The ASQ Certified Construction Quality Manager Handbook

Govind Ramu and Tariq Masud, Editors

ASQExcellence
Milwaukee, WI

Published by ASQExcellence, Milwaukee, WI

Produced and distributed by Quality Press, ASQ, Milwaukee, WI

Publisher's Cataloging-in-Publication data

Names: Ramu, Govindarajan, 1966-, editor. | Masud, Tariq, 1965-, editor.
Title: The ASQ certified Construction Quality Manager handbook / Govind Ramu and Tariq Masud, Editors
Description: Includes bibliographical references and index. | Milwaukee, WI: ASQExcellence, 2026.
Identifiers: LCCN: 2025949955 | ISBN: 9781636941912 (hardcover) | 9781636941929 (pdf) | 9781636941936 (epub) Subjects: LCSH Construction industry--Management. | Project management | Construction industry--Quality control. | Architecture--Management. | Total quality management. | BISAC BUSINESS & ECONOMICS / Industries / Construction | BUSINESS & ECONOMICS / Project Management | BUSINESS & ECONOMICS / Quality Control Classification: LCC HD9715.A2 A77 2026 | DDC 624.068--dc23

ASQ and ASQExcellence advance individual, organizational, and community excellence worldwide through learning, quality improvement, and knowledge exchange.

Bookstores, wholesalers, schools, libraries, businesses, and organizations: Quality Press and ASQExcellence books are available at quantity discounts for bulk purchases for business, trade, or educational uses. For more information, please contact Quality Press at 800-248-1946 or books@asq.org.

To place orders or browse the selection of all Quality Press titles, visit our website at: http://www.asq.org/quality-press.

30 29 28 27 26 GP 7 6 5 4 3 2 1

ASQ
Quality Press
600 N. Plankinton Ave.
Milwaukee, WI 53203-2914
Email: books@asq.org
Excellence Through Quality™

ASQE™
Inspire Excellence

I thank my mom, Vasantha, for her blessings. My dad, Ramu, endured personal hardships and made sacrifices, traveling across India to ensure the rest of the family could sustain itself.

Thanks to my wife Anitha, for her encouragement and sacrifice of countless evenings and weekends for the entirety of this commitment since 2021. Thanks to my daughter Vibha, who inspires me with her passion for the environment, sustainability, and social justice. Thanks to my son Vivek, for his unconditional love. Without their support, my leadership in ASQ CCQM certification, BOK development, and this handbook would not have been possible.
—Govind Ramu

The writing of this book took an extraordinary amount of my personal time. I must thank my family for their understanding: my wife Ambreen, my children M & M (Mona and Maaz), and Pookie, our cat who kept me company during all those long nights.

My heartfelt thanks to my incredible parents (may God bless their souls). They were the constant sources of inspiration and motivation for my personal and professional growth throughout my life. I am sure they must be proud to see me carrying on with the family legacy of writing books.

This handbook is dedicated to my family and to my parents.
— Tariq Masud

Table of Contents

List of Tables and Figures

Acronyms

AASHTO—American Association of State Highway and Transportation Officials
AHJ—Authority Having Jurisdiction
AI—Artificial intelligence
AIA—American Institute of Architects
ASQ—American Society for Quality
ASTM—American Society for Testing and Materials
BCP—Business continuity plan
BoK—Body of Knowledge
C2C—Cradle-to-cradle
CAR—Corrective action replied
CCQM—Certified Construction Quality Manager
CQM—Construction Quality Manager
CMAR—Construction manager at risk
CMFF—Construction manager for fee
CoA—Certificate of authorization
CoC—Certificate of compliance
COPQ—Cost of poor quality
CoQ—Cost of quality
CPFF—Cost plus fixed fee
CSAT—Customer satisfaction
CSR—Corporate social responsibility
CTQ—Critical to quality
DB—Design-build
DBB—Design-bid-build
DBE—Disadvantaged Business Enterprise
DBFM—Design, build, finance, and maintain
DBIA—Design-Build Institute of America
DCR—Daily construction report
DEI—Diversity, equity, and inclusion

DFOW—Definable features of work
DIR—Daily inspection report
DOT—Department of Transportation
DQM—Design quality manager
ENERGY STAR—ENERGY STAR
EOR—Engineer of Record
EPA—US Environmental Protection Agency
EPC—Engineering, procurement, and construction
EPCM—Engineering, procurement, and construction management
ESG—Environmental, social, and governance
eVA—(Not an acronym, but a system used by state agency and local government buyers in Virginia to announce bidding opportunities)
FAI—First article inspections
FAT—Factory acceptance test
FIDIC—International Federation of Consulting Engineers
FMEA—Failure mode and effects analysis
FTA—Fault tree analysis
GDs—General damages
GMP—Guaranteed maximum price
HSE—Health, safety, and environment
HVAC—Heating, ventilation, and air conditioning
IDIQ—Indefinite-delivery, indefinite-quantity
IPD—Integrated project delivery
ISI—Institute for Sustainable Infrastructure
ISO—International Organization for Standardization
ISO 14001—International Organization for Standardization 14001
ISO 9001—International Organization for Standardization 9001
IT—Information technology
ITB—Invitation to bids
ITP—Inspection and test plan
IWBI—International WELL Building Institute
JOC—Job order contracting
JV—Joint venture
KPIs—Key performance indicators
LADs—Liquidated and ascertained damages
LDs—Liquidated damages
LEED—Leadership in Energy and Environmental Design
LLC—Limited-liability company
MBDC—McDonough Braungart Design Chemistry

MBE—Minority business enterprises
MEP—Mechanical, electrical, and plumbing
MOU—Memorandum of understanding
MRR—Material receiving report
NCR—Nonconformance report
O&M—Operation and maintenance
P3—Public-private partnerships
PERT—Program evaluation and review technique
PMBOK Guide—Project Management Body of Knowledge Guide (from Project Management Institute)
QA—Quality assurance
QC—Quality control
QMP—Quality management plan
QMS—Quality management system
QV—Quality verification
RFB—Request for bids
RFI—Request for information
RFP—Request for proposals
RFQ—Request for qualifications
RFQ—Request for quotation
RFx—A collective umbrella term in procurement where x refers to a variable for a request (e.g., RFQ, RFP)
SBSD—Small business and supplier diversity
SMART—Specific, measurable, achievable, relevant, time-bound
SME—Small and medium-sized enterprises
SME—Subject matter expert
SOV—Schedule of values
SOW—Statement of work
SWaM—Small, Women- and Minority-Owned Businesses
SWOT—Strengths-weaknesses-opportunities-threats
T&M—Time and material
TAB—Test and balance
TOR—Terms of reference
UN—United Nations
USDOE—US Department of Energy
USGBI—US Green Building Institute
VDOT—Virginia Department of Transportation
WBS—Work breakdown structure
WIP—Work in place

Preface

Have you ever embarked on a home renovation, only to find yourself entangled in a web of budget overruns, frustrating delays, unforeseen code violations, and persistent quality issues? If such challenges can plague a modest kitchen remodel, imagine the complexities and stakes involved in multimillion-dollar construction projects like community homes, highways, bridges, semiconductor fabrication plants, nuclear facilities, or vast data centers. In these colossal undertakings, the financial and economic implications of quality failures are staggering, alongside critical concerns for human safety and environmental sustainability.

The global construction industry stands as a titan, boasting a market capitalization of $11.5 trillion and experiencing a robust 7.3% year-over-year growth. Yet this impressive scale is overshadowed by significant "cost of poor quality" (COPQ), a pervasive issue that translates into substantial financial losses, frequent safety incidents, and a considerable environmental footprint. Construction activities, from material choices like cement and chemicals to every stage of the project life cycle, contribute a staggering 37% of global emissions, exacerbating the urgent challenge of climate change.

For too long, quality in construction has largely been a reactive exercise, heavily reliant on inspections and checklists. While checklists are invaluable tools for verification, their effectiveness is severely limited when applied in isolation, without the foundational support of comprehensive quality assurance and broader quality management principles. This reactive approach often leads to the discovery of issues far too late in the project life cycle, driving up costs and eroding efficiency. The industry's interchangeable use of "QA/QC," often without a clear understanding of the distinct definitions provided by standards like ISO 9000, further highlights a systemic lack of strategic quality planning. Even within major construction organizations, corporate quality functions frequently operate in silos, disconnected from the realities of site project teams, which leads to fragmented and inconsistent quality practices.

A critical void has long existed: the absence of a formal, recognized certification for construction quality managers. Historically, experienced inspectors or project superintendents were often promoted into quality management roles without adequate training or foundational knowledge in holistic quality management. This often meant they continued to apply the inspection-centric skills that earned them their promotion, rather than embracing the strategic planning and deployment of a comprehensive quality management system.

The ASQ Design and Construction Division (DCD) recognized this profound industry gap. Our journey to address it began as early as 2004 with an initial proposal for a construction quality manager certification. Despite two unsuccessful attempts in 2004 and 2014, the DCD's unwavering commitment to elevating quality in construction persisted. In 2021, we returned to the drawing board, incorporating invaluable feedback from diverse stakeholders. This renewed effort saw the establishment of a dedicated volunteer role for a Body of Knowledge (BoK) and Certification Chair, spearheading an initiative that quickly garnered the support of over 30 international construction industry experts.

The overwhelming demand for such a credential was unequivocally affirmed by our membership. A survey revealed that 92.1% of 266 respondents "strongly agreed" or "somewhat agreed" that an ASQ Certified Construction Quality Manager (CCQM) certification would add significant value to the industry. Furthermore, an astonishing 96.9% of 258 respondents indicated they would either take the exam themselves or recommend it to their colleagues and organizations.

Why is the ASQ CCQM certification so vital now? The construction industry in the United States alone employs over 10.8 million people. Despite its immense size and rapid growth, there has been no formal qualification or certification that addresses "broader quality management" specifically tailored for construction quality managers. A standardized BoK for quality management, universally applicable across the industry, simply did not exist. This certification fills that critical void, providing a much-needed pathway for quality professionals from other industries and those in non-quality titled roles to transition into and meet the unique demands of the construction sector. It will elevate the perception of the quality profession within construction and attract new talent, ensuring that quality considerations are integrated across the entire construction life cycle, including critical aspects of safety and sustainability.

The ASQ CCQM directly addresses the pervasive risks associated with productivity losses, ineffectiveness, inefficiencies, high cost of poor quality (COPQ), and safety concerns that plague the industry. By fostering a deeper understanding and application of quality management principles, this certification promises tangible improvements in project outcomes.

The development of the ASQ CCQM Body of Knowledge was a meticulous balancing act. It was designed to be agnostic to most verticals within the construction industry—from commercial and residential to heavy civil, industrial, and environmental projects—ensuring broad applicability. The BoK is neither too niche nor too generic, making it relevant for both seasoned professionals and those with significant experience but perhaps lacking formal quality training. It is structured to be teachable at community colleges and universities, fostering a new generation of quality-focused construction professionals, and is applicable internationally across all geographical regions. Crucially, it incorporates safety and sustainability in relation to quality planning and embraces new and upcoming technologies and innovations.

The finalized BOK for the CCQM, developed through ASQE's rigorous certification development process, is now available on the ASQ website. It is thoughtfully designed with sections dedicated to each construction life cycle phase and a section addressing essential CCQM skills. The Bloom's Taxonomy cognitive levels were carefully integrated to ensure the BoK emphasizes significant application, analysis, and creation skills where appropriate. Eligibility for this certification requires eight years of on-the-job experience in one or more areas of the CCQM Body of Knowledge, with three years of that experience in a decision-making position.

The ASQ Certified Construction Quality Manager Handbook has been developed in the same sequential order as the BoK, from Sections I to VIII (reflected in Chapters 1–8). It is more than just an exam preparation guide; it is an indispensable desk reference for every construction professional. It encapsulates the comprehensive BoK, providing a practical framework for applying quality management principles in real-world scenarios.

Whether you are a project manager, engineer, construction manager, procurement manager, commissioning manager, superintendent, craft professional, developer, or owner, this handbook will equip you with the knowledge and tools to proactively manage quality, mitigate risks, and drive superior project performance. It will serve as your go-to resource for understanding and implementing the best practices outlined in the ASQ CCQM Body of Knowledge, empowering you to make informed decisions that impact productivity, efficiency, safety, and sustainability.

To enhance the value of this handbook, each chapter includes essay-style questions for instructors and professors to assign to students in construction quality management courses at community colleges and universities. For those preparing for the ASQ CCQM exam, it's important to note that the exam is a single-part, 165-question multiple-choice test, offered exclusively in English. Of these, 150 questions are scored and 15 are unscored. The total appointment time is four and a half hours, with the exam itself lasting 4 hours and 18 minutes.

By introducing more construction professionals to ASQ and its certification offerings, the ASQ CCQM not only addresses a critical industry need but also opens doors to further professional development through other ASQ certifications.

In summary, the ASQ CCQM is a groundbreaking credential, the first of its kind in the global construction industry to cover such a comprehensive Body of Knowledge and launch a formal certification for construction quality managers. As the industry increasingly recognizes the benefits of hiring and empowering quality managers through this certification, we anticipate significant improvements in productivity, efficiency, and a substantial reduction in COPQ and safety incidents. This will lead to enhanced compliance with sustainability goals, ultimately delivering profound financial and economic benefits to organizations, clients, and the construction industry at large. Should this ASQ certification and accompanying handbook contribute to a global enhancement in the competency of construction quality managers, resulting in even a 1% reduction in the annual cost of poor quality, the economic impact would be approximately $115 billion. Consider the profound implications of such an achievement on a large scale.

We are proud to present *The ASQ Certified Construction Quality Manager Handbook*, confident that it will become an essential companion for your professional journey, transforming how quality is understood, managed, and achieved in construction projects worldwide.

—Govind Ramu and Tariq Masud, Editors

Acknowledgments

Thank you to eight key contributors, who were instrumental in drafting entire sections and multiple chapters to assist the Editors in preparing this handbook for publication. They contributed their extensive knowledge and experience from diverse construction verticals and other industry sectors in quality management. The contributors represented a diverse range of geographies, genders, construction verticals, and experience levels, spanning from small to large-scale projects and encompassing all phases from planning through maintenance.

CONTRIBUTORS

Ben Dupslaff is a Minneapolis-based mission-critical professional with 15+ years in business, construction management, engineering, commissioning, and process improvement. He holds an MBA, ASQ CMQ/OE certification and authors deconstrategy.substack.com.

Zakariya S. Al-Helal is a construction quality specialist with expertise in quality and technology management. He worked in public and industrial mega construction projects, and chaired and served on several technical committees. Zakariya holds a master's in strategic quality management, and is an ASQ CQE and Project Management Professional (PMP).

Aram Gharapetian, P.E., ASQ CMQ/OE, CQE, CQA, CCQM, is a Quality Engineer with 20 years of experience in design and construction in the New York City Metropolitan Area, primarily on large infrastructure projects. He holds a B.S. in Electrical and Computer Engineering from Rutgers University in New Brunswick, New Jersey.

Abbigail Meah-Ali is a professional with over 24 years of expertise in quality and construction project management. Ms. Ali is a certified trainer. Her certifications include PMI-PMP, ASQ CSSBB and Certified Quality Engineer (CQE), among others. She is a member of the Chartered Quality Institute (CQI), Project Management Institute (PMI), Institute of Internal Auditors (IIA) and International Project Management Association (IPMA)-USA.

Dennis Murphy PE, ASQ CMQ/OE, is a senior quality manager with Kiewit Infrastructure Co. from Sterling, VA. Dennis is a licensed professional engineer in Washington, DC, Maryland, and Virginia, as well as an ASQ Certified Manager of Quality and Organizational Excellence. Dennis has over 25 years of construction industry experience, and has successfully managed QA/QC on multiple complex construction projects.

Micah Stanton, with 20+ years of experience transitioning from aerospace to construction, has held roles from superintendent to director of quality. He's led diverse projects including housing, healthcare, and data centers. Passionate about team building, he drives continuous improvement through mentoring, strategic questioning, and minimizing waste.

Jesse Fox's 38+ years construction career has focused on QA/QC, building commissioning, and risk management. Jesse currently serves as a senior quality/commissioning group manager for Suffolk Construction. He has extensive experience and knowledge on Quality Management Systems, as a Certified Commissioning Provider and Lean Six Sigma Black Belt.

Collin Sutt is a senior organizational excellence and quality consultant in the construction industry, and has nearly 25 years of experience in the construction and manufacturing industries. He has been a state level Baldrige Examiner, and is a senior member of ASQ, the past chair of ASQ's Design and Construction Division, and holds ASQ CMQ/OE, CSSGB, and CRIS certifications.

REVIEWERS

Thank you to Willis H. Thomas, PhD, PMP, CPT, project manager, Golden Ratio Management Corporation.

We extend our special gratitude to Ben Dupslaff for his exceptional contributions, significantly enhancing the value and completeness of this handbook through the extensive number of chapters authored within the given time frame. We also wish to acknowledge Aram Gharapetian for not only authoring his assigned chapters but also for offering his technical expertise with a peer review of several chapters and for providing comprehensive technical comments to elevate the quality of the content.

We extend our gratitude to the following construction professionals who contributed to the initial development of the ASQ CCQM Body of Knowledge from March 2021 to April 2022: Amanda Skillern, Anita Lidbury, Bill Riley, Careese ACHAT, Collin Sutt, Corey Zussman, Dan Domalik, Don Archer, Eileen Gee, Essam El Nahrawy, Jeffery L Williams, John F Mascaro, Larry Owen, Loan Ngo, Nick Shah, Osmond Ikwueme, Phil Munro, Ray Crawford, Raymond Giles, Rick Simon, Robert Achilles Sellen, Robert Beaver, Robert Connors, Sal Calvo, William Hall, and Zakaria Alhelal.

We express our gratitude to the ASQ Design and Construction leadership team (2021–2022), specifically the late John F. Mascaro, Ray Crawford, and Collin Sutt, for their foresight in establishing a construction quality certification.

—Govind Ramu and Tariq Masud, Editors

Govind Ramu spearheaded this program as the ASQ's Design and Construction Body of Knowledge and Certification chair from March 2021 until the certification's launch in February 2025. His responsibilities included monthly discussions regarding BoK inputs with over 30 international volunteers, presenting the BOK at ASQ World conferences, supporting the endeavor to secure ASQ approval, contributing to exam development, and program managing the ASQ CCQM handbook. For the efforts and accomplishment, ASQ Design and Construction division honored Govind Ramu with their inaugural John Mascaro Service Award, citing, "This award is presented to Govind Ramu, in recognition of his significant contributions to the ASQ Design and Construction Division's initiative of establishing a certification for Construction Quality Managers. Govind has demonstrated exemplary professionalism and service in helping achieve the division's strategic objectives in the development of the CCQM certification program, and related Body of Knowledge."

ADDITIONAL THANKS

I want to thank my co-editor Tariq for his passion, significant support, and commitment during this entire collaboration. Thanks to all contributors and reviewers of this handbook. Thanks to ASQE staff and DCD leadership team for their professionalism and support for this project. I am also grateful to the late John Mascaro, past DCD chair. Without his leadership and passion, we would not have launched the ASQ CCQM exam.

—Govind Ramu

I want to express my gratitude to my co-author Govind Ramu as well as ASQE for giving me the opportunity to contribute to the industry through the writing of this handbook. It was an honor to work on this project and share the knowledge and experience that I have gathered over my professional career, and for that I must also thank so many from whom I have learned all that I know.

—Tariq Masud

CONTRIBUTORS' ACKNOWLEDGMENTS

The American Society for Quality and Quality Press would like to thank the Quality Press Peer Review Committee for its invaluable volunteer participation and contributions to this work. Without our volunteers' subject matter expertise, time, and passion for creating content, none of our efforts would be possible.

QUALITY PRESS PEER REVIEW COMMITTEE MEMBERS

Scott A. Laman, Chair
Peter Pylipow, Vice Chair
Melvin Alexander
Lance Coleman
Ahmad Elshennawy
Gary Jing
Trevor Jordan
Jane Keathley
Mary McShane-Vaughn
Jayet Moon
G.S. Sureshchandar
Tiea Theurer

Chapter 1

Precontract Phase

Project Delivery, Contract, and Payment Models

A quality team is an indispensable part of any construction project; one of its major roles is to ensure the construction project meets client expectations. Thus, an understanding of the project client is required to better deliver quality. Additionally, in dealing with the client it is essential for the quality manager to understand how the project is contracted and is to be delivered. The quality manager and their team are expected to be important members of the project from its beginning to its completion.

PROJECT DELIVERY METHODS

> Identify various types of clients (e.g., private, commercial, federal, state, municipal, and international) and their use of project delivery methods, such as Engineering, Procurement, Construction Management (EPCM), Design-Bid-Build (DBB), Design-Build (DB), Design, Build, Finance, and Maintain (DBFM), Construction Manager at Risk (CMAR), and Construction Manager for Fee (CMFF). Understand project lifecycle phases (e.g., project initiation, project planning, project execution, monitoring & control, and project closeout) align with the project delivery method. (Understand)
>
> Body of Knowledge I.A.1

Types of Clients in the Construction Industry

Construction projects range from those required for individual housing to those constructed for a region or even a nation, such as highway systems. The type of client and consequently the quality requirements for such a wide range of clients vary significantly from one industry to the next, and are also different based on the region and physical location. The requirements are influenced by several factors such as regulatory compliance, technological advancements, global supply chain dynamics, economic conditions, internal processes, cultural norms, and preferences of the clients, to name a few.

Also, client involvement in quality management of the projects can range from those who are fully vested in quality to those who consider quality as just another implied feature of the end product while the project works its primary targets of on-cost and on-time delivery. Depending on their focus on quality, their industry standards, and their preference for a particular project delivery type, different clients choose to deliver the construction projects through a variety of project delivery methods.

Lastly, from the clients who rely on their own quality resources to verify the quality of the project to the clients who hire independent quality consultants to perform the quality functions to those who depend on the hired design and construction firms and allow them to perform their own quality reviews and testing, the construction industry has a very wide variety of the operators, clients, and performing organizations that deliver the construction projects in a variety of ways. One common way to categorize the construction industry clients is based on the end use.

Classification of Construction Projects Based on End Use

Construction projects are often categorized based on their scale, structure type, and end use. The most common classification is the "project end use categorization," which refers to classifying the projects based on their intended final purpose or function. This classification groups them according to what they will be used for once completed.

The main types of construction projects based on end use are: commercial, residential, mixed-use, industrial, and heavy civil; and public or private clients. The end use of these projects dictates the design requirements, regulations, and potential environmental impacts. Also, the same scope of work that is performed for a public client may include different quality, environmental, and legal requirements than if the same project is performed for a private developer. Likewise, construction contractors generally specialize in a particular type:

- *Commercial clients*: These are organizations or individuals that need a construction project for running their business. Examples include restaurants, hotels, shopping malls, and so on.
- *Residential clients:* These are usually individuals who need a construction project unrelated to any business. Examples include construction work for their own homes or the home of a family member.
- *Mixed-use clients*: Mixed-use developments are more popular among private developers who prefer to diversify the end users and reduce their financial risk by having multiple types of end users occupy the same building. Examples include structures that have office,

residential, and commercial end users all inhabiting different sections of the building (e.g., shops, residential apartments, and offices in the same building).

- *Industrial clients*: Industrial construction includes those projects that are built for industrial use. Examples include manufacturing plants, warehouses, power plants, factories of many different kinds, chemical processing facilities, and oil refineries. The clients for these projects could come from different industry sectors such as agriculture, manufacturing, construction, transportation, and communication sectors. These types of projects usually require specialized engineering, along with quality specifications and other regulations that may not be required by other construction projects.
- *Heavy civil clients*: Heavy civil construction includes large-scale engineering projects typically associated with infrastructure or public works. Examples include transportation systems (e.g., construction of highways, bridges, tunnels, railways, airports), or utilities (e.g., water and sewage systems, dams, ports, or communication and power distribution networks) and similar large-scale public works projects. These projects are also called "horizontal construction" because they are typically longer than they are tall. Projects that are taller than they are long are called "vertical construction."
- *Public versus private clients*: The last broad-based classification of clients should be made based on the ownership; namely, the clients could be public or private. The public sector comprises organizations that are controlled and managed by the government. On the other hand, the private sector clients include businesses that are owned and managed by private individuals or entities. It is important to note that there are also some public sector entities that are owned by investors who represent a partnership between public and private clients (these projects are termed public-private partnerships [PPP or P3]). As such, there is a variety of possible ownership on the construction project that may have full or partial ownership rights over the constructed facilities.

Some of these clients are institutions on national, regional/state, or local/municipal levels; examples of public projects may include public hospitals, schools, and government buildings. There is significant overlap between the ownership of commercial and institutional construction, and it can greatly impact the quality of the projects because the design requirements, building materials, and equipment needs for the different owners can vary significantly.

The Different Project Delivery Methods

There are many approaches to achieving successful project delivery. These approaches or delivery methods describe how the project is designed and built and the responsibilities of the involved parties such as the project client, designer, and contractor.

Some of these delivery methods as well as their benefits and risks are as follows:

Engineering, Procurement, and Construction Management (EPCM)

Engineering procurement construction (EPC) or engineering procurement construction management (EPCM) is a project delivery method that involves the coordinated management of design, procurement, and construction. The EPC or EPCM contractor is usually responsible for the fully functional project; this means it includes the primary operational phases of the project life cycle, namely design/engineering, procurement, and construction phases.

Although both methods (EPC and EPCM) are described here as if they are the same, they differ slightly, primarily in terms of the risk that the EPC team may take on with respect to construction. Whereas the typical EPC team takes on complete risk for all three phases of design/engineering, procurement, and construction, the EPCM team may not take the risk for the construction (the owner takes that risk).

The primary distinction of the EPC method of project delivery is that a single team is selected to provide comprehensive management services for all three phases of the project (except that for the EPCM, they may not actually perform the construction work, instead managing the construction on behalf of the client/owner).

Benefits:

1. EPC is a commonly used method in large-scale projects, especially where complex engineering or proprietary technology is involved. EPC provides a single point of contact for project-related services and manages the project from start to finish. This streamlines all processes much better than a multiorganizational team that consists of personnel from different organizational cultures, goals, and objectives.
2. EPC projects offer more cost certainty as all costs for the project are estimated by one organization. This reduces the risk of cost overruns for the owner.

Risks:

1. With one organization performing both design and construction of the project, the expertise and technical knowhow can be limited when compared with the technical expertise of a firm that specializes in design or construction services alone.
2. Potential for conflict of interest is greater when the construction is performed by the same firm that designs the project. The firm may prioritize its own interests over those of the owner/client and require a more stringent oversight by the owner.
3. Less flexibility in design changes is usually characteristic of the EPC delivery method. This is a drawback for those clients and projects where evolving project conditions may necessitate changes in the design or construction.

Design-Bid-Build (DBB)

Design-bid-build is the most traditional method of project delivery; it involves a design firm and a contractor working for the client under separate contracts. This method includes three distinct phases: design, bidding (procurement), and building (construction).

The client first enters into a contract with the designer. Once the design is completed and agreed on with the client, a request for bid is sent out to potential contractors. The client then selects a contractor firm and forms another separate contract with them. The client may hire the designer as a consultant to oversee the construction activities on its behalf.

Benefits:

1. Clear role definition and responsibilities.
2. Greater control by owner: Both the designer and the builder (contractor) are contracted directly with the owner, resulting in the highest control of the project for the owner.
3. Higher transparency due to public bidding.
4. Competitive pricing.
5. More flexibility and customization for the owner.

Risks:

1. Longer duration due to separate, unrelated contracts for design and build.
2. Less cost-effective than the DB method despite its competitive pricing, since the DBB takes longer.
3. Potential for more conflicts between design and construction since having different contracts means the two do not work directly together, which causes time delays as well as conflicts in the approaches of the designer and the construction contractor.
4. Greater risk for the owner and less for the two performing organizations, the designer and the contractor. Having more control over the project also means the risk is the responsibility of the owner.

Design-Build (DB)

In the design-build method of project delivery, the owner makes only one contract to deliver the project from design to construction, that is, combines the design and construction phases into a single contract. Usually, the DB projects include a limited liability company (LLC) created by the joint venture of a design firm with a construction contractor firm. This single entity then manages both design and construction throughout the project. The DB contracts are typically considered "best value" contracts, meaning that they prioritize the most advantageous combination of price, quality, and other factors in the DB's team rather than simply the lowest price alone. To achieve the best pricing from the potential DB teams, most owners perform the preliminary design, which allows for more accurate pricing proposals by the DB proposers. Like all project delivery methods, there are benefits as well as risks associated with this delivery method.

Benefits:

1. Faster project completion, due to single point of contact—one team.
2. Usually a cheaper option than DBB and most other delivery methods, as it takes much less time to finish.

3. Better collaboration among the stakeholders.
4. Offers good accuracy in pricing and more reliable budgets because DB pricing is based on preliminary (i.e., 30%) design and is not like some other delivery methods that must base their pricing on 0% design completion.

Risks:

1. Lesser control of owner.
2. Potential for higher cost due to lack of competitive bids.
3. Higher potential for conflict of interest.
4. Higher level of trust needed to work with the newly created entity, which may consist of performing organizations that are adversaries in their pursuit of other projects.
5. Lack of culture in a limited liability company. The DB team is really a fusion of cultures, as the team consists of not a single organization but includes multiple personnel with different goals, objectives, and vision. Because of this, the DB teams may have internal challenges that could hinder the overall performance and ability of the project team to achieve the project goals and objectives.

Design, Build, Finance, and Maintain (DBFM)

DBFM is a project delivery method that typically involves a private sector consortium responsible for designing, building, financing, and maintaining the infrastructure asset over the contract term. This method brings together a public-private partnership (PPP) and integrates the design, construction, financing, and maintenance of a project into one contract. This method is preferred in the private sector, especially for those projects where the public entity agrees to give its control over the asset/facility under a long-term agreement to the consortium that finances the project.

Benefits: DBFM offers many of the same benefits as the design-build projects, plus some more:

1. Results in potentially higher cost savings through more optimized design and construction processes because the DB organization is selected for the long haul, which also brings more vesting of the interests of the performing organizations as they know they are part of the project for a long duration.
2. Incentivizes the DB team to develop better, low-maintenance solutions.
3. Allows the public sector to transfer the risks to the private sector, which is the biggest advantage.

Risks:

1. Complex to set up and manage.
2. Potential cost overruns due to unforeseen design challenges.
3. Project delays possibly caused by external factors.

4. Continued financial stability of the DB contractor that is responsible for both design and construction.
5. Lesser control for the owner.
6. Potentially negative impacts on project's completion and quality if the DB encounters financial difficulties.
7. Potentially higher expense for the taxpayers over the long term than traditional project delivery models because the private sector firms often seek to recover their costs plus a profit margin through their fees during the postconstruction operations and maintenance phase.

Construction Manager at Risk (CMAR)

This method of project delivery is preferred by owners who want to shift much of the risk and responsibility involved in the construction project to a construction manager (CM) firm that takes the responsibility for building the project at a fixed maximum price. Because the CM takes on this risk, the method of delivery is called construction manager at risk. In this method, the owner hires a firm as the construction manager. This firm also acts as the general contractor for the project of the firm that is at risk; they are responsible for delivering the project within their proposed guaranteed maximum price (GMP) and are liable for any cost overruns.

Unlike the DB project where the design is performed by the designer of the DB team, in the CMAR method of project delivery, the design is developed initially by a design firm that is hired by the owner, who develops the design to a level that the contractor is better able to propose the GMP. This initial design is prepared only to get a more accurate estimate of the project, which is then further designed and built within the proposed maximum price estimate.

Benefits:

1. Owner gets cost certainty.
2. Owner transfers the risks to the CM. As a result, the owner has the least amount of financial risk on the project.
3. The CM confirms the fixed price halfway through the design phase, and then it is their responsibility to get the project completed within that fixed price. The advantage here is that the CM does not have to predict the maximum price based on an incomplete (conceptual) design but that the guaranteed price is based on a more dependable preliminary design. Further, the direct oversight of the design firm enables a tighter control of the design as well as construction services.
4. If the CM is able to complete the project within the target price, the profit is usually shared between the owner and the CM.

Risks:

1. Construction manager at risk (CMAR) assumes the financial responsibility for any costs that exceed the GMP agreed on with the owner. Note that the GMP is based on a very defined scope. Any change in this scope can result in claims and increased cost.

2. Project cost may not be determined until later.
3. If the contractor bids come in higher than anticipated, then the CM firm must bear the financial burden of the project.
4. An inexperienced CM could result in problems in both the design and construction portions of the project, leading to fundamental problems with the end-product quality, disputes among stakeholders, and ultimate project failure.
5. The owner relies significantly on the performance of the CM.
6 Although schedules can be somewhat compressed because of early collaboration with the CM, the CMAR method may not allow for quick project turnaround like a DB.
7. The CMAR is constantly under pressure to deliver quality as well as achieve the cost targets. There is the possibility of less-than-ideal quality of the project, and if the price target is not being achieved, quality may be affected or compromised.

Construction Manager for Fee (CMFF)

In the CMFF method, the CM firm acts primarily as a consultant, advising the owner on the construction process without taking on significant financial risk for the project cost. The CM does not perform construction but instead performs the oversight/management during the construction phase of the project. The CM firm is primarily hired during the construction phase of the project, and usually has no role for the design part of the project. Typical construction management fees range between 3% and 5% of the total project cost. The fee charged for the CM services typically includes preconstruction planning, coordination, and management throughout the construction phase.

Unlike the CMAR project, the scope of the CM firm is only limited to the oversight of construction services; they do not carry the risk of completing the project within a guaranteed maximum price.

Benefits:

Owner gets additional consulting from an expert consultant firm to oversee the work of the builder. Because the CM firm has previous similar work experience, they become the subject-matter expert consultants for the owner. This results in:

1. Better construction performance
2. Reduced project costs
3. Improved efficiencies
4. Better project timelines
5. Better risk management
6. Better quality control (and quality assurance as well where provided)

7. Improved communication among stakeholders
8. More expertise in navigating complex construction processes

Risks:

1. The CM does not assume any financial responsibility for the cost overruns; the risk still belongs to the owner.
2. The CM firm has almost no direct interaction with the design firm.
3. The role of CM may cause conflicts because several of the responsibilities that otherwise belong to the construction contractor are handled by the CM firm that manages the contractor firm. A higher/better definition of roles and responsibilities is required for smooth operations between the contractor and the CM.

Job Order Contracting (JOC)

Job order contracting is an indefinite-delivery, indefinite-quantity (IDIQ) project delivery method. This means that multiple projects can be completed over the life of one long-term contract (also considered for a lot of projects under one program).

This method is usually the choice of those clients that have a lot of related (or sometimes unrelated) projects they want to issue for an undefined period of time; it is commonly utilized where it is difficult to determine the precise quantities of services or supplies that the owner may require during that period. This is a common delivery method of procuring construction services, repairs, and other supplies for the federal services, but it is not as commonly used in the private sector.

Benefits:

1. Flexibility for the owner (usually the federal government) to procure for varying needs that cannot be accurately predicted upfront.
2. Reduced paperwork.
3. Stronger long-term relationships with the owner (usually a federal department).

Risks:

1. Uncertainty for the contractor about the amount of work they may receive.
2. Typically long lead times for approval of IDIQ contracts.

Integrated Project Delivery (IPD)

IPD is one of the newer construction project delivery methods where the key parties that are involved in the design, fabrication, and construction aspects of a project are all joined together under a single contract agreement. IPD aims to maximize efficiency and collaboration, as it integrates all the project participants (owner, designer/architect, and the contractor) through all phases of design, fabrication, and construction.

Benefits:

1. Enhanced collaboration among all project stakeholders.
2. Improved communication.
3. Reduced waste through early problem identification.
4. Better cost control.
5. Faster project completion.
6. Increased innovation in the team.
7. Shared risk and responsibility.
8. Focus on achieving the best possible project outcome for all stakeholders, with everybody working as a team throughout the design and construction phases.

Risks:
As much as it aims to produce great collaboration among all stakeholders, integrated delivery is not without its challenges. The risks include:

1. High upfront cost as a result of engaging all the performing organizations at once.
2. Complex contractual agreements.
3. Heavy reliance on trust and collaboration between all parties.
4. Potential for slower decision-making due to consensus-based approach.
5. Difficulty finding suitable partners who are willing to fully embrace the collaborative model.
6. Impracticality of IPD for simpler projects because trust takes a longer time to develop, and in situations where high levels of trust are not readily established between the parties, the slow start can negatively impact the project teams.

Selection of the Delivery Method

The method of project delivery is typically chosen by the client and must be carefully made based on factors like project complexity, budget constraints, timeline, risk tolerance, owner's involvement, and team collaboration. The chosen method will have its benefits as well as some risks, and the impact on project team composition, quality, schedule, cost, and risk can be significant.

The clients need to understand the basic principles of these delivery methods, their advantages and disadvantages, and choose the method that is right for their projects. The description of various methods of project delivery is provided above for the understanding of a Certified Construction Quality Manager (CCQM). The selection of the right method is typically not within the scope of a quality manager's direct role; usually this decision is made by the senior management, but having a quality manager involved with the decision is useful and shows management's interest in quality as well as the empowerment they may offer to the quality management of the project.

Project Life Cycle Phases

The ASQ Body of Knowledge for the CCQM examination describes the life cycle of a construction project as consisting of the following seven phases:

1. Precontract
2. Planning
3. Design
4. Procurement
5. Construction
6. Commissioning and closeout
7. Operations and maintenance

Because each phase represents a distinctly different aspect of work, it is best to break down the construction project in these seven phases of the project life cycle.

Precontract Phase

The first phase of a construction project's life cycle is the precontract phase. In this phase, the stakeholders determine how the project will meet the business needs as well as the objectives for which the project is initiated. Some of the roles in this phase belong to the owner, and others belong to the remaining stakeholders of the project. It is essential that all stakeholders and the project team are aligned and understand the project success criteria. The alignment of stakeholders increases the likelihood of completing the project deliverables throughout its life cycle. This phase is also termed the "initiation phase" by many. As part of this phase, the project sponsor authorizes the project to proceed, and the project's charter, its policy, and its objectives are finalized. For the construction (and project) management, the primary deliverable of this phase is the project charter, which establishes the authority of the project manager and documents the project milestones and its success criteria. Approval of the charter signals the advance of the project to the next phase, which is the planning phase. The key aspects of the precontract phase include the following:

1. Project delivery, contract, and payment models
2. Contractual requirements
3. Bid evaluation and finalization

Planning Phase

In its simplest definition, the planning phase is where the project manager and the team create the plans to achieve the project's goals. The project solution is developed in this phase to the level of detail where the steps that are necessary to meet the project's objectives are laid out.

The planning process should begin with the identification of the strategic goals and objectives, followed by the planning of a quality management system that would sustain the quality effort. Lastly, the planning phase must include resource planning, which includes the identification of tasks and resource requirements along with the strategy for producing them. The following are the important subelements of planning for a project. (Note: This list is not all-inclusive.)

1. Strategic quality planning
2. Planning the quality management system (QMS) and the quality audit program
3. Resource planning

The primary deliverables of the planning phase of a construction project's life cycle include the project plan (also sometimes called the project management plan), which usually contains several supplementary plans and a detailed work breakdown structure (WBS) or task list with duration and start and finish dates (linked with baseline schedule). Another deliverable of the planning phase includes the baseline schedule along with the project's critical path. In addition to the mobilization and demobilization planning, the resource plan must identify who does what work, at what time, and if any special skills are required. The budget planning specifies the costs; the procurement plan focuses on vendors and subcontracting; and the risk management plan identifies the possible risks and how to handle them.

Design Phase

The design phase in a construction project life cycle is the stage where comprehensive design plans for the project infrastructure are developed. This is the first of the two execution phases of the project life cycle (the other one is the construction phase). Design includes the drawings (previously called blueprints), the cost estimates, detailed specifications, and calculations. The final design, which is also called the "construction documents," lays the groundwork for the entire construction process before actual building begins. Many in the construction industry also refer to the design phase as the "preconstruction" phase. The design phase may be described in terms of design inputs, design process, design outputs, the quality control (QC) aspects of the design, and the quality assurance (QA) aspects of design (also sometimes called "quality monitoring").

Following are the substages of the design phase:

1. Design inputs
2. Design process
3. Design outputs

It should be noted that QC in design and QA should be implemented across all three subphases of design.

Procurement Phase

The procurement phase of the construction project's life cycle is the administrative phase of the construction project's life cycle, where all of the necessary materials, equipment, and services that are required for the construction process are acquired—that is, all the steps involved in obtaining goods (e.g., materials and components) or services (e.g., equipment, labor/personnel, and specialty resources) that are needed for the project. Some of the activities in the procurement phase include the implementation of quality and risk on the procurement of the material and personnel resources that are needed for the project.

Usually, the procurement of parts (components and materials) and sometimes labor or professional services such as testing laboratories are procured from the suppliers (of materials) or vendors (of the components and assemblies). Some of the aspects of project work that may be performed during the procurement phase are listed here. (Note: This list is not all-inclusive.)

1. *Quality requirements for procurement of suppliers/vendors*: The quality requirements for the vendors are defined and communicated during the procurement phase of the project.
2. Determination of standard reporting (quality and management reports).
3. Resource requirements (for suppliers/vendors).
4. Risk management; initial risk assessments related to the vendors and suppliers.

Construction Phase

There are two execution phases in the life cycle of a project: design and construction. It is in the construction phase that the physical building or development of the project takes place. Most projects start their construction phase with preconstruction activities such as site preparation, and equipment and personnel readiness inspections, and then proceed to the actual construction where the installation of the materials and components takes place according to the definable features of work (DFOW) that are typically included in the construction contract.

The DFOWs are tasks that are separate and distinct from each other; they include control requirements and work crews that are unique to those tasks. In other words, the DFOWs are the distinct, separate tasks of the project and ones that can be inspected for their quality and compliance with the project's requirements and specifications. It is the DFOW that makes construction integral to quality management. Because of the DFOW, it becomes possible to inspect construction and determine if the produced work is in compliance with the contract requirements as well as with the standards of excellence.

The construction phase of the project life cycle requires the deployment of the construction plan, as well as the application of the construction quality control (CQC) and quality assurance (QA) to the construction activities. Following are the subphases of the construction phase:

1. Construction plan deployment.
2. Process management.
3. Destructive and nondestructive tests.
4. Compliance management. Notice that compliance management generally represents the QC activities during the construction phase, which are

aimed to ensure that the constructed activities comply with the product as well as process requirements.

5. Quality monitoring and reporting. Notice that this represents the QA for the construction phase.

Commissioning and Closeout Phase

In a construction project, "commissioning" refers to the process of systematically testing and verifying that all building systems are functioning as intended according to design specifications, while "closeout" marks the final stage of the project, signifying the completion of construction activities and the transition to the building's operational phase, including handing over the project to the owner and final documentation.

Both stages require formal documentation that must be completed for the immediate release, as well as for the permanent records to the project's document control repository. The quality manager must ensure that the aspects of commissioning and closeout are addressed before the project is closed out for good.

1. Impact assessment and compliance
2. Required documentation
3. Records of nonconforming work
4. Quality deliverables verification
5. Final certifications
6. Final turnover and closeout documentation
7. Warranties (both explicit and implied)

Operations and Maintenance Phase

Once the project is completed for construction, it may be turned over to the owner (or sometimes to the owner's representative) to operate and maintain it. There are two things that are more significant in the operations and maintenance phase:

1. *Sustaining the quality management program requirements*: The quality program for the operations and maintenance phase is slightly different from the production phases as most of the emphasis now is not on producing the design and construction but on the maintenance of existing facilities. However, sustaining the quality control and quality assurance requirements must remain essential in this postconstruction phase of the project as well.

2. *Business continuity*: Business continuity refers to the ability of an organization to maintain essential functions during and after a disaster or unexpected disruption. This should include planning for the continuity of services, even in cases of unplanned events. The example of planning for extreme events such as COVID-19 represents an extreme

event, and preparations must be made to handle the project during the uncertain times. A project's resilience is significant during this phase of its life cycle.

CONCLUSION

The role of quality managers on construction projects is a function of quality planning, quality assurance, and quality control activities. In order to establish and implement the required quality on the projects, it is essential that they are familiar with the factors that can significantly impact the quality of their projects.

Knowing the differences between various client types, the various project delivery methods, and the multiple phases of the construction projects allows the quality managers to understand their role and the extent of their responsibilities. It helps them better manage and resolve quality issues when they occur and allows them to develop and maintain effective relationships with the various stakeholders, all of which is essential to collaborate and manage the quality of the projects.

QUESTIONS

1. What factors could cause a change in project requirements?
2. How can these changes impact the quality requirements of the project?
3. Why is customer satisfaction important for future work between the consultant and the client? What can the project quality manager do to enhance it?
4. What are the types of construction projects? Give an example for each.
5. Explain the pros and cons of the provided delivery methods.
6. Give examples of the main activities that project quality managers perform in the various phases of a project's life cycle.
7. Compare and contrast the role and risk allocation between a construction manager at risk (CMAR) and a construction manager for fee (CMFF) project delivery method. How might the owner's level of involvement and risk exposure differ between these two approaches?
8. Explain how the "end use" classification of a construction project can significantly impact quality requirements. Give examples illustrating how regulatory compliance, technological advancements, or environmental considerations might differ between residential and industrial projects.
9. Analyze the potential challenges of implementing an integrated project delivery (IPD) model. Considering the risks associated with IPD, under what specific project conditions or constraints would you recommend against using this delivery method? Justify your reasoning.
10. Describe the relationship between the project life cycle phases and the effectiveness of quality management. How does the focus of quality control (QC) and quality assurance (QA) activities shift from the design phase to the construction phase?
11. Discuss the potential implications of selecting the wrong project delivery method. Focusing on the design-bid-build (DBB) versus design-build (DB) approaches, outline a scenario where the choice of one over the other could lead to significant budget overruns, schedule delays, or compromised project quality.

CONTRACT TYPES

> Understand various types and sources of contracts (e.g., Public-Private Partnership (P3), Design-Build Institute of America (DBIA), American Institute of Architects (AIA), International Federation of Consulting Engineers (FIDIC), lump sum, subcontract, consultant agreement, consultant staffing agreement, joint venture contracts, and purchase orders) along with their quality requirements. Recognize contract terms and conditions (e.g., liquidated damages [LDs], liquidated and ascertained damages [LADs], and general damages [GDs]). (Understand)
>
> Body of Knowledge I.A.2

A construction contract is a legally binding agreement between the project owner and the contractor that explains the work that the contractor is required to perform and the payment method. It defines a project's schedule, budget, quality requirements, and other necessary aspects of the project. Projects by definition are unique endeavors, and they vary greatly in terms of complexity and size. Thus, there are several types of construction contracts, and each has its advantages and disadvantages for the stakeholders involved.

To clarify the BoK language about contract administration, readers should first review the project delivery methods discussed earlier. This section focuses on contract types based on legal structure, including individual contracts, partnerships, subcontracts, joint ventures, consultant agreements, staffing agreements, and purchase orders. Later in the chapter addresses payment-model contracts, such as lump sum, unit pricing, time and material, and cost-plus.

Practitioners should clearly distinguish among the three buckets of contract types–delivery methods, legal contract structures, and payment models–to avoid any confusion.

Extensive documentation and standardized guidelines regarding these classifications are maintained by several authoritative industry bodies. Primary resources include the Construction Management Association of America (CMAA), the Design-Build Institute of America (DBIA), the American Institute of Architects (AIA), and the International Federation of Consulting Engineers (FIDIC) (see Glossary). Additional regulatory information and classification details can be sourced from the Office of Contracting and Procurement in Washington, D.C.

Individual Contract

A construction contract is a legally binding agreement between any two parties that are involved in a construction project. Such a contract can be between the project owner and the designer/architect, or between the project owner and the general contractor. Sometimes, the owners directly sign a contract with a subcontractor firm that is not the prime general contractor, or else directly contract with the suppliers of materials and services. Each one of these contracts is between two parties and must outline the scope of work, payment terms, timelines, and responsibilities of each party.

Individual contracts can be classified into many different types based on the method of payment and the terms of payment. The classification of contracts between the parties based on payment methods is described in "Payment Models and Methods" later in this chapter.

Partnership Contract

A simple partnership contract is where two parties choose to have shared control and decision-making in the project. Usually, as part of the partnership, the partners contribute capital, skills, or other resources and share in the profits, losses, and liabilities of the business. Unless otherwise specified, in a simple partnership contract, each partner is entitled to an equal share of the profits and losses. Partnerships can be structured as general partnerships, limited partnerships, or limited liability partnerships, depending on the desired level of liability protection for the partners.

Joint Venture Contracts

A joint venture (JV) agreement is a legal contract between two or more parties who agree to collaborate on a specific project or venture for a limited time. A JV contract is also a partnership contract but is only for a limited time, whereas a typical partnership contract is for unlimited time duration.

Under a JV agreement, each party contributes something of value to the venture, such as money, expertise, people, or equipment. In return, each party shares the profits or losses of the venture, according to the terms outlined in the contract agreement.

For construction projects, JV contracts are usually made when two or more construction contractors choose to pool their resources and share the risks and potential profits. This makes it possible for the general contractors to participate in bidding on a project that may be beyond their individual capacity.

The JV firms form a new and separate legal entity to bid for the project. If the JV is successful in its bid, it executes the construction not as individual construction companies but as the JV firm. As in any partnership, it is important to clearly state the purpose and scope of the agreement, contributions, roles, responsibilities (responsibility of work management, quality management, etc.), ownership and liabilities, intellectual rights, termination terms, and mechanisms for dispute resolution.

Sometimes joint ventures occur on the owner's side, not the contractors, and in this case the owners need to be clear about the quality requirements that need to be implemented for the project. The nature of business, quality standards, and personnel experience of each side can vary; and the engaged owners should design their requirements taking these differences into consideration in a way that aligns with the project scope and supports its success.

There are several ways that JV contracts differ from the simple contracts of individual firms:

1. *Profit Sharing*
   In a JV, the parties typically agree to split the profits and losses of the venture according to a predetermined formula, such as a percentage of

capital that each party brings to the table. Notice that this is different in a simple partnership contract where the contracting firm is entitled to the entire profits and losses occurring from their contract with the owner.

2. *Joint Ventures Managed Differently than the Individual Contracts*
   In a single-party contract with the owner, the management and operation of the project is the responsibility of the representative project manager of the performing organization. However, in the joint venture contracts, the JV parties frequently retain control over their respective areas of expertise and decision-making authority. In other words, not all decisions regarding the project are made by the JV partners in equal capacity—in many cases, there is a leading partner that is based on their controlling position.
3. *Liability in JV Contracts*
   Because of the way the joint ventures are managed, the liability of each party is also limited. Each of the JV partners is responsible for its own actions and is liable only for its own share of the venture's debts and obligations.
4. *Subcontracting*
   In this arrangement, the primary general contractor authorizes another firm (a subcontractor) to undertake part of the project work that it had secured with the owner. It is widely used in construction contracts because of the nature of the industry, which is diversified by type, size, complexity, and advancement in materials and methods that must be used for construction projects.

Subcontracting makes it economical for the primary contractor to utilize specialized labor and equipment as needed to complete the project activities and also transfer the risks to the subcontractor. It is especially common in the construction projects where complexity of the scope requires a specialized skill set and subject-matter expertise that the general contractor may not have. The subcontractors are hired by the project's general contractor, who continues to have overall responsibility for project completion and execution within its stipulated parameters and deadlines. The subcontracts should ascertain the exact scope and the quality of work required, and avoid overlapping in work items. This is in addition to other necessary conditions such as the schedule of the construction and the cost of the finished work. In many public infrastructure projects, the owners may require that any subcontractors that are to be hired by the general contractor must be pre-approved by the owner (client) before they may be assigned their scope of work.

Consultant Contract (Agreement)

A consultant contract, also known as a consulting agreement or professional services agreement, is a legally binding document that outlines the terms and conditions of the professional services that are provided by independent consultants to a client. Such a contract includes the scope of work, compensation, and duration for the contract. In the construction industry, consulting contracts vary a lot and are generally offered to individuals or firms that offer expert advice and subject-matter experience for highly specialized technical tasks. With their knowledge and past experience in similar work, the consultants usually help the projects avoid costly pitfalls and enhance the project's effectiveness by mitigating the risk factors. In most cases, consultants are hired for specific tasks with well-defined deliverables and time commitments, and are required to provide ongoing reports on the progress of their work and deliverables. A consulting agreement can be more expensive than a longer-duration contract, but the high cost may be offset by the invaluable guidance and expertise of the consultant and the practical solutions to the problems that would otherwise take much longer for inexperienced teams working on the same challenges. Consultants may be hired to work on any aspect of the construction project where dedicated help is warranted, such as in the scope, function, safety, quality, scheduling, budget, or any other niche area of expertise that is not frequently offered by the general contractors' teams.

Consultants are generally selected based on their technical qualifications and reputation. Some examples of the work they do include: (1) estimating the total cost of the project, (2) creating the budgets, (3) selecting subcontractors, (4) administering the construction contracts, and (5) ensuring that contractors stay within the approved budgets. The typical roles for which dedicated consultant services are sometimes requested include project managers, designers, architects, quantity surveyors, and service engineers.

Sometimes, hiring an individual consultant with subject-matter expertise is a relatively less expensive and more beneficial option than accepting the services of a less-than-experienced alternate that the general contractor can provide from their internal team.

Consultant Staffing Contract

Consultant staffing agreements are usually made with staffing firms that provide construction projects with various types of temporary and sometimes permanent employees. Their construction recruiters usually understand the nuances of the construction industry staffing and can meet the hiring needs of the projects quickly and effectively. These firms usually operate on a commission or a fee schedule that varies with the types of roles that the staffing agency may fill (temporary, contract, or permanent) and based on the nature of the work.

The key elements of a consultant staffing agency contract would include:

1. Parties involved
2. Scope of services provided
3. The roles the staffing firm may fill (temporary, contract, permanent, temporary-to-permanent)
4. Duration of the contract
5. Fees and payment terms, including how the agency would be compensated for the services provided by the staffed person (hourly rates, project fees, or a percentage of the employee's salary, and payment schedules)

6. Onboarding protocols
7. Termination clause and compete versus noncompete clauses that restrict the provided staff from working for a competitor or starting a similar business for a specific time and geographic area after working on the current contract

Purchase Order (Short-Term Contract)

Whereas all contracts are made for generally longer-term and more complex agreements, a purchase order is a short-term contract for a specific purchase and outlines the details of a transaction. Generally, the difference between all other types of contracts and a purchase order is that the purchase orders are used for transactions whereas all other contracts are used for the procurement of items or services over an extended period of time (generally greater than one year). Also, most contracts include renewal options, but the purchase order is only a commercial document that becomes a contract on acceptance by the supplier of the product or service. Clarity on the acceptance terms of the purchase order is crucial in order to create its legal standing.

Other Contract Types

A nonbinding form of contractual arrangement is called the memorandum of understanding (MOU). An MOU outlines the general intentions and key points of a future agreement but is not a legally binding agreement with specific obligations and consequences for noncompliance.

An MOU usually defines the broad understandings that are agreed on between the two parties, but it is not required to be formally accepted. It is normally considered a stepping stone toward a contract. An MOU is typically used to state an intention to negotiate and to agree to a contract at some later date. As such, it is not as detailed as a contract.

Breach of Contract Terms and Conditions

Managing the contract is crucial for every project to achieve success and minimize the disputes and damages. However, any breach of the contract can result in damages that can be incurred by one or more of the parties involved. Following is a summary of the potential legal consequences of breach of contracts, that is, what may occur if one of the two contracting parties fails to fulfil their obligations as outlined in their legally binding agreement.

Liquidated Damages (LD) or Liquidated and Ascertained Damages (LAD) Clause (in the Contract)

Liquidated damages are the predetermined sum of money that is specified in a contract to compensate for losses incurred when a party breaches the agreement, particularly when actual damages are difficult to ascertain. In construction contracts, this typically manifests as a fee per unit of time (the "liquidated amount") in the event of a missed schedule milestone such as substantial completion. An LD clause may be used in the construction contracts, for example, if the owner must get compensated if the prime contractor (or their subcontractor) is delayed in finishing the work of the contract.

Because the exact amount of damages due to a contract breach can be difficult to quantify in ongoing construction, a contractual clause for liquidated damages provides a mechanism to calculate the damages (or estimate of actual losses). It fixes the sum of damages or the rate at which damages can be assessed, in advance should a breach of contract occur.

Other terms for liquidated damages include liquidated and ascertained damages (LAD), pre-estimated damages, and stipulated damages. The "ascertained" part in the LAD emphasizes that the damages are predetermined and agreed on, which avoids the need to prove actual losses.

The biggest benefit of the LD clause is that it encourages the involved parties to comply with their contractual obligations, because they know that the damages clause can be enforced.

Also, the contractor can use a clause to factor the price of its exposure (in case of contract breach) into the bid. Also, in case of delay, if the contractor is asked to accelerate the project, they can use this clause to compare "the cost of accelerating and completing work by the schedule" against "the amount of liquidated damages."

One special provision to note is that the justice system (courts) enforces the LD clauses but may not enforce a penalty clause. This can be, for example, when the rate that is stipulated is extravagant and unconscionable in comparison with the greatest possible loss from a particular breach of the contract.

General Damages Clause (in the Contract)

General damages (GDs) are more commonly recognized for breach of contract, including delayed completion. In case a contractor or subcontractor completes the work after the contractually required date, they would generally be liable for GDs that include actual losses caused by that delay. This liability, however, would be subject to usual contractual damages requirements such as foreseeability and the duty to mitigate loss.

It requires specific drafting in the contract for the LDs to apply; and in the absence of LD provisions, GDs will apply to late completion. Therefore, GDs are the legal "default" LD provisions stated in the contract.

CONCLUSION

To summarize, in construction contracts it is important to have clear, sufficient, and exact terms in the agreement between the owner and the contractor that leave no room for interpretation and future disputes. A well-written contract serves as a system for communication and protection of rights, setting out the obligations and responsibilities of each party. This should help in eliminating unnecessary changes, reducing risks, and disputes, and overall enhancing the project's success by its delivery with the required quality, cost, and schedule.

For any project, the terms of the contract are legally binding requirements that can expose the business to risk. It is essential that the legal requirements of the contract are managed effectively. A quality manager should understand the contracts and their terms in order to manage risks and facilitate effective quality control and quality assurance efforts. This knowledge also helps the quality manager to ensure compliance throughout all phases of the project and assist with continual improvement efforts, all of which ultimately improves the collaboration between the contract parties and stakeholders, while also supporting the compliance and governance of the contract.

QUESTIONS

1. What is the difference between individual contract and partnership contract?
2. What is the difference between individual contract and joint venture contract?
3. Why is subcontracting common in construction projects? How does this practice impact the quality of projects?
4. What attributes will you consider as a project quality manager in the qualification and selection of a consultant to perform technical inspection for the project activities?
5. What attributes will you consider as a project quality manager in the qualification and selection of a third-party testing laboratory?
6. How are joint venture contracts in construction projects formed? Why?
7. How are liquidated damages different from general damages? Does this have an impact on project quality? Why?
8. Beyond the shared financial risks, how do the organizational structure and decision-making process in a joint venture contract differ from that of a simple partnership contract in construction projects, and what are the potential implications of these differences on project execution and quality?
9. Discuss how the inclusion of a preapproval requirement for subcontractors by the project owner in public infrastructure projects might impact the risk allocation and management strategies of the primary general contractor.
10. Given that both LDs and GDs address breach of contract, explain a scenario where a project quality manager might advocate for explicitly including a detailed LD clause instead of relying on the default GD provisions, and what considerations should guide the determination of the LD amount?

PAYMENT MODELS AND METHODS

> Understand the risks associated with and the differences between payment models (e.g., unit rates and reimbursable costs) and methods (e.g., Guaranteed Maximum Price (GMP). (Understand)
>
> **Body of Knowledge I.A.3**

Payment methods wield significant influence over project execution, impacting cash flow, contractor finances, and ultimately, project productivity and performance. Efficient and timely payments foster smoother operations and positive owner–contractor relationships, while delays and disputes stemming from contractual variations can strain contractor finances, potentially leading to project delays and dissatisfaction. Therefore, understanding the nuances of various payment models and methods is crucial for project quality managers, as this knowledge aids in planning quality activities and incentivizing timely defect resolution, ensuring project completion.

Payment methods significantly impact the work performed on any project. They influence the project cash flow and the finances of the contractor, and because of that may have a significant impact on the productivity and performance of the project. Efficient and timely payments lead to smoother execution and better relationships between the owners and the contractors, while delays and disputes that sometimes occur because of the different types of contracts between the parties can cause financial strain on the contractors. This sometimes results in project delays or other dissatisfaction between the parties.

It is important for the project quality managers to understand the various payment methods for projects and how the payment schedules and payment methods could adversely affect the quality of work that is performed. Although quality managers are not directly responsible for the project budgeting, understanding a contract's payment type is useful, as it helps them plan the quality activities better. The knowledge of the payment method used by the contract and its payment schedule familiarizes the quality manager with the need for timely resolution of the defects so that the contractor may get paid for the completed work. If used constructively, this knowledge can be used to incentivize the contractor to address the nonconformances quickly. It can help avoid unnecessary disputes and ensure timely project completion as well.

Below are some advantages and disadvantages of the most common payment methods used in the industry.

Lump Sum—Also Called Fixed-Price Contract

Billing with a lump sum is when the two parties agree on a fixed price for the whole job before the work begins. The contractor can ask for the payment in advance or based on project completion assessed periodically, typically monthly. It is always best to set clear payment schedules that coincide with project milestones or specific work stages. For lump-sum jobs, the contractor and the client must agree to the total cost before the work begins. In this method, the project cost to the owner is known, so the contractor may not have to produce too many backup documents to claim the costs; only a standardized report of what was accomplished and the evidence of the delivery of work.

There are two common payment schedules that are used for construction invoicing for lump-sum contracts:

1. *Progress payments based on completion of milestones*: These lump-sum payments are based on reaching certain predefined milestones. They are partial payments but are made only at defined stages of completion of the project (e.g., payment would be made at 30% completion, 60% completion, 90% completion, or any other preagreed format). This method of progress payments is not very common because contractors prefer ongoing partial payments that correspond to the amount of work they complete, even if it does not reach a scheduled milestone, which could happen for multiple reasons.

2. *Progress payments based on percentage of completed work*: In this method of lump-sum payments, the payment is not based on reaching specific milestones but billing incrementally as the job goes along. The payments may be made on a monthly basis, or biweekly or weekly basis, as

agreed. On the exact date, the portion of work completed to date is used to determine how much to bill. On the projects that include it in the contract, only the work that is approved by quality control as having met the quality requirements (i.e., no deficiency) may be paid.

When Does a Lump-Sum Contract Make the Most Sense?

The lump-sum contracts are best suited to projects that have a simple and well-defined project scope. They also work best for those projects that work to a single bottom line (profit or loss). Such contracts are good for short-term projects and cases where the long-term cash flow maintenance is less of an issue. Another advantage of the lump-sum payment method is that usually they are very easy to prepare and there isn't much backup documentation required.

Risks of Using the Lump-Sum Method

The lump-sum method of payment may be simple, but it poses risks to both owners and contractors. The risk to the owners in the lump-sum method of payment is that sometimes quality issues may be found when the contractors rush to finish the work so as to meet the deadlines and get paid. This sacrifices the quality of work and creates unnecessary follow-up and corrective actions. Also, in the lump-sum method of payment, the contractors may take advantage of the gaps in the scope of work and claim additional scope and change orders, thereby passing the unexpected expenses to the client/owner. Lastly, once the contractor is paid in lump sum, they do not have much incentive to follow up and complete the work that was found deficient or lacked quality.

For the contractors, the risk is that if the scope was defined well and if the contractor fails to estimate the costs correctly, then all additional costs that may be incurred to complete the work that was scoped out must be borne by the contractor. The negative outcome of this is a poor relationship with the owner, or worse, it leads to litigation and claims where the two parties struggle over the scope of work after the fact.

The best thing to do for the lump-sum method of payments is to ensure that the contract documents have complete and clear quality requirements, and also to perform regular quality inspections (audits) to confirm compliance with these requirements. The lump-sum contracts require clear and well-defined scope of work, accurate cost estimates, and preferably progress milestone payments (instead of single lump-sum payments). These are some of the good practices for lump-sum contracts that can help both the client and the contractor reduce the payment risks and achieve better project outcomes.

Unit Pricing Contract

A unit pricing model is one where the contractor breaks down a project into individual, measurable units of work or materials, with a predetermined price for each unit. The contractor breaks down the project price into blocks (units) of repeatable work and then charges for each unit that is completed.

For example, a contractor may determine that they can charge a unit price of $800,000 per km of installed pipeline. So, if and when the scope of the project changes and an additional km of pipeline needs to be installed, the contractor can just bill for the additional units upon completion. Each "unit price" includes all costs, overhead, and profit for completing the specific work unit. The amount of the contractor's profit in a unit price contract depends on how efficiently they complete the work.

When Does a Unit Pricing Contract Make the Most Sense?

The unit pricing method is especially helpful in situations where the needed units of work are unknown, such as the removal of soil during site preparation. This method can also be used in combination with any other billing methods. Unit pricing can make the project expenses very clear to the contractor and the client because it assigns a fixed price per unit, which helps in efficient budget management. This method is also helpful in projects that may include some flexible (extra) work, such as additional work of the same type, in which case the contractor only adds units of work to the invoice and gets paid.

Risks of Using the Unit Cost Method

A risk in the unit pricing contract is that it requires a detailed breakdown of the price that the contractor is charging to build it. If the contractor is unable to correctly determine the unit price of his work, such as not knowing the number of units that are worked, then poor pricing of the units can destroy the contractor's profit. Also, in cases where the nature of work is unique such that unit price is hard to determine, then the unit pricing method may be difficult to implement and may result in losses for the contractor.

For the client/owner, the risk is that in this method of payment, the client may find it difficult to know the total price of the project until the work is complete.

Time and Materials (T&M) Contract

In this method, the payment of invoices is based on the quantity of labor hours and the cost of the materials for the project. The contractor includes its profit margins in the hourly labor prices that are quoted to the client.

This type of contract is best used when the scope of work is expected to change or is not clearly defined. It gives the client flexibility in paying a set amount for the materials and time taken to complete the work, even when each aspect of the work is not completely known at the beginning. This lets the two parties get the work started with ease, with quickly established contract terms and conditions. The contractor does not have to estimate the total cost of the project; they get paid for their time to install each unit of materials as the project progresses.

For such projects, the client can set a maximum number of labor hours in the contract terms (using a clause similar to a not-to-exceed condition). In case the contractor exceeds the specified labor hours, the additional cost cannot be billed to the client, until a change order is issued by the owner. This is a mechanism for the client to cap the cost of the project in the T&M contracts.

A T&M contract should be structured in a way that lets the contractor bill for a sufficient amount of money to cover its fixed costs. This is because the contractors do need to cover their fixed costs even if the variable quantities may underrun (i.e., the situation where the actual amount produced or delivered is less than the quantity that was originally ordered or estimated).

When Does a T&M Contract Make the Most Sense?

This method is best used for small contracts and for self-performed work. It is preferred when the scope of work is expected to change or is not clearly defined, or where the lack of clarity in scope does not enable the performing organization to predict the time or material that may be required to complete the task. The method is also most applicable for short-term tasks but should not be the preferred method of payment for longer-duration projects (which should be better scoped out and priced ahead of the actual work).

Risks of Using the T&M Method

Because there can be no established budget for applying the T&M method, the total cost in this contract can escalate and go far beyond the initial estimates. Another risk for the owner is that the contractor may not have an incentive to prioritize efficiency, as they can keep operating and keep getting paid for all of their time and materials. Because of this risk in the T&M method, the owner must implement an extensive (quality) oversight process that would actively track the materials and work hours that are used/claimed for the project. The level of client involvement in the oversight of a T&M contract is significantly more than what is required in a fixed-price contract, as they must establish such a mechanism to oversee the time spent and materials that the contractor claims they are installing and therefore charging.

Because there are a lot of unknown variables in the projects that force owners to use a T&M contract, there are some potential disadvantages for the contractor as well. It is important for the contractor to carefully consider the advantages of T&M and determine whether or not the advantages outweigh the risks. Their risks can include cash flow issues and a potential for disputes with the owner over the costs. If the project duration is long, it is best for both parties to take time to do further research about the project's conditions and consider a fixed-price contract instead.

Cost-Plus Contracts—the Three Subtypes

Another payment model is known as cost-plus. Cost-plus construction contracts stipulate that the contractor be reimbursed for the project costs (cost) and in addition receive a fee (plus) that is usually some percentage of the cost. In these contracts, the contractor bills the owner/client for all the costs that go into the project as well as a fee for their overhead and profit. This fee can be a fixed amount set at the beginning of the contract or a percentage of the contract cost. As such, there are three variations of the cost-plus contracts:

a. Cost-plus fixed fee with no maximum

b. Cost-plus fixed fee with GMP

c. Cost-plus fixed percentage

When Does a Cost-Plus Contract Make the Most Sense?

Cost-plus contracts make the most sense for unclear scopes and requirements, and in longer-duration projects where many changes may happen (such as prices of the materials or labor costs or other reasons). These contracts offer the contractor the financial incentive to control costs and keep them low. At the same time, they give the client more visibility and details about the incoming costs and percentage profits that are made by the contractor. These contracts are designed to protect the contractor more than the owner, as the contractor can offload much of its financial risk to the client. The client takes the risk that the contractor may work to maximize its profit.

Also, the administrative process of this method of payment is demanding for both the contractor and the client. The contractor is required to provide documents (evidence) detailing all the costs incurred such as worker time sheets, material invoices, rental agreements, subcontract agreements, material contracts, and so on; the owner needs to diligently review and approve all such records.

Risks of Using the Cost-Plus Method

While the cost-plus method can be a necessity for projects that do not have a clear and well-defined scope, the expenses require a well-managed control system by the client to avoid escalation in the cost and unnecessary or not-required project features. For example, detailed quantity takeoffs become essential before releasing funds to the contractor. The risk of the use of cost-plus method is on the owner/client, and not as much on the contractor.

Following is a description of each of the three types of cost-plus fixed-fee contracts:

1. Cost-Plus Fixed-Fee with No Maximum Price

A "cost-plus fixed-fee contract with no upper limit" means the contractor is reimbursed for all actual costs incurred, plus a fixed fee for their services, with no cap on the total project cost. There is no predefined maximum amount the project can cost; the client agrees to cover all costs, even if they exceed initial estimates.

When does CPFF with no maximum make the most sense?

From the owner's perspective, the owners/clients find the CPFF contracts practical only when the project scope is not well-defined or when they prioritize quality and flexibility over strict cost control. From the contractor's perspective, this method of contracting is advantageous only in those projects that have uncertain scope or where it is very difficult to accurately estimate the costs. CPFF with no cap guarantees a profit to the contractor regardless of the project's cost fluctuations. However, the benefit is that because the contractor is paid for all costs, the likelihood of achieving a higher quality of work is more as they would not be cutting corners and rushing to finish the project and move on.

What are the risks of using the CPFF with no maximum contract?

Obviously, the biggest disadvantage is to the owner/client. This is because in such a contract, there is no guaranteed maximum price defined. As a result, the cost incurred by the contractor could exceed the budget of the owner/client. For this reason alone, there is always a higher likelihood of disputes between the owner and the contractor over the allowable costs and the contractor's expenses. Authorizing a CPFF contract with no upper limit is a decision that must be made by the owner/client after a lot of deliberation.

2. Cost-Plus Fixed Fee with GMP

The cost-plus fixed fee with GMP is commonly called the GMP method. This method of contracting is used to set an upper limit to the cost of completing a project. While the lump-sum method allows contractors to keep the additional profits in case they can complete the project cheaper than expected, the cost-plus fixed fee with GMP allows the client/owner to keep the savings. In this contract type, the contractor absorbs the additional costs beyond the guaranteed maximum price for the project. A GMP contract passes all the project's financial risks to the contractor.

When does CPFF with guaranteed maximum price make the most sense? This method is most useful in those cases where the scope of work is incomplete, unclear, or not detailed before bidding, or there is a work delay. For the client, the GMP method may result in disputes with the contractor particularly for unforeseen or unknown items.

There is a slight variation in these types of contracts as well, where some contracts allow for a contingency fund to cover the unexpected costs; this is allowed for some changes that are not in the original scope of work. Doing this creates a contract of cost-plus fixed fee with a guaranteed maximum, but with room for the unexpected.

What are the risks of using the CPFF with guaranteed maximum price contract? Whereas the GMP contract incentivizes the contractors to manage their costs effectively and complete the project within the agreed-on budget, it also may increase the possibility of the contractor wishing to complete the work within the approved budget and overrule legitimate quality concerns because they may delay the project or add costs.

It is therefore essential that all the GMP contracts establish greater independence of the quality manager from the production/construction function. In the contracts that are CPFF with GMP project arrangements and where the quality control and quality assurance personnel are not independent from the span of control of the construction manager, the nature of such a contract arrangement generally tends to impact negatively on the quality of work performed by the contractor.

3. Cost-Plus Fixed Percentage

In this type of agreement, a contractor is reimbursed for all of their actual project costs, plus a predetermined percentage of those costs as their profit or fee. The fixed percentage of the total project costs is supposed to cover any overheads, and other indirect costs, plus offers the contractor greater opportunity to make profit than most other contract types do.

When does a cost-plus fixed-percentage contract make the most sense? This method is typically used when the contractors are able to convince the client/owner of the high level of risk included in the project scope. For example, when there is a lot of ambiguity in the project scope or when the scope of work is subject to change, then the contractor may request the client/owner allow for the flexibility in project management that the contractor must plan for. These contracts also make sense in those long-duration projects where the estimated costs of materials may have significant fluctuations over the years, for example, leaving the contractor with limited profits or sometimes resulting in losses. Having a provision for cost plus a fixed percentage of the costs to be given in addition to the cost compensates for the additional risk the contractor is willing to take in such projects.

What are the risks of using the cost-plus fixed-percentage contract? Similar to the cost-plus fixed-fee contract, the cost-plus fixed-percentage contract also puts the risk of potential cost overruns on the shoulder of the client/owner. The contractor may erroneously feel they are not incentivized to control the costs because they are awarded a percentage of the costs in addition—the more their costs, the more would be their profits. For example, the contractor may procure materials that are above the required quality standard and still transfer these costs to the client.

Because of the potential for very high profits through a fixed percentage of the additional costs, this contract type requires a lot of oversight by the client. It is fair to allow such a contract only when the client/owner has the resources to appoint a strong and independent quality verification team that represents the interests of the client/owner, who must keep a very close watch on the work that is performed by the contractor.

CONCLUSION

All of the above-mentioned types of contracts have their advantages as well as their shortcomings. There is no one set of conditions that would qualify one contract payment type to be more useful than others. Also, the usefulness of one contract type over another becomes a subjective opinion that is based on the perspective from which it is viewed—what is considered an advantage for the owner may not be seen the same way by the contractor firms.

Their competing priorities, their resource constraints (including the availability of oversight resources to oversee the quality performance of the contractors), and the inherent complexities and risks of each project must be the basis of determining the most appropriate contract payment type. A quality manager should be familiar with the details of the contract type selected for their project, in addition to the delivery method chosen for the project delivery. Knowing so enables the quality manager to facilitate the project to achieve its collective goals and objectives.

Navigating the complexities of payment models and methods requires careful consideration of project specifics, stakeholder priorities, and inherent risks. No single approach fits every scenario, as the suitability of each model hinges on factors like scope clarity, risk tolerance, and resource availability. Quality managers must possess a thorough understanding of these payment dynamics to effectively facilitate project objectives, ensure timely defect resolution, and foster positive relationships between owners and contractors. Ultimately, informed payment strategies are integral to successful project execution and achievement of desired outcomes.

QUESTIONS

1. What are the risks of using a lump-sum payment method for the owner, and for the contractor?
2. When is it appropriate to use the unit pricing method (give example). Also, when is it not appropriate to use the unit pricing method?
3. What is the difference in the payment approach between the unit pricing method and the time and materials (T&M) method?
4. What are the various types of cost-plus methods of payment? What are the risks associated with each?
5. What are the quality risks that can result from a poor payment system or slow payments?
6. How can a clear payment schedule in the contract protect the rights of the client and the contractor?

7. Beyond just price predictability, how does the choice between a lump-sum contract and a T&M contract directly impact the dynamic of risk allocation between an owner and a contractor, and how might this allocation affect quality oversight strategies?
8. Consider a project with a largely defined scope but potential for minor scope variations throughout the project life cycle. Discuss the rationale for and against using a unit pricing contract versus a cost-plus contract with a guaranteed maximum price (GMP), emphasizing the mechanisms for cost control in each approach.
9. In situations where a cost-plus fixed-percentage contract is deemed necessary, what specific safeguards can a client implement to mitigate the risk of inflated costs and ensure that contractor profits are reasonably aligned with project efficiency and quality?
10. How might the dependency or independence of the quality management team from the construction manager impact the effectiveness of different payment models, particularly in preventing quality compromises due to financial incentives or pressures inherent in those models?

PAYMENT DOCUMENTATION REQUIREMENTS

> Understand payment documentation requirements, such as schedule of values, percent completion determination (cost complete vs. work complete), retainage, billing methods, and billing terms. (Understand)
>
> **Body of Knowledge I.A.4**

In a contract, the payment terms and documentation are very important; they should clearly outline payment schedules, milestones, and methods in order to avoid financial disputes. Payment applications are instruments used by contractors to request progress payments throughout the project, and they include detailed documentation about what work was done, which materials were used, and what payment is due. A contractor typically submits the applications every month to reflect work completed within that month, and the client is contractually required to make the payment within a specified period. Payment applications submitted by the contractor to the client are a standard part of any construction project. They help the client identify and monitor the costs and validate payments. However, they can be incomplete and error-prone; for example, containing duplicative charges, inaccurate data, or inflated prices. This results in over or under charges that can put any of the parties at risk. Therefore, the payment applications should be organized and supplemented with accurate and proper documentation.

In a payment application, the contractor is typically required to show the status of the contract sum to date, the total dollar amount of work completed and stored to date, the amount of retainage, the total of previous payments, a summary of change orders, and the amount of current payment requested. The documents that are typically included in an application payment are:

- *Payment application form*: A wide variety of templates are available for the forms.
- *Schedule of values*: A list of every work item along with each item's cost.
- *Change orders*: A written agreement between the client and the contractor that amends an existing contract (e.g., to change the contract amount, work requirements, or schedule).
- *Invoices from suppliers or vendors*: Invoices of materials procured during the current pay period but that have not been paid yet.
- *Receipts*: Purchased materials with paid invoices.
- *Visual documentation*: Photos, drawings, or other visual proof of work completion or materials currently stored.

Below are some good practices that help the two parties validate costs, mitigate financial risks, and maintain a good relationship.

Schedule of Values

Every payment application should have a well-organized and detailed breakdown summary of items being built. The schedule of values (SOV) is a list of every work item on a construction job, along with each item's cost. It depicts the amount of work completed and requested for payment, in addition to the percentage completed. The client and contractor should develop a mutual agreement about a well-organized SOV outline to avoid complications in processing payments, such as difficulties in verifying the percentage of work completed or inaccurate inputs. Establishing the SOV upfront helps in managing expectations among the parties and minimizes disputes.

It is critical to have organized line items in the SOV and to make sure all subcontractors and suppliers have their lines on the SOV. For instance, if five or six contractors are executing a site work, the SOV should show their separate charges and how they ladder back up to the requested payment. In addition, documentation of labor, when applicable, helps to make sure the payment requested correlates with the hours worked. As the project progresses, this will result in easy and more accurate validation of the charges and claimed percentage of work completion. This in turn should improve the cash flow and help in better schedule control. Furthermore, fees and markups should be separated in the SOV and have their lines, which makes it easier to validate costs and figure out discrepancies.

Percent Completion Determination

The contract document usually defines the process for submitting a payment application, including the documents to be submitted and the submission deadline. A payment application, including the schedule of values, shows the percentage of each item of work that a contractor has completed, the amount to be paid, minus any retainage that the client is withholding. The contractor submits the payment application before the deadline of each pay period for the work completed and materials purchased within that period. Because each payment application builds on the one before it, the contractor needs to be accurate and verify the figures before submitting.

Retainage (Holding Back Payment)

A retention clause in the contract allows the client to withhold a percentage of the payment until project completion to ensure all work is performed to the required standard. In some countries, the term "retainage" is used. These clauses can be useful for ensuring quality; however, they can significantly affect the cash flow of the contractor. Thus, the retention clauses must be fair and reflect the value of the work still to be completed.

Billing Methods

Different construction contracts are used for different project delivery methods; consequently, there are several construction contract pricing structures. The most common are:

- *Time and material (T&M) contracts:* The payment of invoices is based on the number of labor hours and the cost of the materials for the project. The contractor includes its profit margins in the hourly prices quoted to the client. It is important to keep a good record of costs, as the client or general contractor may want to review them before payment.
- *Lump-sum contracts:* The price is a set amount. A lump-sum contract is typically used when the total costs for the project can be easily estimated. In small projects, a subcontractor may bill with a simple invoice at the end of the project. In larger projects, progress billing is used to give the client more control over the work as it progresses and to provide ongoing payments to subcontractors. In lump-sum contracts, generally, there is not much backup documentation required for the billing.
- *Cost-plus contracts:* The contractor bills the owner for each cost that goes into the project, in addition to a fee for the contractor's overhead and profit. The fee can be a fixed amount set at the beginning of the contract or a percentage of the contract cost. Documentation is required for the client to verify the expenses.
- *Unit price contracts:* Contractors break down project prices into blocks (units) of repeatable work and then charge for each unit completed. This is a straightforward billing method except when there is disagreement about how many units are completed. The contractor provides documentation to prove the amount of work completed.

Each billing method has its own advantages and disadvantages, and it is important to make sure the chosen one is appropriate for the specific needs of the project and its risk profile.

Billing Terms

The project client must outline the payment application and process requirements in the contract before the project's outset to properly manage expectations and provide complete clarity for the involved parties. Also, the contractor should read the contract closely before submitting any application and ensure what is submitted meets the requirements. The contractor should be as thorough and detailed as possible when applying for a payment because if a deadline is missed or a required document is missing, the application could simply be rejected. Mistakes in the application mean payment delays, which can impact the cash flow. In some cases, the contractor may even have to wait until the next pay period to apply again. The more evidence the contractor provides to back up the application request, the more likely the client approves it quickly and releases the payment. Remember, contractors get paid for what they claim *and support* with documentation, and this requires a complete and accurate payment application.

CONCLUSION

Accurate and thorough payment documentation is essential for successful construction projects. Clearly defined payment terms, meticulously prepared schedules of values, and comprehensive supporting documentation minimize disputes and ensure smooth financial transactions. Understanding various billing methods and the implications of retention clauses enables both clients and contractors to manage expectations and mitigate risks. By adhering to these best practices and maintaining open communication, all parties can foster a positive working relationship and facilitate timely, equitable payments, ultimately contributing to the project's overall success.

QUESTIONS

1. What are the documents submitted in a typical payment application?
2. What is a schedule of values? Why is it important?
3. How is a retention clause in a contract useful for the project's quality?
4. Why are there various contract pricing structures?
5. Give examples of reasons that cause clients to reject payment applications.
6. Compare and contrast the documentation requirements for billing under a time and materials (T&M) contract versus a lump-sum contract. Explain why these differences exist and what risks they mitigate for each party involved.
7. Discuss the potential impact of an inaccurately prepared schedule of values (SOV) on project cash flow and schedule control. Give examples of how errors in the SOV can lead to disputes and delays.
8. Analyze the dual purpose of retainage clauses in construction contracts. Explain how they can be beneficial for the client in ensuring quality and the potentially negative consequences for the contractor's financial stability.
9. Describe the difference between "cost complete" and "work complete" in the context of determining percent completion for payment applications. Why is it crucial to have a clearly defined method for this determination in the contract?
10. Explain how proper visual documentation, such as photos and drawings, can support a payment application and why it might be particularly critical in cases involving specific billing methods like unit price contracts or disputes over work completion.

Contractual Requirements

SCOPE, SCHEDULE, AND RESOURCE PLANNING

> Define a project's scope components (e.g., boundaries), develop the schedule and resource plan, track milestones, and identify roles and responsibilities in terms of quality of different entities (e.g., trade partner, contractor, owner, and quality manager). (Analyze)
>
> Body of Knowledge I.B.1

Every project is defined by a scope that needs to be completed within a certain time frame and within a specified budget. Making changes to the scope without proper control leads to scope creep; missing project deadlines can lead to missed market opportunities and dissatisfied stakeholders; and increased costs cut into organizational profits and might hinder future investments. For a successful project, disciplined scoping, scheduling, resources planning, and budgeting are essential.

Define Project Scope

A project scope statement is a foundational document that sets the stage for successful project execution. It defines the project objectives, deliverables, boundaries (what is part of the project and what is not), constraints, and acceptance criteria. The constraints may include any physical, legal, or policy issue that limits the options of the project team or constrains their efforts in completing the project. Constraints may or may not change during the project. If they do change, they may lead to opportunities for the project team, but they may also pose new risks for the team to overcome.

A well-defined scope is crucial for aligning the project activities with the organization's goals and managing the stakeholders' expectations. On the other hand, a poorly defined scope increases the risk of a project delay, exceeding budget, or both. This makes documentation and approval of the scope a necessity before proceeding with subsequent phases of the project.

The project scope statement sometimes gets confused with a project scope of work. A scope of work is an agreement of work—typically between the client and the contractor—that details the work to be performed, including products, project timeline and milestones, and progress reports. Breaking the scope into a structure from large to small components—work breakdown structure (WBS)—assures that all the scope activities are identified and defined within a common framework. Utilizing a WBS is an effective means to subdivide the project phases into manageable elements that describe the scope of work within each element.

Changes in projects happen, and consequently, the management of the scope should include a formal change-control process to manage changes and ensure the project can still be delivered on schedule and within budget.

Develop the Schedule and Resources Plan

Scope, time, and cost are project constraints that are interdependent. For example, a reduction in the scope reduces a project's duration and schedule. The time represents the total schedule duration for the project from its initiation to closure. It also includes estimates of the work activities' duration, their dependencies, and determination of the project critical path. The critical path determines the minimum time for the project completion. Accelerating the project schedule typically requires addition of resources or reduction in scope. Since the WBS is used to identify all the project activities, different scheduling techniques are used to determine the total duration of the project. There are many tools to plan a project schedule and track and manage the timelines such as Gantt charts, network diagrams, and program evaluation review technique (PERT) analysis.

Managing a project schedule tends to be a bigger challenge than managing cost and it is probably the hardest element to manage in a construction project. Therefore, during bidding it is important to focus on the contractor's capabilities to deliver the project in the necessary time frame. However, the client should plan for schedule delays and extensions, and not agree to something that is not realistic or cannot be met without cutting corners. This would negatively impact project quality and increase risk. The project manager is expected to complete the project deliverables within the agreed schedule dates. If the project is running late, the client needs to be informed to decide how to proceed. If the opposite happens, the client should also be informed as other projects could be started earlier. Contractors often complain that the lack of timely decision-making during construction directly and adversely affects overall progress. Thus, timely decision-making by all parties helps the project move smoothly and avoid delays. While clients could have the right to change the requirements (e.g., request new better materials), the impact of these changes on the schedule is to be considered.

Once the project scope is defined and the WBS is created, the cost for each stage of the project life cycle can be determined. This is by specifying the resources needed, including labor, materials, machinery, and so on, which informs the project budget. At this stage, project estimators work closely with contractual professionals to develop accurate estimates for the costs associated with the project. The costs can be: (1) direct costs that occur in a project and are attached to specific activities, such as the cost of raw materials, labor, and supplies; or (2) indirect costs that are in support of the project, such as rents, salaries of the administrative staff, and cost of utilities. Also, they can be classified as (1) fixed costs, which are those that don't change throughout the life cycle of the project (e.g., setup costs and rental costs); or (2) variable costs that change due to the amount of work performed (e.g., hourly labor wages). Sunk costs are another type that describe an investment that has already been incurred and cannot be recovered (e.g., marketing, and research).

Identify Roles and Responsibilities

Depending on the size, type, and risk of the project, the following key entities are generally involved in a project's construction and its quality management.

Client

This is usually the owner of the project and the entity responsible for the project's future operation and maintenance (O&M). The client ensures all parties are contractually engaged to plan, design, construct, commission, and operate the project; are qualified; and have properly defined roles and responsibilities. The client also provides the necessary funding to achieve the project objectives in a timely manner and makes regular payments according to the contractual provisions and progress of the work.

Designer

The designer is responsible for preparing the design packages (drawings and specifications) for the project; this can also include preparation of the bidding documents. The designer also interacts with the consultant, as necessary, to assist in design-relevant matters during the construction, such as reviewing contractor submittals that have design relevance to ensure they are consistent with the design intent. In some cases, the designer is retained to provide the functions of the consultant.

Contractor

The contractor is expected, throughout the project duration, to have sufficient resources available to perform and complete the work in accordance with the agreed schedule and budget. Also, the contractor is directly responsible for the quality of the project and should describe the measures used to perform the work in method statements. In addition, the contractor inspects and tests the work to verify compliance with the design and quality requirements. The contractor is responsible for employing competent staff with experience working on similar types of projects. Competent and experienced personnel will aid the team in identifying when variations to specified procedures are necessary.

The role of a quality manager in a construction project is crucial, as the manager ensures the project adheres to rigorous quality standards throughout its life cycle. The responsibilities include overseeing contract development, strategic quality planning, resource management, documentation, and project closure. The manager is expected to apply principles to plan, manage, and control the quality of the project, thus ensuring the final deliverables meet stakeholders' expectations and industry and quality requirements.

Trade Partner

A trade partner is a specialized contractor or firm that provides expertise in a particular trade or field such as electrical work, plumbing, or HVAC. The high level of skill and precision in their work better ensures that the project meets industry standards and regulations, and leads to higher-quality outcomes and more efficient processes. They can also have access to the latest technologies and techniques in their trades, which enhances the overall project performance. The main difference between trade partners and subcontractors is that trade partners are onboarded early in the design stage, and they participate in the development of execution strategies, project schedule, and cost. Subcontractors are usually hired at a later stage, after the execution plan is prepared.

Consultant

The consultant is responsible for construction oversight and, typically, quality assurance. They also provide the client with contract management support, ensuring an effective link between the owner and the contractor. The consultant ensures that the construction work is carried out in accordance with the contract design and specifications. The team typically includes full-time specialists, quality control inspectors, and quantity surveyors. They review and approve the contractor's quality management plan and monitor its implementation and effectiveness throughout the project. The consultant reviews the contractor's test records for compliance with project design and specifications, as well as reviewing and approving the contractor's method statements before construction. In addition, they liaise with the designer to resolve design changes during construction. The consultant should also have the capability to undertake independent inspections and tests necessary to confirm construction compliance to the requirements.

CONCLUSION

Effective scope, schedule, and resource planning are the bedrock of successful project execution. By meticulously defining project scope, developing realistic schedules, and strategically allocating resources, stakeholders can mitigate risks and ensure desired outcomes. Clear communication, robust documentation, and a formal change-control process are essential for navigating complexities and adapting to changes. Ultimately, adhering to these fundamental principles empowers project teams to deliver quality work within specified timelines and budgets, satisfying all stakeholders and securing future opportunities.

QUESTIONS

1. Why is the documentation of a project scope important? What should be covered in the document?
2. How is a project scope statement different from a project scope of work?
3. What is work breakdown structure (WBS), and why is it important?
4. Why should the client be informed about project delay or early completion?
5. How are the cost and schedule of a project determined?
6. What are method statements, and how are they used by contractors?
7. Beyond merely listing them, explain how the interconnectedness of scope, time, and cost constraints can create a ripple effect on a project, using a specific hypothetical scenario to illustrate your point.
8. Compare and contrast the roles of trade partners and subcontractors as outlined in this section, focusing on their impact on project planning versus project execution phases.
9. The section mentions the importance of a formal change-control process. Describe a situation where a seemingly minor change request could significantly impact

multiple aspects of the project (scope, schedule, resources) and what steps the project manager should take.

10. Discuss the significance of identifying both direct and indirect, as well as fixed and variable costs, during the resource planning phase. How does understanding these different cost categories contribute to effective project budgeting and risk management?
11. Considering the varying perspectives of the client, contractor, and consultant, how might potential conflicts arise regarding schedule management, and what strategies can be implemented to mitigate these conflicts and ensure timely project delivery?

QUALITY MANAGEMENT INPUTS

> Determine and analyze quality management inputs (e.g., Certificate of Authorization (CoA), input for proposals, and Request for Proposal (RFP)) to adhere to a project's quality standards. (Analyze)
>
> **Body of Knowledge I.B.2**

Adhering to a project's quality standards necessitates a comprehensive understanding and analysis of various quality management inputs. These inputs, such as the Certificate of Authorization (CoA), proposal components, and the request for proposal (RFP), form the bedrock of successful project execution. By meticulously examining these elements, project quality managers can ensure alignment with requirements and expectations. This section delves into the intricacies of these critical inputs, elucidating their significance and the role they play in maintaining quality throughout the project life cycle.

The project quality manager needs to determine and analyze quality management inputs to adhere to a project's quality standards and requirements. Such inputs may include but are not limited to a CoA, input for proposals, and an RFP.

Certificate of Authorization (CoA)

The practice of professional engineering covers several activities, such as planning, designing, composing, evaluating, advising, reporting, directing, or supervising works that require the application of engineering principles. Individuals or entities that engage in the practice of these professional engineering activities are typically mandated to apply for a permit to practice. Safeguarding life, health, property, the public welfare, or the environment requires the application of engineering principles.

A CoA allows an individual and a business entity to offer and provide professional engineering services to the public. These CoAs are distinct from licenses issued to individuals to practice professional engineering. Therefore, an individual or an entity cannot offer to the public, or engage in the business of providing to the public, services that are within the practice of professional engineering without a certificate of authorization. The CoAs are issued by government organizations that license professional engineers, maintain professional standards, and set codes of professional conduct and ethics.

As mentioned, an individual who advertises and promotes (personally or through an entity) as offering professional engineering services is required to obtain a CoA. Additionally, professional engineering services provided to the public through the sale of a product—custom designed or an original (not off-the-shelf)—require a CoA. The certificate is required for offering professional engineering services on a part-time and volunteer basis as well. In such cases, the individual should avoid potential conflicts of interest with the employer and provide the client with a written statement about the nature of employment and the attendant limitations on the services to the client. Additionally, a company that does not hold a current CoA that is in accordance with a Professional Engineers Act is not authorized to advertise its services as "engineers." Furthermore, the holder of a CoA is typically required to obtain insurance for professional liability.

Input for Proposals

A proposal that is professionally written can make the difference between winning and losing a construction contract; this is especially true when the price is not the only focus of the client or when there is significant competition. Therefore, it is crucial for the contractor to develop a well-written and thorough proposal that is customized to the client's needs.

Contractors usually participate in various bids to fill their pipeline and secure future work. To do so, they submit a project proposal, and in that document, they try to stand out from competitors that are bidding for the same project. In the proposal, the contractor should work to accurately inform the client about the project requirements, its timeline, cost, and other conditions necessary to complete the construction. The document usually starts with a cover letter, company profile, and project overview. In its preparation, the contractor evaluates the project's objectives and specifications. By evaluating the scope, it comes with a cost estimate that factors in the direct and indirect costs, such as subcontractor quotes, profit markups, and overhead. The proposal should also contain the quality standards to be applied. In addition, the document is expected to provide the client with a breakdown of the schedule phases, preconstruction and construction responsibilities, key team members, relevant experience, and examples of previous projects. The contractor may also attach supporting documents to the proposal, such as team member resumes and insurance documents. Depending on the contract delivery method (lump-sum, cost-plus, etc.), there could be more than one proposal for different phases of the project, such as both an engineering and construction proposal.

A proposal document should contain the following at a minimum:

- The names and contact information of the client and contractor
- The location of the project

- Detailed description of the scope of work (possibly referencing drawings or blueprints)
- Quality standards
- A schedule, cost, and payment plan

Additionally, the proposal may contain change order policies and relevant warranty information. Among the above list, it is crucial to have a clear, comprehensive, and specific scope of work. This is to avoid potential ambiguity and change orders afterward. If the work description (e.g., in a construction drawing) is not specific, the product would not be known until it is completed. This may result in customer dissatisfaction and making changes that increase costs and result in delays.

Request for Proposal (RFP)

A request for proposal (RFP) is a formal document used to solicit bids from qualified contractors for a specific project. The document outlines the project's details such as scope, objectives, budget, duration, and evaluation criteria. It enables the client to clearly communicate the requirements and expectations, and invites the contractors to submit detailed proposals to qualify them for the job.

The competitive nature of RFPs ensures the client receives better competitive offers that allow for informed decisions when selecting a construction partner. The process also promotes transparency, fairness, and ultimately contributes to the success of the project.

In the realm of construction, there are two crucial components in the RFP: (1) a request for qualifications (RFQ) and (2) a request for information (RFI). Each type serves distinct purposes.

The request for qualifications is a preliminary document used by a client to evaluate and shortlist potential contractors based on their qualifications. It establishes a pool of qualified contractors that can later be invited to submit detailed proposals and is used as a tool to streamline the selection process by ensuring only qualified contractors bid for the project. The document focuses on a contractor's capabilities, past performance, financial stability, and other qualifications relevant to the job. It is concerned with the contractor's ability to successfully undertake the project rather than the specific project details. In the RFQ, the contractor provides information about the company's history, relevant experience, certifications, and key personnel.

The request for information is another tool used to clarify uncertainties or gather information that might impact the bidding process; however, it is not a solicitation for bids or proposals. An RFI is valuable when project requirements or specifications are unclear, and more information is needed for accurate proposal submissions. For example, common phrases such as "or equivalent" or "to match existing" in the specifications might lead to disputes over whether a contractor's choice is equivalent to the specified product and matches the client's expectations. In this case, the contractor can apply an RFI to clarify the requirement and remove the ambiguity. Generally, a contractor might submit an RFI to seek clarification on project requirements, technical specifications, or scope ambiguities in the RFP. On the other hand, a project client benefits from the RFI by receiving insights into potential issues early on, so adjustments can be made before the formal proposal submission. Therefore, the RFI is a mechanism to effectively promote communication between the client and the contractors that contributes to the project's success.

CONCLUSION

A comprehensive understanding and thorough analysis of quality management inputs, such as the CoA, proposal inputs, and the RFP, are essential for upholding project quality standards. These elements collectively provide a framework for project execution, ensuring alignment with regulatory requirements, client expectations, and project objectives. By diligently addressing each input, project quality managers can mitigate risks, foster clear communication, and ultimately drive project success, delivering outcomes that meet and exceed stakeholder needs.

QUESTIONS

1. What is the purpose of the Certificate of Authorization (CoA)? Who issues them and why?
2. In what situations does it become crucial for contractors to develop a thorough proposal customized to the client's needs?
3. When does it become possible to have more than one proposal for a project? Give examples.
4. What is the difference between request for qualifications and request for information in the RFP process?
5. Which document is concerned with the contractor's ability to successfully undertake the project rather than the specific project details?
6. Beyond simply offering engineering services, what specific scenarios require an individual or entity to obtain a CoA, and what are the potential consequences of not doing so?
7. Considering the various components of a proposal, how does a contractor effectively balance providing detailed information with avoiding potential ambiguities that could lead to future disputes or change orders?
8. Explain the interconnectedness between RFQ, RFI, and RFP in the context of construction projects. How does each stage contribute to the overall success and efficiency of the bidding process?
9. If a contractor identifies an area in the RFP where the specifications are unclear, what proactive steps can they take beyond submitting an RFI to ensure their proposal accurately reflects the client's intended needs and minimizes risks?
10. How might the requirements for a CoA influence the content and quality standards presented in a contractor's proposal, particularly in terms of liability and insurance?

STAKEHOLDERS

> Identify and acknowledge internal and external stakeholders and engage them. (Apply)
>
> Body of Knowledge I.B.3

Stakeholders in a construction project are individuals, groups, or entities that have an interest or a stake in the project. They can influence or be influenced by its outcomes, and their involvement is important to project success. The stakeholders can include owners, contractors, architects and designers, inspection agencies, regulatory authorities, investors, end users, local community, and the general public.

The expectations from the project and the requirements of all stakeholders are unlikely to coincide, and some of them will seek to influence the project to meet their own requirements. This pressure or influence often leads to changes that increase the project's complexity and risks. However, if stakeholder concerns are not addressed, the project is also unlikely to deliver optimum value for all involved.

The identification of a stakeholder can be complex. For example, identification of the client is sometimes difficult, as there may be a designated single point of contact who is not really the "client" but rather is a representative of the client's organization. Also, within the client organization there often is a range of individuals with very different "stakes" in the project. These individuals with differing needs result in a "multiheaded" client, which makes decision-making a complex process (as questions cannot be answered directly by the designated representative). The representative needs to negotiate with the various stakeholders within the client organization before providing a response. When trying to meet the diverse goals of your multiheaded client, it can be difficult to find the right solution that satisfies those individuals and prevents negative impacts to the project.

Consequently, in construction projects it is important to strike the right balance between stakeholder involvement and isolation of the project to achieve delivery on budget and on schedule and also maximize benefits for the stakeholders. It is also essential to gain a clear understanding of who the stakeholders are, what their expectations are, and what motivates them. This process is called stakeholder analysis.

Identification of Key Players

Stakeholder analysis is a strategic process used to identify, understand, and assess the interests and expectations, motivations, and influence of the stakeholders that can affect or be affected by a project. The analysis is conducted to better understand and gain insights into the dynamics and relationships among the project stakeholders so project managers can make informed decisions and develop effective strategies for engagement. This analysis extends beyond the hard-and-fast technical specifications of the project, and it is primarily communication-oriented—the project manager spends a significant amount of time communicating and clarifying the requirements of a variety of project participants.

Each participant, or stakeholder, has many interests that change during the different phases of the project, which can result in changes to the technical requirements that had been assumed to be stable. There are also several nontechnical requirements that need to be addressed for the project to be a success, such as a team member's requirement to know the project goals and the team's role throughout all project phases. This is in addition to the financial sponsor's requirement (sufficient confidence at the beginning of a project that the money will be spent effectively) and the end user's requirement (the resulting delivered product will be functional and will meet the expectations).

Stakeholder influence is often felt in the early stages of the project, when the project is still flexible and can be more easily changed. Once the project starts progressing, it takes on more momentum, and thus the effort to stop or alter its direction is far greater. While stakeholder influence may decrease at the start of construction, it will often increase again as the project nears completion and handover.

Acknowledging Internal and External Stakeholders

Project stakeholders can be broadly classified into two groups: internal and external. External stakeholders are individuals or organizations that are not part of the client organization but have an interest in the project. Internal stakeholders can be anyone from within the organization, such as the users of the project or related marketing, IT, human resources, or other departments or employees that can affect or influence the project.

For publicly funded projects, the number of stakeholders can easily reach several dozen or more. Examples of main stakeholders include:

- Funders (e.g., government departments, grant providers, or partners)
- Users (e.g., passengers for a transportation project)
- Regulatory authorities (e.g., planning authorities involved in rail projects)
- Those affected by the project (e.g., neighbors living near the network system)
- The press and media (a significant group that can greatly influence the public perception of the project)

Stakeholders' Analysis and Engagement

In stakeholder analysis, often a range of techniques or tools are applied to identify and understand the needs and expectations. Also, understanding the attributes, relationships, and interfaces among the project advocates and opponents assists in strategically planning the project. Typically, the project management team begins with managing and then influencing the stakeholders' expectations to ensure a successful project.

The analysis is best accomplished before a project is initiated or at its beginning phase. Because the analysis can involve sensitive information, the facilitator should manage unproductive interests and hidden agendas when discussing stakeholders. Additionally, the team members should have sufficient trust among themselves, carefully reveal issues, and deal with potentially undiplomatic information.

Table 1.1 Stakeholder analysis table.

Stakeholder	Interests	Impact	Priority
Client	Avoid liability Reduce pollution	Mid High	1
Sponsor	Expand market	Mid	3
Inspection agency	Comply to standards and specifications	High	2
Project management	Avoid going over budget	Low	4
Local municipality	Protect the environment	Mid	5

1=highest priority

Source: Ramu, Govind. "Lineup of Players," *Six Sigma Forum*, 16:1, 23–25.

Below is a simple three-step approach to conducting stakeholder analysis and engagement: (1) identify project stakeholders, (2) assess stakeholder importance and influence, and (3) define level of influence.

Identify Project Stakeholders

To be classified as a stakeholder, the person or group must have some interest or influence that could impact the project. A brainstorming session with appropriately selected members and a facilitator can be conducted to initially consider all the stakeholders and possibly drop some in later stages of the analysis. To get a more powerful understanding of needs and expectations, it is helpful to be specific and identify stakeholders by name rather than generic terms such as customer, sponsor, owner, and so on. Sometimes, it is difficult to determine who is considered truly inside or outside the project context.

The stakeholders can then be listed in Table 1.1 with their key interests, potential impact on the project, and priority in relation to other stakeholders. The analysis should take a dynamic view and consider changes in these parameters over the life of the project.

Assess Stakeholder Importance and Influence

A stakeholder with high influence can control key decisions within or over the project. Such influence is usually derived from hierarchical, economic, social, or political position or power. Also, influence can be derived from individuals with personal connections to others of influence, or those with expert knowledge, and so on. The project cannot be considered successful if needs, expectations, and issues of stakeholders who are deemed "important" are not addressed. For example, the users of the project's product are typically considered of high importance.

Define Level of Participation

Following the effort to understand the stakeholders, the next step is to assess their level of participation and information needs. All the stakeholders will not be involved in all aspects of the project or in all its life cycle phases. The previous steps should help identify potential groupings of stakeholders; those individuals should receive similar project information. A participation matrix (a variation of a typical RACI matrix, as shown in Table 1.2) can be used for strategizing the involvement of stakeholders. The matrix shows the role of each stakeholder in the life cycle stages of the project.

Table 1.2 Stakeholder engagement matrix.

	Inform	Consult	Involve	Collaborate	Support
Initiation	Stakeholder B Stakeholder C		Stakeholder C	Stakeholder A Stakeholder B	
Planning	Stakeholder A		Stakeholder C		Stakeholder C
Execution		Stakeholder B	Stakeholder D	Stakeholder A	
Control				Stakeholder C	Stakeholder B
Closing	Stakeholder E	Stakeholder D	Stakeholder B	Stakeholder A	

Source: Ramu, Govind. "Expert Answers," *Quality Progress*, 54:12, 8–9.

The tools in Tables 1.1 and 1.2 should assist in analysis and understanding of the involved project stakeholders and their needs, influence, importance, and participation throughout the project life cycle. These tools can also be presented in different formats or creatively combined and used to better manage stakeholders.

CONCLUSION

Effective stakeholder analysis and engagement are paramount to project success. By meticulously identifying, assessing, and managing stakeholder expectations and influences, project managers can navigate complexities and mitigate risks. Striking a balance between involvement and strategic direction ensures the project stays on track while maximizing benefits for all parties. Employing tools like stakeholder analysis tables and engagement matrices facilitates informed decision-making and fosters clear communication, ultimately leading to the delivery of projects that meet diverse needs and achieve optimum value.

QUESTIONS

1. How can the key stakeholders of a project be identified?
2. What should be done in case there is a conflict of interests among stakeholders?
3. What are nontechnical requirements that need to be addressed during this stage? Give examples.
4. How different is a stakeholder's influence at the start of a project compared to during the project execution?
5. How can you encourage stakeholder engagement?
6. What is the difference between stakeholder influence and importance? Give an example.
7. How does the concept of a multiheaded client impact the stakeholder analysis process, and what specific strategies can a project manager employ to navigate this complexity?
8. Explain why stakeholder influence tends to fluctuate throughout a project's life cycle, with a specific focus on the early stages and near completion, and discuss the implications of these fluctuations on project management strategies.
9. Beyond simple identification, how does the determination of stakeholder "importance" differ from the assessment of stakeholder "influence," and why is it crucial to distinguish between these two factors in project planning?

10. Using the concept of the stakeholder engagement matrix (Table 1.2), describe a scenario where a stakeholder might transition between different levels of participation throughout the project, and explain the reasons behind this shift.
11. Given that sensitive information may be involved in stakeholder analysis, what ethical considerations and communication strategies should a facilitator employ to manage potentially unproductive interests or hidden agendas while still ensuring necessary information is gathered?

LOCAL CONTRACT OPPORTUNITIES AND CONTRACT VALUES

> Understand the impact of local contract (e.g., DBE, SWaM, local small businesses and suppliers) opportunities and contract values (e.g., allocation percentages to local resources) in relation to the quality of a project. (Understand)
>
> **Body of Knowledge I.B.4**

Supply chain resilience is a key strategic objective and a national security concern for countries, especially with the increasing global interconnection and trade. Thus, several countries have developed programs for local content and have established policies to support small and medium-size enterprises, and disadvantaged business enterprises to contribute to the development of the national economy. These local content programs encourage government organizations and government-owned companies to purchase from private sector service providers and suppliers based on the size of their contribution to local content. This enhances the competitiveness of local industries and drives innovation and confidence in quality of local manufactured materials and designs. By building local capabilities and creating opportunities for private sector players across the value chain, countries aim to build a regional and global position in the business market.

Examples of Local Contract Opportunities

In the United States, for example, the State of Virginia has an agency called the Virginia Department of Small Business and Supplier Diversity (SBSD), which is dedicated to enhancing the participation of small, women-, and minority-owned businesses in Virginia's procurement opportunities. Its mission is to provide education and assistance, economic opportunity, and access to capital for small businesses in order to generate employment and economic growth. The SBSD is responsible for administering two certification programs: (1) the Small, Women- and Minority-Owned Businesses under Virginia's "SWaM" Procurement Initiative and (2) the Disadvantaged Business Enterprise (DBE) Program. The second is a federal certification program under the U.S. Department of Transportation (DOT). The SBSD also provides support to certified businesses, in cooperation with other agencies and departments, through business development and procurement advocacy programs.

The SWaM is one example of initiatives that governments create to enhance business opportunities for SME or minority-owned companies. A SWaM vendor is a business that has been certified and listed in the SWaM Vendor Directory, and the Vendor Directory is the listing used by state procurement officials to locate SWaM-certified vendors. The certification is granted for a five-year term that can be renewed. These initiatives help in creating robust systems for evaluating and monitoring SWaM-procurement plans and at the same time bring greater focus to the value of supporting small businesses. The DBE program allows small businesses and businesses owned by historically disadvantaged groups to certify and compete for projects that are federally funded through the Virginia Department of Transportation (VDOT). This includes rail, airport, seaport projects, and public transit projects. It also includes women-owned and minority-owned businesses. Becoming SBSD-certified in Virginia means the vendor is certified by the state. Similarly, obtaining DBE certification means the vendor is certified to participate in the federal program for transportation-related projects.

In addition to the above-mentioned governmental programs and initiatives, governmental organizations, and sometimes national corporations, set forth a certain percentage goal of purchases from SMEs and DBEs to enhance local content participation in the procurement activities. Furthermore, SMEs and DBEs can register in systems such as an electronic system of Virginia (eVA), which is used by more than 13,000 state agencies and local government buyers to announce their bidding opportunities, get quotations, and place orders for goods and services. By registering with eVA, the firm joins a community of more than 45,000 businesses offering quality goods and services.

Benefits and Risks of Using Local Suppliers

The procurement of local products and materials has been an important way of supporting local economic growth, and this is especially true in the construction industry. Explicit targeting of local content in construction projects can support local content efforts. For example, when constructing a building, its imported components such as roof materials (e.g., pressed aluminum sheeting), light and electrical fittings, kitchen units, flooring tiles, doors, and so on, all can be targeted for supply and installation by local firms. Imported materials and products typically provide little employment for importers, suppliers, and transportation companies in the importing country. On the other hand, the use of local materials and components creates direct employment within the country. The increased local revenue achieved through purchasing local goods can be used to improve personnel training, machinery, and infrastructure, leading to increased economic competitiveness and productivity, and better quality of life in the long term.

Furthermore, when components, such as tiles, fail or break, it would not be difficult to replace or repair them compared to imported products. Inability to replace the tiles, for example, can result in significant waste as wholesale replacement may be required. Specifying local materials helps ensure that replacement components can be more easily sourced. This prolongs the life of a building, reduces waste, and minimizes costs associated with refurbishment. Compliance with best-practice health and safety standards can also be more readily confirmed with local products through local certification and checks. And large-scale transportation of materials and products has a range of negative impacts including consumption of fossil fuels, noise, air pollution, and road accidents. Sourcing materials and products locally reduces these negative impacts.

It should also be mentioned that locally manufactured materials and products may not achieve the required levels of quality as imported goods. This can be caused, for example, by limitations in local skills and equipment. This situation can be addressed by enabling local manufacturers to access training and improved equipment, providing access to state funding for these activities, or by support with technical expertise. Additionally, local manufacturers may not be able to produce materials and goods in the required quantities; avoiding this situation could require planning and incorporation of lead-in times to enable scale-up production and stockpile supply needed for large projects. Furthermore, governments can help SMEs by ensuring healthy competition in the local industry and providing set policies to ensure monopolies are avoided. Governments and big companies have capacities to develop local content manufacturers with limited capabilities, and they should aim for efficient manufacturing with high quality and competitively priced local products.

QUESTIONS

1. Why do governments give attention to and develop programs to support local content in projects?
2. How do government certification programs support local content?
3. How can corporations support local content?
4. Give examples of how local content can be applied in the construction of a residential building.
5. What advantages can local content bring to local consumers?
6. How can a project benefit from the use of local materials and services?
7. Beyond economic growth, what are two strategic or national security implications of prioritizing local content as described in the document?
8. Explain the interplay between state-level and federal-level certification programs like SWaM and DBE in the context of local content and project funding in Virginia.
9. The document mentions both advantages and potential limitations of using locally manufactured materials. Describe a situation where the long-term advantages might outweigh the initial challenges and how those challenges can be mitigated.
10. How does the eVA system contribute to the broader goals of local content initiatives beyond just announcing bidding opportunities, according to the document?
11. Synthesize the document's information to explain why a government might choose to implement both percentage goals for SME/DBE procurement and provide support for training and improved equipment for local manufacturers.

PROCUREMENT AND CONTRACTING METHODS

> Understand procurement methods, such as sole source, single source, competitive bid, lowest bid, and technical bid, and contracting methods, such as general contracting and subcontracting. Identify and understand the best value proposal. (Understand)
>
> **Body of Knowledge I.B.5**

In the complex area of procurement, understanding the nuances of sourcing and contracting is paramount. This section delves into the various procurement methods, from competitive tendering to sole and single sourcing, each with distinct advantages and drawbacks. We explore the RFx process, including RFIs, RFQs, and RFPs, which are crucial for gathering information, estimating costs, and defining project requirements. By analyzing these methods, this section aims to equip procurement professionals with the knowledge needed to make informed decisions, optimize value, and ensure transparency in the acquisition of goods and services.

Procurement Methods

In procurement, sourcing revolves around identifying and evaluating potential suppliers for products and services, and choosing a supplier that gives the best value. The use of competitive tendering is generally the preferred sourcing option. However, there are times when alternative sourcing options are required.

Competitive Tendering

A sourcing organization may solicit open bids via a public notice, which is known as *competitive tendering*. There are two main forms:

- *Open competitive tendering* occurs when an organization invites companies to respond to a need via public notice.
- *Closed competitive tendering (or restricted tendering)* occurs when an invitation to bid is issued to predetermined or restricted suppliers.

The process promotes an invisible form of competition between the bidders that results in the procurer gaining value for money. It provides transparency and helps avoid the potential to favor a specific supplier. Despite its positive attributes, the process is relatively expensive and slow. It also discourages open communication between the procurers and the sellers, stifles value chain innovation, and lacks flexibility in relation to potential price.

Two-Stage Tendering

This method has two procedures under it, each with a two-stage process. In the first procedure, the procurement department receives a proposal with two envelopes. One contains the technical proposal, and the second contains the financial component. First, the bidder submits a technical proposal highlighting its solutions to fulfill the requirements specified by the procurement team. The proposal is scored by the project team based on the relevance of the solution. Once an agreement is reached, the bidder is invited to submit its financial proposal for further discussion.

In the second procedure (the second approach), a partial proposal is submitted instead of the full technical proposal, which gives room for customization. Once a bidder is selected, it submits the complete technical proposal together with the financial component. The financial proposal is opened only if the technical proposal is evaluated. The combined evaluated score of the technical and financial components is the deciding factor for choosing a bidder.

Sole Source

A sole source procurement can be defined as procurement through solicitation of a proposal from only one source, where a contract is entered into without a competitive process. This is based on the justification that only one known source (or supplier) exists and can fulfill the requirements. States generally do not permit noncompetitive procurements; however, exceptions are allowed where competition is not feasible. For solicitation of proposals from only one source, one or more of the following circumstances must apply:

- The item or service is available only from one source.
- The public exigency or emergency for the requirement does not permit a delay resulting from competitive solicitation.
- After solicitation of a number of sources, the competition is determined to be inadequate.

An agency requirement for a particular proprietary product or service, good or service "uniqueness," or preference for a brand name product might not justify a sole source procurement. Some examples of sole sourcing procurement may include system or product maintenance (e.g., additional licenses, updates, or replacement parts), information technology, and laboratory equipment.

Processing a sole source procurement may require publishing a "notice of intent to sole source" to the public. This posting allows potential vendors to view and indicate interest in bidding. In such a situation, the procurement official can cancel the sole source and consider the other sources in a new competitive bidding process and withdraw the intent to sole source. In some other practices, an appeal or protest process related to sole source procurements is allowed. The appeal/protest can be sustained if the intended vendor for the sole source was indeed not the only vendor who can provide the service or commodity. If the appeal is sustained, the agency is denied permission to proceed with a sole source; if the appeal is denied, the sole source procurement moves forward.

Single Source

In this method of sourcing, a supplier (or a contractor) is purposefully chosen when there are other suppliers (or contractors) available for the product or service. Usually the single-sourced supplier has a proven track record and comparatively offers the lower cost without compromising product quality. In the context of a developing economy, the supplier can have the potential to contribute to local content (which results in job creation, economic development, etc.). When there are multiple sources, organizations have a relatively strong negotiating power advantage and can avoid single sourcing even in a crisis. This sourcing method has disadvantages, such as blatant overpricing or increased risk in supply disruption (e.g., in case of supplier bankruptcy, inability to deliver, or factory fire).

Single and sole sourcing, for example, can offer benefits that in some cases exceed the advantages associated with competitive tendering. However, they require justification for being used as methods of acquisition, and there should be governance mechanisms to ensure each of these strategic sourcing alternatives are transparent and follow the correct processes.

Procurement professionals and decision-makers involved in overseeing the acquisitions also need to be aware of the specific differences between the options and their impact on the project requirements.

Request for X (RFX)

Request for X, or "RFX," is a term that encompasses all requesting processes in the procurement process. Request for quotation (RFQ), request for information (RFI), and request for proposal (RFP) are all part of the RFX process.

- *Request for information (RFI)* is a request to gather existing data about a potential supplier before formally sourcing its product or service. It helps, for example, in evaluating potential suppliers' capabilities and avoiding noncompliance issues.
- *Request for quotation (RFQ)* is a request to a supplier asking to give an estimate about the costs for the completion of a product or service. It includes a comprehensive price quote, submission deadline, and payment terms.
- *Request for proposals (RFP)* is also called a request for procurement. At this part of the procurement process, the manager announces intentions of creating a project that needs the product or service. An RFP provides a comprehensive overview of the intended project and its requirements, and solicits supplier bids.

CONCLUSION

The use of subcontract arrangements in the overall procurement strategy is widespread within construction. For example, general contractors often solicit their own tenders from subcontractors they need to hire to complete specialized jobs in projects. The general contractor then combines subcontractor proposals to prepare the tender and submit it to the client. Consequently, mastering the procurement process is important for general contractors and subcontractors alike.

Understanding the distinctions between procurement methods, from competitive tendering to sole and single sourcing, is critical for navigating the complexities of acquisition. The RFx process, encompassing RFIs, RFQs, and RFPs, serves as the backbone for gathering necessary information and securing the best value. As procurement professionals, it is essential to recognize the advantages and disadvantages of each method, ensuring transparency, efficiency, and compliance in all transactions. Mastery of these principles not only optimizes resource allocation but also strengthens organizational relationships with suppliers and stakeholders, ultimately driving project success.

QUESTIONS

1. What are the types of tendering? How are they used?
2. What makes it necessary in some situations to use a sole source?

3. How can a sole procurement source be managed to enhance transparency?
4. What is a single source, and how is it different from a sole source?
5. What does RFx include in a procurement?
6. Beyond cost considerations, what strategic advantages might a developing economy seek to gain by utilizing single sourcing, and how might these advantages be measured or evaluated?
7. Explain the nuances between the two distinct procedures within the two-stage tendering method. How does each approach mitigate potential risks or address different project complexities?
8. While the document acknowledges the disadvantages of competitive tendering (e.g., cost, time), how might a procurement team proactively address these limitations to still leverage the benefits of competition?
9. The section mentions that "an agency requirement for a particular proprietary product or service . . . might not justify a sole source procurement." Under what specific conditions would such a requirement actually justify sole sourcing?
10. Discuss the inherent conflict that can arise when "preference for a brand name product" is cited as justification for sole sourcing, and how a procurement professional should navigate this conflict to maintain transparency and fairness.

DELEGATED DESIGN

> Apply delegated design or specialized engineering to transfer specific design responsibilities for a project to a contractor or subcontractor and understand the associated risks. (Apply)
>
> **Body of Knowledge I.B.6**

Delegated design is a crucial aspect of modern construction, shifting specific design responsibilities from the project owner and architect to the contractor. This approach allows for specialized expertise to be integrated into the project, particularly for complex systems like fire suppression or curtain walls. Understanding the nuances of delegated design is essential for all parties involved (as shown in Table 1.3), as it entails unique contractual obligations, risk allocations, and professional responsibilities. Properly navigating this design methodology ensures a successful project outcome, while neglecting its complexities can lead to disputes, delays, and liabilities.

Contractors are not expected, and may not be legally allowed, to provide professional services that constitute the practice of engineering or architecture unless such terms are specifically stipulated in the contract documents. Delegated design is a form of design collaboration where the contractor takes responsibility for design of an element or a portion of the project. It can be used, for instance, in the design of retaining walls, fire suppression systems, curtain walls, and other building components. To perform the work activity, the contractor may hire a consultant to perform delegated design services on a project. The "contractor" can be a general contractor, subcontractor, or construction manager; whereas

Table 1.3 Summary of the roles in a delegated design.

Project party	Function
Client	Delegates the design of systems, materials, or equipment to the contractor and its appropriately licensed design professional. This is through the drawings and specifications prepared by the architect. The architect and contractor rely on information and services provided by the other, through the client.
Architect	Specifies the performance and design criteria that the contractor is required to achieve.
Contractor	Provides design services through an "appropriately licensed design professional."
Licensed design professional	Uses its own signature and seal on design documents produced for the project.

the "consultant" is expected to be a design professional licensed to perform the specified design services. The terms "contractor" and "consultant" are used flexibly because numerous parties in the project could hire someone, or be hired, to conduct the delegated design services. The agreement that defines the portion of the design to be performed by the consultant can be (1) a stand-alone agreement or (2) used as a complementary document to many client/contractor or client/construction manager agreements.

In a delegated design, it is the contractor's responsibility to develop design details for elements of the project as part of its scope of work. These responsibilities are associated with design criteria and performance specifications. The process of a delegated design is communicated through shop drawings from the contractor and its design agency to the architect and client. The client and architect have responsibilities when design is delegated. They are expected to specify the performance and design criteria that the delegated design team must satisfy to properly execute the work. Design criteria are usually specific, such as a requirement to use a certain product in the project or a specific dimension to be met. Performance criteria are used to inform the contractor of what is to be met for proper execution of the work. It is then up to the contractor to decide how the set criteria are ultimately achieved. Generally, the contractor relies on the adequacy and accuracy of the provided performance and design criteria of the contract documents. The owner and architect, on the other hand, rely on the adequacy and accuracy of the services and certifications that are provided by the contractor and its delegated design team.

It is the contractor's responsibility to have all the design services provided by an appropriately licensed design professional. This design professional signs and seals the project-related drawings, shop drawings, calculations, specifications, certification, and other submittals. The architect reviews the contractor's shop drawings and other submittals; and the responsibility for the delegated design still rests with the contractor and its licensed professional. The architect reviews the shop drawings and submittals to check their conformance with the design concept and information given in the contract documents.

There are several key features (and also risks) of delegated design services. Here are some tips for navigating the risks:

- *Understand the roles and responsibilities:* The performance criteria of a delegated design are usually established by the design professional's specifications. It is the responsibility of the design team to ensure adequacy of the performance criteria. The contractor's responsibility is to achieve the portion of the design delegated to it (e.g., fire protection system). This design needs to conform to the performance criteria, applicable building code requirements, and the other applicable standards. Clearly defined roles between the design team and contractors helps reduce conflicts and ensure smoother project execution.
- *Address the delegated design services properly in the contract:* An agreement that includes delegated design services should address (1) the contractor's scope of work, (2) delegated design responsibilities, (3) the approach to exchange and review design information, (4) contractor's compensation, and (5) requirements for professional liability insurance that is to be obtained by the contractor.
- *Recognize legal and technical consequences:* When the contractor performs delegated design work on a project, the action could erode the ability to rely on the client's implied warranty of adequacy of the plans and specifications.
- *Consider insurance:* The contractor may incur liability for the design delegated to it; therefore, contractors should consider procuring professional liability insurance when providing delegated design services. Each of the parties in the contract should consider appropriate insurance coverage for their respective disciplines and contractual and professional responsibilities.

CONCLUSION

Delegated design, while offering specialized expertise and streamlining processes, introduces a complex web of responsibilities and potential risks. Clear communication, meticulously defined contractual agreements, and appropriate insurance coverage are crucial to mitigate these risks. All parties involved—client, architect, contractor, and licensed design professional—must fully understand their roles and obligations to ensure project success. By adhering to these principles, the complexities of delegated design can be effectively navigated, leading to a well-executed project that meets performance and design criteria while minimizing disputes and liabilities.

QUESTIONS

1. What could be examples of delegated designs in a project to the construction of a sports stadium?
2. What are the roles of a contractor in a delegated design?
3. How are shop drawings used, and why are they developed?

4. What risks can be associated with this design approach?
5. What should be included in a delegated design services agreement?
6. Beyond simply stating performance criteria, how can an architect proactively minimize potential conflicts between design intent and contractor implementation in a delegated design scenario? Consider the interplay between specifications and shop drawing reviews.
7. Discuss the implications of a contractor engaging in delegated design on the owner's implied warranty of the adequacy of plans and specifications. How might this shift in responsibility impact potential legal disputes?
8. Analyze the dynamic between the "flexibility" in the use of the terms "contractor" and "consultant" and the requirement for a "licensed design professional" to sign and seal documents. What potential challenges arise from this dynamic, particularly in larger or more complex projects?
9. Considering Table 1.3, how does the flow of information and responsibility impact the client's overall project risk in a delegated design model compared to a traditional design-bid-build model?
10. Evaluate the significance of including specific requirements for professional liability insurance in the delegated design services agreement. How does this contractual stipulation address and mitigate the inherent risks of delegated design for all parties involved?

Bid Evaluation and Finalization

PREQUALIFICATION PROCESS

> Analyze the role of audits in the prequalification process, align the prequalification process to meet organizational needs, and explain the prequalification process for short-listed bidders. (Analyze)
>
> Body of Knowledge I.C.1

An audit is a vital component of any quality management system (QMS) and a key requirement of the ISO:9001 standard. Audits play a crucial role in pinpointing potential problems and proactively preventing nonconformities. In construction, audits serve various purposes, including prequalification audits. These assess the management systems of subcontractors, suppliers, and designers before contracts are finalized. Examples of audits in construction projects include:

- Third-party audits for certification purposes when an accredited third-party organization is invited to conduct audit at a construction project or company to verify conformance to the ISO 9001 (Quality Management System) or ISO 14001 (Environmental Management Systems) and issue certification.
- Audits by the client or its representative to ensure all contractual requirements regarding the QMS of the project are in place. This

makes the construction industry one of the industries where the "customer" can perform audits to the "product" during its production.

- Internal audits by the parent company, where the quality department of the parent company (contractor or other) audits a specific project team. These audits are conducted to monitor the performance of the project according to the company procedures. They are also used in construction joint ventures, where each partner monitors the project's system and its performance.

Prequalification Criteria

Prequalification is a systematic qualification process that can be used before inviting a request for bids (RFB) in the procurement of goods, works, or services. There are various approaches for the prequalification, such as requiring prequalification (or registration) with the organization to be completed before contract award (upfront or in parallel with a tender process), and only shortlisting prequalified companies. The process is separate from the bid evaluation procedure, which concentrates on the price and merits of the bid itself. Prequalification is typically used to select potential contractors and suppliers for large-scale contracts, contracts with complex and highly specialized services, or when the cost of preparing detailed bids is high and could discourage participation. Prequalification can be applied in other circumstances, such as when a large number of contracts are let on a "slice-and-package" basis.

Prequalification serves several objectives, such as:

- Ensuring that invitations to bid are extended only to qualified bidders
- Preventing unqualified bidders from winning the bid as a result of superficial evaluation
- Encouraging leading contractors or suppliers to bid
- Enabling insufficiently qualified bidders to avoid the expense of bidding
- Allowing well-qualified firms to price their bids more competitively with the knowledge that they will only be competing with other qualified bidders
- Giving an incentive for potential bidders to form a joint venture
- Limiting the number of bidders to a manageable size for the procuring entity

Prequalification also allows purchasers to:

- Assess the interest generated by the project among qualified firms and make any necessary adjustments in the procurement process or conditions of contract (e.g., payment terms and completion times)
- Reduce the amount of work and time involved in evaluating bids from unqualified contractors
- Encourage local firms to form joint ventures with other local or international firms to benefit from their resources and experience

- Reduce problems associated with low prices submitted by bidders of doubtful capability

On the other hand, prequalification has its disadvantages:

- It might increase procurement lead time.
- It requires review of all prequalification applications, whereas post-qualification requires the review of the qualifications of, normally, the lowest evaluated bidder.
- Collusion, and the possibility of price-rigging, can get easier, because of the limited number of identified bidders.
- Subjective judgement by evaluators when applying the prequalification criteria, and the discretionary rights by the employer, can provide opportunities for deviations from the expected high standards of ethics and impartiality in prequalifying the applicants. Therefore, prequalification should be made under transparent conditions to better balance some of these advantages.

Suppliers and contractors subjected to this procedure are assessed against mandatory criteria that can include:

- Eligibility (nationality, conflict of interest, bank eligibility)
- Historical contract nonperformance (history of nonperforming contracts, suspension based on execution of bid/proposal, pending litigation and litigation history, environmental and social performance and history)
- Financial situation and performance (financial capabilities, average annual turnover)
- Experience (general experience, specific experience)
- Past performance (number of similar contracts, timeliness of delivery)

The prequalification criteria are used for judging the qualifications of the applicant, and they should be established in a manner that would result in a yes or no answer when determining whether an applicant meets the criteria. Excessively "soft" or inapplicable criteria can result in too many unqualified applicants and lead to deterring qualified firms from bidding or obtaining a low bid from a marginally qualified contractor or supplier. A high level of qualification requirements limits the number of bidders, while the opposite would allow unqualified bidders to pass through. Applicants that substantially meet the set qualification requirements are then invited to submit a bid/proposal. Therefore, the qualification requirements are to be set up carefully for each project.

The prequalification documents can be prepared at the same time as the prequalification criteria. However, the criteria are not to be changed during the prequalification evaluation process. The documents enlarge on information provided in the notification advertisement, such as:

- Scope and description of the proposed procurement
- Major quantities of the work

- Main quantities to be procured
- Location of the work
- Procurement scheduling of goods or works
- Payment terms
- Price adjustment provisions
- The language and governing law of the contract
- Other information to enable bidders to assess their interest and respond appropriately

Aligning the Prequalification Process with Organizational Needs

The next activity in the prequalification procedure is to evaluate the alignment of the submissions with organizational needs, as communicated through the prequalification criteria and documents. The evaluation procedure includes two stages:

1. *Preliminary examination*, to assess whether the submitted documents and their formats are in conformity with the requirement. In case of no submission or missing required documents deemed as important, the applicant becomes disqualified at this stage.
2. *Qualification evaluation*, to examine whether the submitted documents are in compliance with the qualification requirement; this is by applying the pass/fail criteria. In principle, an applicant is to be disqualified if it does not satisfy the qualification requirements, and the criteria set up beforehand are to be applied during the evaluation. The method is to be based on absolute, not comparative, evaluation.

In case the applicant for prequalification is a joint venture (JV) that is formed by two or more firms, the conditions applying to JVs are to be stated clearly in the prequalification documents and to subsequent bidding by the JV. For example, a partner of the JV may not be allowed to submit or to participate with more than one bid. Additionally, the resulting contract is to be legally binding on all partners jointly, and a JV agreement is to be submitted together with the bid. (Or, a letter of intent to execute a JV agreement in the event of a successful bid can be signed by all partners and submitted with the bid, together with a copy of the proposed agreement.) There are many other details to be considered in the situation.

A *slice-and-package procedure* is used when a large homogeneous project is sliced into smaller similar contracts that are bid simultaneously to attract the interest of both small and large firms. Firms offer their bids on individual contracts (slices) or on a group of similar contracts (packages). The award is then made to the combination of bids offering the lowest cost. Slices that comprise a number of similar construction units together in a small area are called "lots," and they are bid concurrently with other similar lots as part of the larger "package." In case the prequalification evaluation includes slice-and-package requirements, the procedure used in the evaluation should be mentioned in the summary, and the summary should also show the aggregate total of contract value for which the applicants are considered prequalified.

Typically, there is no limit on the number of successful applicants prior to the evaluation of prequalification. All applicants that meet the specified prequalification criteria are allowed to bid. However, it is important to set the criteria to allow a reasonable number of applicants to get qualified. An excessive number of applicants is not preferable, as it increases the burden of the procuring entity when evaluating the bid proposals.

After evaluating the applications received, the evaluation committee should prepare a report that addresses each of the pass/fail criteria set in the documents. The report should also explain the disqualification of applicants that failed to meet the criteria. The procurement entity then promptly notifies each supplier or contractor who participated whether they have been prequalified or not. Promptly after the notification, the entity invites bids from all prequalified applicants.

Principles in Evaluation of Bids

The prequalification documents should provide a clear basis to evaluate prospective bidders, following an objective process and based on fair and transparent criteria. The sought documentation should always be relevant and clearly stated, and not impose an excessive burden of preparation or paperwork, so the prospective bidders are not deterred. In addition, there are certain principles and practices to be clearly understood and observed in the bid evaluation process, such as:

- *Prompting of public bid opening*: The time for bid opening is to be the same as the deadline for bid submission or promptly thereafter. This is to preserve the integrity of bid submission, which promotes transparency at the very start of the bid evaluation process.
- *Confidentiality of procedures*: After the public opening of the bids, information relating to the examination, clarification, and evaluation of bids, and recommendations concerning awards, are to be kept confidential and not communicated to any person not officially concerned with these procedures until the notification of the contract award to the successful bidder.
- *Clarification and modification of bids*: A bidder is not permitted to modify its bid after the bids have been opened. Only clarifications that do not change the substance or price of the bid may be requested or accepted by the contracting agency. The request for clarifications and the response are to be sent in writing or by acceptable electronic means.
- *Rejection of bids*: All valid bids received are to be evaluated in accordance with the criteria, methods, and procedures provided in

the bidding document. Only those bids that do not substantially meet the specifications, bidder qualifications, and other requirements of the bidding document can get rejected.

- *Currency of bid evaluation*: Bid prices are to be expressed into a single currency for evaluation and comparison purposes.
- *Rejection of all bids*: All bids rejection is justified when the bids submitted are not substantially responsive or where none of the bids received meet the specifications, bidder qualifications, and other requirements—or in case there is evidence of lack of competition. Where all bid prices have substantially exceeded the cost estimate, a negotiation can be made with the lowest evaluated bidder for a reduction of the bid price. In case no satisfactory contract can be concluded, rebidding can be initiated with appropriate modification of the scope of the contract.

CONCLUSION

The prequalification process is a critical gatekeeper in procurement, ensuring only capable entities proceed to the bid stage. Its meticulous application—encompassing audits, defined criteria, and alignment with organizational needs—streamlines selection and minimizes risks. While presenting challenges such as increased lead time and potential collusion, its benefits in securing qualified bidders and fostering transparency are paramount. By adhering to clear principles and objective evaluations, organizations can leverage prequalification to achieve project success and uphold high ethical standards in contracting.

QUESTIONS

1. What type of audits can be applied in a construction project? How is this different from other industries?
2. When is the prequalification process required?
3. What practices are used by companies for prequalification?
4. How is prequalification different from bid evaluation?
5. What is "slice and package," and why is it used?
6. Beyond initial screening, how does the prequalification process contribute to long-term relationship management with contractors or suppliers?
7. Explain the inherent tension between transparency and confidentiality in the prequalification and bid evaluation process, and suggest a method for mitigating potential conflicts.
8. Considering the potential for increased procurement lead time, under what specific project circumstances would the benefits of prequalification unequivocally outweigh this disadvantage?
9. Discuss how the use of slice-and-package bidding strategy can impact the prequalification criteria and subsequently the type of bidders who apply.
10. Analyze why subjective judgment in prequalification evaluation, although a known risk, is difficult to completely eliminate, and propose one practical strategy to minimize its impact.

CRITERIA DEVELOPMENT

> Review the criteria for assessing and comparing bids (e.g., dollars, schedule, lead times, workforce, products, assemblies, safety, sustainability) to integrate the bid criteria with quality requirements, critical to quality (CTQ) concepts, and examine bid criteria for the specific contract type in use. (Analyze)
>
> Body of Knowledge I.C.2

Evaluating bids and finalizing contracts are critical phases in any procurement process. Establishing clear, valid, and objective criteria ensures fair treatment of all bidders and maximizes the likelihood of selecting a supplier who can successfully fulfill the contract requirements. This section delves into the intricacies of developing effective evaluation criteria, ranging from mandatory pass/fail standards to weighted scoring systems that balance cost with quality. Understanding these methods is essential for any organization aiming to achieve the best value for money and minimize risks in their procurement activities.

Evaluation criteria are the standards against which bids are compared. They can be categorized into (1) mandatory criteria, (2) weighted criteria, and (3) weighted criteria with mandatory elements.

1. *Mandatory criteria* are straightforward methods where bids are rated as pass/fail, responsive/nonresponsive, or comply/noncomply. They are used in evaluation for goods procurement and could be used for the procurement of services and infrastructure works. The mandatory criteria are the first criteria against which bids are evaluated to eliminate those that do not conform to these requirements. Thus, careful consideration should be made when using mandatory criteria about whether the requirement in question is significant enough to merit rejecting a bid regardless of its other benefits.

2. *Weighted criteria* can be measured in terms of degree of responsiveness. The scale used for measuring the degree of responsiveness depends on the procurement method and category of procurement. Weighted criteria are usually used in the evaluation of services.

3. *Weighted criteria with mandatory elements* have defined mandatory minimum requirements and are measured above that minimum requirement. For example, a requirement may be set for a consultant to be skilled in at least two types of work activities, and a rated score may be assigned for capabilities, if the additional capabilities add value to the requirement.

The client defines evaluation criteria based on key project factors, such as cost, technical requirements, experience, project schedule, past performance, and compliance. Different weights are assigned for each evaluation criterion to reflect its relative importance. For example, technical requirements might carry more weight than cost. Furthermore, the client can request clarifications regarding any ambiguities. Based on the evaluation results, negotiations will be initiated to finalize terms with the top-ranked bidder.

The solicitation documents are expected to clearly state the evaluation method (instructions to bidders) and the evaluation criteria. This is to be decided before proposal submissions and is not to be changed from what was originally described in the solicitation documents. Thus, it is important that the procurement officer gives appropriate thought to the selection of the method and the criteria. The purpose of establishing the evaluation criteria is to:

- Develop the ability to compare the merits of offers.
- Minimize the risk of selecting a subcontractor or supplier not able to perform the contract satisfactorily.
- Identify the best match to the project requirement.

The solicitation documents should create a level playing field where all potential suppliers receive the same information at the same time, and the evaluation criteria are measurable and provide equal opportunities for all suppliers to be awarded the contract. In addition, the commercial risks are to be fairly distributed between the contracting organization and the selected supplier. Furthermore, the requirements are to be clearly expressed in terms that encourage sufficient responsive offers to best meet the organization's needs and to avoid spending unnecessary time providing explanations.

Characteristics of Successful Evaluation Criteria

The evaluation criteria spell out the rules by which the offers are compared and ensure fair treatment of all bidders. Evaluation criteria have four elements: They should be valid, objective, measurable, and impartial.

- *Valid criteria*: The criteria being evaluated are critical for the successful execution of the contract. Criteria that do not clearly relate to the requirements in the solicitation are probably not valid and are based on what is desirable rather than a need for the project. If it is reasonable to assume that all suppliers will score high or well on a certain factor, then it would not be a valid significant differentiator. For example, certain language capabilities could be a valid factor to select suppliers because of the country where the assignment will take place. Valid criteria result in an assignment of a higher score to suppliers that actually have a relatively high probability of successfully performing the work.
- *Objective criteria*: The criteria ensure that offers are evaluated consistently, even if they are evaluated by different evaluators, and that a third party can ascertain that the supplier has certain features. An example of this is the capability of the supplier to provide after-sales service or maintenance parts for the constructed project in the country.
- *Measurable criteria*: The criteria must allow evaluators to assign rankings or make pass/fail judgments effectively. The criteria can be

assigned points against degrees of achievement (e.g., two points per year of experience or one point for each similar project in the region) up to a set maximum of points.

- *Impartial criteria*: The criteria must not give an unfair advantage to one supplier over another and must not be tailored around the attributes of favored suppliers or products. For example, if a supplier has 15 years of experience in offering a specific service in the country, it could be inappropriate to use this number as a pass/fail minimum criterion. It will exclude other established suppliers that have a more limited track record of dealing with the country, and these suppliers perhaps are successful in delivering a number of even larger services and to various buying organizations. The criteria should be impartial and facilitate the inclusion of new suppliers with reliable performance records.

In addition, "open" and "'closed" questions can be used appropriately when designing tender questions.

- *Closed questions* (only possible answer is yes or no) are useful for mandatory questions and helpful in clarifying whether a supplier will do something specific now or after contract award (e.g., sign up to particular terms or standards). These types of questions should remove subjectivity in evaluation.
- *Open questions* are helpful to establish how bidders will meet quality requirements. The questions should still be clear as to the nature of the information sought and how responses will be evaluated.

Setting the evaluation criteria effectively requires time and expertise. Market testing can be helpful in overcoming any preexisting bias and ensuring that criteria are fit for purpose. It is also good practice to peer-review quality criteria.

By reaching the solicitation stage (i.e., notifying prospective suppliers of the organization's need to procure certain goods, services, or works), the evaluation criteria identified should be developed and formulated to obtain the best value and inform the potential bidders regarding:

- Administrative rules for the solicitation process, which include the evaluation criteria and the method.
- Functional and performance requirements, and physical characteristics of the required goods, services, or works. This includes a clear statement of minimum quality requirements (schedule of requirements and technical specifications).
- Commercial, legal, and other conditions (proposed contract general and special conditions).

Integrating the Bid Criteria with Quality Requirements

The mandatory content of solicitation documents varies among organizations based on the procurement method and complexity of the requirement. It is a business practice to have separate standard solicitation documents for each procurement method (e.g., RFQ, RFP). In some organizations, the selection of a particular standard solicitation document automatically implies specific general conditions, contractual terms, and so on. Therefore, it is important to ensure that all parts of the standard document fit the requirement (e.g., appropriate specific conditions of contract). The practice of applying standard documents ensures consistency from one transaction to another and streamlines the process. However, these standard solicitation documents need to be customized to a certain degree to appropriately reflect the specific nature of every project. In the case of an RFP, for example, the document structure typically includes some or all of the following:

- Letter of invitation
- Instructions to bidders
- Schedule of requirements
- Technical description (specifications, terms of reference [TOR], SOW including design/drawings, and technical specifications for works and materials)
- General conditions of contract
- Special conditions
- Sample contract and offer submission form

The specifications represent a description of the technical requirements for a material, product, or service, but can also relate to the requirements for services (terms of reference) or work (statement of work). Specifications, terms of reference (TOR), and SOW constitute the technical basis for the solicitation and the evaluation of offers. They determine whether suppliers satisfy the requirements as stipulated in the solicitation documents. They are the "heart" around which the eventual contract is written and administered afterward. These requirements influence the entire lifespan of the procurement (planning, sourcing, solicitation, bid evaluation, review, award, contract performance, completion, and post-contract evaluation). Therefore, clear and appropriate descriptions of the procurement requirements are of utmost importance.

A clear and detailed description of the performance is expected to prevent misunderstandings and disagreements with suppliers/contractors at the time of contract execution. Ambiguity in performance requirements can lead to increased costs, as suppliers may have to factor a contingency buffer into their offer that will then be reflected in the price. Additionally, it is important to clearly state the delivery date for goods or starting/completion dates of services or works where there are specific requirements or when time is critical. Also, when procuring goods, the destination and mode of transport should be included, and when procuring services, the exact location where the services should be provided is to be specified.

The best value for money in the context of offers evaluation means that price alone is not always the only criterion factored into the evaluation. Other criteria such as quality, availability, sustainability, time, compliance, risks, and cost of maintenance and support also determine the best return on investment in the procurement of goods, services, or work. The evaluation criteria and method provided in the solicitation document should enable the bidders to understand the focus of the evaluation (e.g., cost together with compliance, or technical performance).

When applying the methods of solicitation—for example, RFP—the evaluation is conducted by a designated evaluation team in accordance with the organizational regulations and procedures. The evaluation team is expected to apply the evaluation criteria and method to conduct a fair, transparent, and unbiased evaluation. To ensure the integrity of the evaluation and to provide a proper audit trail, all stages in the evaluation process should be documented in writing in an evaluation report, which subsequently forms the basis for the recommendation of award.

Bid Evaluation Methods

One or a combination of both evaluation methods can be used:

1. Lowest-priced technically compliant acceptable offer method

When using this method, the award of a contract is made to the supplier whose offer has been evaluated and determined as both:

- Responsive, compliant, and technically acceptable
- Offering the lowest cost/price

Responsive, compliant, and technically acceptable means fully meeting the specifications, TOR, and SOW required, or reaching a predefined threshold of a maximum attainable score based on the requirements stated in the specifications, TOR, or SOW.

The level of responsiveness, compliance, and acceptability can be measured using straightforward pass/fail criteria or by more complex approaches like a weighted scoring system. In principle, an offer that "fails" is disqualified, while an offer that greatly exceeds the requirement is ranked the same as one that just meets that same requirement.

There is a difference between price and cost; price is the amount of money paid to procure a certain product or service, while cost is the amount of money paid plus other related cost factors such as freight costs, operational costs, disposal costs, and so on, over its lifetime or for the duration of a contract or project. This is also called the "total cost of ownership."

This method of evaluation is typically used in the case of invitation to bids (ITBs), and in a simpler form is also used for RFQs. The essential characteristic of the ITB is that the technical requirement can be clearly and completely specified in a qualitative and quantitative manner. Also, the award is normally made based on lowest cost among commercially and technically compliant bids. In the RFP the proposals are submitted against requirements that cannot be fully defined and/or quantified at the time of solicitation, and where also innovation and the expertise of the supplier is being sought—often in the form of performance requirements.

2. Weighted scoring method (combining cost and quality)

This scoring method is typically used in the case of RFPs, when the award of a contract is made to the supplier whose offer has been evaluated and determined as:

- Responsive, compliant, and technically acceptable
- Having received the best combined score out of a predetermined set of weighted technical and financial criteria specified in the solicitation

This method can be used for more complex procurement where:

- Evaluation criteria cannot easily be quantified
- Different aspects of value (e.g., cost/price and quality) are to be considered and balanced
- Different types of scales are to be used for various factors

This method is typically used for procurement of services where the relative importance of each evaluation criterion needs to be weighted. It also can be used for the evaluation of offers for complex goods, services, and work that require the evaluation to be based on criteria other than price and when it is difficult to evaluate an offer on the compliant/noncompliant scale only.

Under this evaluation method, price or cost is one of the evaluation criteria. For straightforward requirements, price can play a significant part in the selection process, provided the technical criteria are met. However, the more complex the requirement, the more significant the end product; and the less comparable the proposals, the less influence price has on the selection of the recommended proposal. In other words, the points allocated for the technical offer are higher than the number of points allocated for the financial offer. By applying this method, the risk of selecting a potentially nonperforming but low-price supplier is reduced, and the potential savings to be achieved by selecting the lower priced offer do not outweigh the possible implications and costs caused by a nonperforming supplier. Thus, typically the financial offer is given 15-50% importance. The technical proposals are opened and evaluated first, and the financial offers are opened only for proposals with technical evaluation scores above a stated threshold. For proposals where the technical offer does not reach the minimum specified score, the corresponding financial offer is not eligible for further consideration. Some organizations open both the technical and financial offers at the same time, but only share the technical proposals with the evaluation team. This enables parallel evaluation of both offers by different teams. After scoring both the technical and financial components, formulae are applied to calculate the total number of points in order to identify the winning proposal. Below is an example, and other formulae can be used depending on the type of procurement.

- *Technical proposal rating = (Total score obtained by the offer ÷ Maximum obtainable score for technical proposal) × 100*
- *Financial proposal rating = (Lowest priced or cost offer ÷ Price or cost of the offer evaluated) × 100*
- *Total combined score = Technical proposal rating × Weight of technical proposal + Financial proposal rating × Weight of financial proposal*

For example, if the weight assigned to the technical proposal is 60% and the weight of the financial proposal is 40%, the total combined score is calculated as:

Total score = Technical proposal rating × 60% + Financial proposal rating × 40%

The formula methodology and scoring approach for bids should be scenario-tested—that is, by using dummy bids (including low bids) and scores to ensure that results are as expected.

When weighting the quality criteria, within the overall quality score there will be several individual quality criteria (or subcriteria), each of which will need to be weighted against the others in terms of importance to the overall contract and its objectives. The weightings should be set on a case-by-case basis, taking account of factors such as the nature of the goods/services being procured and the objectives of the procurement. Each evaluation criterion must be distinct from each other to avoid double counting. Additionally, it is possible to subdivide an evaluation criterion into lower-level subcriterion (with subcriterion weightings).

CONCLUSION

The development and application of robust evaluation criteria are paramount to successful procurement. By employing a clear understanding of mandatory, weighted, and combined methodologies, organizations can ensure fairness, transparency, and optimal value. Diligent planning, clear documentation, and meticulous adherence to predefined criteria empower the selection of the most suitable supplier. This comprehensive approach not only mitigates risks but also fosters long-term successful contract performance, ensuring the delivery of quality goods, services, or works that align with the organization's strategic objectives and requirements.

QUESTIONS

1. What types of evaluation criteria can be used to evaluate and compare bids?
2. Why are the solicitation documents expected to clearly state the evaluation method and the evaluation criteria?
3. What are the characteristics of successful evaluation criteria?
4. What is meant by impartial criteria? Give an example.
5. How can quality requirements be integrated into the bidding criteria? Give an example.
6. What is the difference between price and cost?
7. Discuss the potential risks associated with relying solely on mandatory pass/fail criteria for bid evaluation in a complex service procurement scenario.
8. Explain how cultural context or geographic location might influence the "validity" of a specific evaluation criterion, and give a practical example.
9. Describe a scenario where a low-priced technically compliant bid might not represent the "best value for money," and justify your reasoning.
10. Given the section's emphasis on clear communication in solicitation documents, how might ambiguity in the "schedule of requirements" or "technical specifications" impact the impartiality and fairness of the evaluation process?

BIDDING PROCESS

> Distinguish the role of quality requirements in the context of different contract types during the bidding process. Identify strategies and techniques for building consensus when finalizing quality requirements. (Analyze)
>
> Body of Knowledge I.C.3

The bidding process is the crucible where project quality is forged. It's not merely about selecting the lowest price; it's about strategically aligning cost with desired outcomes. This section delves into the nuanced interplay between financial considerations and quality standards, exploring evaluation models that prioritize value for money. We'll dissect various bidding procedures, from single- to two-stage approaches, and examine how to foster innovation while managing risk. By understanding these critical elements, organizations can ensure their procurement decisions deliver not just affordability but also excellence.

Balancing the Importance of Cost and Quality

The bids evaluation model should reflect the desired business objectives and outcomes of the procurement. Therefore, it should consider a balance of price and quality, and how they will be evaluated. At the core of the bidding is the list of relevant and measurable criteria against which each bid will be evaluated. The criteria should be proportionate to the subject matter of the contract, its value, and its objectives, and should not be taken from similar or previous programs without proper consideration as to whether they are proportionate and relevant to that particular procurement. Additionally, the criteria should not be based on price alone, should include the consideration of qualitative criteria, and should be clear and unambiguous.

Value for money means securing the best mix of quality for the least outlay over the period of use of the procured goods or services. Thus, to achieve the desired business objectives and outcomes of the procurement, it is essential to effectively determine quality through the bid evaluation process. For example, a 70/30% quality/price weighting sends a clear message to suppliers that quality is significantly more important than price. However, the relative weightings could be undermined by the nature of the criteria. For example, if the quality criteria are designed so most bidders tend to receive the same score, the procurement may still be decided on price (even if quality has a higher weighting). For evaluating the price, finance teams are to be consulted on the most appropriate evaluation pricing methods considering whole-life cost. An effective model balances price and quality by asking if paying a higher price for higher quality is justified, and if the weighting reflects the characteristics of the product or service.

Sequential evaluation of price and quality is another alternative to relatively weighting price and quality. It involves setting a minimum quality score threshold, and bidders are evaluated against the published quality criteria. Only bidders that meet or exceed the quality threshold are taken forward to the next stage of the evaluation (the price evaluation), and those that do not meet the quality threshold are rejected. This approach is most effective when there is a clear definition of minimum requirements that the client wants the project to meet. There is also an additional option where the cost element takes the form of a fixed price or cost, and bidders compete on quality criteria only.

With all bids, it is important to verify that the solution is underpinned by a robust costing model, and bidders should have evidenced how their pricing relates to their proposals on quality, confirming the threshold quality level can be delivered at the price bid.

Deliverability and Innovation

The delivery risk should be taken into account by considering it in (1) an evaluation of risk within the quality evaluation or by applying (2) a separate scored and weighted section on risk. Such a section provides a greater focus on risk and is more useful for complex procurements where an assessment of delivery risk is critical. Including the evaluation of risk within the main evaluation can be more appropriate for lower-value/low-complexity requirements. The evaluation of the delivery risk should be transparent and clear to bidders.

Quality may also refer to innovation, as adopting innovative solutions and emerging technologies improve ways of working and achieve better project outcomes. To drive innovation into the bids, consider:

- An open technical requirement with clearly defined outcomes. Overly prescriptive requirements can prevent suppliers from proposing new and innovative solutions for delivering better outcomes.
- Removing the cap on cost, as this allows suppliers to propose innovative solutions that may add or drive additional value.
- Even where innovation is not the primary goal of a particular procurement, the evaluation criteria should not discourage suppliers from offering solutions with additional benefits that are not directly linked to the primary product or service.

Bidding Procedures

Under open tendering, the procuring entity may choose:

1. *Single-stage bidding procedure*

- *One-envelope bidding procedure*: The bidders submit bids in one envelope containing both the price and the technical proposal. The bids are opened at the date and time advised in the bidding documents. The bids are then evaluated, and the contract is awarded to the lowest-evaluated bidder that meets eligibility and qualifications requirements.
- *Two-envelope bidding procedure*: The bidders submit two separately sealed envelopes simultaneously. One envelope contains the technical bid and the other the price bid enclosed together in an outer single envelope. First, the technical bids are opened at the date and time advised in the bidding document, and the price bids remain sealed. The technical bids are evaluated, and no amendments or changes to the bids are permitted. The objective of this step is to evaluate the technical bids without reference to price. Bidders that do not conform to the specified requirements may be rejected as deficient or nonresponsive bids. Second, bidders who have been declared substantially responsive (meeting the minimum qualification requirements) are invited to submit financial or price proposals together with the revised technical proposals in compliance with the acceptable technical standards. The documents at this stage should

include required changes to the technical bids of each bidder to comply with the final requirements (together with amendments to the bidding document, if any). Sufficient time should be given for the bidders to incorporate the changes required in their technical proposals and prepare price proposals. Next, the price proposals and revised technical proposals are evaluated, and the contract award is made to the lowest substantially responsive bidder.

2. *Two-stage bidding procedure*

The procuring entity prepares a first stage-bidding document with functional performance specifications (rather than detailed technical specifications). In response, bidders offer unpriced technical proposals. The suppliers' qualifications are assessed (if prequalification has not been done), technical evaluation of proposals is conducted, and suppliers are precisely informed about what must be done to make their bid technically responsive. Memoranda of changes are shared with each bidder, and addenda could be prepared to the bidding documents, including revisions to the technical requirements made in the light of the technical evaluation. Next, bidders offer amended bids containing their final technical proposal and a financial proposal. The procuring agency evaluates the proposals (both technical and financial) according to the specified criteria and method in the bidding documents.

The two-stage process has advantages that include the ability of the procuring agency to interact extensively on technical matters with bidders, which is not permissible in the single-stage process. By doing this, the agency can learn from the market and adapt its requirements. It also allows stating the requirements in more general functional or conceptual terms than the detailed functional and technical requirements that are necessary to carry out the single-stage process. This reduces the burden of preparing detailed functional and technical requirements, and encourages suppliers to offer innovative solutions.

Abnormally Low Bid

This is a bid with a price so low as to make the procuring agency question whether it is correct and/or deliverable. Where a bid appears to be abnormally low (e.g., more than 10% lower than the average of all bids), the bidder is requested to explain the proposed price. The agency could only reject the bid where the evidence supplied by the bidder does not satisfactorily account for the low price; the purchaser will be guided by the procedures described in the issued bidding document in handling an abnormally low bid and has guidance to determine if a bid is abnormally low.

Applying the Evaluation Model

Received bids are to be evaluated according to the published criteria and evaluation methodology. It is critical that the scores awarded are supported by records to provide justification for them; this helps demonstrate the process is consistent and can be used as a basis for debriefing unsuccessful bidders. During the evaluation, evaluators could encounter points in bidder responses that are incomplete or erroneous, or where specific documents are missing. Where this occurs, the procuring agency may ask the bidder to submit, supplement, clarify, or complete the relevant information, provided this is done in full compliance with the principles of equal treatment and transparency. Clarification questions are to be sent in writing to the bidder, who is expected to respond in writing within an appropriate time frame set by the procuring agency. It should be noted that clarification is not an opportunity for bidders to improve their bids. Evaluators should not raise clarification questions directly with bidders (as the proposed clarification could go beyond what is permitted). Clarification questions are to be submitted to the procurement or commercial team, who will review and, if appropriate, send them to the bidder. Also, the procuring agency could consider whether any of the questions need to be reviewed by legal first to ensure they are allowed. Clarifications to bidders' responses are then shared with all evaluators evaluating that particular area of the bid, not just the evaluator raising the clarification questions.

Once individual scoring is complete, evaluators send their scores and evaluation reports to a moderator for review. A moderation meeting is then held to agree to a single consensus score. This is particularly important where evaluators' scores are different for the same bidder. The moderator is responsible for leading the moderation meeting where evaluators explain their scores and the reasons, asking evaluators about any apparent errors or discrepancies identified and whether adjustments need to be made to their scores and evaluation reports, and leading evaluators to agree to a single justifiable consensus score in case there are different scores to the same response (which should not simply be an average). Once the moderation process is completed, this is the end of the scoring part of the evaluation process. Evaluators should not discuss their scores with other evaluators even after the completion of the moderation process.

Bids may be scored identically or near-identically, which can be mitigated through good-quality evaluation criteria and an effective scoring approach; however, this possibility can never be entirely eliminated. It may therefore be sensible to include an optional final evaluation stage that would be triggered where two or more bids are awarded a final score that is the same or within a specified range. The process to be followed in such circumstances needs to be clearly set out as part of the published evaluation methodology.

Finalizing the Award Process

Once the award decision is approved internally, the next step is to notify successful and unsuccessful bidders in writing. There is a mandatory standstill period that commences once the procuring agency has notified unsuccessful bidders. This notice is often referred to as the "standstill letter," and it contains certain prescribed information such as details of the winning bidder, their own and winning bidder's scores, and the characteristics and relative advantages of the winning bid based on the evaluation. Good feedback to bidders can be useful to them by helping them understand what they did well and what they could have done better, with points to consider in the future. The feedback to the unsuccessful bidders should be based on evaluators' and moderators' notes, and it should be sufficient to let each bidder understand why they received the score they did.

At the end of the evaluation process, the procuring agency produces an evaluation report to demonstrate that the evaluation has been performed properly and to provide evidence supporting the scores given to each bid.

CONCLUSION

A well-structured bidding process is the foundation of successful project execution. By meticulously defining quality requirements, strategically balancing cost and value, and implementing transparent evaluation methods, organizations can ensure that procured goods and services meet their intended purpose. Clear communication, robust documentation, and a commitment to fairness are vital. This section has illuminated the complexities and critical nuances of bidding procedures, equipping stakeholders with the knowledge to navigate these processes effectively and achieve optimal project outcomes.

QUESTIONS

1. How can the importance of quality be emphasized during the project bidding?
2. Give an example that shows how the relative weightings can be undermined by the nature of the criteria.
3. How can sequential evaluation be used to balance price and quality? When is this approach appropriate?
4. Why is a moderation meeting required?
5. How can a bidder submit clarification questions? How should the request be handled by the procuring agency?
6. How can contractors, through the bidding process, be encouraged to innovate? How does this improve the quality of the project?
7. Beyond simply stating quality is important, how can a procuring agency structure its bid evaluation model to practically ensure that quality takes precedence over price, even with weighted scoring systems? Explain with reference to potential pitfalls mentioned in the text.
8. Compare and contrast the advantages and disadvantages of the two-envelope bidding procedure versus the two-stage bidding procedure. Under what specific project circumstances would you recommend one over the other? Justify your choice.
9. The section mentions that "clarification is not an opportunity for bidders to improve their bids." However, it also states that "memoranda of changes are shared with each bidder" in the two-stage bidding procedure. How can these seemingly conflicting statements be reconciled? Explain the boundaries and ethical considerations involved.
10. Discuss the potential impact of an "abnormally low bid" on project quality and risk. What measures can a procuring agency take to both investigate such bids fairly and protect the integrity of the procurement process?
11. Explain the significance of "consensus scoring" during the evaluation process. What role does the moderator play in ensuring fairness and accuracy, and what are the potential consequences of failing to achieve a true consensus?

Chapter 2

Planning Phase

Strategic Quality Planning

A robust project quality strategy is the cornerstone of successful project delivery. It serves as the guiding framework, integrating stakeholder intent, procurement considerations, contracting approaches, and regulatory compliance into a cohesive plan. By proactively addressing quality throughout all project phases, from precontract planning to turnover, organizations can mitigate risks, enhance collaboration, and ensure stakeholder satisfaction. Establishing clear key performance indicators (KPIs) and securing management commitment further solidifies the strategy, and drives accountability and continuous improvement, which helps the project achieve excellence.

A project quality strategy is the framework guiding decisions that ensure project success. It integrates stakeholder intent, procurement considerations, contracting approaches, and regulatory compliance. The strategy must be proactive, evolving alongside project phases, from precontract planning to project turnover. By securing management commitment and defining KPIs, a project ensures that quality objectives are effectively implemented and maintained.

It is important to note that once formulated, the quality strategy becomes the basis of quality planning, which includes the plans for both quality control (QC) and quality assurance (QA). Note that strategy is at the higher level and consists of the logical choices that the organization/project may consider; it carries a longer-term perspective. Quality plans that support the quality strategy, on the other hand, are developed to include the action steps (who, what, when). Both are important, with the quality strategy as the precursor to the quality plans. The description of quality assurance is provided later in this chapter ("Quality Assurance"), and quality control is defined in "Quality Control."

PROJECT QUALITY STRATEGY DEVELOPMENT

> Formulate a comprehensive project quality strategy, integrating stakeholder requirements (stated and implied), procurement requirements, contracting approaches, and regulatory compliance based on Authority Having Jurisdiction (AHJ), and securing management commitment. Establish key performance indicators (KPIs) to align with each element of the quality strategy. (Create)
>
> **Body of Knowledge II.A.1**

The components that enable the development of the quality strategy include:

- Vision
- Mission
- Purpose
- Values

The vision, mission, and purpose are the precursors of strategy and are based on the values of an organization. The *mission* describes the clear objective that the organization or the project intends to achieve. The *vision* is what the organization aims to achieve in the future. The *mission statement* describes the actions and methods that must be considered to achieve the vision. In other words, the vision describes the "what," the mission describes the "how," and the purpose describes the "why" of quality. When the vision and mission are well-defined, the next thing to consider is the purpose of the project. When the purpose of undertaking the project is known, it is easier to develop the project quality strategy and easier to keep it consistent with the vision, mission, as well as the purpose. Following the development of the quality strategy, the project must perform quality planning, which includes both quality control assurance (QA) as well as quality control (QC) planning.

This "implementation of quality on a project" is characterized by a similar structured guidance that is provided for the different levels of the organization's operations. It begins with the vision and mission of the project, which should describe why the project is being performed. While the vision and mission describe the desired purpose and the benefits of the project, it is the next tier of guidance where the quality strategy describes how those goals and objectives may be achieved.

The quality strategy of the project is the high-level narrative that identifies the quality priorities for the project and thereby sets up the direction of the quality effort for the project. It guides the project's journey from its beginning to its desired future state and also describes the quality processes that would be used to achieve the desired state.

The strategy is reviewed regularly and may be modified only if the project believes that it no longer accurately reflects the project's commitment to the quality ideals it describes. Examples of the conditions for modifying the strategy include significant changes in the project's strategic direction, in market conditions, or in the regulatory requirements that apply to the project. If a change in quality strategy is warranted, approval of the change should be obtained from all stakeholders who approved the original quality strategy.

A clear definition of the purpose in terms of the project's goals and objectives becomes the starting point of quality planning, as it helps develop the project's quality strategy through a structured approach.

Each construction project generally proceeds through the following phases:

- *Precontract*: This stage generally confirms the project's purpose.
- *Planning*: In the planning phase, the project establishes stakeholder intent and the regulatory requirements.
- *Design*: In the design stage, the focus of quality strategy is to integrate constructability and quality verification early.
- *Procurement*: The quality strategy during this phase must be to select materials and labor based on capability, not just cost.
- *Construction*: During the construction phase, the quality strategy must implement robust quality assurance and quality control mechanisms.
- *Project turnover and closeout and transition to owner operations*: The focus of quality strategy during the closeout phase is to verify compliance with stakeholder expectations and achieve client acceptance.
- *Operations and maintenance*: The quality strategy during the operations and maintenance (O&M) phase focuses on ensuring quality operations for the rest of the operational life cycle.

The level of detail will be different for each project, scaling as needed based on the team's experience and project complexity. The developed list is a means of breaking the project up into manageable chunks. (These "chunks" are sometimes referred to as "definable features of work.") Each item on this list will proceed through both quality assurance and quality control activities, guiding the team toward the achievement of quality on the project: delivering stakeholder intent.

Creating the quality strategy for any project requires completion of the tactics shown in Table 2.1. The quality strategy is developed when the Why, What, When, Who, and How questions are effectively answered.

"Quality" itself has varying definitions—including those noted by Deming, Juran, Crosby, Feigenbaum, and ISO 9000:2015—and is subjective to each stakeholder.

There are two facets to the definition of quality:

1. Quality of process
2. Quality of the delivered product

Table 2.1 Project quality strategy outline.

Idea	Tactic
WHY is a strategy required?	Strategy is required because it helps to keep the quality planning consistent throughout the project. To develop the quality strategy, first define the purpose (of the project) and break it down to the specific objectives. Ensure that the strategy is consistent with the vision, mission, and values of the involved organizations. (Note: Vision, mission, and values are precursors to strategy.)
WHAT needs to be completed?	Next, outline the project priorities and identify the tasks that must be completed. The lists of priorities and tasks are essential to develop the (tactical) quality plans, which include the quality assurance (QA) and quality control (QC) plans. Leverage the concepts previously stated (stakeholder intent, experience and lessons learned, critical path activities, and the drawings and specifications) as well as the project's risk profile.
WHEN will each action be completed?	Add the time dimension to the project's objectives. This helps describe the order in which the strategy and its associated tactics must be implemented.
WHO will complete each task?	Outline the responsibilities for each action. Clearly articulate who will complete what according to the timeline.
HOW will the quality objectives be realized?	Illustrate what "good" looks like. Ensure all stakeholders understand what success means. This means identifying the "acceptance criteria," which is used by the QC and by the QA to ensure that the quality objectives are realized.

Project teams must achieve both quality of *process* and quality of *product* (the completed structure). The project quality strategy must outline the process by which the product will be delivered, the process must be followed by the design and construction teams. The purpose of quality assurance auditing, then, is to confirm that there was indeed quality of process, and also that quality control helped achieve the quality of product.

The development of quality plans is the responsibility of the performing company assigned the task of completing that phase of the construction project based on the quality strategy. For example, if the project is issued in the traditional design-bid-build (DBB) format, it becomes the responsibility of the design firm to develop the quality plan for the design phase. Later on, when the project for the construction is awarded to the general contractor, it is essential for the contractor firm to develop the quality plan for construction. The quality plans developed for the design phase and the ones developed for the construction phase would be consistent only if the quality strategy of the project remains consistent.

Integrating Stakeholder Requirements

Stakeholders' stated and implied needs must be reconciled into the project's quality strategy. Early collaboration is critical to ensure that these expectations are defined, documented, and integrated into design and construction processes.

Quality begins with the first stakeholder conversation. In his book, *Out of the Crisis*, Deming notes that quality starts with intent, and that intent is translated by designers into project implementation. Intent takes many forms and is dependent on the stakeholder, and every project has multiple stakeholders. The quality strategy is the means by which stakeholder intent is understood and translated into the project down to those executing the work. Expectations are derived from intent. *Intent* as a noun means *purpose*. As a verb, intent means *showing eager attention*.

The project team must proactively extract and document expectations and intent through:

- Stakeholder meetings to define quality objectives
- Experience-based adjustments informed by lessons learned
- Critical path identification to prioritize quality measures
- Detailed drawing and specification analysis

Teams need additional context to understand what's truly important from the stakeholder perspective. What is the *purpose*, what is the *intent*, and where should each team member *show eager attention*?

The project must achieve consensus among its performing organizations on the quality objectives. The role of the quality manager is to facilitate the definition of the quality goals and objectives and ensure stakeholder alignment. Expectations must be reconciled with the final scope of the project along with its quality objectives via a stakeholders' meeting (or equivalent). To deliver quality in design and construction, stakeholder intent needs to be managed and understood. This must occur both at the beginning of the project and as new stakeholders come on board. This includes the designer/architect, general contractor and its subcontracting team, as well as the operations and maintenance staff. When selected, stakeholders must contribute to keeping the quality strategy relevant and applicable. Doing so not only reaffirms the project's commitment to quality, but also ensures that all stakeholders remain committed to the same quality goals and objectives, regardless of their own priorities for the manner in which those quality goals may be achieved. It is the primary goal of the CQM to ensure this happens.

Procurement Requirements

A quality strategy development needs procurement requirements for the project. The project supply chain includes physical materials and contracted labor. Subcontracted workers contribute 85% of a project's labor, and design team members are also part of the supply chain. Price influences both material and labor, and balancing price and performance according to stakeholder intent is essential.

Procurement decisions directly impact quality. Awarding contracts based solely on cost increases the risk of quality failures. A comprehensive quality strategy demands that subcontractors and suppliers be selected based on qualifications, track record, and compatibility with project needs. Though cost certainty is a critical constraint on every project, the project quality strategy must outline a clear process for selecting materials and labor that focuses on qualifications beyond price alone. This balance is achieved through early collaboration during quality assurance—the planning activities *before* the work starts.

Vendors and manufacturers who are part of the supply chain must be included in the project's quality strategy development. In the early phases of the project, identify suppliers willing to work as part of the project team to ensure their products are compatible with others. Ensure they are willing to participate in preconstruction and preinstallation meetings to verify the installation sequence prior to mobilization. Early input with contracting teams reduces the potential for substitution requests and changes in general after design completion.

Contracting Approaches

The quality strategy of a project can be affected significantly by the contracting approach that the project adopts. There are several different ways in which a project may be delivered. Every project team makes a strategic decision to deliver the project through a specific contracting approach and delivery method. The impact of the delivery method and contract type becomes significant because the different methods may shift the primary responsibility for quality among the owner, the designer, the contractor, or sometimes a construction management firm.

Table 2.2 shows the five typical contract types, provided here for reference only. More detail on contract types is included in Chapter 1.

There are multiple different approaches for project delivery; the decision of which contracting approach is best suited for the project must be made by the project management based on factors such as the budget constraints, project complexity, required schedule, owner risk tolerance and expertise, the desired level of control over the project, and the available expertise of the design and construction teams.

There is no one correct approach; however, similar contracting approaches for design and construction should be used for both design and construction if the contracts are independent. Designer and constructor contracts must be compatible and facilitate collaboration. The CQM and project team must identify the collaboration opportunities, regardless of how the project is contracted. In addition to the contracting approach, a project's payment method adds another layer of complexity.

Table 2.2 Contracting approach and project/pricing advantages.

Contracting Strategy	Ideal Project Conditions	Pricing
Lump Sum, Fixed/Stipulated Price	Clear scope of work	Fixed price for entire project
Time and Materials (T&M)	Unclear scope of work, smaller projects	Material costs at predetermined/hourly rates
Unit Price	Repeatable units/scopes of work with unknown quantity/duration	Fixed unit costs
Guaranteed Maximum Price (GMP)	Known scope, fixed budget, uncertainty/high potential for change, schedule driven/early start required	Maximum cost identified, cost plus a percentage or fee (may include stipulated cost cap)

- *Fixed-price (lump sum)*: Best for well-defined projects but may limit flexibility
- *Cost-plus and GMP*: Allow for quality-based selection but require stringent oversight
- *Design-build and CMAR*: Facilitate early contractor involvement, improving constructability and reducing quality risks

Each delivery method for construction projects offers advantages as well as limitations. The type of delivery method used can impact the project's quality planning. Table 2.3 shows the various contract delivery method types with a high-level comparison of the risks of lesser collaboration between the various phases of a construction project.

With integrated project delivery (IPD), the entire project team is under one contract before the design phase begins, and there is shared risk and reward among all stakeholders. In design build (DB), the design and construction team are under one contract as one entity entirely responsible for design and construction. (Design build can be designer- or contractor-led.) Under construction manager at risk (CMAR), the construction manager acts as the stakeholder's representative and controls design, construction, and associated information flow. On DBB

Table 2.3 Contracting delivery methods.

Project Delivery Method	Contract Delivery Method Used	Risk of Lack of Collaboration
Integrated Project Delivery Method	The entire project team is under one contract (designer/architect, contractor, owner).	Risk and reward are shared. Integrated projects offer the most collaboration between all stakeholders.
Design-Build (DB) Method	Design and construction are under the same contract; they usually work for the owner but could also work for the concessionaire. (Note: There are several variations in this delivery method.)	Usually DB projects are contractor-led, but they can be led by the design teams. More collaboration is usually possible between the various phases of the construction project life cycle.
Construction Manager at Risk (CMAR)	The construction manager is hired as the representative of the client/stakeholder; the CMAR oversees the design, construction, and associated information flow.	Because construction is linked with design through the construction manager, this method forces a higher level of collaboration. The risk of lack of collaboration belongs to the construction manager, not to the owner of the project.
Traditional Design-Bid-Build	The owner has direct control of design; the construction contract is issued separately to the contractor. This is a traditional project that is run in phases; a construction contractor may be hired much later in the project life cycle.	There is the least amount of collaboration between the designer and the construction contractor in this method of project delivery. The risk of lack of collaboration belongs to the owner.

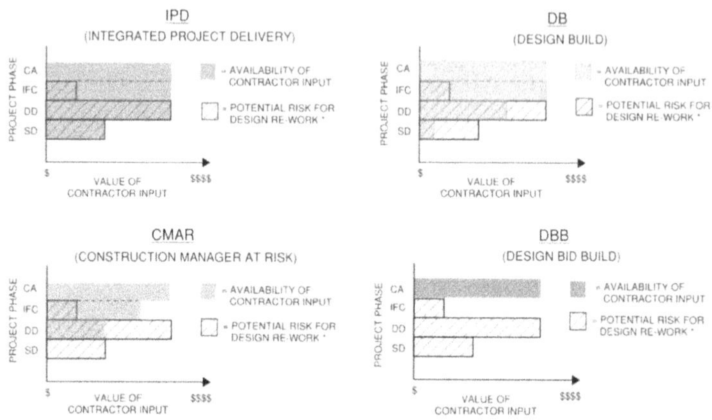

Figure 2.1 Value of contractor input.

projects, the owner has direct control of the design and construction team via two independent contracts. (See Chapter 1, "Project Delivery Methods," for more information.)

In Figure 2.1, the hash-marked bars represent the time when critical decisions are made, a visual representation of risk to be mitigated. This is when collaboration is most important. Collaboration opportunities—specifically, what information is needed, by whom, and at what time—need to be identified early, regardless of the contracting approach or delivery method. Collaboration opportunities must be created intentionally to prevent gaps in quality planning and execution.

Regulatory Compliance

Another critical element needed for planning a quality strategy for a project is the need for regulatory compliance. The requirements of the clients and any other Authority Having Jurisdiction (AHJ) must be included as part of quality planning. The AHJ is a critical external stakeholder with both stated (building code) and implied (code interpretation) intent. Key to understanding these requirements is the AHJ's early involvement in the project. AHJs dictate project compliance through building codes and safety regulations. Compliance strategies should not only meet baseline requirements, but also anticipate interpretations and expectations specific to the jurisdiction. Early AHJ involvement mitigates risks of late-stage rework and delays.

Early understanding of the following is essential related to the AHJ:

- Roles in the regulatory system/jurisdiction
- Who needs to review what and by when
- What permits are needed
- Specific interpretations of the code

- How to look up specific codes and references to prepare for AHJ conversations
- The AHJ's specific concerns or items about which they are overly sensitive. (Note: These sensitivities may carry more weight in the interpretation than another team member's understanding of the code.)
- Special inspections required and who needs to hire the associated inspectors
- Who has relationships within the jurisdiction to attain the needed information

Additional proactive compliance measures include:

- Early consultations with AHJs to clarify interpretations
- Integration of special inspections into the quality plan
- Clear documentation of compliance paths to expedite approvals

Figure 2.1 shows where these issues would significantly impact the design and permitting process, where those specific code interpretations should be known. It is too late to learn of a contentious interpretation after the building is nearing substantial completion or framing/structural components are already in place. The general contractor is the link to the local code official who can help the team understand how the code will be interpreted.

Securing Management Commitment

Another significant aspect of strategic quality planning is to secure the commitment of the management for the chosen strategy. Management must foster a culture where quality is integral to decision-making. Without leadership endorsement, quality initiatives often become secondary to cost and schedule pressures. Only management can create a psychologically safe environment for communication. A safe atmosphere must exist for transparent communication to occur. Collaboration will not happen with isolated communication. Management at each organization involved in the project has the power to enable this type of environment to exist. Once the environment and collaborative culture are created, the team can work within them.

Management commitment is demonstrated by:

- Prioritizing quality over short-term cost savings
- Creating an environment of psychological safety for quality concerns to be raised
- Supporting early collaboration among all stakeholders

Management must also commit to collaboration even if the contracting approach or delivery method does not provide an ideal framework for it—and that collaboration *actually happens* once the opportunities are identified. This means a commitment to the planning effort.

Establishing Key Performance Indicators (KPIs)

A quality strategy is the project's overall approach to achieving its quality goals. It is essential that as part of the strategy, the measurable metrics that must be used to track progress toward the achievement of the goals and objectives are defined. Key performance indicators (KPIs) are the performance metrics that act as triggers for action, informing a team of poor performance. Leading indicators are metrics used to identify proactive actions, signaling a potential problem. They measure behavior *prior* to the start of the work, a prediction of future performance—measurements of quality assurance. Lagging indicators are metrics that reveal themselves *after* a problem. They measure risks not being mitigated after the work starts, reflecting past performance—measurements of quality control. KPIs drive accountability by measuring both leading and lagging indicators of quality performance.

There is no normalization between any two projects. Each project is unique with no consistent data set from project to project, which is why KPIs for one project do not apply directly to another. There is some consistency, project to project, within a specific sector where cost of poor quality (COPQ) can be measured; but it is nonetheless a forecast, and forecasts are inherently inaccurate.

Examples of leading and lagging indicators include:

Leading indicators (predictive):

- Frequency of design coordination meetings
- Completion rates of preinstallation checklists
- Timeliness of requests for information (RFIs) and submittal reviews

Lagging indicators (reactive):

- Number of punch list items per project phase
- Defect rates in completed work
- Warranty claims and rework costs

KPIs must align with stakeholder expectations and project priorities to ensure quality objectives are being met efficiently.

To establish KPIs for the project, the project team must quantify all stakeholder intent—in addition to the critical physical elements of the project requiring eager attention (e.g., definable features of work and design intent)—and establish a success metric for each. KPIs must measure both the "good" (positive) and the "bad" (negative). Positive metrics are additional motivators to validate the right behaviors driving toward high customer satisfaction (CSAT), while negative metrics identify risks that need to be mitigated or where corrective action is required.

There are diminishing returns for attempting to measure everything, capturing data on all potential data points, and being 100% accurate. Construction is unlike manufacturing in that it is difficult to measure specific data points, yet the impact is important to understand. With much of today's software, automatic reports can be generated or set up to showcase this information.

The perfect list of KPIs is unattainable, as is identifying a KPI for each element of the project. Project teams must select KPIs that drive behavior for specific project goals and stakeholder expectations. These are the KPIs that truly matter.

Planning the Maintenance of the Quality Strategy

The final element to a comprehensive project quality strategy is periodic review, maintenance, and continual improvement. For projects that are of a longer duration and complex in nature, it is good practice to review the strategy at least on an annual basis. Such a review is aimed at substantiating the progress that is made in implementing the quality strategy and its supporting quality plans. It must be noted that although it is not preferred, the quality strategy may need to be modified due to major changes in the project's conditions, such as market shifts or major changes in the regulatory requirements.

When the project conditions change significantly while the scope does not change, it becomes essential to revisit the quality strategy to ensure it is still relevant and achievable using the subsequent quality plans. If it is determined that the goals are not achievable using the existing strategy and its subsequent quality plans, then that may warrant modifications in the strategy as well as the quality plans. The purpose should be not to change the goals but the means to achieve them.

CONCLUSION

A successful project quality strategy identifies the key collaboration points, regardless of the contracting approach or delivery method, and ensures that the collaboration occurs. Early stakeholder involvement is essential—from AHJ and contractor input in the early design phases to design handoff of information to the contractor at the start of construction. Assigning KPIs that drive these collaborative behaviors is key to delivering stakeholder intent.

By managing quality proactively at each project phase, teams can mitigate risks, improve collaboration, and ensure successful project outcomes. In an industry driven by efficiency and performance, a well-crafted quality strategy is the foundation for delivering excellence.

QUESTIONS

1. What is the difference between quality assurance and quality control?
2. Why is it important to have early input in the design from material and product fabricators?
3. What is a critical element of design and construction team contracts?
4. What is a common problem with stakeholder input? Give an example of how it can be solved.
5. Give an example of a leading and lagging indicator.
6. Explain how different project delivery methods (e.g., design-bid-build, design-build, CMAR) impact the level of collaboration between stakeholders and how this can affect the project's quality strategy. Give examples.
7. Discuss the relationship between "stakeholder intent" and the development of a project quality strategy. How does a project team effectively elicit and manage both stated and implied stakeholder expectations throughout the project life cycle?
8. Describe the importance of integrating procurement requirements into the project quality strategy beyond simply selecting the lowest bidder. How can early vendor involvement and qualification-based selection contribute to achieving project quality goals?

9. Differentiate between leading and lagging key performance indicators (KPIs) in the context of project quality. Give two examples of each, and explain how they can be used to drive continuous improvement and accountability.
10. Explain how a project's quality strategy must be a dynamic and evolving document. What factors might necessitate a modification of the quality strategy and what steps should be taken to ensure alignment and approval of these changes among stakeholders?

ENVIRONMENTAL, SOCIAL, AND GOVERNANCE (ESG) PRACTICES

> Apply sustainable practices and programs (e.g., LEED®, ENERGY STAR®, and Cradle to Cradle Certification), social responsibility, and governance standards to enhance the overall quality and long-term viability of projects. (Apply)
>
> **Body of Knowledge II.A.2**

In today's increasingly conscious world, environmental, social, and governance (ESG) practices have transcended traditional business considerations to become fundamental drivers of project success and organizational viability. Evolving from concepts like the triple bottom line and corporate social responsibility, ESG now encompasses a multifaceted approach to sustainable and responsible operations. This section delves into the intricacies of ESG within the construction industry, exploring its historical development, core components, and practical application in achieving both environmental sustainability and social equity through effective governance. Understanding ESG is no longer a niche expertise but a necessity for quality managers and stakeholders alike.

Historical Perspective of ESG

Evolving from the triple bottom line concept, in 2004, the term ESG, or environmental, social, and governance, was coined by the United Nations Global Compact. The UN Global Compact was a nonbinding pact that encouraged businesses and firms worldwide to adopt sustainable and socially responsible policies, and to report on their implementation. In addition to environmental sustainability and social responsibility, the ESG concept suggested that good governance should make up the third aspect of the new initiative.

The components of ESG incorporate many socioeconomic factors, most significantly, protecting the environment, addressing climate change and impacts, improving labor conditions and human rights, and enhancing diversity. ESG has integrated all previous concepts and initiatives that were related to its stated goals for good business practices. Over time, it has become increasingly important to the ever-expanding customer base, as new customers, particularly those that operate in the public domain, are beginning to carry much greater influence over purchasing (procurement) and other aspects of the project work. They often use ESG metrics to make their decisions about the projects and the direction of their businesses.

ESG and the Construction Industry

For the construction industry, the most significant ESG initiatives manifest on the environmental aspects of ESG (with social and governance aspects increasingly gaining attention). There are many construction projects that are undertaken with the goal of reducing their carbon footprints, emphasizing energy-efficient buildings, reducing their water usage, and in general engaging with more sustainable sourcing of their materials. ESG is increasingly becoming a measure of responsibility for the infrastructure industry organizations and all those that are involved in design and construction.

Although ESG represents a subject-matter expertise (SME) that is sometimes considered a niche, a Certified Construction Quality Manager (CQM) is expected to have at least a minimum level of understanding about these three aspects of business. The quality manager must learn about ESG and the connection between it and the quality of the project work. In the sections below, we will review all three aspects of ESG practices in greater detail.

Environmental Sustainability (E in ESG)

The first component of ESG is "environmental sustainability." For the construction industry, this aspect of ESG has gained the most attention. It is important for construction projects to construct buildings that are sustainable over a longer term. Sustainability was defined by the United Nations Brundtland Commission in 1987 as "meeting the needs of the present without compromising the ability of future generations to meet their own needs."

The environmental sustainability initiatives of contractors' parent organizations and their construction projects must aim to reduce the impacts of high carbon emissions, resource depletion, and waste generation of the building materials. The objective should be to utilize more sustainable material options instead of typical traditional materials. For example, preferred options may include the use of recycled materials, responsibly sourced wood, and cross-laminated timber, all of which can reduce the overall environmental footprint of construction projects.

The construction industry is profoundly influenced by environmental factors, as it utilizes a lot of building materials such as wood, steel, concrete, aluminum, and others, all of which can have significant environmental impacts due to the energy-intensive processes involved in their production.

Like quality management, where the quality of the process is measured differently from the quality of the product, the environmental sustainability of construction projects must also be viewed in terms of two aspects of sustainability:

- *Process-based sustainability*: Sustainability of the process looks at the efficiency, effectiveness, and fairness of the management practices that are used by the project to implement sustainability.

- *Product-based sustainability*: Sustainability of the product measures the life cycle impacts of the products that are used by the project (and measures the impact of sustainable products used in all phases of the project, not just the construction phase).

To measure the implementation of sustainability in organizations, there are currently several well-known certifications in the construction industry. These certifications have been initiated by two different organizations, although both address process as well as product sustainability. The evaluation of the environmental sustainability of the constructed work as well as the sustainability of the materials and components that are used in construction is part of both certifications.

Process-Based Environmental Sustainability

Process-based sustainability is meant for the "built environment," or human-made conditions. This is the environment where we spend most of our time: buildings, villages, cities, public infrastructure, transportation, highways, and other urban spaces. The term has been used in architecture, urban planning, public health, sociology, and anthropology and is increasingly becoming the center of attention for designers and constructors. With the increasing focus on social responsibility, communities everywhere are expecting that civil infrastructure should be environmentally sound and sustainable, offering society the opportunity to improve their quality of life without degrading the quantity, quality, or availability of natural and social resources.

Currently, there are two leading associations in the United States that are at the forefront of promoting sustainability of processes and have created technical measurement scales and frameworks that quantify the sustainability performance of projects and organizations. These are the U.S. Green Building Council (USGBC) and the Institute for Sustainable Infrastructure (ISI). The technical frameworks that have been issued by these leading organizations for sustainability are used extensively by the construction industry.

The framework issued by the USGBC considers sustainability in construction and awards points corresponding to a certain level of sustainability efforts in the design and building of sustainable infrastructure. The framework used by ISI is slightly different, as it utilizes a combination of technical measurement of process-based sustainability with the physical health of the people that inhabit that facility.

All sustainability frameworks, although slightly different from each other, offer a structured way to measure and assess an organization's sustainability efforts. Selecting which sustainability framework should be considered by an organization or a project to quantify their sustainability efforts is a decision that management must make. The right answer may be found within the scope and preference of the project stakeholders; one certification is not superior to the others; because of their focus on slightly different sustainability indicators, one may better serve different end users than another.

The frameworks issued by USGBC as well as by ISI have four levels by which any building/structure (project) is evaluated (Table 2.4). The more points it earns,

Table 2.4 Four levels of sustainability certification.

1. Level 1 Rating/Status—VERIFIED/CERTIFIED—First level of sustainability certification
2. Level 2 Rating/Status—SILVER—Second level of sustainability certification
3. Level 3 Rating/Status—GOLD—Third level of sustainability certification
4. Level 4 Rating/Status—PLATINUM—Highest level of sustainability certification

the higher the classification of that structure with respect to sustainability and resilience.

The organizations that aim to achieve these environmentally green and sustainability targets for their infrastructure usually have qualified personnel who are SMEs in various areas and use their expertise to design and construct sustainable infrastructure. A description of the three frameworks for process-based sustainability follows.

Leadership in Energy and Environmental Design (LEED) Certification

The USGBC is a nonprofit organization that supports the development of healthy and resilient communities through the "transformation" of the built environment. USGBC is best known for its development of the Leadership in Energy and Environmental Design (LEED) green building rating system. LEED certification stands out as one of the most globally recognized green building certification programs for the built environment.

To achieve LEED certification, the design for a new building must apply for the certification that is the most applicable to its scope. Existing infrastructure can also be evaluated for how closely it meets the standards of excellence in each category of the LEED standards. Called "the project," the applicant building (or a new design) earns points if it is evaluated that it adheres to the prerequisites and credits that address the best practices for carbon, energy, water, waste, transportation, materials, health, and indoor environmental quality. Projects go through a verification and review process by the certification body of the USGBC and are awarded points that correspond to a certain level of LEED certification.

There are nine categories that form the framework of the LEED evaluation of the buildings and infrastructure.

1. *Integrative process*: Focuses on an approach to building systems and equipment that is comprehensive, considers interrelated issues, and looks for synergies among systems and components.
2. *Location and transportation*: Grades the location of the building where it makes most sense and not on land that is unsuitable for development.
3. *Sustainable sites*: Awards credits for protecting natural habitat, keeping open spaces, dealing with rainwater, and reducing light pollution and heat island effects.
4. *Water efficiency*: Covers water use reduction, rainwater management, and the employment of alternative, non-potable water sources for appropriate end uses.

5. *Energy and atmosphere*: Gives credits when the design of the building effectively handles energy usage and mitigates its effect on the atmosphere.
6. *Materials and resources*: Describes the environmental criteria to follow in selecting construction materials and the methods employed to manage construction waste.
7. *Indoor environmental quality*: Covers the overall air quality control and assessment, tobacco smoke control, acoustics, low-emitting materials, construction air quality management, thermal comfort, and lighting, and gives credits if the design considers these elements.
8. *Innovation*: Awards extra credit points to projects that incorporate innovative and sustainable building features that exceed the LEED rating system requirements or thresholds.
9. *Regional priorities*: Awards credits to the projects that focus on local environmental, social equity, and public health priorities.

The buildings and infrastructure (usually new designs) are assessed and evaluated by the USGBC's third-party certification program, which uses LEED standards for the design and construction of high-performance buildings and infrastructure. Credit points are scored against each noted category for adhering to the best practices of design and construction and addressing the environmental and sustainability targets of carbon emissions, energy, water, waste, transportation, materials, health, and indoor environmental quality.

Envision Certification

The Institute for Sustainable Infrastructure (ISI) is another organization that has made significant contributions to the environmental sustainability of infrastructure projects. Its framework similarly encourages systemic changes in the planning, design, and delivery of sustainable, resilient, and equitable civil infrastructure, and issues the ENVISION certification for the processes of the construction projects in the built environment. The sustainability rating system or framework issued by ENVISION is slightly different from that of LEED, but it is similarly aimed to help the projects through quantifiable metrics for process-based sustainability. The ENVISION certifications are also based on the achievement of points against the ENVISION framework.

This certification requires the applicant project to provide evidence of compliance in five categories of the Envision-SP Certification.

1. *Quality of life*: Assesses how the project enhances public health, improves safety, and minimizes the negative impacts of construction through its focus on well-being, mobility, and community.
2. *Leadership*: Evaluates how the project and its leadership offer the synergies of commitment, collaboration, teamwork, and stakeholder engagement to make it more sustainable and resilient. This subsection includes a holistic view of the project and its full life cycle in terms of three subareas: collaboration, planning, and economy.

3. *Resource allocation*: In this category, the Body of Knowledge (BoK) covers three sub-areas of Materials, Energy, and Water. Includes the evaluation of knowledge of the assets that are needed to build the infrastructure and maintain it.
4. *Natural world*: This category examines the project site, conservation, and ecology and the applicant's knowledge of the ecosystem in which their project is situated.
5. *Climate and risk*: This section covers two sub-areas, namely emissions and resilience of the project. The changing future operating conditions and their understanding and mitigation of the negative impacts are evaluated under this category of the BoK.

Obviously, there cannot be a one-size-fits-all certification that addresses all sustainability targets of the built environment. Organizations must determine which standard/framework suits their organizational sustainability goals most closely, and consider the recognition that one standard may have in any physical location (e.g., Europe, North America) versus another. Both LEED and ENVISION are world-recognized standards for environmental sustainability that are slightly different in their application.

There are several other organizations working for conservation and environmental sustainability, but the focus of the work by these two associations has created more quantifiable metrics than most others.

WELL Certification

To address the social aspect of sustainability, a new certification framework was introduced in 2013. This certification, called the WELL Certification, was introduced by the International WELL Building Institute (IWBI). This is a "well-being focused standard," which emphasizes the concept of constructing "healthy buildings" and is commonly called the next generation of sustainable infrastructure. This concept of healthy buildings was intended to support the physical, psychological, and social health and well-being of the people who work and live in them.

Thus far, this standard has had more success mostly outside North America. According to one report, by 2016, over 200 projects in 21 countries had adopted the WELL certification.

The process for on-site assessments and testing for the certification is called "performance verification." On-site measurements are taken for various air and water quality parameters, as well as sound and light levels. It is a distinct process from traditional building commissioning and ensures that the building performs as intended, according to WELL requirements.

Performance verification is completed by an authorized WELL Performance Testing Agent, who usually spends one to three days in the building to validate its documentation and complete a series of performance tests, spot-checks, and measurements covering all WELL concepts. Testing is completed according to IWBI's sampling protocols in the WELL Performance Verification Guidebook.

Although sustainability is typically considered a niche area in construction, it may be useful for a quality manager (such as CQM) to further their knowledge about sustainability because this is the direction of future construction. This is because sustainability can be considered another feature of the project (or an organization) that promotes longer-term viability of the processes and systems and ultimately leads to higher-quality end products and processes.

Product-Based Environmental Sustainability

In addition to process-based sustainability frameworks, sustainability frameworks also exist at the product level. Sustainability that is intended for materials and components is called product-based sustainability.

Several certification programs issued by various organizations act as benchmarking ratings for materials and components used in construction projects. Product-based sustainability certifications provide several benefits; the most significant is increased consumer trust and confidence that comes from using materials and components that are more compliant with safety, quality, and sustainability standards.

Construction projects that differentiate themselves by using well-known certified products and materials can build more customer loyalty and access new markets that value certified products and components. Using such products enables constructed projects to achieve their LEED and ENVISION ratings more easily. Following is a list of the more well-known certification frameworks for determining the environmental sustainability of materials and components.

Energy Star

Energy Star certification is an energy efficiency program that is administered by the US Environmental Protection Agency (EPA) in partnership with the U.S. Department of Energy (USDOE). As of 2024, more than 75 product categories are considered eligible for the Energy Star rating label. These products include appliances, electronics, lighting, heating and cooling systems, and other items that are utilized mostly in residential buildings.

This program is a widely recognized scoring program for energy efficiency in products and materials that are used in building construction. There are more than 100 different metrics for the performance, including KPIs for energy use and costs, water use and costs, waste and materials, and greenhouse gas emissions.

Cradle to Cradle (C2C)

The C2C-certified products program was introduced in 2005 as a proprietary trademark of the consultant firm MBDC (McDonough Braungart Design Chemistry). This is a product assessment and certification standard that certifies safe, healthy, and sustainable products. The benefit of the C2C standard is that it promotes product development in a cyclical way so that all resources may be used effectively. The goal is to avoid mixing biodegradable products with oil and chemical-based products.

Social Practices (S in ESG)

Like environmental sustainability, social aspects of construction management have similarly evolved over time. Social practices encompass the human dynamics, ethical considerations, and community impact of project execution. In the modern history of project and construction management, the role of social practices can be traced back to the 1980s when the concept of corporate social responsibility (CSR) was adopted by some organizations; this started public awareness of the social responsibility of corporate businesses.

However, the impact of social responsibility and the lack of its effectiveness was highlighted through the globally publicized scandals of Enron, Tyco, and WorldCom, all of which happened within a span of a few years (around 2000). The common theme in terms of the failure of the companies' social responsibility in these scandals was flawed corporate reports. The reports overemphasized public relations but lacked hard evidence of success through measurable performance indicators and quality processes. Following these scandals, corporate standards such as Sarbanes-Oxley (in 2002) and ISO 26000 (2010) were introduced to provide guidance to business organizations on how to operate in socially responsible ways.

As a result of these standards and best practices, the concept of social responsibility is now a lot more mature. Business organizations, especially public ones, are now required to offer more than just public relations statements; some industries are pushing for mandatory ESG reporting that includes quantifiable and tangible results that must correspond to publicly defined, specific goals.

In the social practices category, the most common and applicable social practices a project may consider include the following (and note that this is not an inclusive list):

1. Diversity in the project teams
2. Equity and inclusion in the project teams
3. Disadvantaged and minority-based businesses
4. Community impact

Diversity in Project Teams

The most prominent and well-known sub-element of the social practices is the concept of diversity, equity, and inclusion (DEI). This concept has its roots in the 1960s civil rights movement, but it has gained more significance in the construction industry in recent years. DEI is a business framework for the workplace that may be applied to an organization or even a project. It is important to recognize that, like any initiative, there are both positive and negative impacts of DEI. The positives include increased innovation, improved financial performance, and a more inclusive workplace that fosters employee engagement and retention. However, the negative impacts of DEI have been the potential backlash that results from the perception of reverse discrimination as well as the potential for lowered standards (which is something that directly impacts quality).

It is important to realize that by itself, diversity in the workplace is an important and useful social concept—but it must be addressed carefully and equitably if it is to be useful and practical. Businesses are realizing that a diverse and inclusive workforce is essential for their success. In the newly conscious landscape of diversity, construction projects must also quantify the measurement of their progress toward their diversity goals.

Diversity in project teams refers to the presence of differences within a given setting. The Oxford English Dictionary defines diversity as "the practice of including or involving people from a range of different social and ethnic backgrounds and of different genders, sexual orientations, age, religious identity, and status." Diversity allows for a wider range of perspectives, experiences, and ideas, all of which usually lead to increased creativity, innovation, better decision-making, and a stronger understanding of different markets, and ultimately contributes to a more productive and inclusive work environment.

Diversity has two fundamental dimensions:

- *Diversity that exists due to internal (personal) reasons which cannot be changed*: Examples of such diversity include age, gender, ethnicity, religion, race, sexual orientation, physical ability/disability, appearance, and language. It is not possible to carry out an intervention on the diversity that happens due to these reasons. The organization must acknowledge and manage it.
- *External reasons for diversity that are not personal*: These are usually factors outside the immediate control of individuals, although it may be possible to modify them. Examples of external reasons include the diversity due to geographic work locations, or varying skill sets, education levels, interests, backgrounds, and experiences. It is possible to balance the external factors that cause diversity through organization-level interventions. For example, although it may be challenging, it is possible to elevate the skill sets of different people through training, or to change their work location, or to invite them to new interests.

It is very important for quality managers to understand diversity in their workplace, as it assists them in implementing quality management systems more effectively when they understand the composition of their projects and the parent organizations. The benefits of a diverse workforce are significant and must be considered for long-term growth, as well as for improved productivity and reduced turnover, both at the project's and the parent organizations' levels.

Equity: Another Social Initiative

Another important social initiative is "equity." *Equity* is the act of making the processes and programs within the organization (or the project) equitable, impartial, and fair such that they provide equal possible outcomes for every individual. Equity involves trying to understand the needs of each person and giving them what they need to achieve an equal opportunity in life. In other words, equity is the "equality of opportunity but one that is based on delivering justice." True equity is therefore merit-based, not loyalty-based or quota-based. When people feel they have been treated equitably, they feel more valued and consequently have a higher commitment to the cause of their organizations.

Minority-Based Social Equity Initiatives
To keep the construction industry equitable, large publicly funded construction projects are required to offer opportunities to Disadvantaged Business Enterprises (DBE) and Minority Business Enterprises (MBE). DBE is a federally mandated program intended to provide contracting opportunities to small businesses that are owned and controlled by socially and economically disadvantaged individuals. The objective of the DBE and MBE initiatives is to allow smaller-scale businesses to participate in government contracts so that they also get fair contracting opportunities and foster inclusive economic participation to socially disadvantaged groups in a competitive market that they would otherwise struggle in.

Governance (G in ESG)

The third and the last element of the ESG is "governance," which relates to the administration of the projects or organizations for sustainable infrastructure and social diversity and inclusion. *Governance* refers to the systems, processes, and practices that an organization (or a project) uses in order to operate. Good governance practices are those that include transparency, accountability, and integrity in their decision-making. Good governance is an essential part of the ESG initiative, because without it, environmental sustainability and social justice cannot be achieved.

Governance is important because it sets the rules, policies, and procedures that help organizations meet business targets along with environmental and social targets. Effective governance makes sure that environmental and social policies are created, implemented, and followed.

An important aspect of good governance includes corporate/business ethics, as well as project ethical guidelines. The governance aspect of ESG is a macro concept; it requires the projects and organizations as well as their stakeholders to think of big-picture goals and objectives, and commit to responsible projects that address climate-related issues, energy consumption, waste, and their policies, risks, and opportunities.

CONCLUSION

The three domains of environmental, social, and corporate governance (ESG) are intimately linked. The most interesting and practical thing about the ESG initiative is that the projects as well as their parent organizations are required to provide quantitative and qualitative disclosures that cover their ESG policies. Unlike in the past, when similar initiatives served mostly as public relations exercises, the ESG-related initiatives of the projects and their parent organizations are now considered the primary drivers for change.

CQM quality managers can only have impact on the continued development of their projects and parent organizations when they understand the ESG initiatives and engage with the relevant aspects of the project and organizational work to achieve efficiency as well as effectiveness.

ESG practices are no longer optional add-ons; they are fundamental to the sustainable success of the construction industry. By embracing the principles behind ESG, organizations can drive positive change, mitigate risks, and foster long-term value creation. Quality managers, like CQMs, play a crucial role in integrating these principles into project workflows, ensuring both efficiency and effectiveness. The journey toward a more sustainable and equitable future requires continuous learning, adaptation, and commitment to these evolving standards, ensuring that both projects and organizations thrive in a world that is increasingly conscious to ESG.

QUESTIONS

1. What are the most common environmental sustainability certifications for the built environment?
2. What are the sustainability certifications that are focused on certifying the products and materials used in construction projects?
3. Which sustainability certification addresses the social aspects of next-generation "healthy buildings"?
4. How many categories form the basis of LEED certification?
5. How many categories form the basis of ENVISION certification?
6. What are the performance indicators for the social practices of organizations that measure the positive impacts on the well-being of the local and global communities?
7. Compare and contrast the LEED and ENVISION certification frameworks, highlighting key differences in their approaches to evaluating process-based environmental sustainability in the built environment.
8. Explain the distinction between process-based sustainability and product-based sustainability in construction projects, providing examples of how each is measured and certified.
9. Discuss the evolution of social practices within ESG, outlining how concepts like corporate social responsibility have changed over time and what factors have driven these changes.
10. Analyze the two fundamental dimensions of diversity in project teams, and explain why understanding these dimensions is important for quality managers implementing quality management systems.
11. Describe how the concept of "equity" differs from "equality" in the context of social initiatives within ESG, and give an example of how equity is addressed in the construction industry through specific programs.

ALIGNMENT OF QUALITY PLAN WITH OTHER PROJECT PLANS

> Create the quality plan, ensuring alignment with the quality strategy and technology strategy. Align the quality plan with other project plans, such as the Health, Safety, and Environment (HSE) Plan, Execution Plan, Communication Plan, project schedule, and Risk Management Plan. (Create)
>
> Body of Knowledge II.A.3

Construction projects are multidisciplinary in nature and are spread across a phased completion schedule that includes different performing teams working on their scope of work, many times quite independently from each other. It is not uncommon on construction projects to find a disconnect where the people working on one phase do not even meet with those working on the next, let alone collaborate. It is partly the sequential and phased nature of construction projects, and secondly the fact that there are different and unrelated performing organizations working on the different phases, that causes this disconnect. The disconnect is so prevalent that some do not even consider design an integral part of the construction project. Choosing to think in micro terms, and only the construction phase, indicates limited thinking that can be changed only when the expanded scope of a construction project is considered.

The CCQM BoK describes construction as a life cycle activity that begins at the precontract stage and proceeds through its developing stages of planning, design, and procurement before the actual construction work may begin. Considering a construction project as not just limited to the construction phase but in terms of its life cycle requires that all the phases of the construction project's life cycle be aligned to the same (project's) goals and objectives. This can be achieved only through good quality planning and creation of a project where synergistic efforts pull it in the same direction, bringing consistency in the quality of work and optimizing the quality performance of all the performing teams involved.

There are multiple commonalities between the various functional disciplines on a construction project that can also be leveraged to bring more consistency in the project work.

What Is Alignment of Quality Plan, and Why Is It Needed?

Construction projects are traditionally inclined toward a more structured, "waterfall" style of project management. The reason for this is the unidirectional nature of work, which is characterized by its linear and sequential nature. In this methodology, each phase of the work must be completed before moving on to the next. Unless the project is extremely unpredictable (which is rare and may warrant the use of agile technologies), most construction projects find the cost of going back, which is considered rework, as enormous. The purpose of alignment of quality planning is to arrange the multiple, often individually incentivized plans into one seamless system that meets the needs of all stakeholders (Griffith).

In this context, the alignment of quality plans becomes the most significant way to ensure that all phases of the construction project remain consistent with their eventual goals and objectives. Even though the earlier phases of the construction project are different in terms of their scope, they are an integral part of the construction project's life cycle. It should be possible to bring a flow to the project life cycle by aligning all its phases and making alignment a force multiplier. In that respect, quality planning and management is the one common thread that connects all seven phases of a construction project's life cycle because, regardless of their different scopes, the commitment to quality should remain consistent across all the phases. Using quality planning for the alignment of the construction project from its initiation to planning, design, procurement, to construction, and even beyond the closeout is the best solution for the project.

The question, then, is at what level/tier of the quality system should the quality alignment be made? Whether they are formal or informal, all quality systems that exist within organizations as well as in the projects operate with a hierarchical tiered structure that exists to ensure that quality issues are handled efficiently and effectively. A tiered structure enables information to be shared with the right people operating at the right levels. Its ultimate purpose is to achieve quality and operational effectiveness. As the more critical quality issues are elevated to the top management, the operational issues of quality may be handled at the quality planning levels, preferably resolving them at the lowest level that is practical.

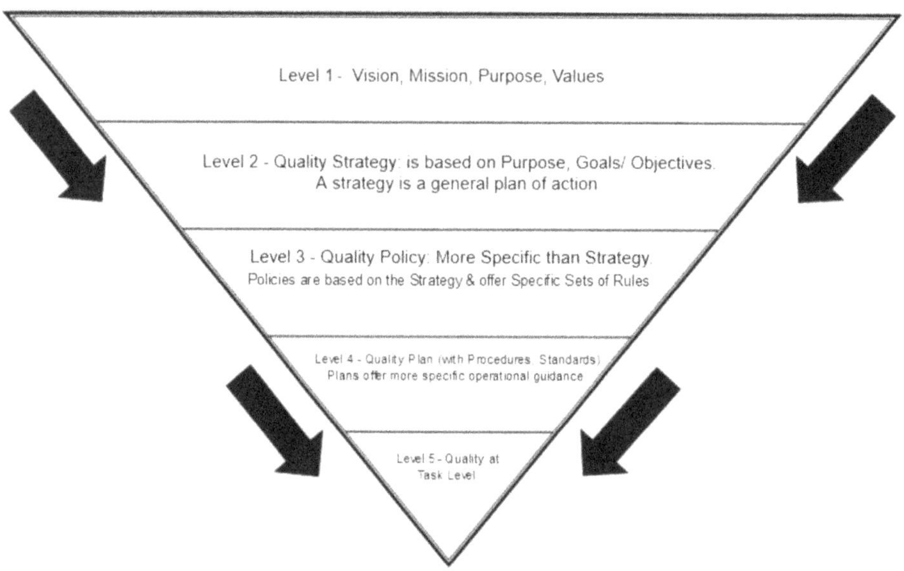

Figure 2.2 Top-down structure of organizations' quality management system.

Figure 2.2 shows the various tiers of a typical quality management system. The CQM BoK recommends that alignment must be planned for the construction project life cycle at the quality plan level, as that is the most appropriate tier of planning where the guidance offered is neither too broad nor too narrowly defined.

At the quality plan level, the planning is specific enough to utilize quality procedures, standards, and quality activities—all of which, once implemented, are able to align the various phases of construction to the common objectives of achieving quality throughout all the phases.

Alignment Versus Integration

The first thing about alignment is that it is not the same as integration. Alignment is the process of ensuring that all performing teams that are working on a project work toward the common goals, whereas integration is the process of combining parts into a unified system. The primary goal in the quality planning of the construction projects is alignment, not the integration of unique and unrelated processes.

The performing organizations that work in the earlier phases of the construction project life cycle (namely precontract, planning, design, and procurement phases) are usually different from the construction contractor. But they are also the stakeholders of the construction project. The construction quality manager must engage with the personnel who are/were responsible for the quality of work in the earlier phases, some of which may have happened before the construction quality manager was hired. In this respect, the design quality manager becomes the single most important person that the CQM must engage with to align the construction quality plan with the design phase. Other personnel who may be similarly responsible for quality in the earlier phases (whether titled or not for the quality role) may also be engaged by the construction quality manager with the goal of achieving the alignment of the quality strategy throughout all phases of the project.

A Project Quality Plan Versus an Organizational Quality Plan

The first challenge of alignment is that the projects need a project quality plan (which should *not* be the organization/enterprise-level quality plan). Project quality plans are different from organizational quality plans. A project is a specific, temporary, time-bound initiative undertaken by a project team, whereas organizational plans are developed at the enterprise level and outline the organization's permanent direction and strategic objectives. Most organizational plans, regardless of department/function, do not scale, nor do they effectively accommodate the context of project work. A compartmentalized corporate culture, if directly translated to the project level, usually creates competition between the functions. It is therefore essential that the project quality plans be written to a level that is appropriate for their scope. In any phase of the construction project, the performing organization must be required to develop a project quality plan that is specific to their work on the project.

Project quality plans can change if they need to because they are more tactical in nature, not strategic. These plans must be written to describe the project circumstances and include information about the project's customers (internal and external). They must also describe the project's technology (e.g., what document control system the performing organization intends to use for their part of work on the project, or their cost management system, etc.).

The alignment between their own organizational quality plan and their project quality plan is the responsibility of the performing organization that develops the quality plan for its phase of the construction project. The quality plan that is prepared for the design phase of the project is the responsibility of the design team. Likewise, the quality plan that is prepared for the construction phase of the project is the responsibility of the construction team.

If the construction quality manager finds that work on the earlier phase is still in process, and that the performing organization never provided a project quality plan for the earlier phase work, then they must request a submittal of a dedicated project quality plan for the performing organizations' scope of work. Also, the project-specific quality plan, when submitted by the earlier-phase performing organization (such as a designer), must be formally approved by the construction project quality manager. The phase-based project quality plan that is issued by the performing organizations should be formally approved only if it is aligned with the quality objectives of the construction project.

Four Alignment Challenges

Typically, there are four challenges to making the necessary alignments to the quality plans, described in detail below.

Align the Quality Plans to the Project Quality Goals

A common failure of alignment in quality planning happens when the earlier phases of the project issue a quality plan that uses their own organization's goals and objectives as the guiding standard. This challenge is slightly different from the earlier described scenario where the performing organizations do not even develop a project-specific quality plan.

In most construction projects, there are multiple performing organizations that may only work on part of the project; for example, there are project teams that only work in the pre-bid phase and are precluded from design or construction. Others may join the project life cycle at the design phase and only perform design. Yet others join during the construction phase, or even later. These performing organizations are all stakeholders of the construction project together with the project's eventual client(s) or owner(s), who are usually considered the primary stakeholders. In addition, depending on the nature and scope of the subcontractors' tasks and the complexity of their work as well as its criticality, the construction project may also request project-specific quality plans to be prepared by the vendors, trade partners, design teams, clients, and other potential stakeholders.

It's important to recognize that quality planning for construction projects will include personnel who belong to different performing organizations, some of whom may be traditional adversaries or competitors. However, for the sake of the project's success, they must all work together as one team, targeting the project's goals and objectives (and not their own organizational objectives). To prevent rewrites or resubmittals, the required formats of writing the project quality plans may be issued, so that the quality plans that are developed by various performing organizations are all consistent and aligned.

At a minimum, the key aspects of quality planning that need to be aligned among all stakeholders should include the processes for quality control (QC) and quality assurance (QA) for all critical aspects of work. The KPIs of project quality must be defined the same way by all performing organizations that work on the construction project throughout its life cycle. All performing organizations must be required to define the critical quality attributes for their phases of work.

Some examples of critical quality include the following:

- In-process defects/defectives
- Customer complaints
- Inspection and checking status
- Nonconformances and corrective actions
- Alignment of data throughout all phases
- Alignment of the format of the quality plans
- Alignment of the project risk throughout the project life cycle
- Alignment of document control throughout the project life cycle

All phases of the project must require the performing organizations to identify the most critical attributes that are applicable to their scopes of work.

Another aspect of alignment of the quality plans is the flow of the documentation from one phase of the construction project to the next. For example, the submittals workflow that transmit the design documents over to the construction contractor, and the reverse of that, such as the description of the construction contractors' role in the constructability reviews of design, are two critical areas for a construction project where the two phases of design and construction are directly connected. The description of these two activities must be aligned in both the design and the construction quality plans.

Another much-needed alignment that is required throughout all phases of the construction project involves the communication protocols for the project. All performing organizations that work on the various phases of the construction project must be required to follow the same communication protocol in all phases of the project.

To achieve alignment between the different quality plans, it may benefit the construction project to consider the design quality manager (DQM) as a representative of the construction quality manager for the design phase. Likewise, all other quality plans that are prepared by other performing organizations, such as subconsultant firms, vendors, surveyors, or any other support functions, may also benefit by considering their quality leads (whether titled for quality roles or not) as an extension of the staff of the construction quality manager.

Align the Quality Plan with Other Functional Plans

The project quality plan needs to be aligned with the other functional plans that are also issued for the project. Examples of such plans include the environmental plans, health and safety plans, document control plans, project execution plans, and subcontractor plans. Instead of leaving them in silos, aligning these plans with the project quality plan creates a consensus-based approach that offers enhanced benefits of consistent oversight in addition to improved efficiency for other related areas and functional disciplines besides quality.

Just as an organization must integrate its individual departmental processes to improve its operational effectiveness and achieve its business strategy goals, the project-level management plan requires similar integration of its health, safety, and environment (HSE), schedule, quality, risk management, and other plans to deliver contractual goals and stakeholder expectations that are common to the project. These functional plans are typically seen as separate, each with their own manual, associated procedures, and implementation plans.

For the alignment of all functional plans, it is important to consider all other related functional plans that are also prepared for the project. The construction quality manager must ensure that such functional plans are all aligned with the project quality plans.

Obviously, the best practice is to think about such alignment during the early planning of the overall project, thereby issuing a consistent format to all the performing companies. Doing so allows all parties to prepare their functional plans in a format that is consistent across all phases of the project. Even though the content of the other plans such as environmental plans, health and safety plans, document control plans, project execution plans, and subcontractor plans for niche services may be slightly different due to their different scopes, once they use the same project-specific quality objectives and strategy as their guide, these functional plans become more aligned with the project quality plans. In all cases where the construction quality manager finds that they are not, then alignment activity is needed.

Alignment of Technology

Use of technology is another significant challenge for alignment. Inefficiencies can be mitigated if the technology used throughout the different phases is consistent (or the same). The most significant alignment in this respect is the description and use of a well-developed, well-communicated centralized document management repository for the project. All phases of the project should be subjected to or aligned to the same document management software for the storage of their records, and it should remain consistent throughout all phases.

Sometimes in construction projects, different repositories are selected for storing the records of different functions; for example, the safety records of the project may not be retained in the project's document management software, which stores the project's quality records. Similarly, other, different software may be used to store, monitor, and retain the records of environmental and sustainability compliance.

Alignment of technology, especially the use of same or interoperable software, saves frustration and the waste of significant time, especially during the closeout phases of the project.

Lastly, it is now possible to use artificial intelligence (AI) to easily cross-reference multiple documents for a structured comparison of data and concepts, as well as compliance with requirements. Two quality plans, one written by the design firm and the other written for construction quality, can be compared to highlight similarities, differences, and even discrepancies, often with a much higher level of accuracy and efficiency compared to manual methods. Obtaining alignment between the performing organizations and functional disciplines is becoming easier with AI.

Align the Organizational Structure of the Project

Another alignment that must be made is in the roles of personnel in quality control (QC) and quality assurance (QA). These roles must be identified for the project life cycle, not just for a singular phase. For many projects and phases, scalability may require that individuals wearing multiple hats fulfill these roles. Role definition ensures that the subsequent person responsible for quality knows who to contact for quality in the previous phases (the construction quality manager would know the quality lead during the design phase, etc.).

At a minimum, the names of personnel who are assigned the lead roles for quality must be included as the key staff for the project in all its phases. Construction projects that are considered small in scale and do not create a role for quality management must still identify the responsibilities for quality management and identify the people to whom those responsibilities are assigned, even if they are not identified as independent quality personnel.

The organization structures of the individual "project-specific performing organizations" should be obtained for all phases of the project (whether designer or contractor), and these should be shared with all the stakeholders.

One of the best approaches to implement quality in construction projects is to prepare a *zipper plan*—a strategy invented for use on complex projects that have multiple organizations working together. It is an excellent tool for project collaboration. The zipper plan connects like-level counterparts across each organization, ensuring smooth communication and seamless integration, much like the teeth of a zipper interlocking to create a unified whole. It provides a direct link between people by identifying an immediate client, the next person who should receive their deliverable, or someone who should be directly involved with its delivery due to their role on that deliverable in a previous phase. Aligning the key people from all aspects of the project at every tier of the project operation fosters better partnership. Figure 2.3 shows an example of a zipper plan.

The zipper plan concept may also be used for all functional disciplines, not just for quality. Following this provision, formal meetings may be arranged to ensure these functional leads of different organizations develop a one-on-one relationship with each other, and their collaboration must continue throughout the life of the project.

Creating zipper plans frequently helps identify gaps in an organizational structure immediately. People come and go on long-duration projects, and sometimes you don't finish with the same teams you started with. The zipper plans

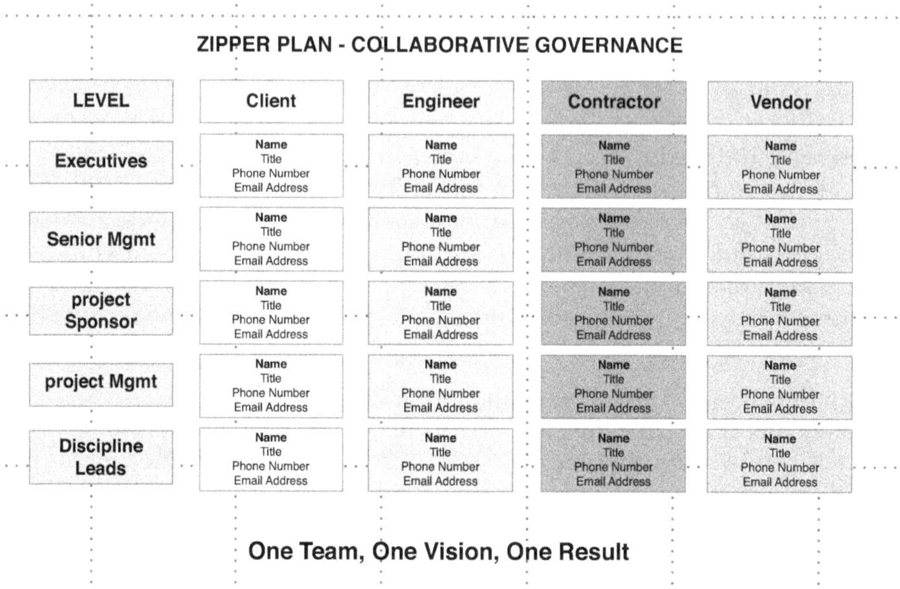

Figure 2.3 Zipper plan.

help identify the need to fill missing roles; if one lead person leaves, their absence is immediately noted, thereby allowing quicker replacements, better engagement, and more sustainable partnerships.

Instead of making an effort to write one project quality plan that integrates the different phases of the construction project and forcing this task over to one person (such as the construction quality manager), it is better to allow the individual performing organizations to create their own quality plans as well as other discipline-specific plans.

This ensures two things: (1) that the plans are created by SMEs for their scope of work and are applicable to their phase of the construction project's life cycle; and (2) it aligns them all to the project's goals and objectives and educates them on their roles in the construction project in terms of the big picture, thereby offering them the best opportunity to achieve the strategic quality objectives of the project.

CONCLUSION

Aligning quality plans across all phases of a construction project is paramount for success. Moving beyond traditional siloed approaches, strategic alignment ensures consistent quality, efficient workflows, and shared objectives among diverse stakeholders. By emphasizing project goals over individual organizational targets, and integrating technology and functional plans, a cohesive and unified project life cycle emerges. Implementing tools like a zipper plan further strengthens collaboration and communication, solidifying quality management as the vital thread connecting all aspects of the construction journey, ultimately driving superior outcomes.

QUESTIONS

1. What are the different types of alignments that are necessary in quality planning?
2. Who is responsible for writing the quality plans for a construction project?
3. Explain the critical distinction between "alignment" and "integration" in the context of quality planning for a construction project, and describe why alignment is the primary goal.
4. Describe the four common challenges to aligning quality plans in a construction project. For each challenge, give an example of how it might manifest in a real-world scenario.
5. Why does the section advocate for project-specific quality plans rather than relying solely on organizational-level quality plans? How does this approach impact the collaboration among different performing organizations?
6. What is a "zipper plan" and how does it facilitate better communication and integration among different organizations working on a construction project? Give an example of how it might resolve a common communication issue.
7. Discuss the role of technology in achieving alignment of quality plans. How can technologies like artificial intelligence (AI) assist in this process, and why is aligning document management software particularly important?

Quality Management Systems (QMS)

Quality management systems (QMS) must be designed during the planning phase. The QMS must consider the applicable industry standards and create project-specific quality plans for the project. The planning for quality assurance (QA) and quality control (QC) for the project must also be done during the planning phase. (Note that whereas planning for QA and QC may be performed during the planning stage, the implementation of QA and QC is performed during each phase of the project life cycle.)

INDUSTRY STANDARDS

> Analyze the quality management system (QMS) to ensure it follows industry standards and understand how it complies with international standards (e.g., ISO 9001, ISO 10006, ISO 14001, and ISO 45001) to deliver quality outcomes. (Analyze)
>
> Body of Knowledge II.B.1

This section delves into the aspects of quality management systems (QMS) and their alignment with industry standards. It emphasizes the importance of analyzing a QMS to ensure compliance with international standards like ISO 9001, ISO 10006, ISO 14001, and ISO 45001.

Understanding an organization's QMS and specific project requirements is crucial for successful project execution. This section highlights the necessity of evaluating and adapting the QMS to meet evolving project needs and conformance requirements.

Analyze the Quality Management System (QMS)

To ensure that your organization's QMS will meet the needs and requirements for a specific project, one must have knowledge of both the organization's QMS and what specific requirements exist for the project. Additionally, one must understand how their organization's QMS and the project's specific quality management plan (QMP) will integrate or align with other systems and plans, both organizationally and from a project perspective.

The creation of a QMS is a strategic decision, and the design and implementation of the QMS are influenced by many factors, including customer requirements, market expectations, industry regulations, and the organization's overall business environment, as well as the risks and opportunities associated with that environment. As such, most organizations are very intentional in how they design and deploy their QMS, ensuring it meets or exceeds the typical owner expectations in their target industry sector(s), as well as any industry or local regulations. It is important to integrate quality strategy into your organization's overarching strategic planning efforts to ensure that these inputs are included in QMS design and improvement.

If quality is integrated into your organization's strategic planning efforts, then your QMS is likely already designed to meet or exceed requirements for any given project that they would typically pursue. But sometimes an organization will pursue projects in new geographic markets or industry sectors, or with new project owners. And sometimes requirements simply evolve faster than an organization adapts its QMS. This is why it is important to evaluate your QMS in terms of each project's specific needs and requirements for compliance, quality, and safety. To effectively evaluate requirements and achieve alignment, one might follow a process as outlined in the following paragraphs.

Identify Applicable Industry Standards and Stakeholder/Owner Requirements

Research and list all relevant industry standards, regulations, and codes that apply to your construction project. This may include international management system standards such as ISO 9001, ISO 10006, ISO 45001, and ISO 14001 for occupational safety and environment, industry-specific guidelines or regulations, and safety codes. This may vary by region, geography, and construction verticals supported.

Understand QMS Requirements

Review the requirements and objectives of your current QMS. Ensure it is designed to meet quality standards applicable to your specific project. There are also helpful guidelines to understand requirements published by the national standards body.

Conduct a Gap Analysis

Conduct a gap analysis by comparing the requirements of the identified industry standards with the project's existing QMS processes and procedures. Identify any gaps or areas where the QMS does not fully address the standards. Gaps may be due to inapplicability of certain requirements or areas not being implemented by the project. So, each gap identified needs to be assessed carefully to confirm if there are valid missing pieces. You may express this gap graphically using a *spider chart*, which provides a visual of requirement areas where the project has significant gaps. More information on how to perform gap analysis is provided later in this chapter ("Resource Planning").

Develop an Action Plan

Based on the gap analysis, prepare a detailed action plan that outlines steps to address areas of nonconformance to QMS requirements. This could be presented to the leadership as a Gantt chart and might include updating processes, controls, creating new documentation, or enhancing training programs—or perhaps the nonconformance areas can be addressed through the deployment of a project-specific QMP.

Implementation of Action Plan

Deploy the changes needed to align your QMS with the project's requirements, and ensure all stakeholders are informed and trained on any new procedures or updates to existing practices by updating and incorporating them into your communication plan and your training matrix.

Internal Audits and Management Reviews

Conduct regular internal audits to assess conformance with both the QMS and project requirements. Ensure these audits are specifically included in the QA plan, and use them to continually check the effectiveness of implemented changes. Present audit findings and the status of the overall project QMS to the leadership team, and obtain commitment for continual improvement.

Continuous Monitoring and Improvement

Establish an internal system for ongoing monitoring of processes to ensure they remain aligned with the project's requirements and deliver quality outcomes. Include these changes as part of regular quality committee review and evaluation. This can involve periodic reviews, employee feedback, and updated training, as well as companywide rollout of these options.

Engage Third-Party Registrars

If certifying an organization or project's QMS through an ISO management systems standard, consider engaging an accredited third-party registrar organization at the earliest opportunity (e.g., program planning) to provide an objective pre-assessment of your QMS's conformance with management system and industry standards. This will help set up a roadmap to registration. Having a third-party registration for an organization's QMS can lend credibility and uncover areas that internal teams might overlook.

Stay Updated

To keep abreast of any changes in industry standards and update the QMS accordingly, regularly review industry publications, join professional organizations, and partake in training sessions to ensure ongoing compliance.

CONCLUSION

Ultimately, the strength of a QMS lies in its ability to adapt and evolve. By periodically analyzing, aligning, and improving our QMS with industry standards and project-specific needs, we can ensure not only conformance but also a commitment to excellence. This proactive approach fosters trust, enhances project outcomes, and positions the organization for sustained success in an ever-changing industry landscape. It is important to remember that quality is not a static goal but a continuous journey of refinement and improvement.

QUESTIONS

1. Beyond meeting basic requirements, why is it important for an organization to integrate its quality strategy within its overarching strategic planning efforts?
2. The document mentions using a "spider chart" in gap analysis. What is the purpose of using a spider chart in this context, and what does it visually represent?
3. What are two scenarios described in the document where an organization's existing QMS might not fully meet the needs of a new project, even if the QMS generally meets industry standards?
4. The document suggests engaging third-party registrars. At what stage of the process is it recommended to engage them, and what benefits does it provide?
5. The document emphasizes that "quality is not a static goal but a continuous journey of refinement and improvement." How does the document suggest organizations ensure this continuous monitoring and improvement of their QMS?

QUALITY MANAGEMENT PLAN

> Create a project-specific quality management plan (QMP) that complies with the appropriate QMS, incorporating policies and procedures (e.g., nonconformance control procedures, inspection and testing procedures, and method statements). Integrate PMBOK® Guide principles and system assurance practices into the development of the QMS and QMP. (Create)
>
> **Body of Knowledge II.B.2**

A *quality management plan* (QMP) is an important component of the quality management system that needs to be developed for effective implementation of quality. Although sometimes mistakenly used synonymously, a QMP differs from a QMS in that a QMP is created to meet the needs of managing the activities for a specific project, while a QMS is a broader system for establishing policies and procedures pertaining to an entire organization's quality and business operations. The QMS is the overarching, structured framework of an organization's policies, processes, and procedures for meeting quality requirements and objectives, whereas the QMP is a specific plan within that system.

Effective project execution hinges on a robust QMP. The QMP for projects is a project-specific blueprint that details how quality standards will be met. This plan incorporates policies, procedures, and practices to ensure deliverables align with established requirements. Multiple QMPs may need to be developed for each phase of the project, especially if these phases are performed by different performing organizations. Each performing organization is responsible for developing a QMP; for example, the design firm must develop the QMP for design, and the construction contractor firm must develop the QMP for construction, and so forth for the other phases.

The general guidelines that describe what the quality management plan should include are usually included in the technical specifications or requirements that are issued for the project.

In this regard, some guidance for good system assurance practices is also available in the Project Management Institute (PMI) Body of Knowledge (BOK), which is published in the *PMBOK® Guide*. By integrating those principles and system assurance practices, and addressing potential gaps of traditional approaches, the QMP may be developed to provide a structured framework for managing quality. Organizations can apply the *PMBOK® Guide* principles and system assurance practices as a guide for the development of their QMP. Table 2.5 shows opportunities for synergy between both best practices.

Structure of a Quality Management Plan

The type of project being constructed, contractual requirements, owner expectations, and/or third-party oversight for quality—all may have an impact on what sections or subsections are included in a QMP. Each phase of the project must focus on the quality management activities that are pertinent to that phase; for example, a design quality management plan addresses quality for design, and the construction quality management plan addresses quality for construction.

All QMPs must include a few standard items and preferably be based on a standard organizational template, which may be modified as needed to meet the quality requirements for a specific project. The common elements that must be in all QMPs should include the project's QA and QC plans, procedures to address nonconformances, project scopes and/or a definable features of work (DFOW) log, inspection and testing plans, testing requirements, and any industry standards and performance specifications that the materials or systems in the project are required to meet.

Creating a project-specific QMP involves outlining how the project will adhere to quality management standards through a set of policies, procedures, and practices designed to ensure project deliverables meet established quality requirements. An example outline of a QMP that ensures alignment with a QMS, incorporates *PMBOK® Guide* principles, and integrates system assurance practices is provided below as a template.

Table 2.5 Integration of QMP in alignment with QMS and PMBOK Guide.

Aspect	*PMBOK® Guide* Integration	System Assurance Integration
Policy and Objectives	Align with customer satisfaction, prevention, and continual improvement.	Include goals for audit compliance, defect minimization, and regulatory adherence.
Organization and Roles	Clearly define quality responsibilities emphasizing management commitment and team involvement.	Assign specific roles for audits, nonconformance management, and verification/validation.
Processes and Procedures	Document key construction activities focusing on prevention.	Embed practices like independent reviews, supplier qualification, inspections withhold points, and configuration management within processes.
Resources	Allocate necessary resources for quality management activities.	Allocate resources for system assurance activities like auditor training and testing equipment.
Documentation and Records	Maintain records supporting evidence-based decisions.	Maintain records demonstrating compliance with system assurance (audit reports, inspection records, test results).
Improvement	Implement continual improvement based on feedback and performance data.	Utilize audit findings and nonconformance analysis for QMS and process improvements.
Objectives (Project)	Define measurable quality goals aligned with QMS and stakeholder needs.	Include project-specific system assurance goals like inspection pass rates and verification completion.
Roles (Project)	Clearly define quality responsibilities for the project team.	Assign specific roles for system assurance tasks like inspections and nonconformance management.
Control and Assurance	Detail inspections, tests, and reviews focusing on prevention.	Specify types and frequency of system assurance activities (site audits, verification, validation, third-party inspections).
Standards and Specs	Reference relevant quality standards, codes, and project specifications.	Ensure these include system assurance requirements (testing protocols, acceptance criteria).
Acceptance Criteria	Define clear and measurable criteria for deliverables.	Link criteria to verification and validation requirements.
Tools and Techniques	Specify quality tools like checklists and control charts.	Include system assurance tools like audit checklists and nonconformance reports.
Change Management	Outline the process for managing changes while maintaining quality.	Emphasize impact assessments and approvals for changes to maintain asset integrity (configuration management).
Communication	Define how quality information will be shared.	Include protocols for reporting audit findings and verification/validation results.
Risk Management	Identify and plan responses to quality risks.	Include risks related to system failures and non-compliance, defining mitigation.

(continued)

Table 2.5 Integration of QMP in alignment with QMS and PMBOK Guide. (*continued*)

Aspect	*PMBOK® Guide* Integration	System Assurance Integration
Audit and Review Schedule		Specify the schedule for planned quality audits and reviews.
Reporting and Metrics	Define how quality performance will be measured and tracked.	Include metrics related to system assurance (nonconformances, audit findings, verification success).
Lessons Learned	Describe how project learnings will be used for future improvement.	Ensure lessons learned related to system assurance are also documented.

Sample Quality Management Plan (QMP) Template for a Construction Project

1. Introduction

The QMP should include an introduction that describes the project phase for which the QMP is being developed; for example, a Design QMP may have a slightly different scope and purpose than a Construction QMP. The following information may be included in the Introduction of the QMP:

Project Title: [Insert project name here]

Project Number: [Insert project number here]

Client: [Insert client name here]

Contractor: [Insert contractor name here]

Date Prepared: [Insert date here]

Revision Number: [Insert revision number here]

1.1 *Purpose*: This QMP outlines the quality management system and procedures that will be implemented for the successful execution of the [Insert Project Name Here] construction project. It defines the responsibilities, authorities, and processes necessary to ensure that the project deliverables meet the specified requirements and achieve the desired level of quality.

1.2 *Scope*: This QMP applies to all activities undertaken by [Insert Contractor Name Here] and its subcontractors involved in the [Insert Project Name Here] project, from commencement to final handover.

1.3 *References*: List any relevant documents referenced in this QMP (e.g., contract documents, specifications, codes, standards, organizational QMS manual).

2. Project Information

2.1 *Project description*: Briefly describe the project and its key components.

2.2 *Key stakeholders*: Identify the main stakeholders involved in the project (e.g., client, architect, engineer, regulatory agencies).

2.3 *Contractual requirements*: Summarize the key quality-related requirements outlined in the project contract.

3. Quality Policy and Objectives

The QMP should identify the quality policy and the quality objectives for the specific phase of the project (which should be consistent with the overall quality objectives of the project). The following may be included:

3.1 *Quality policy*: State the project-specific quality policy, aligning with the contractor's overall quality policy and emphasizing commitment to customer satisfaction and continual improvement. (*PMBOK® Guide* principles: Customer Satisfaction, Continual Improvement).

Example: [Insert Contractor Name Here] is committed to delivering high-quality construction that meets or exceeds the client's requirements and all applicable standards and regulations. We will achieve this through a proactive and systematic approach to quality management, focusing on prevention and continuous improvement.

3.2 *Project-specific quality objectives*: Define measurable quality objectives for this project. These should be SMART (specific, measurable, achievable, relevant, time-bound).

Examples:

Achieve a [specify percentage]% pass rate on all mandatory inspections.

Maintain zero major nonconformances throughout the project life cycle.

Achieve [specify percentage]% client satisfaction based on post-completion surveys.

Complete all verification and validation activities as per the schedule. (*PMBOK® Guide* principle: System Assurance).

4. Roles and Responsibilities

The QMP should identify key project team members and their specific responsibilities related to quality management.
Project manager: Responsible for overall quality management on the project.
Quality manager/quality control (QC) engineer: Responsible for implementing and monitoring the QMP, conducting inspections, and managing nonconformances. (*PMBOK® Guide* principle: System Assurance.)
Site superintendent: Responsible for ensuring quality workmanship onsite and adherence to procedures.
Foreman/supervisors: Responsible for ensuring the quality of work performed by their teams.
Other relevant roles (e.g., safety manager, environmental manager): Clearly define the authority and reporting structure for quality-related matter. (*PMBOK® Guide* principles: Management Responsibility, Engagement of People.)

5. Quality Control Activities

The QMP should describe the specific quality control activities that will be implemented throughout the project life cycle. The following are some of the most pertinent elements of QC that must be included in the quality management plan, although some of the noted sub-elements (such as the ITP) may not apply for all situations. Focus on prevention over inspection. (*PMBOK® Guide* principle: Prevention Over Inspection.)

5.1 *Control of documents and records*: Outline the procedures for managing project documents and records to ensure they are controlled, current, and readily available. (*PMBOK® Guide* principle: System Assurance: Configuration Management).

5.2 *Control of purchased materials and services*: Describe the process for ensuring the quality of materials and services procured from suppliers and subcontractors, including supplier qualification and inspections. (*PMBOK® Guide* principle: System Assurance: Supplier Quality Management.)

5.3 *Inspection and testing plan (ITP)*: Refer to a separate detailed ITP document (if applicable) or outline the key inspection and testing requirements at various stages of the project (e.g., material receipt, foundation, structural, architectural, MEP).

- Specify the type of inspection (e.g., visual, dimensional, functional).

- Identify the party responsible for each inspection.

- Define the acceptance criteria for each inspection.

- Include hold points where work cannot proceed without inspection and approval. (*PMBOK® Guide* principle: System Assurance: Verification.)

5.4 *Control of nonconforming products/work*: Define the process for identifying, documenting, evaluating, and resolving nonconformances. Include procedures for rework, repair, or rejection as well as instructions for the disposition of nonconformances. (*PMBOK® Guide* principle: System Assurance.)

5.5 *Calibration and control of monitoring and measuring devices*: Outline the procedures for ensuring that all measuring and testing equipment is properly calibrated and maintained.

6. Quality Assurance Activities

The QMP should also describe the planned quality assurance activities that are aimed at providing confidence that the QMS is functioning effectively.

6.1 *Internal quality audits*: Outline the schedule and scope of internal audits to assess compliance with the QMP and identify areas for improvement. (*PMBOK® Guide* principle: System Assurance: Independent Reviews and Audits.)

6.2 *Management reviews*: Describe the process for periodic reviews of the QMS by project management to ensure its continued suitability, adequacy, and effectiveness. (*PMBOK® Guide* principle: Management Responsibility.)

6.3 *External audits*: Specify any planned external audits by the client or regulatory bodies. (*PMBOK® Guide* principle: System Assurance: Independent Reviews and Audits.)

6.4 *Verification and validation*: Detail the activities planned to ensure that the constructed product meets specified requirements (verification) and fulfills its intended purpose (validation). (*PMBOK® Guide* principle: System Assurance.)

7. Standards and Specifications

The QMP should list all applicable standards, codes, and project specifications that will be used as the basis for quality requirements.

Examples: [Insert relevant building codes], [Insert relevant industry standards (e.g., ASTM, ISO)], [Insert project specifications section numbers].

8. Acceptance Criteria

The QMP should define the criteria that will be used to determine the acceptability of project deliverables. This should be clear, measurable, and aligned with the contract requirements.

Examples: Compliance with specified tolerances, successful completion of required tests, client approval of mock-ups.

9. Quality Tools and Techniques

The QMP should identify the quality tools and techniques that will be used on the project to support quality management activities.

Examples: Checklists, control charts, cause and effect diagrams, Pareto charts, flowcharts.

10. Change Management

The QMP should outline the process for managing changes to project requirements and ensuring that the impact on quality is assessed and controlled. This should include procedures for identifying, documenting, reviewing, approving, and implementing changes. (*PMBOK® Guide* principle: System Assurance: Configuration Management.)

11. Communication Plan

The QMP should describe how quality-related information will be communicated among project stakeholders. This includes reporting on inspection results, nonconformances, audit findings, and quality performance metrics. (*PMBOK® Guide* principle: Relationship Management.)

12. Risk Management (Quality Related)

The QMP should identify potential quality risks that could impact the project and outline the planned mitigation strategies. (*PMBOK® Guide* principle: System Assurance: Risk Management.)

Examples: Material defects, poor workmanship, inadequate supervision, delays in inspections.

13. Audit and Review Schedule

The QMP should provide a schedule for planned internal and external quality audits and management reviews. (*PMBOK® Guide* principle: System Assurance: Independent Reviews and Audits.)

14. Reporting and Metrics

The QMP should define the key quality performance indicators (KPIs) that will be tracked and reported on the project.

Examples: Number of nonconformances, inspection pass rates, client satisfaction scores, number of corrective actions completed.

Specify the frequency and format of quality reports. (*PMBOK® Guide* principle: Evidence-Based Decision-Making.)

15. Lessons Learned

The QMP should describe the process for capturing and documenting lessons learned related to quality management throughout the project life cycle. These lessons will be used to improve future projects and the organization's QMS. (*PMBOK® Guide* principle: System Assurance.)

16. Appendices

The QMP should include any supporting documents, such as:

- Inspection and testing plan (ITP)
- Sample inspection checklists
- Nonconformance report form
- Supplier quality requirements

The QMP sample template described above features common elements that may be included in the quality management plan to ensure that project quality management is systematically addressed and is in alignment with the broader organizational QMS. Each section would require tailoring to the specific project and industry requirements for optimal effectiveness.

CONCLUSION

Developing a QMP requires many inputs; you'll incorporate contractual requirements regarding performance specifications, along with industry standards for performance and testing methodologies. Additional elements may include plant visits, third-party testing, required certifications or qualifications, and so on (based on the life cycle phase of the project). The QMP may also include policies surrounding the handling of cost of quality (CoQ), with instructions on how those costs are to be captured. (See Chapter 8 for further information on CoQ.)

It is also extremely valuable to leverage the organization's knowledge management systems for best practices and lessons learned when developing a QMP. This ensures that the project team can learn from past projects beyond the limits of the project team's experiential knowledge. Incorporating lessons learned and best practices into a project's processes and tools ensures that benefits are gained through the leveraging of that knowledge.

To ensure continual improvement of the QMP (and the QMS overall), it is also a best practice to have scheduled intervals to collect feedback from the project team and all involved stakeholders, to analyze the effectiveness of the QMP, and to make modifications and course corrections as recommended.

QUESTIONS

1. How does the Quality Management Plan (QMP) differ in scope and purpose from the Quality Management System (QMS)? Explain how they relate to each other within an organization.
2. The section mentions the integration of *PMBOK® Guide* principles and system assurance practices. Give two examples of how each of these is integrated into the development of a QMP, as shown in Table 2.5.
3. What are some of the key components that should be included in the structure of a QMP, and why are they important for project success?
4. Explain the importance of "prevention over inspection" in quality control activities, as mentioned in the QMP sample template. Give an example of how this principle can be applied in a construction project.
5. Why is it important to incorporate "lessons learned" into the development and continuous improvement of a QMP, and how can this be effectively achieved?
6. Explain the distinction between a QMS and a QMP, and discuss why an organization might choose to use PMBOK Guide principles in their QMP to improve effectiveness in modern construction management.
7. Describe the key components that should be included in a comprehensive QMP, and explain how factors such as contractual requirements, owner expectations, and third-party oversight influence the structure and content of the QMP.
8. Outline the steps involved in creating a project-specific QMP, and explain the significance of incorporating elements such as QA/QC plans, nonconformance procedures, inspection and testing plans, and industry standards into the document.
9. Discuss the importance of leveraging organizational knowledge management systems and incorporating lessons learned when developing a QMP. Give an example of how lessons learned can be integrated into a specific component of the QMP, such as the definable features of work (DFOW) log.
10. Explain the role of continuous improvement and feedback mechanisms in refining a QMP and the overall quality management system. How can scheduled intervals for collecting stakeholder feedback and analyzing plan effectiveness contribute to successful project outcomes?

QUALITY ASSURANCE (QA)

> Develop and implement quality assurance (QA) practices to verify the project meets quality specifications and is compliant with applicable standards by establishing standardized quality policies and procedures. (Create)
>
> Body of Knowledge II.B.3

In the realm of project management, ensuring quality is paramount. This section delves into the crucial distinctions between QA and quality control (QC), two integral components of a robust quality management system. While often used interchangeably, QA and QC serve distinct yet complementary functions. QA is process-oriented, focused on the processes to prevent defects, establishing standards and procedures to guarantee consistent quality. Conversely, QC is product-oriented, emphasizing inspections and testing to verify that the final deliverables meet specified requirements. Understanding these differences is essential for project success.

Defining Quality, Quality Assurance, and Quality Control

A project-level quality plan has two categories of tasks: quality assurance and quality control. Assurance and control are separate functions that complement one another. As mentioned previously, ASQ defines quality control as "the part of quality management that is focused on fulfilling the quality requirements of the product." ASQ defines quality assurance as "the part of quality management that is focused on providing confidence that quality requirements will be fulfilled." So, in simple terms, QC is focused on the product quality, and QA is focused on the process quality.

Any physical element of a constructed project will first have an *assurance* task, where the work is planned, with a corresponding *control* task, which validates its completion according to the client intent. For example, QA develops a standard checklist for a process, while QC leverages that checklist to inspect the product. Table 2.6 shows several ways in which the two concepts differ.

Quality assurance is also the effort of *prevention* of future quality problems by planning and organizing the project effectively. These planning activities, as shown in Table 2.6, will vary from project to project and will depend on the current phase of the project. It is the responsibility of the construction quality manager to outline what they are. Project team staff also need to understand their responsibilities for those tasks and receive the necessary training to handle them.

QA and QC processes complement each other, and they typically happen independently of each other—many times in succession. QA begins at the start of the project phase's work (during the planning stage) where the processes that QC must follow are determined. Once the actual work is done, QC is completed via the checking and inspection processes (defined by QA during the planning phase). Lastly, QA audits assure that the QC activities were performed exactly as planned.

Some examples of QA activities performed during the different phases of a project life cycle are noted in Table 2.7 (this list is not all-inclusive).

Stakeholder expectations are also defined and finalized by QA in each phase before the work starts. Those expectations are then continuously validated during QC. These expectations—along with the specifications and standards associated with the project—serve as the foundation for quality that is implemented through QC and later verified by QA in the form of QA audits.

Using QA in the Planning Phase to Plan the Quality of Work

Not all project elements require the same rigor of QA planning. Quality assurance creates the framework that identifies which planning activities need to occur and how to achieve the balance between quality and all other project-related plans. (See "Alignment of Quality Plan with Other Project Plans" in this chapter.) The incorporation of lessons learned, best practices, and other requirements is also a

Table 2.6 Quality assurance and quality control.

Quality Assurance (QA)	**Quality Control (QC)**
Focuses on process quality	Focuses on product quality
Uses the process specifications to establish the processes for quality	Uses the product specifications to measure the QC characteristics of the product
Establishes the standards	Inspects against (verifies) the standard
Major challenge: Define the processes that QC must follow.	Major challenge: Assist QA in defining a good QC inspection and test plan for the product-based testing/checking.
Goal of QA: Ensures processes produce consistent quality	Goal of QC: Ensures product meets required specifications and standards
Concept of big-Q—strategic	Concept of little-Q—tactical
QA includes more intangibles.	QC includes more tangibles
QA thinks on a spectrum (it is not binary); performance of QA may be measured on a scale (percentage).	QC thinks in binary terms; QC must always be measured on pass/fail basis.
Rule for QA: QA must *not* do QC's work; QA's role is to ensure that QC is effectively done.	Rule for QC: QC must correct the noncompliance/deficiency, and recheck to pass. QC cannot leave a fail unaddressed.
QA's focus is on preventing the defects by defining and implementing processes to ensure the quality of the product.	QC's focus is on detecting/finding defects by inspection and testing. QC verifies that the product is meeting the standard that is/was established through quality assurance. If not, then implement corrective action.
Culture: Proactive—(QA is leading indicator)	Culture: Reactive—(QC is lagging indicator)
QA person must understand general principles of QC (but may not be SME in the discipline) and should understand principles of QA.	QC should be done by an SME in that discipline; QC personnel may or may not be an expert in QA.
QA is typically sampling-based (QA only samples a limited percentage of QC).	QC may be sampling-based, if possible; sampling may be normal, tightened, reduced (it is based on levels of inspection severity).
Typical tools for QA: Process audits, flowcharts, training programs, preventive actions	Typical tools for QC: Checklists, inspections, sampling, control charts
QA plans for the QC and creates the QC plan along with specific ITP's (Note: ITPs are for construction phase only.)	QC follows the QC plan and performs the inspections and reviews according to the QC plans.
QA includes front-end planning and later back-end quality assurance that QC is done (uses audits)	QC follows the lead that is provided to it by the QA plan (QC is back-end work by its nature).
Success of QA depends on QC following the plans correctly.	Success of QC depends on good QA planning (QC needs good ITPs).
Continuous improvement in QA: The QA uses feedback from QC to improve processes and standards, preventing future issues.	Continuous improvement in QC: The QC reports defects to QA for root cause analysis. Continuous improvement may happen if the lessons learned are implemented to improve the QC processes.

Table 2.7 Typical project phases and QA planning activities.

Project Phase	Example QA Activities
Precontract	Development of basis of design
Planning	Confirmation of stakeholder intent
Design	Outlining of the inspection and test plan (ITP)
Procurement	Specifying manufacturer requirements
Construction	Preinstallation/preparatory meetings
Project Turnover and Closeout	Closeout planning meetings
Operations and Maintenance	Development of operations and maintenance (O&M) manuals

Table 2.8 QA and QC for air handling units example.

Air Handling Units (AHU)	
Quality Assurance	***Quality Control***
Establish what project elements need collaboration and planning and how that collaboration will be facilitated.	*Establish how the execution resulting from collaboration and planning will be validated.*
Holding a preconstruction meeting with the mechanical subcontractor	First work inspection
Defining the intended function of the AHU	Delivery inspection
Outlining the test and balance (TAB) plan	Conducting testing, adjusting, and balancing (TAB) according to the plan
Establishing the Commissioning Plan	Performing functional testing to validate operation against the design and stakeholder intent

QA activity that must be timely and appropriate (See Chapter 5). Table 2.8 illustrates how QA and QC activities and tasks may be planned during the planning phase of a project.

Examples of QA Failures

Listed below are some examples of QA failures during the design and construction phases of the project life cycle.

Design phase:

- *Requirements management failures*: Requirements management is a front-end QA planning activity that can be implemented as a QA process to identify, document, analyze, and manage the stakeholder needs throughout a project or product life cycle. Wherever it is used as quality planning tool, it enables the organizations and projects to ensure that the products or services they are delivering would meet the requirements of the stakeholders. Requirements management can and should be scalable to the size and the scope of the projects. Mostly performed during the design phase, if this design quality assurance activity is not performed correctly, then the results show in the form of poor QC of the design. Poor requirements management is a QA failure, which subsequently results in QC failures.

- *Design quality staffing issues*: Resourcing of design quality assurance staff is a common challenge on many construction projects. Where there is no independent quality assurance staff available to perform the QA, the design quality assurance activity is usually not performed. Informal design quality reviews without a defined QA process for verification of comment incorporation is a common failure of QA in the design phase of the projects.
- *Software validation failures*: Software validation is another QA activity. Using design software that is not validated for the design application where it is used causes design quality issues. In such cases, the design could be wrong despite good QC checking.
- *Design quality assurance that is focused on catching incidents of failures in the QC*: It is not the responsibility of QA to perform the work of QC. Quality control checking must be performed to catch and correct the design's quality incidents. Quality assurance must only sample the design work that has been checked by QC. Also, QA should be focused on "process quality" of design; its primary aim is to identify any systemic deficiency in design. The role of QA is not to focus on the quality incidents (because that is the role of QC).
- *Design QA that is checklist-focused*: The design QA that is solely dependent on checklists only tends to verify what is included in the checklist. As a result, it fails to do effective QA because it does not audit what is not included in the checklist. Checklists tend to limit the scope of QA and should be used with a caution about their limitation.
- *Poor design change process*: Value engineering changes that are made to the final design by the designer must be planned and issued out as a formal engineering change. Such changes preferably must be issued through QA workflows that are formally managed by the document control. In cases where design changes are done manually, failures in design QA are more common.
- *Quality assurance that is taking the responsibility for "correction"*: Correction is the responsibility of QC, not of QA. QA (such as the design QA manager or the CQM) should only direct the QC checkers to perform QC corrections—QA should not perform QC checking nor perform corrections.

For more detailed information about design QA, see Chapter 3.
Construction phase:

- *QA that performs QC checking of the construction work*: This is a very common failure in construction projects. QA should let construction

QC inspectors perform the inspection work and not try to perform their work for them. QA's role is to verify that QC is correctly performed as it was planned, not to perform QC inspections.

- *QA that writes off nonconformances to quality incidents and not to the failure of the QC process*: This is another common QA failure. The purpose of QA is to think bigger—on the systemic processes, and not on individual incidents. It is not the responsibility of the QA manager (such as a CQM) to issue nonconformances to the product quality incidents. If QA must issue any nonconformance, it should only be written to the QC process that was unable to identify the QC failure. Identification of the nonconformance by QA is warranted only if systemic failures are identified; the incidental failures must be redirected to QC staff, and they must get them corrected (and report to QA when correction is completed).
- *QC is lapsed*: The lapse of quality control reporting is a failure of QA. The QA manager (e.g., a CQM) should ensure that quality meetings are performed as scheduled and that ongoing quality reports (daily, weekly, monthly, etc.) are also completed on time and are effective.
- *Inspection and test plan (ITP) is not correct*: The development of the ITP is typically not the responsibility of QC inspectors; mostly they just use it to perform their inspections. The preparation of the ITP is a higher-level activity. It should include the QA manager (such as a CQM), who must coordinate with the designer to prepare the ITP that QC inspectors may use during the construction inspections. If the ITP is not correct, it is considered a failure of QA.
- *Document control failures in construction*: When document control failures occur, wrong final design drawings may reach the field. Document control failures in construction can result in multiple subsequent failures of quality, but these can all be prevented through effective QA planning and audits.

Quality assurance principles must also be implemented throughout all other phases of the project life cycle, even though design and construction are the two major phases that include production of deliverables. (For more detailed information about construction QA, see Chapter 5.)

Incorporating Lessons Learned

Incorporating lessons learned from team experience is another essential part of QA. The team must be able to plan accordingly to eliminate the risk that those same issues from previous projects will occur again.

The CQM must understand at what point in QA for each project phase each specific lesson learned item needs to be discussed. (See Chapter 5 for more information.) Go over lessons learned and other collaboration points when the discussion is most effective—some will be helpful during precontract, while others must be reviewed during procurement.

CONCLUSION

The project team on a construction project has both assurance and control responsibility. A dedicated QA manager is generally the key driver of QA activities. If there is no dedicated quality professional assigned to this role, then the project team must collectively determine how to execute QA activities for the scopes that they manage. Regardless of whether quality assurance consists of dedicated team members or not, the project staff needs to be trained and made knowledgeable about QC as well as QA. According to their roles, all project team members must be capable of understanding how to create processes that will achieve the stated quality objectives for the project.

QUESTIONS

1. Explain the difference between quality assurance (QA) and quality control (QC) in the context of a construction project, providing an example of a QA activity and a contrasting QC activity for the same project element.
2. According to the text, how does the timing of QA activities differ from the timing of QC activities within a project phase? Elaborate on why this sequence is important for effective quality management.
3. Discuss the role of lessons learned in the QA process. At what stages in the project life cycle are these lessons most effectively incorporated, and why?
4. The document highlights several QA failures during the design phase. Describe two of these failures and explain how they can lead to QC issues during later project stages.
5. What are the responsibilities of a QA manager (e.g., CQM) in relation to QC activities, as described in the text? Specifically, what actions should QA not take, and why is this distinction critical?

QUALITY CONTROL (QC)

> Develop and implement quality control (QC) measures such as an inspection and test plan to ensure compliance with requirements. (Create)
>
> **Body of Knowledge II.B.4**

QC is a critical aspect of quality management, distinct yet intertwined with QA. While QA focuses on processes to prevent defects, QC is centered on verifying product quality against established standards. Essentially, QA sets the stage and QC ensures that performance meets it. This section of the QMS delves into the definitions, differences, and applications of QC, particularly within design and construction phases of the project life cycle. Understanding QC's role, from inspections to documentation, is essential for achieving successful project outcomes and fulfilling quality requirements across any organization.

Defining Quality Control

QA and QC are two aspects of quality management. Often used interchangeably, the two terms actually are focused on two different aspects of quality. While some of the QA and QC activities are interrelated, they have different purposes. As defined by ASQ, "Typically, the QA activities and responsibilities cover virtually all of the quality system in one fashion or another, while QC is a subset of the QA activities." Further, ASQ describes QC as the "part of quality management which is focused on fulfilling the quality requirements." To achieve this goal, QC acts as the inspection aspect of quality management.

Quality control applies to all organizations, regardless of whether they are product-based or service-based. It aims to ensure that the outputs meet the necessary specified standards and requirements. In the context of a construction project, the goal of QC is to ensure that the end product (constructed infrastructure) meets the specified quality standards, conforms to design specifications, and complies with regulatory requirements. To achieve this goal, QC planning must be performed for all phases of the construction project life cycle, not just for the construction phase.

How Is QC Different from QA?

As discussed previously, quality control and quality assurance are two different aspects of quality management. They are different in terms of their focus, and because of that, the scope of each is different, even though there are some commonalities between the two. Also, it must be noted that there can be other elements in the quality system of an organization (or a project) that might not be specifically covered by QA/QC activities and responsibilities but may involve QA and QC.

In *The Management of Quality in Construction*, J. L Ashford defines quality control as eliminating nonconformance. This is one definition of quality control, but there are several other definitions as well. Another definition of quality control is given by the quality guru Joseph Juran; "Quality Control is the regulatory process through which we measure actual quality performance, compare it with quality goals, and act on the difference."

In all definitions of quality control, the one common element is that it is focused on the measurement of the quality of the "product." ASQ defines QA as "the part of quality management that is focused on providing confidence that quality requirements will be fulfilled."

So, in simple terms, QC is focused on the product quality, QA the process quality. There are several other ways the two concepts are different from each other. (Table 2.6 describes the differences between QA and QC.)

Because of the inherent differences between QA and QC, it is advisable to keep the two groups independent. Even though they may be related and frequently overlap, keeping their roles independent helps ensure that there is no conflict of interest on the project. If the same person who represents QA also performs QC, they could be checking their own work, which always becomes a source of bias and conflict of interest.

Why Is the QC Plan Developed by the QA Group?

The way QC relates to QA is very important to understand. Ideally, the QC plan is developed by the QA group because the focus of QA is typically larger than that of QC. (Whereas QC is interested mostly in the quality of the product, the QA is interested also in the processes that must be established first if the quality of the product is to be achieved.) Their bigger-picture perspective, particularly their emphasis on preventing defects from occurring, allows QA to include more comprehensiveness in the inspection and testing plans (ITP), which are part of the quality plan that they prepare for the project's execution phases. QA achieves this goal by working in tandem with the QC group as well as with the project's design group. Together as a result of this collaboration, QA outlines all of the QC inspections and tests that must be performed during the execution phases of the construction project (i.e., for both the design QC and construction QC planning).

Organizations that do not create the separation of QC and QA functions are advised to consider the different focus of the two quality concepts. Creation of a QA function, even if that may consist of a single person in the QA role, helps improve the organizations as well as projects.

QC Planning for Design Versus Construction Phase

It is a best practice to recognize and ensure that the QC plan for the design phase should be developed separately from the QC plan for the construction phase. These two plans must be developed by the respective performing organizations.

In other words, the design QC plan must be developed by the design QA (i.e., by the QA team of the designer). Likewise, the construction QC plan must be developed by the construction QA (i.e., by the QA team of the construction contractor).

In those projects and organizations that do not differentiate between the two functions, the QC planning is typically less effective or insufficient. The best practice is to have independent functions of QC and QA for all phases of the project life cycle.

Quality Control for Design

In many construction projects, the design team is hired long before the construction contractor is brought onboard. To ensure design quality, the QC processes for design checking must be developed by the design quality team. The design quality plans should include the processes for design QC and QA. In all those projects that are delivered in the design-bid-build format, or those where the construction contractor is not hired until later, the quality planning for design must become the responsibility of the design team. Some of the most common failures of design quality are noted below. Attention to these during the planning phase helps mitigate or eliminate the design quality control problems significantly (this list is not all-inclusive).

- Intradisciplinary (i.e., within the discipline) design QC checking is not performed correctly.

- Interdisciplinary (i.e., across the discipline) design QC checking is not performed correctly.
- Design quality control staff is not qualified/trained. This happens when a design is checked by staff that is not qualified to review that design (e.g., a civil design is checked by an electrical designer, who is not a subject-matter expert in the civil design discipline).
- Design plans are not checked against design calculations. The result is poor/wrong cross-references in the design plans (which cannot be easily identified).
- There are design input errors. The typical result of this common QC failure is a flawed design that will not meet the design requirements.
- Constructability reviews are incomplete or not performed at the right time. This is a QC failure. Typically, a design must be developed to the 60% design stage before constructability review comments are practical and useful.
- Document control is lacking during the design phase. The document control failures such as naming or wrong version numbering causes comment incorporation in the wrong versions of the design files. This results in wasted time and rework.
- As-built design drawings processes not following the same rigor of QC as the original drawings is another common failure of design QC for many projects.

These are only some of the most common design QC problems that are noticed on complex designs. (For more detailed information about design quality control, see Chapter 3.)

Quality Control for Construction

Like the development of a QC plan for the design phase, a QC plan must be developed for the construction phase as well. The construction quality plan should include the processes for both construction QC control and QA. (For further information about construction QC, see Chapter 5.)

A construction QC plan must address two things:

1. General QC requirements (process-specific)
2. Specific QC requirements (product-specific)

General QC requirements may include information regarding the time and scheduling of the QC inspections. In this regard, the construction quality plan must include the procedures for three types of QC inspections that must be performed for all construction projects: (1) preparatory inspections, (2) in-process inspections, and (3) final inspections.

Other QC inspections that pertain to construction quality may include some or all of the following (notice that some of these are performed on the construction field activities, whereas others are not):

1. *Unit (component) or system testing*: Where applicable, the construction QC plan must also include unit testing and system testing. The term "unit testing" is borrowed from the software industry; like software testing, the unit testing for a construction project must include testing the individual components, or "units." These tests may be functional, to ensure the components and assemblies to be installed are working as intended according to the functional and nonfunctional requirements and specifications. (See more detailed information about the related aspects of unit and system testing in Chapter 5.)
2. *Material and factory (source) inspections*: A factory acceptance test (FAT) is a test or a series of tests that are performed on equipment (usually assemblies or parts that must be installed as part of the construction). As the name indicates, FATs are performed on-site at the supplier's facilities (factory) and aim to ensure the product quality as well as the process that manufactures those parts and equipment.

   In addition to FATs, another significant QC test for the suppliers of the construction projects is first article inspections (FAI). These inspections are also sometimes called "benchmark inspections."

   In FAI testing, the new or modified production process of the supplier is inspected to ensure it is producing conforming parts and component assemblies that meet the manufacturing specifications detailed in the technical or engineering drawings. FAIs are performed at the suppliers' facilities and should be included as an essential (though separate) part of the inspection and test plan for the construction projects. (More specific and detailed descriptions of FATs and FAIs are provided in Chapter 4.)
3. *Work-in-place inspections*: As applicable, the construction QC plans may include ongoing formal or informal inspections that may be conducted as the work is constructed/installed.

Specific QC Inspections in the Construction Phase

The specific QC inspections that may be required for the construction projects will always be different from one project to the next. Those that must be performed for the construction pertain to the "quality of the product" being constructed. Specific construction quality requirements are determined based on the project's definable features of work (DFOW). Each applicable DFOW is a task separate and distinct from other tasks and has separate control requirements; the inspection and test plan (ITP) must include specific information on what needs to be checked for each DFOW.

An ITP is the most significant component of the QC plan for the construction phase. It outlines the product-specific QC inspections and tests that must be performed in the field to verify that the completed product meets the quality standards and requirements. Its primary purpose is to outline all the QC inspections and tests that must be performed at various stages of construction. These

QC checks and inspections verify the compliance of the constructed work (or its components) to the product-specific quality requirements. ITP also serves as a roadmap for QC activities throughout the product's life cycle. Lastly, it is important to note that ITPs may be developed for the construction activities, as well as for preconstruction activities.

Depending on the scope of the construction work and its definable features, there may be several ITPs that need to be developed. Creating a single ITP can become too complicated; it may become overly generic and impractical. Each ITP should include product-specific QC inspections to be performed during the construction phase.

It is important to note that the ITP is prepared only for the construction QC plan; design QC is addressed in a different manner. (See Chapter 4 for more specific details about the application of QC in the design and construction phases.)

Common QC Failures in the Construction Phase

Some of the most common failures of construction quality are noted below. Attention to these during the planning phase helps mitigate or eliminate the common construction QC problems significantly (this list is not all-inclusive).

- *Missed QC inspections*: Missed daily QC inspections, incomplete inspections, incomplete reporting, or late reporting.
- *Construction QC staff issuing "deficiencies" instead of "nonconformances" where warranted*: This is a very common construction QC failure across the construction industry. This failure is primarily the result of poor definition of what is a quality problem and what is not. When the QC inspection is performed, often the construction contractor agrees that the work completed is deficient but claims that it will be corrected. Since they intend to correct it, the QC inspection should not classify the work as nonconforming but as deficient. They put the deficient item on a punch list and continue with the construction, claiming that the punch-list item will be addressed later.

  The schedule determines whether the QC inspection should identify the work as deficient or nonconforming. As a rule, if the deficiency can be fixed without affecting the time (schedule) or the budget (cost) of the project, it may be called a deficiency. However, if the failed inspection cannot be corrected without impacting time or budget, it must be issued as a nonconformance. This is one of the most significant QC issues of the construction industry, and it should be discussed by all stakeholders with a consensus answer formally achieved and communicated. This may require management-level intervention and involvement of QA (e.g., a CQM).
- *ITP is not correct:* If this happens, then QC inspections are not correct. This is another common construction QC problem.
- *QC witness point and hold point inspections are not defined in ITP:* The result is that these inspections are not performed by QC staff.

- *Witness and hold points are defined, but QC is not informed about the construction activity.* The result is missed inspections. Once again, this is a QC failure, but it cannot be resolved by the QC inspection staff alone—management or the QA manager (e.g., a CQM) must enforce QC inspections.
- *QC inspectors are not qualified for the discipline in which they are performing QC inspections.* Lack of training for QC inspectors is a common construction QC failure.
- *Document control failures in construction*: If wrong drawings are accessible in the field, it may result in multiple quality failures. Also, the failure of good document control in the construction phase can result in multiple failures down the line in the project life cycle.
- *Construction QC is not independent from the construction contractor staff*: If both QC inspectors and the construction/production staff are reporting to the same supervisor, the production staff may be checking their own work. This can create a conflict of interest. Once again, this QC failure requires management intervention or the involvement of QA (e.g., a CQM).
- *Construction QC inspectors are not monitoring the as-built changes in the field*: As-built drawings prepared by the designer may include wrong information.

(For more detailed information about construction QC, see Chapter 5.)

Documentation and Recordkeeping of QC in Construction

Documentation of QC inspections is the responsibility of quality personnel (QC inspectors and reviewers as well as QA personnel). By itself, document control is both a QC and a QA function because QA personnel typically establish the procedures and processes for effective document control. And then QC personnel (inspectors, supervisors, and SMEs who perform the construction QC) record the evidence of QC in the form of completed tests and their pass/fail status. Lastly, QA personnel (e.g., QA auditors, which may include the CQM) perform audits to ensure the QC inspections were performed as planned. The support discipline of document control must collaborate with both QC and the QA personnel. (Detailed guidance on what to include as the quality documents and quality records for design and construction phases, respectively, is provided in Chapters 3 and 5.)

CONCLUSION

Quality control is not merely a checklist exercise; it is a fundamental process, ensuring adherence to standards and achieving desired outcomes. The distinction between QA and QC is critical, with each playing a vital yet separate role. From planning in the design phase to execution in construction, robust QC measures are imperative. Proper documentation, independent functions, and clear ITPs are essential for effective QC. Ultimately, diligent QC, coupled with continuous improvement, guarantees quality outputs and minimizes errors throughout the project life cycle.

QUESTIONS

1. Explain the distinct roles of quality assurance (QA) and quality control (QC) within a project, highlighting why maintaining independence between the two groups is considered a best practice.
2. Discuss the differences between the QC plans for the design phase and the construction phase, and why these plans are developed by different entities.
3. Describe the significance of inspection and test plans (ITPs) in the construction phase, including who develops them and why they are considered the most significant component of the QC plan for construction.
4. Analyze the potential consequences of failing to distinguish between "deficiencies" and "nonconformances" during construction QC, and how this issue might require intervention from higher-level management.
5. Compare and contrast the "tangible" aspects of QC with the more "intangible" aspects of QA, providing examples to illustrate these differences.

Resource Planning and Analysis

Effective project execution hinges on meticulous resource planning and analysis. This section delves into the critical processes of identifying, evaluating, and securing the necessary people, tools, and technology for project success. Understanding the intricacies of task allocation, individual capabilities, and timely availability is paramount. Resource planning starts out with the assessment of resources. The roles of quality personnel and internal stakeholders and their training needs must be defined. Resources for third-party inspections as well as the measurement system of the project organization must be reviewed. Also critical is provision of the quality management tools to ensure that the project has the resources it needs to succeed.

RESOURCE ASSESSMENT AND MOBILIZATION

> Identify the specific quality resources and timely availability needed for a project and develop a mobilization plan including allocation and a gap analysis of current and necessary resources. (Create)
>
> Body of Knowledge II.C.1

This section emphasizes developing a robust mobilization plan, including allocation strategies and gap analyses, ensuring that the right resources are in place at the right time to deliver quality outcomes and stay within project scope, budget, and timeline.

As a project comes to life, individuals from each participating organization come together to form the project team. The life cycle of this team shifts throughout each phase of the project—from precontract to operations and maintenance—as tasks and responsibilities are handed off from one individual to another. These processes, tools, and technology, and the people, or team members, associated with them are the resources needed to accomplish the project. A plan must be in place to guide the project team, outlining:

- What task needs to be done?
- Who will complete the task?
- What tool or software will be used to complete it?
- What process and project phase is the task part of?
- Are the tools capable of completing the task?
- Are the assigned project team members trained and capable of completing the task sufficiently?
- Are the assigned people, tools, or software available at the right time?

In addition to individuals coming together to form a project team, various organizational cultures and requirements also influence the tasks to be executed. As roles and stakeholder responsibilities are outlined, it is essential to remember that individuals, not just the corporate organizations involved, form the team. While company policies are to be followed and integrated at the project level, the project-level tasks require the most focus and planning for success.

Roles and Stakeholder Responsibilities

At the project level, the tasks, tools, and team members responsible for executing them must be used efficiently. Effective planning of these resources is directly correlated to the level of quality achieved and the ability to deliver the project on time and budget. Team roles and responsibilities—tasks to be completed as part of the project's processes—come from multiple sources:

- Tasks associated with the team member's specific role
- Organizational policies, plans, and procedures
- Contract documents
- Client or operator requirements and expectations
- Experience and lessons learned

There are tasks related to quality for each of those sources (and the list is not all-inclusive). Once the responsibilities are outlined, resource identification and assessment can begin.

Resource Qualification, Capacity, and Training Needs

When project team members are assigned, not all have the necessary experience or training to complete a task. Also, not all technology or software can perfectly facilitate the task. A lack of specific knowledge, experience, or tool functionality can become part of a project's risk profile and—if related to a tool—can be tied back to risk assessment and appropriately mitigated.

It is essential for the CQM—or in the absence of a dedicated quality team member, the project team's shared responsibility—to ensure each team member assigned to a task is able to perform it—understanding, for example, if a project requires mechanical, electrical, plumbing (MEP) experience, or if an external subject-matter expert (SME) is necessary to support structural steel coordination. Identifying and evaluating missing key skill sets are critical skills for a CQM.

Resource Availability

Identified resources may not be available at the specific time the task is to be completed. Design often lacks constructability input due to little or no availability of an SME with construction experience. Even during the early project phases, resources must be available to support the QA and QC activities.

Understanding the facets of the project quality plan and how they align with the project schedule will ensure collaboration happens between resources and that the necessary tasks get completed. A tool that identifies roles and responsibilities, the RACI matrix, is shown in Table 2.9.

External Resources and Experience

When resources such as people or technology are assigned to a project and may not be able to perform certain functions, external resources are needed. For example, Table 2.9 shows a project team with a dedicated CQM. Many project teams do not have this luxury, and thus the quality tasks must be shared among them. The organizational quality leaders may have to train and guide the team members in carrying out these specific tasks when direct project team resources are unavailable.

Data, Reporting, and Measurement Systems

In addition to people and process-related resources, there are reporting requirements and data needs for the project. The project team must understand what resources will be used to report on leading and lagging indicators to measure project performance to KPIs, objectives and key results (OKRs), and client intent goals. KPIs are specific, quantitative metrics that monitor ongoing operational performance, mostly on day-to-day and tactical actions, goals, and objectives. On the other hand, OKRs are more aspirational in nature and are typically used by those organizations and projects that intend to measure progress toward strategic change and innovation.

Gap Analysis

Gap analysis is a planning process typically used to compare a current state to a desired future state. The difference between the two states is called the "gap," which is typically described in terms of the improvements required to achieve the desired state or as the potential risks that may exist in the present state that need to be addressed. Gap analysis can be performed to find the gaps in resources, performance, skills, technology, or any other areas of interest. It takes the form of an audit, which produces a report that allows the team to determine what actions should be taken to address identified gaps. The goal is to assess current and available resources and determine what needs to be done to ensure that the project team is sufficiently resourced with qualified staff and other technical resources.

The steps that must be included in a gap analysis audit are more or less universal and do not vary too much. However, the tools that may be used could differ significantly. Also, depending on the nature of the gap that is to be identified and the audience for which the analysis is performed, the gap analysis may be performed either at a high level or at an operational/tactical level. The scope of the audit must first be defined to ensure it addresses the needs of the gap analysis. The scope should detail what is included and excluded in the analysis to collect the right information.

For example, the CQM BoK is intended for and focused on the dedicated CQM. However, many projects do not have the requirement or budget for this role. (This does not excuse the team from having a quality plan or establishing QA and QC processes.) Performing a gap analysis on the resources of the project can help identify the personnel who are available and assigned the roles of quality, which helps the project team determine which resources are not staffed. This enables optimal resource allocation and reduces costs by preventing the possible overuse of resources. It also improves management decision-making as well as the quality of strategic planning.

The Problem of Constrained Resources

Because quality-related processes and tools are susceptible to becoming isolated (as discussed earlier in this chapter), resources can be strained as the various project-level plans (safety, environmental, quality, etc.) compete for attention. A project team may think it needs separate resources or staff to manage the project quality plan—or other plans. However, when integrated effectively—and ensuring quality is the foundational plan on which other plans are built—resources can be shared and project operational costs may be put to more efficient use.

An unfortunate reality of the industry is that not all projects are staffed appropriately during each phase. Underperforming resources, such as untrained team members or broken software, can constrain others as they try to pick up the workload. Effective resource planning helps the entire team as members communicate in an open and transparent manner about what needs to be done and what support is required.

CONCLUSION

The construction process is inherently interdependent; the performance of any single contractor at a jobsite relies on the successful work of others. If this expectation is established up front as part of the culture at the project level, then the constraining of resources can be mitigated, if not avoided.

Robust resource assessment and mobilization are foundational to project success. Identifying qualified personnel, appropriate tools, and timely availability is crucial. Thorough gap analyses of resourcing and proactive training address potential shortcomings and mitigate resourcing risks. By establishing clear roles, responsibilities, and communication channels, teams can avoid resource constraints and optimize performance. Recognizing the interdependence of team members and the importance of integrated planning ensures efficient resource utilization, fostering a project environment that is conducive to quality outcomes, timely delivery, and budgetary adherence.

QUESTIONS

1. Why is outlining who is responsible for each task important from a quality perspective?
2. What is a key requirement of a team member when they are assigned as a subject-matter expert (SME) to a project?
3. What must the quality champion confirm for each team member when they are assigned a specific task?
4. Explain how different organizational cultures and individual responsibilities impact resource planning for a project, and why focusing solely on corporate policies may be insufficient for ensuring quality task execution.
5. Describe the significance of a "gap analysis" in the context of resource assessment and mobilization. How does it contribute to risk mitigation and project success?
6. Discuss the challenges associated with "constrained resources" in project execution and how integrating the quality plan with other project-level plans can address these challenges.
7. Why is understanding both leading and lagging indicators crucial in resource planning for data, reporting, and measurement systems, and how do these relate to KPIs, OKRs, and client intent goals?
8. Explain the interdependence of contractors on a jobsite and how this dynamic relates to resource availability and the overall project culture, as mentioned in the concluding section of the document.

ROLES OF QUALITY PERSONNEL AND INTERNAL STAKEHOLDERS

> Ensure all relevant quality personnel and internal stakeholders understand the quality plan and have the necessary knowledge and skills to carry out their responsibilities. Formulate the roles and expectations of quality personnel, contribute to the personnel qualification process, and ensure all personnel understand their roles and responsibilities. (Create)
>
> Body of Knowledge II.C.2

Effective quality management in construction hinges on clearly defined roles and responsibilities for all involved. This section explores the essential roles of quality personnel and internal stakeholders, emphasizing how these roles lay the foundation for project success. By formulating clear expectations, ensuring personnel qualifications, and employing tools like the RACI matrix, organizations can mitigate risks and enhance collaboration. This section will detail the responsibilities of key quality professionals and demonstrate how proper planning can lead to efficient execution and achievement of quality objectives throughout the project life cycle.

Roles of Quality Professionals in the Project Life Cycle

Effective quality management in construction projects begins with well-defined roles and responsibilities for quality personnel and internal stakeholders. During each phase of the project, it is essential to ensure that all individuals involved in QA and QC understand the project's quality plan (see the sections "Project Quality Strategy Development" and "Alignment of Quality Plan with Other Project Plans") and possess the necessary knowledge and skills to execute their responsibilities. Establishing clear roles, setting expectations, and verifying personnel qualifications help create a strong foundation for project success.

For each project phase, there are both QA and QC functions that must be set up along with the processes that must be followed to perform QC and QA (see "Quality Assurance [QA]" and "Quality Control [QC]"). There also needs to be separation of roles and expectations for the quality professionals. Generally, there are three types of quality roles:

- General quality leadership (high-level)
- Quality assurance personnel
- Quality control personnel

These roles will exist at both the corporate and project level. General quality leadership is represented by the project manager, the project principal, and senior leadership that has the overall responsibility for the quality as well as the delivery of the project. The role should be discussed for every project, regardless of its size and scope. Even small projects that cannot afford dedicated resources for QA and QC must include a discussion of quality and its importance.

The roles for QA and QC may be dedicated resources or additional responsibilities for personnel with other roles on the project. Assigning dedicated resources to QA and QC roles significantly increases the emphasis on these roles, as those resources can provide greater focus and expertise as well as accountability, which elevates the importance of the quality. Rightsizing the quality effort with dedicated QA and QC staff is a strategic management decision that must be made based on the size, scope, and significance of the projects. When projects are complex and high-risk, with mandatory compliance requirements, it serves to have dedicated resources for QA as well as for QC. Note that even those roles who are not dedicated to a quality group or function—or do not have a quality title—still have a responsibility to understand the quality plan, objectives, and expectations. This includes those installing the work and those supervising the work in any project phase.

Another important aspect of quality resourcing on construction projects is that many times, the projects that are issued as design-bid-build do not have a

CQM until a later stage when the construction contractor is hired. In all such cases, it is advisable to identify a dedicated person for the implementation of quality; the person assigned as the quality manager for design then becomes the single most useful person that the CQM may engage with to align the quality planning of construction with the earlier design phase.

Formulate the Roles and Expectations of Quality Personnel

The first step in resource planning is to clearly define the roles of quality personnel and internal stakeholders. Each role should be tailored to the project's scope, complexity, and regulatory requirements. Key roles in the quality management process include:

- *Quality manager*: Oversees the entire quality management system (QMS), ensures compliance with industry standards, and leads quality initiatives.
- *Quality engineers/inspectors*: Conduct inspections, verify compliance with specifications, and document QC activities. These roles are most often associated with the QC function.
- *Project managers*: Integrate quality processes into overall project planning and ensure alignment with contractual obligations.
- *Construction supervisors*: Implement QC measures on-site and coordinate with quality personnel to address nonconformities.
- *Document control specialists*: Manage and maintain quality documentation, ensuring accessibility and compliance with recordkeeping standards.
- *Internal auditors*: Perform process and system audits to evaluate the effectiveness of the quality management framework.

Provision for quality roles should be made by all performing organizations, and the effort can be scaled according to the size and criticality of the projects. Quality roles specifically must be defined, even if no position can be created for it. For example, project team members who have not performed the work but are familiar with the function can fulfill a quality role as long as they are guided and mentored accordingly. They also do not need to be given a formal quality title.

By clearly defining these roles, organizations can prevent overlapping responsibilities, reduce inefficiencies, and ensure accountability in quality management.

Contributing to the Personnel Qualification Process

Once roles are established, it is critical to ensure that all quality personnel and stakeholders have the appropriate qualifications and training. This involves:

- *Identifying required competencies*: Each role should have clearly outlined qualifications, including certifications, experience, and technical skills.

- *Providing training programs*: Regular training sessions should be conducted to keep personnel updated on best practices, new technologies, and evolving industry regulations.
- *Conducting competency assessments*: Periodic evaluations are necessary to verify that personnel meet the necessary standards and can effectively perform their duties.
- *Encouraging continuous learning*: A culture of continuous improvement should be fostered through mentorship programs, workshops, and professional development opportunities.

A primary responsibility of the CQM is to offer training on the quality plan and help all project team members and stakeholders understand the quality expectations and requirements. Ensuring that personnel are well-trained enhances project efficiency, minimizes errors, and improves overall quality outcomes.

Management Tools and Infrastructure

The risk of quality deliverables being missed can be mitigated by the CQM creating a matrix that helps communicate the roles and responsibilities. The *roles and responsibilities (RACI) matrix* is an effective tool for managing and understanding the roles of all internal and external stakeholders and project team members. It identifies who is executing what process, whether they are qualified, and associated milestones.

Subcontractors, architects, key members of the design team, and clients should also be included. A RACI matrix, as shown in Table 2.9, should be created during each major project phase to help ensure resources are coordinated. One or multiple team members could be assigned a task. The RACI matrix also identifies any gaps in resources if tasks need to be completed but are unassigned.

- R: Responsible—*Responsible* for executing the task.
- A: Accountable—*Accountable* to ensure the task gets done.
- C: Consulted—*Consulted* as part of the task being executed; a stakeholder.
- I: Informed—*Informed* on the task for context.

The RACI matrix is a sample tool, but others of a similar nature outlining who is doing what and their specific level of involvement need to be created within the context of the project team and the organizations involved. Following the RACI matrix, the CQM should create a stand-alone "roles and responsibilities" document for each individual team member. This could serve as an additional guide for each team member to promote their awareness of the necessary quality tasks and how they can help the project achieve its quality objectives.

Internal Stakeholder Engagement

Quality management is not limited to dedicated quality personnel—it also involves internal stakeholders such as project executives, procurement teams, and

Table 2.9 Sample roles and responsibilities matrix.

Task	Role				Tool/Technology
	Project Manager	Quality Manager	Superintendent	Subcontractor	
Mobilization					
Conduct preconstruction meetings for each scope of work	R	C	A	I	Technology or tools that will facilitate the action.
Conduct preinstallation meetings for trade coordination	I	A	C	R	Technology or tools that will facilitate the action.
RFI management	R	I	C	C	Technology or tools that will facilitate the action.
Submittal management	R	C	A	R	Technology or tools that will facilitate the action.
Closeout					
Inspections	I	A	C	R	Technology or tools that will facilitate the action.
Commissioning	I	A	C	R	Technology or tools that will facilitate the action.

subcontractors. Effective communication and collaboration among these stakeholders ensure that quality objectives are aligned with overall project goals. Key strategies for stakeholder engagement include:

- *Quality planning meetings*: Regular meetings should be held to discuss quality expectations, potential risks, and corrective action plans. (See Chapter 5 for more information.)
- *Clear documentation and guidelines*: Quality requirements should be documented in accessible formats and shared with all relevant personnel.
- *Feedback mechanisms*: Open channels for reporting issues and suggesting improvements should be established to promote proactive problem-solving. (See Chapter 5.)

CONCLUSION

Due to the increasing complexity of many modern projects, larger teams are needed to manage quality requirements. By fostering strong engagement and communication among stakeholders, construction projects can maintain high-quality standards and improve collaboration across teams. Each project stakeholder will employ individuals assigned to cover the interests of that stakeholder. It is necessary for the CQM to implement a structure to understand who is accountable for the quality deliverables.

Establishing well-defined roles, ensuring personnel competency, and engaging internal stakeholders are critical steps in the planning phase of quality management. A structured approach to resource planning enables construction teams to execute quality processes effectively, comply with industry regulations, and achieve project success. By prioritizing these elements, organizations can build a solid foundation for continuous improvement and long-term quality excellence.

Establishing well-defined roles, ensuring personnel competency, and engaging internal stakeholders are critical steps in the planning phase of quality management. A structured approach to resource planning enables construction teams to execute quality processes effectively, comply with industry regulations, and achieve project success. By prioritizing these elements, organizations can build a solid foundation for continuous improvement and long-term quality excellence. Fostering strong internal stakeholder engagement and communication ensures high-quality standards and improves collaboration across all teams.

QUESTIONS

1. What are three key elements of the RACI matrix?
2. What are the primary responsibilities of the CQM for aligning roles and responsibilities?
3. What is another tool for ensuring team members understand their roles and responsibilities related to quality?
4. Explain the difference between general quality roles and dedicated quality roles (assurance or control) at both the corporate and project level in construction quality management. Give examples of each.
5. Describe how the RACI matrix helps mitigate the risk of missed quality deliverables, and why it's important to include stakeholders beyond the internal quality team in this matrix.
6. Discuss why simply defining quality roles is insufficient and why contributing to the personnel qualification process is equally critical for effective quality management in construction projects.
7. Analyze why roles and responsibilities documents for individual team members should be created in addition to the RACI matrix, and explain how these documents contribute to achieving project quality objectives.
8. Elaborate on the significance of stakeholder engagement in quality management beyond dedicated quality personnel, and describe how strategies like quality planning meetings, clear documentation, and feedback mechanisms contribute to high-quality standards.

TRAINING NEEDS IN QUALITY MANAGEMENT

> Identify a project team's quality-specific training needs to achieve project deliverables, develop a training plan, deploy the training, and evaluate the effectiveness of the training. (Create)
>
> Body of Knowledge II.C.3

Ensuring project quality hinges on a well-trained team. This section delves into the crucial process of identifying, developing, deploying, and evaluating quality-specific training. Recognizing that training directly impacts quality, we will take a structured approach, aligning with ISO 9001 standards and instructional system development models. By analyzing project requirements, roles, and individual skill sets, we can pinpoint necessary training, from general awareness to trade-specific expertise. This proactive approach fosters competence, mitigates risks, and ultimately leads to successful project deliverables.

For many quality professionals who may not be involved in training and development aspects within their organization, it may seem strange to have this area of responsibility rolled up under the purview of a quality manager rather a learning and development specialist. However, when you consider the direct impact training can have on quality, you'll understand why, as a quality manager, you want to be involved with your training programs, even when there are dedicated learning and development professionals within the organization. That's precisely why training is an aspect of the ISO 9001 standards for quality management systems.

Per *The ASQ Certified Manager of Quality/Organizational Excellence (CMQ/OE) Handbook* (Fifth Edition), "planning for training should closely resemble the planning done for projects."

For example:

- A need is identified.
- An objective, outputs, and outcomes are stated.
- Stakeholders (trainees) are identified.
- Development work to be done is laid out (work breakdown structure).
- A timetable for development, delivery, and evaluation is set (Gantt chart).
- Resource requirements are specified (trainers, facilities, equipment, materials).
- Applicable metrics are established.
- A budget is developed.

- The plans are approved.
- The training program development is monitored and measured.
- Training is delivered.
- Outputs are evaluated (ability to perform task).
- Outcomes are evaluated (training objective achieved).
- Lessons learned are reviewed.
- Documentation of the development and delivery process, outputs, and outcomes is added to the organization's knowledge database.

The ASQ CMQ/OE Handbook framework provides a guide that aligns well with the four-phase process detailed in this section, intended to help a CQM address training needs in quality management.

The ADDIE Model

One of the most widely applied models used in the development of training programs is the *ADDIE model* (see Figure 2.4), which was developed at Florida State University in the 1970s. The name is an acronym for the five phases it defines for building training and performance support tools: Analysis, Design, Development, Implementation, and Evaluation. Note that some variations of this model exist, with slight differences, but all generally consist of the same five-stage framework.

While it is similar to the traditional framework, in one variation of the ADDIE model, evaluation is emphasized throughout each of the cycles (Figure 2.5).

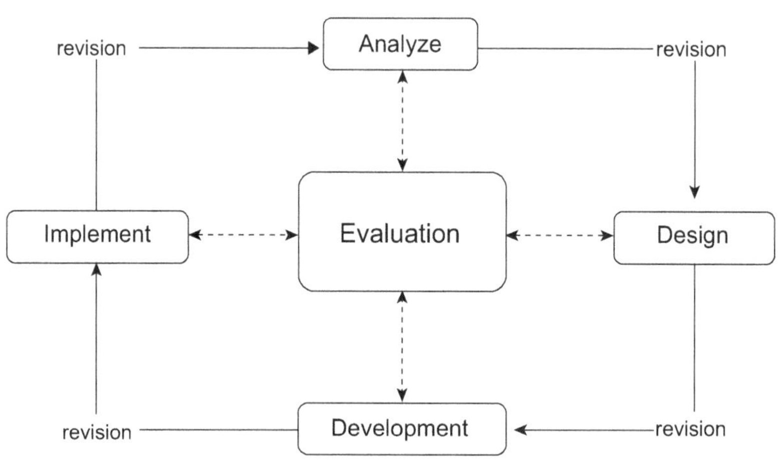

Figure 2.4 ADDIE model phases.

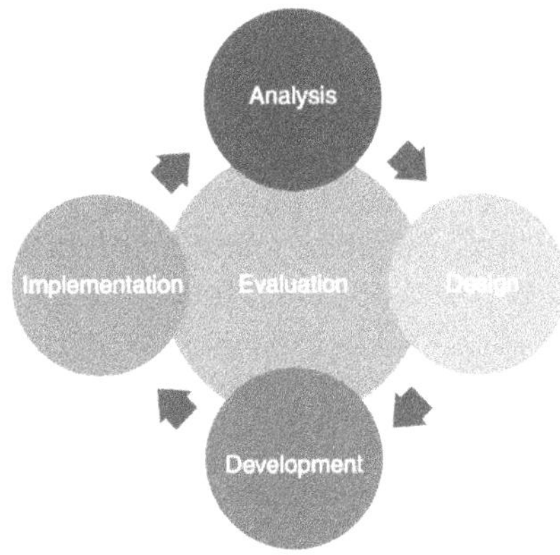

Figure 2.5 ADDIE model of training evaluation cycle.

Identifying Quality-Specific Training Needs

Once a resource assessment for the project is completed and project team members have been assigned appropriately in a clear RACI matrix (see Table 2.9), the CQM should have awareness and a clear understanding of:

- Who will be involved in the project?
- What are the team member qualifications and current skill sets?
- What roles will they fill for the project?
- What tasks will need to be completed relating to quality?
- What tools, software, and processes are required to complete the tasks?

As previously mentioned, not every project will have a full-time dedicated quality professional assigned to it, let alone a team of quality personnel to address all of the processes and procedures impacting and relating to quality. Furthermore, even when there is an entire team of quality professionals on staff, there are typically certain roles and responsibilities for specific tasks or scopes that still become assigned to various other project team members that relate to, or impact, quality. For example, on some large, complex projects, submittal reviews might be the responsibility of a member of the quality team, while on smaller projects, this task might be handled by project engineers. Therefore, whoever is championing the quality training efforts for the project must have a good understanding of the contract's quality management requirements, the organization's QMS, the project's quality plan, and how all of the various roles within the RACI matrix impact or interrelate with processes impactful to quality, and the knowledge or skill sets required for each to be accomplished.

The development of a training matrix at the organizational level can also be valuable, as it can aid in easy determination of who has already acquired specific skills or completed various training.

When determining the training needs, there are two kinds of needs:

- *Mandatory needs:* Those that satisfy any mandated requirements expected by management, the customer, a governing body, or a standards organization, such as ISO. Examples of mandatory training in quality must include the understanding of the QMS, the quality strategy, the quality plans, and the procedures for QA as well as QC. The project teams also need to be trained in the basic expectations of quality management from all employees, regardless of their title and nature of work.
- *Expressed needs*: Those intended to improve a deficiency in a particular competency. Quality training for specific competencies may be applicable to some, but not to all. Those who may need specific training in quality tools may be identified based on their roles and responsibilities as they pertain to QA or QC of the project deliverables. For example, the quality training needed by an electrical inspector may be different from that of a civil inspector because the nature of their testing may be completely different. Many organizations identify the trade-specific needs of their QC inspectors and include them in the training plans and the training matrix. For organizations that have an ISO 9001:2015-compliant QMS, there are specific requirements within clause 7.2 Competence that must be met. It should be noted that in the 2015 update to ISO 9001, the word "training" was removed and replaced with "competence." This change highlights that training is simply one facet of developing or enhancing competences needed for personnel.

An individual's competencies are determined by a combination of education, training, and/or experience that would be appropriate to their particular role. While education and training are often treated synonymously, they are actually quite different in that education involves increasing one's knowledge and information, while training involves the development or refinement of a skill or skill set. However, along with experience, they make up our competencies.

To meet the expectations of ISO 9001:2015 requirements surrounding competencies, an organization must assess the required competencies of all individuals who can have an impact on quality outcomes and ensure that they have acquired the appropriate level of competencies needed to successfully fulfil that role.

"Quality is impacted by everyone in the organization" is a core principle of modern quality management, most notably total quality management (TQM). It holds that every individual, from top leadership to frontline employees, influences the overall quality of a product or service. This mindset is a very helpful way to look at training needs identification, as some positions and titles may not seem obvious to include until further analysis is conducted. For example, one might overlook a jobsite's administrative assistant when evaluating quality-related competencies. However, if they cannot do their job proficiently, then their missteps in performing clerical or administrative duties could certainly have a downstream impact on the project's quality or performance.

Another consideration is what general training or awareness must be provided to trade partners, or third-party agencies or consultants. Are there policies, processes, or tools that they need to be made aware of? Consider all stakeholders and groups that will be involved in the project and evaluate if there are any specific competence needs that may need to be addressed.

The ISO 9001 section titled *Awareness* explicitly requires that anyone doing work under the organization's control be made aware of quality policies and objectives, and their potential impacts on performance and quality. Given that emphasis, at the very least, the creation of a brief quality orientation might be an important tool to be deployed for all personnel that enter the jobsite, providing them with an overview of critical aspects of the QMS and site-specific quality plan, and so on—thus providing and creating awareness for all stakeholders involved in the project.

Having established the stakeholders that will need specific training, the optimal delivery methods for the training may be determined along with the required resources. Resource consideration for training should include budgetary concerns, the need for justification of the cost and usage, and the time involved in the training sessions. All these considerations can have an impact on the design of the training plan.

Design and Develop a Training Plan

After working through the needs analysis and determining what training needs exist (and for whom), the next step is the development of the training plan. Taking input from needs assessment, determine the optimal delivery method for each training to be included in the training plan, along with required content, to meet the desired learning outcomes. Next, establish a work plan to show all the tasks that will need to be done to accomplish the development, deployment, and evaluation of the training. An effective work plan also requires a schedule to create accountability and to show the constraints and deadlines that need to be met. This is particularly important if training requires specific resources that necessitate advanced scheduling. Therefore, specify what resource needs exist, such as facilities, trainers, special equipment, and/or training materials.

Once the work plan has been developed, along with scheduled milestones and goals, establish a process that tracks the progress of the development and deployment of the training program, along with the evaluation of the effectiveness of training conducted. Consider which KPIs and metrics will be important, where these inputs will come from, and how they will be measured. Additionally, the plan should explore the opportunities to benchmark these metrics. Can any be compared to other projects? Are there other outside organizations that have established best practices and lessons learned that the project can benefit from?

Organizations are now more aware of the links between training and quality, and that training should be a proactive part of quality management; therefore, projects should be allotted a budget for training. If so, work through the costs of the training recommended by the needs assessment, ensure that the program can be deployed with the allocated budget, and adjust accordingly. Where a training budget does not exist, convey the importance of the training to management by leveraging the needs assessment as justification.

With the budget established and a clearly defined training plan, the training plan must be submitted to management for formal approval and then proceed to implementation. The developed training program should include metrics that evaluate the effectiveness, efficiency, and outcomes of the program, and help align it with business goals of the project and the organization.

Implementation: Training Plan Deployment

Once the training program has been designed and developed, and has gone through some preliminary beta testing to ensure that it is effective, it is time to deploy the program to its intended audience. A *training matrix* is a visual tool, often a table or spreadsheet, that can be used to track, manage, and plan employee training and skill development. A training matrix is helpful to track who has completed the recommended or required training and identify who needs to be included in communications regarding specific training.

Leveraging this information, the training plan may be communicated to the intended audience. The plan will include training requirements or recommendations, along with the dates of the training and the process for completion. Most well-resourced organizations have online training portals with such information for their employees and external stakeholders (where required). If deadlines are established for the completion of training, then this must be communicated. The person responsible for the training program must ensure the training dates are met. Holding individuals accountable for meeting the training deadlines helps create a culture where learning and development are priorities. It is also helpful to be clear about the intended outcomes associated with each training module, as it provides the target audience with an understanding of the benefits of completing that training.

Throughout the deployment phase, create a culture where learning and development are important by rewarding training completion, celebrating goals being met, and providing team members adequate time to complete their training, even while juggling competing priorities.

Improvement: Evaluating Training Plan Effectiveness

Evaluating the effectiveness of training helps determine how useful the training is and how it can be improved in the future. The following may be considered for training evaluation:

- Evaluate outputs (ability to perform tasks).
- Evaluate outcomes (training objective achieved).
- Review lessons learned.
- Add documentation of the development and delivery process, outputs, and outcomes to the organization's knowledge database.

While the program is being deployed, collect information regarding the outcomes of completed training to evaluate and improve training as needed. Is the intended audience better prepared to perform specific tasks upon completing the training? Were the objectives of the training met? Were any lessons learned that can help improve the project deliverables? These are important questions to analyze that will help you understand the effectiveness of the training program, as well as how to improve it. The feedback cycle should leverage the plan-do-check-act (PDCA) framework for continual improvement throughout the duration of the training plan.

Once training is completed, a postmortem meeting should be conducted, documenting the development and delivery processes, and the outcomes, along with any lessons learned that will be incorporated into the organization's knowledge management system to leverage in the future.

A Structured Approach to Training

For a construction project team's quality-specific training needs to achieve project deliverables and to develop a training plan, a structured approach is recommended. The two major aspects of the structured training include:

1. Identification of the project team's quality-specific training needs
2. Development of the training plan

Identify the Project Team's Quality-Specific Training Needs

This crucial first step involves understanding what knowledge and skills the team requires to meet the project's quality deliverables.

Analyze Projec t Documents

- *Contract documents*: Review the contract for specific quality clauses, standards referenced, inspection requirements, and acceptance criteria.
- *Project specifications*: Thoroughly examine the technical specifications for materials, workmanship, testing, and performance requirements.
- *Drawings and plans*: Understand the design intent and quality expectations embedded in the project's visual representations.
- *Quality management plan (QMP)*: If one exists, review the defined quality processes, procedures, and any identified training needs.
- *Regulatory requirements*: Identify all applicable building codes, safety regulations, and environmental standards that have quality implications.

Identify Key Project Deliverables and Associated Quality Standards

- List the major deliverables of the project (e.g., foundations, structural framework, MEP systems, finishes).
- For each deliverable, identify the specific quality standards, codes, and specifications that apply. This will highlight the technical knowledge required.

Define Roles and Responsibilities

Clearly define the roles of each team member and their specific responsibilities related to quality. This will help pinpoint role-specific training needs. Examples:

- *Project manager*: Overall quality oversight, QMP implementation
- *Quality manager/QC engineers*: Inspection planning, execution, reporting, nonconformance management
- *Site superintendent*: Ensuring quality workmanship on-site
- *Foreman/supervisors (by trade)*: Ensuring quality of work within their specific trade
- *Craftspeople (by trade)*: Performing work according to quality standards
- *Inspectors (internal and external)*: Verifying compliance with quality requirements

Conduct a Skills Gap Analysis

Assess the current knowledge and skills of the project team members against the identified quality requirements. This can be done through:

- *Interviews*: Talk to team members and their supervisors to understand their experience and perceived training needs.
- *Questionnaires*: Use structured questionnaires to gather information about their understanding of quality standards and procedures.
- *Review of past performance*: Analyze performance data from previous projects to identify recurring quality issues that might indicate training gaps.
- *Observation*: Observe team members performing tasks to identify areas where their skills could be improved.

Identify Specific Quality Training Needs

1. *General quality awareness*:
    - The importance of the project QMP
    - The organization's quality policy and objectives
    - The cost of poor quality
    - The importance of adherence to procedures and specifications
    - Basic principles of quality control and quality assurance
    - Reporting quality issues
2. *Trade-specific quality requirements (not everyone needs specific quality training; it may be specific to the applicable trades)*:
    - *Concrete works*: Proper mixing, pouring, curing techniques, testing procedures (slump, compression), and reinforcement placement

- *Structural steel erection*: Bolting and welding procedures, dimensional tolerances, and inspection requirements
- *Masonry*: Mortar mixing, brick/block laying techniques, jointing, and waterproofing
- *MEP (mechanical, electrical, plumbing)*: Installation standards, testing and commissioning procedures, and code compliance
- *Finishes*: Application techniques for plastering, painting, flooring, etc., and quality standards for materials

3. *Inspection and testing procedures (training to ensure consistency in measurement methodology)*:
   - How to perform specific inspections (e.g., visual, dimensional, or nondestructive testing)
   - Proper use of inspection tools and equipment
   - Understanding acceptance criteria and tolerances
   - Documentation of inspection results
4. *Nonconformance management*:
   - How to identify and document nonconformances
   - The process for raising and resolving *nonconformance reports* (NCRs)
   - Understanding root cause analysis and corrective actions
5. *Quality documentation and recordkeeping*:
   - Proper procedures for completing inspection checklists, test reports, and other quality-related documents
   - Importance of accurate and timely recordkeeping
6. *Communication and reporting*:
   - Effective communication of quality issues and findings
   - Understanding quality reporting requirements

Develop a Training Plan

Once training needs are clearly identified, the next step is to develop a structured training plan (see Table 2.10). This plan will outline how the training will be delivered and managed.

Key Elements of a Training Plan

Training objectives: Clearly state what the participants should be able to do after completing the training. These should directly address the identified training needs.
Target audience: Specify which roles or groups of team members will attend each training session.

Table 2.10 Training plan example.

Training Topic	Target Audience(s)	Objectives	Content Outline	Training Method(s)	Duration	Schedule	Trainer(s)	Evaluation Method(s)
Project quality management plan (QMP) overview	All project team members	Understand the purpose, structure, and key requirements of the project QMP.	Introduction to the QMP, key sections, individual responsibilities, reporting procedures.	Classroom session	1 hour	Week 1	Quality manager	Attendance, Q&A
Concrete placement best practices	Concrete crew, foreman, QC engineer	Apply best practices for concrete placement to ensure quality and minimize defects	Preplacement checks, pouring techniques, vibration methods, finishing, curing requirements	Workshop, OJT	3 hours	Week 2	Concrete specialist	Observation, checklist
Understanding welding standards (AWS D1.1)	Welders, welding foreman, QC inspector	Interpret and apply the relevant requirements of AWS D1.1 structural welding code	Review of key clauses, weld symbols, acceptance criteria, inspection procedures	Classroom session	4 hours	Week 3	Certified welding inspector	Written test
Nonconformance reporting and resolution	Project manager, QC engineers, site superintendent, foreman	Understand the process for identifying, documenting, and resolving nonconformances effectively	Definition of nonconformance, NCR form completion, escalation procedures, root cause analysis basics, corrective action implementation	Classroom session	2 hours	Week 4	Quality manager	Case study analysis

Training content (modules): Outline the specific topics that will be covered in each training module. Be detailed and relevant to the objectives.
Training methods: Determine the most effective methods for delivering the training. Consider:

- *Classroom sessions*: For theoretical knowledge and discussions
- *On-the-job training (OJT)*: Practical training at the worksite under supervision
- *Workshops*: Hands-on activities and skill development
- *Online modules/e-learning*: For flexible and self-paced learning
- *Toolbox talks*: Short, focused sessions on specific quality topics
- *Demonstrations*: Showing correct techniques and procedures

Training schedule: Develop a timeline for when each training session will be conducted. Consider the project schedule and the availability of team members. Prioritize training based on project phases and immediate needs.
Training resources: Identify the resources required for each training session.

- *Trainers*: Internal experienced personnel, external consultants, or equipment suppliers
- *Training materials*: Handouts, presentations, videos, manuals, or checklists
- *Equipment and tools*: For practical demonstrations and hands-on training
- *Training venue*: Classroom, site area, or online platform

Training budget: Allocate the necessary budget for training activities, including trainer fees, materials, and any travel or accommodation costs.
Training evaluation: Define how the effectiveness of the training will be evaluated. This could include:

- *Quizzes or tests*: To assess knowledge acquisition
- *Practical assessments*: To evaluate skills application
- *Feedback forms*: To gather participant satisfaction and identify areas for improvement
- *Observation of on-the-job performance*: To see if the training has led to improved quality in their work

Recordkeeping: Establish a system for documenting all training activities, including attendance records, training content, and evaluation results.

CONCLUSION

Establishing a robust training program is paramount for ensuring project quality. By meticulously identifying needs, developing tailored plans, deploying training effectively, and rigorously evaluating their impact, organizations can cultivate a competent workforce. This proactive approach not only fulfills ISO 9001 requirements but also fosters a culture of continuous improvement. Ultimately, investing in quality-specific training translates to enhanced project outcomes, mitigated risks, and sustained organizational excellence. The lessons learned and documented throughout this process will serve as invaluable assets for future endeavors.

QUESTIONS

1. How do you define "competence" in the context of ISO 9001:2015, and why is this distinction important for quality management?
2. Describe three examples from the ADDIE framework for each phase.
3. Explain why a jobsite's administrative assistant might be included in quality-related training, even if their role doesn't seem directly tied to technical quality control.
4. How does the section differentiate the ADDIE model from other training frameworks, and why is that difference significant?
5. Differentiate between "mandatory needs" and "expressed needs" in the context of quality training, and give a brief example of each within a construction project scenario.

THIRD-PARTY INSPECTION AND OUTSOURCING

> Identify the need for third-party inspection services as applicable and develop criteria for selecting and integrating outsourced quality services into the project. (Create)
>
> Body of Knowledge II.C.4

In the construction industry, the strategic engagement of third-party inspection services is crucial for ensuring quality control and addressing specialized needs. This section delves into the planning and analysis of resource allocation, focusing on the identification, evaluation, and integration of outsourced quality services. It examines the "make or buy" decision, weighing the short-term costs against the long-term benefits of in-house versus third-party expertise. Understanding when and why to utilize external providers, along with selecting the right partners, is essential for project success.

It is common practice for general contractor firms to hire third-party contractors to provide labor or materials for tasks that are unique, are of short duration, and that require special skills not typically offered by the general contractors. Planning for the third-party resources is the process of strategically identifying, evaluating, and selecting external service providers and vendors to fulfill those specific needs of the project that the general contractors typically do not offer. In the context of the project life cycle, these services are used during the construction phase of the life cycle and are sometimes hired for the post-construction (O&M) phase as well. Third-party quality inspections fall into the same category of special services for the construction projects, though there are exceptions.

Make-or-Buy Decision Criteria for Third-Party Inspection Services

Typically, third-party inspection services are procured for QC and less frequently for QA. This is because QC services are required for the short duration of those specialized tasks, whereas the QA functions of quality planning, compliance evaluation, and lessons learned are best suited for the in-house QA staff (such as the CQM) of the contractor firm to perform.

The "make or buy" decision about the use of third-party inspection services, which is typically made during the planning phase of the project's life cycle, weighs short-term affordability against the long-term benefits of performing these services in-house. This final decision falls under the purview of the project's procurement function and may take place during the procurement phase, but the planning phase is where the "need for the outsourcing" is typically identified. Note that sometimes third-party inspection is mandatory. For example, some projects require International Electrical Testing Association (NETA)-certified third parties to do wire testing. Using third-party inspection services versus offering these services in-house is dependent on frequency, uniqueness, and independence/impartiality.

Frequency (or Infrequency) of Inspections

The most notable reason for choosing to outsource third-party inspection services is the frequency (or infrequency) of required testing. In cases where there isn't a high volume of consistent production/construction, the contractor will usually consider outsourcing the QC inspection and testing services. Conversely, the project team should consider building an in-house inspection team if there is a high volume of consistent production and a long-term need for inspection. Using third-party inspection services generally has advantages in terms of cost-effectiveness only when used for a short duration and infrequent need. Due to the higher cost of such services, economy of scales becomes negative the longer you use third-party quality inspection services.

Uniqueness of the Required Inspections

Typically, QC personnel in the construction industry are trained in the general trades, which include disciplines such as civil, structural, electrical, mechanical, plumbing, welding, and others. If the project's tasks, scope, and criticality is such that its inspection needs are typical, then the standard inspection and testing that is typical to the general construction trade disciplines is sufficient. In such cases, the economies of scale support developing an in-house QC inspection team.

However, most general construction firms do not consider it worthwhile to carry specialized expertise in niche areas and instead utilize subcontractor firms to perform the work and ensure its compliance. If inspection services are needed for an activity that is very specific in nature, or unique, it may require SMEs to perform the tests and inspections to review the materials, products, and their performance.

Special inspections warrant the use of outsourcing the work to third-party inspection firms that can offer those services; they have the personnel who are trained and usually qualified through professional certifications to perform such QC inspections as well as the equipment that is necessary to perform the tests for the niche areas.

To determine which project elements warrant special inspections, the project team must examine the design specifications and identify the technical areas where third-party inspection services make the most logical sense. Note that in some situations, special inspections may be required by code. Such inspections are crucial to ensure that specific technical standards and quality benchmarks are met.

Independence and Impartiality

The most compelling reason why third-party inspections are used in construction projects is their ability to offer objectivity in QA. As external entities, third-party QC inspection firms offer an independent, unbiased perspective and provide a fresh and impartial evaluation of the constructed works, materials, products, services, and processes.

If a project has the need to collect independent assessments on the quality of the products and processes, it may consider procuring third-party inspection services. Their impartiality and the ability to scale inspections based on a project's special needs are additional reasons to outsource such inspections.

Independent third parties provide an unbiased assessment of the project's adherence to technical and quality standards. Unlike internal inspections, which might be influenced by the project's stakeholders, third-party inspectors have no vested interest in the project's outcome. Their impartiality ensures that inspections are objective and more reliable.

Criteria for the Selection of Third-Party Inspection Services

The following are the recommended criteria for selecting the right third-party inspection service provider for the project needs.

1. *Proximity to the Project Site*:

Third-party inspection services that are local to the project's physical location offer the most utility because of many contributing factors:

(a) they are usually better known to the other stakeholders of the project than any out-of-town testing firm;

(b) they can offer quick turnaround, in terms of both the work they are tasked with and their resources;

(c) they may have an institutional knowledge of the area, having worked on some of the previous projects in or around the physical location of the project;

(d) using them may be regarded as a positive impact on the part of the project in terms of supporting the local economy; and

(e) they usually offer a reduced environmental footprint due to lower travel distances, which is typically the reason for their more efficient, time-sensitive turnaround that out-of-town inspection services cannot provide.

There are some disciplines of construction where it is strongly recommended to utilize local area third-party suppliers and inspection services, like surveying and geotechnical. In both disciplines, such services may offer extensive past experience working in the local area.

2. *Capability (Defined in Terms of Equipment and Personnel)*:

The capability of a third-party inspection service should be measured on the basis of their proposed scope of work. At minimum, the following may be considered as elements that help determine the capability of the third-party inspection firm:

- *Competence of the QC inspectors*: The competence/qualification of their inspectors must be evaluated. The term "competence" emphasizes the person's ability to apply knowledge and skills in the required discipline effectively; the term "qualification" is usually linked with the formal recognition of the person's knowledge and ability from a recognized training source.

  Before selecting third-party QC services, the project team needs to predetermine which disciplines (trades) need to be supported by discipline-specific QC technicians. It is best to use the project's *definable features of work* (DFOW) as the starting point and determine which of the scope elements would require specialty inspections. Information about the DFOW is usually included in the technical specifications.

- *Test equipment of the third-party inspection service*: The second criterion for the selection of third-party QC inspection services is the equipment of the inspection/testing firm. Typically, the scope of testing for standard horizontal construction projects includes testing for soils, concrete, masonry, welding, materials, and so on. The scope of testing for vertical construction projects, on the other hand, typically includes testing for interior construction elements, which may include electrical inspections for circuits, cables, boxes, heating, ventilation, air conditioning, and several other elements that qualify more as specialty construction.

  While evaluating the third-party inspection firm, their testing equipment must be assessed for their ability to perform inspection and testing for the niche areas that are defined in the scope of the project work.

- *Calibration program for the test equipment*: Another important decision criteria is the quality of a service's calibration program. *Calibration* describes the level of precision and traceability of testing equipment. Typically, the better-performing testing organizations are certified for their calibration programs by external accredited organizations.

3. *Technical and Administrative Performance*:

Another major criterion for the selection of one third-party inspection service provider over others is their quality performance. The evaluation of quality performance may be measured against two categories:

- *Technical performance*: The performance of a third-party testing firm must be evaluated against testing that is performed by their

QC technicians. Track the accuracy of their inspections and measurements, which is a factor of the technical competence of their personnel as well as the accuracy of their equipment. Also, because the decision to select the right third-party firm is made in the early planning stage, contact previous clients and provided references of the potential outsourcing firms to verify their experience with the firms' quality of work, reliability, delivery timelines, their communication, and their overall performance.

It is in this respect that the CQM may oversee the technical performance of the third-party inspection firms. Their compliance adherence will show how well their inspections complied with the technical provisions and industry standards for their previous clients.

We recommend using a qualitative assessment template for the technical evaluation, where the quality manager (or their field quality representative) may use their subjective judgment and base the selection on predefined measurement variables for technical performance. Some factors that may be used to evaluate the technical performance of third-party inspection providers include (but are not limited to) the following:

- Number of inspections/tests performed (over a finite period)
- Complexity of testing and number of special inspections performed (examples include concrete special inspections for mix design and batching, welding of structural elements, sensitive equipment commissioning, and building envelope commissioning)
- Compliance checking—a statistical comparison of test sample results pulled for simultaneously prepared test specimens but tested by an independent testing firm
- On-time availability of their inspection staff (measured against the schedule)
- Missed inspections
- Number of qualified technicians (as per qualification requirements)
- Depth of roster (backup QC inspectors, when required)
- Number of noncompliance issues: The number of recorded issues with the performance of the third-party firms
- Follow-up of issues and timely closeouts
- Proactive identification of potential problems

Create a weighted scoring matrix to score these technical elements and arrive at an educated decision on the selection of potential third-party inspection firms.

- *Administrative performance*: Like any other vendor, the administrative performance of third-party testing firms also affects the quality of their service experience and should be considered when making the best decision for the selection of these firms. Some aspects of administrative performance may also be considered as the criteria for the selection of the outsourced third-party inspection and testing firms.

- *Customer service performance*: Research the customer service performance of their staff as well as their collaboration with other performing organizations. In particular, if the specialty inspection firm has previously worked with the client, their feedback about the customer service experience is very useful.

It is a common practice to first ask the client if they are familiar with a third-party inspection firm that performed well on some previously completed project. Collecting such information helps to hire a known and familiar firm as the specialty third-party contractor because their positive experience with the client is an additional reason to select them.

Role of the CQM in the Outsourcing Decision for Third-Party QC Inspection and Testing

The role of the CQM in the outsourcing decision for third-party QC inspection and testing is dependent on the nature of the project delivery method. For the traditional phased project delivery method, such as design-bid-build projects, the construction contractor is not hired during the early planning phase of the project. In such cases, the decision of third-party QC inspection services is also not made during the planning phase, and the CQM can have no input. However, if the project is delivered through the progressive design-build method, the CQM may be hired at the time when this decision is made. If available, the CQM may be included in the decision of procuring these services and facilitate the planning and procurement of the third-party inspection services. For all other methods of project delivery, the CQM may assist in procurement only if they are already hired at the time when these decisions are being made during the project life cycle.

The CQM usually understands the project's technical requirements, which can be a great help in putting together a weighted decision matrix for the selection of the third-party inspection services. If the levels of acceptable performance are clearly defined, and the decision criteria factors (capability, cost, quality, and compliance) and potential risks are included in the weighted matrix according to the priorities of the project, such a matrix could be utilized by the procurement team to onboard other subcontracted services as well.

CONCLUSION

In summary, the effective utilization of third-party inspection services is essential for maintaining quality and managing specialized project needs in the construction industry. By carefully identifying the need, evaluating potential providers, and considering factors such as frequency, specificity, and independence, projects can strategically leverage external expertise. The "make or buy" decision must balance cost-effectiveness with long-term project goals. Ultimately, choosing the right third-party partners ensures adherence to standards, provides impartial assessments, and contributes significantly to overall project success and client satisfaction.

QUESTIONS

1. What is the best reason to outsource quality control inspection services?
2. What factors identify the need for third-party inspection and testing services?
3. What should be the criteria for selecting the best third-party inspection service for the project?
4. What factors measure the capability of a third-party inspection service?
5. What factors represent the quality performance of a third-party inspection service?
6. What key considerations would lead a project manager to choose third-party inspections over building an in-house inspection team, and how might these considerations change during different phases of a project?
7. Beyond simply checking for defects, what additional benefits does the use of independent third-party inspection services provide to a construction project, particularly concerning stakeholder perception and regulatory compliance?
8. The section mentions both "capability" and "quality" when selecting third-party inspection services. How are these two criteria distinct, and what specific factors should be examined to evaluate each?
9. How does the role of the Construction Quality Manager (CQM) vary in influencing the selection and utilization of third-party inspection services, depending on the project's delivery method and the timing of contractor involvement?
10. The document highlights the importance of "proximity" when selecting a third-party inspection service. Explain why geographic location is a significant factor, and provide at least two disciplines where this consideration is particularly critical.

MEASUREMENT SYSTEMS ANALYSIS

> Assess the measurement system and its requirements to verify if equipment accurately and reliably measures specific parameters. Ensure calibration and certification requirements are met and that all equipment, including equipment used by subcontractors, is accurate, precise, and compliant with relevant standards and regulations. (Evaluate)
>
> Body of Knowledge II.C.5

Accurate measurements are the bedrock of successful construction projects, ensuring quality, safety, and adherence to specifications. The reliability of these measurements hinges on the robustness of the measurement system itself. This section delves into the critical processes of assessing and verifying measurement equipment and procedures. Ensuring that every tool, from tape measures to complex laser surveying instruments, is calibrated and used correctly is vital. Without a thorough understanding of measurement system analysis (MSA), you risk inaccuracies that can lead to costly errors, delays, and potential safety hazards.

Assessing the Capability of a Measurement System

Assessing the measurement system involves evaluating the entire process used to obtain measurements, not just the individual piece of equipment. The goal is to understand the sources of variation and ensure the system is capable for its intended purpose.

Key aspects of ensuring that the system is capable include the following:

- *Defining the measurement task*: Clearly understanding what needs to be measured (e.g., length, elevation, strength, temperature, moisture content), why it's being measured (e.g., compliance with specs, progress payment, safety check), and the required level of accuracy and precision
- *Identifying system components*: Recognizing all elements contributing to the measurement result
- *Equipment*: The specific tools used (tape measures, total stations, lasers, concrete test machines, scales, thermometers, pressure gauges, etc.)
- *Operator*: The person taking the measurement and their skill/training
- *Procedure*: The documented (or standard) method for taking the measurement
- *Environment*: Site conditions (temperature, humidity, vibration, dust, lighting) that might affect the measurement
- *Software*: Any programs used for data collection or analysis, integration with equipment hardware
- *Evaluating suitability*: Determining if the chosen equipment and procedure can meet the required tolerance or specification limits. For instance, using a standard tape measure might be unsuitable for high-precision steel fabrication alignment where a total station is required.
- *Measurement system analysis (MSA) concepts*: While formal MSA studies (like gage repeatability and reproducibility/GR&R) are more common in manufacturing, the underlying principles are applied less formally in construction:
  - *Repeatability*: Can the same operator get consistent results using the same equipment to measure the same item multiple times? High variation indicates issues with the equipment or operator technique under specific conditions.
  - *Reproducibility*: Can different operators (inspectors) get consistent results using the same equipment to measure the same item? High variation here points toward differences in operator training or technique.
- *Accuracy*: Accuracy describes how close a measurement is to its true value. Does the measurement system consistently measure higher or lower than the true value (as determined by a traceable standard)?

The prerequisite of accuracy is that the equipment used must be calibrated. Accuracy in construction projects is crucial and relies on comparing measurements against consistent reference points, known as "datums." These datums must be established using the same measurement system to avoid discrepancies. For instance, if you're measuring distances on a site, all measurements should be based on the same baseline to ensure they are accurate and comparable. Accuracy is dependent on checking against consistent datums using the same measurement system with calibrated equipment.

- *Stability*: Does the measurement system remain consistent over time? This is addressed by regular calibration and verification.

  The CQM is required to understand how each element of this system works together and how it needs to be monitored to ensure reliable results. This can be challenging when projects scale in complexity, with many simultaneous processes happening at once, carried out by multiple stakeholders. Continuous monitoring is necessary to ensure that the measurement system will produce consistent accurate measurements.
- *Assessing procedures and training*: Reviewing measurement procedures for clarity and correctness; ensuring operators are properly trained on both the procedure and the use of the specific equipment
- *Documentation review*: This includes checking the records of calibration, of verification, and the records of operator training

Calibration

Besides assessing that the measurement system is capable, it is also important to ensure the measurement equipment is currently accurate and reliable. A piece of equipment capable of performing the intended measurement may become inaccurate and unreliable due to the conditions in which it is used, or because of poor equipment handling. It is therefore essential to perform accuracy and reliability checks on the measurement equipment before it is used to ensure the equipment is performing as expected.

Importance of Calibration

Calibration is the process of configuring an instrument to provide a result for a sample within an acceptable range. It involves comparing the measurements of the instrument with a known standard or reference. This ensures that the instrument's readings are accurate and reliable.

Using calibrated equipment is essential to maintain accuracy. Calibration ensures that the tools and instruments used for measurements provide correct readings. Without calibration, even the most precise instruments can give erroneous results, leading to potential issues in the construction process. Calibration involves comparing the measurement device against a known, traceable standard (a standard whose calibration is linked back to national or international standards, like NIST in the United States) and adjusting it or documenting its deviation.

When equipment is calibrated, it means that its measurements are aligned with the standard values. This is crucial because even minor deviations can lead to significant errors in construction projects. For example, if a laser level is not calibrated correctly, it might show a level surface when in reality it's sloped, which will lead to structural issues.

Potential Issues Without Regular Calibration

Without regular calibration, instruments can drift from their accurate settings due to wear and tear, environmental factors, or other effects of time. This drift can cause:

- *Measurement errors*: Inaccurate readings that can affect the quality and safety of the construction
- *Rework*: Incorrect measurements that might necessitate redoing parts of the project, leading to delays and increased costs
- *Safety risks*: Structural integrity that might be compromised and pose safety hazards

Example of Uncalibrated Equipment Used on Construction Site

Imagine a construction site where a theodolite (an instrument for measuring angles) is used to lay out the foundation of a building. If this theodolite is not calibrated, the angles might be slightly off, leading to a foundation that is not perfectly square. This small error can cascade through the project, causing misalignment of walls, floors, other structural elements, and even the skin of the building. The failure to use a properly calibrated theodolite for the layout of these foundations could result in millions of dollars of costly rework and the project falling behind schedule. To avoid such problems, the CQM must ensure that all such measuring tools and devices are properly calibrated prior to their use.

What Does Calibration Entail?

In practice, calibration involves three things:

- *Setting a schedule*: Choose regular intervals for calibration based on the manufacturer's recommendations and the frequency of use. The frequency of calibration interval depends on the following:
  - Manufacturer recommendations
  - Project specifications or contractual requirements
  - Frequency and severity of use
  - Environmental conditions (harsh environments may require more frequent calibration)
  - History of the equipment (if it often drifts out of calibration)
  - After any suspected damage (e.g., being dropped) or repair

- *Using standards and accredited sources*: Use reference materials or standards that are traceable to national or international standards. It is important to use accredited calibration laboratories. ISO/IEC 17025 accredited laboratories provide higher confidence in the traceability and quality of the calibration.
- *Documentation*: Keep records of calibration results to track the performance of the instruments over time. Maintain calibration certificates that clearly show traceability, the date of calibration, the next due date, the standard used, the results (as-found and as-left conditions), and the environmental conditions during calibration. Equipment should be clearly labeled with its calibration status.

Role of the CQM and Measurement Systems

The CQM plays a pivotal role in ensuring that all elements of a measurement system work together seamlessly. The CQM needs to understand the challenges of complex projects and the need for continuous monitoring of the measurement and test equipment (M&TE). To tackle these challenges, continuous monitoring of the measurement system is essential. This involves:

- *Creating a formal system for the M&TE used on the project*: The CQM must ensure that the measurement systems for the contractor firm as well as the subcontractors is organized around a formal system that ensures regular audits to ensure the calibration status of the equipment. By continuously monitoring the M&TE, a CQM can ensure that the project maintains high accuracy and quality, even as it scales in complexity.
- *Ensuring that subcontractors on the project also have an M&TE program:* The CQM must ensure that subcontractors working on the project also have a program for the measurement and test equipment that they utilize. The subcontractors' quality plan must be required to include specific information about their M&TE program. Ensure subcontracts clearly state the subcontractor's responsibility to use only calibrated and compliant equipment and to provide proof of calibration on request. The CQM must also require formal submission of the M&TE list of their equipment with current calibration certificates before their equipment arrives on-site and is allowed to be used. It is a best practice to include the clause allowing the CQM or their delegate to audit and inspect the MT&E records of the subcontractors. The CQM must treat the measuring and test equipment of the subcontractor the same as the main contractor.
- *Receiving inspections:* Verify calibration status (check labels, review certificates) for any new M&TE brought to the site (whether it's the contractor's own or that of subcontractor) before it is released for use.
- *System for handling nonconformances to M&TE*: Enact a system to ensure that the nonconforming equipment is repaired, recalibrated, or replaced, and that proper documentation is obtained before it's

returned to service. Document the entire nonconformance and corrective action process. If the use of nonconforming measuring and test equipment is found through an audit or through alternate sources, the CQM must initiate an impact assessment to determine the potential impact of any work performed that used the nonconforming equipment since its last valid calibration. Follow-up may require retesting or remeasuring previously accepted work.

- *Regular field audits/inspections of the contractor's own M&TE system*: The CQM or delegated quality personnel must conduct regular site inspections specifically targeting M&TE. Ensure that the required field verification checks (e.g., two-peg test for levels) are being performed and documented.
- *Regular audits of subcontractors' M&TE systems*: The CQM must conduct periodic audits on subcontractors' operations, focusing on their adherence to M&TE control procedures outlined in the project quality plan and their own quality plan. The CQM must ensure that M&TE records of the subcontractor are readily retrievable for verification, audits, or investigation if measurement discrepancies arise.
- *M&TE log/register:* Establish a system for collecting, reviewing, and filing all calibration certificates. The CQM must check the physical equipment against the MT&E log/register and verify that the calibration stickers/labels are present, current, and legible. The CQM must randomly request and review calibration certificates. Observe equipment usage to ensure proper handling and application. The M&TE log must include the following:
  - Correct equipment identification (must match the equipment)
  - Calibration dates and due dates
  - Traceability statement (e.g., to NIST)
  - Accreditation of the calibration lab (if required)
  - "As-found" and "as-left" data (which show if adjustments were needed)
  - Indication of pass/fail or compliance status
- *Continuous improvement of the measurement system:* The CQM must initiate a continual improvement program that ensures regular communication with users of the M&TE, communicating the importance of using calibrated equipment and adhering to procedures during project meetings (e.g., at kickoff or weekly quality meetings). All issues, such as any nonconformances and their corrective actions, must be formally reviewed and approved by the CQM.
- *Training:* A continuous improvement program may also include ongoing and as-needed training to the contractors' own staff and relevant subcontractor staff who use critical M&TE. The staff must be trained in the proper use, handling, and storage of the measuring and test equipment.

By implementing these measures systematically, the CQM creates a framework of control, ensuring that all measurements influencing project quality, safety, and compliance—regardless of who performs them—are made with accurate, precise, reliable, and properly certified equipment traceable to relevant standards.

CONCLUSION

A robust measurement system analysis is fundamental to the success of any construction project. By prioritizing accurate equipment, thorough procedures, and comprehensive training, CQMs can minimize errors and ensure compliance with the requirements of a measurement system that manages the measuring and test equipment for the project. Regular calibration, verification, and diligent record-keeping are crucial for maintaining the reliability of the measurement system. As projects become more complex, these practices become even more essential for stakeholder alignment and consistent outcomes. Ultimately, a well-managed measurement system not only avoids costly rework and delays, but also ensures the safety and integrity of the final constructed infrastructure.

QUESTIONS

1. Describe how environmental factors can impact measurement accuracy in construction, and give two examples of mitigation strategies a construction quality manager (CQM) might implement to address these issues.
2. Explain the difference between "repeatability" and "reproducibility" in the context of measurement system analysis (MSA), and why is it important to assess both in a construction setting?
3. Discuss the significance of calibration traceability to national or international standards (like NIST) in ensuring the accuracy of measurement equipment on a construction site.
4. Beyond just providing calibration certificates, what other proactive steps should a CQM take to manage subcontractor-owned measurement and testing equipment (M&TE) effectively?
5. How does a comprehensive M&TE register contribute to preventing measurement errors and ensuring quality in complex construction projects with multiple stakeholders?

QUALITY MANAGEMENT TOOLS AND INFRASTRUCTURE

> Identify and assess quality tools (e.g., data analysis tools, statistical process control, dashboards, construction pre-control checklists, and budgets) and quality infrastructure essential for effective quality management. (Evaluate)
>
> Body of Knowledge II.C.6

Effective quality management in design and construction relies on a combination of robust tools and established infrastructure to facilitate their use. Effective integration of these tools and their infrastructure ensures project teams meet their quality objectives, deliver on their client's intent, comply with regulatory requirements, and improve productivity.

Quality Tools and Phases

Effective quality management in construction projects requires the strategic use of quality tools and infrastructure across all project phases. As discussed earlier in this chapter in the section "Project Quality Strategy Development," there are general phases for every design and construction project:

1. Precontract, predesign, and programming phase
2. Planning phase
3. Design and preconstruction phase
4. Procurement phase
5. Construction and construction administration phase
6. Commissioning and transition to owner operation
7. Postconstruction, operations, and maintenance phase

Within each phase, there are multiple tools available in the industry; see Table 2.11. (Note that this list of tools is not all-inclusive.) The CQM who may be a quality champion must evaluate which tools will ensure the specific project's quality objectives are met and that the technological infrastructure is in place to facilitate their use. (Though not shown, the incorporation of lessons learned into each tool of every phase is essential.)

The project phases will have tasks that fall under either QA or QC, which form the backbone of quality management. (The differences between QA and QC have already been described in this chapter.) Robust QA and QC systems integrate tools like checklists, testing equipment, and software, supported by well-trained personnel. Evaluating these systems requires analyzing their efficiency, accuracy, and adaptability to project demands.

This flexible framework ensures quality management at every phase of the construction life cycle, fostering project success and long-term stakeholder satisfaction.

Data Analysis Tools

Data analysis tools provide the foundation for monitoring and improving construction quality. Tools such as spreadsheets, business intelligence, building information software programs, and project management software programs enable project teams to collect, organize, and analyze vast amounts of data. These tools can help teams identify trends and areas of concern, and evaluate project performance against established metrics. For instance, tracking defect rates or material compliance over time enables proactive interventions, reducing the probability of

Table 2.11 Example quality tools by phase.

Phase	Quality Tools
Predesign and programming	• Data analysis tools
	• Quality assurance tools
Design and preconstruction	• Risk management tools (e.g., FMEA)
	• Statistical process control (SPC)
	• DFX (design for excellence; may include design for constructability, maintainability, serviceability, experiments, etc.)
	• Dashboards
	• Budgets
Procurement	• Supplier evaluation templates
	• Performance dashboards
	• Quality assurance tools
Construction and administration	• Construction quality management software
	• Statistical process control (SPC)
	• Inspection and test plans (ITPs)
	• Key performance indicators (KPIs)
Transition to owner operation	• Pre-punch/punch/completion lists
	• System testing/commissioning
Postconstruction, operations, maintenance	• Maintenance management systems, condition monitoring tools, feedback mechanisms

rework. Evaluating tools involves examining their usability, scalability, costs, and compatibility with existing systems, ensuring they support data-driven decision-making effectively.

Each organization involved in a project has its own set of tools, and the project itself may have specific tools noted in the contract. The quality champion's role is to understand the context of each organization and their available tools, along with the project's quality objectives, and evaluate which tools will ensure achievement of those goals. (See Table 2.13 for the definition of an effective quality tool.)

Statistical Process Control (SPC)

Statistical process control (SPC) uses statistical methods to monitor and control construction processes. Control charts, for example, visually represent process variations, distinguishing between natural process variations and nonrandom variations (issues) that require corrective action. By identifying anomalies early, SPC helps maintain process stability and ensures consistent quality. When evaluating SPC, critical factors include its ability to adapt to project-specific requirements and the team's proficiency in interpreting statistical outputs. Training workers and supervisors to use SPC effectively is an integral part of embedding this tool into quality management practices.

Dashboards

Dashboards offer a centralized, real-time overview of quality metrics, progress, and compliance. These visual interfaces consolidate data from multiple sources, such as site inspections, material tests, and safety audits. A well-designed dashboard allows stakeholders to assess project health at a glance, enabling quick corrective actions. Evaluating dashboards requires analyzing their functionality, customization options, and ability to integrate with other systems. High-performing dashboards provide actionable insights while minimizing information overload.

Dashboards facilitate SPC evaluation. Triggers can be implemented into the dashboard software that act as upper and lower control limits, signaling when a team should take corrective action. SPC can be conducted at both the organizational level and the project level. Lessons learned from other projects delivered by each organization will help identify upper and lower control limits for the project's KPIs. Leading and lagging indicators must also be identified so the team can be proactive in preventing significant quality issues.

For example, a potential metric could be the number of defects. A general contractor may understand that, for healthcare projects under 100,000 square feet, no project in their history of delivering similar projects exceeded 25 defects in any given week. If there is a sudden increase in defects—or more than 25 defects are identified—it may signify significant problems that warrant immediate attention. A sudden increase in defects, even if still below 25, may be a trigger for the CQM to act; the project team should not wait until the number of defects exceeds 25.

As with all metrics, context is essential. In this example, a sudden increase in defects may mean either:

- The project team members are doing their jobs—they are conducting and documenting robust inspections.
- The number of defects truly is much higher than on prior projects, and further investigation is necessary to identify the cause(s) of the increase.

While the goal for the number of defects should be zero, a low number of defects could also be both *not* a cause for concern *and* a cause for concern. Either the actual number of defects is indeed low or the project team members are not out conducting their QC tasks, so they are not identifying work being installed incorrectly.

The CQM must be able to understand from the data what is happening. These situations are very contextual, and critical evaluation is needed to fully understand and take action.

Budgeting Tools

Quality in construction is intrinsically linked to financial planning and control. Each organization has its own means for establishing budgets and controlling them. Budget management tools are part of the financial management systems of the organization. They enable accurate cost estimation, tracking, and reporting.

These tools help ensure resources are allocated efficiently, minimizing waste and ensuring that quality objectives are met within budget constraints.

Evaluate budget tools by looking at their accuracy, transparency, and integration with other QMSs—as well as other project plans. Tools that provide real-time updates and predictive insights are particularly valuable for maintaining financial and quality performance.

Construction Checklists

Construction pre-control checklists are proactive tools that ensure readiness before mobilization. They are used during the early stages of a construction project to ensure all necessary conditions, materials, and processes meet the required specifications and standards before work begins. These checklists help prevent errors (error-proofing, e.g., *poka yoke*), rework, and delays by verifying the foundational elements of the project are in place and compliant with regulatory and project-specific requirements.

These checklists cover critical areas such as material quality, site preparation, worker certification, and equipment functionality. By systematically addressing potential risks, these tools help prevent defects and delays. The effectiveness of pre-control checklists depends on their comprehensiveness, ease of use, and alignment with project-specific requirements. Regularly updating and customizing checklists ensures they remain relevant and effective across diverse projects.

While the term used here is "construction pre-control checklist," there are many tools used in the industry to facilitate the same intent under a different name. Again, the CQM must evaluate which tools they have at their disposal within the bounds of the project's specifics, and put them into action. For example, there are many tools that could satisfy a construction pre-control checklist. A preconstruction meeting agenda with a subcontractor is a checklist for a conversation that goes over the critical items necessary to plan prior to mobilization and meets the intent of what is being discussed in this section. Table 2.12 illustrates the essentials of any tool used to meet the intent of a construction pre-control checklist.

Table 2.12 is not all-inclusive; the general idea is to review and confirm the project's preparedness to deliver the project prior to mobilization.

Quality Infrastructure: Standards

Quality infrastructure provides the framework necessary to support and sustain the use of the project's quality tools. It encompasses the organizational structures, processes, and resources essential for implementing and maintaining effective quality management systems.

Not only must the quality tools meet the requirements noted in Table 2.13, but they must also fit within the processes being used on the project. Establishing and adhering to standardized processes, such as ISO 9001 quality management system requirements' international standard, ensures consistency and reliability across projects.

These standards provide a benchmark for evaluating performance and identifying areas for improvement. Organizations must regularly review and update these processes to reflect industry advancements and project-specific needs.

Table 2.12 Construction pre-control checklist essentials and intent.

Element	Intent
Verification of materials and equipment	• Confirm that all materials meet specified standards (e.g., strength, grade, type).
	• Check that equipment is operational and suitable for the intended tasks.
Review of plans and documents	• Ensure design drawings and technical specifications are up-to-date and available.
	• Confirm that permits, licenses, and approvals are in place.
Site readiness	• Verify site conditions, including soil preparation, grading, and access.
	• Check for proper marking of utilities, boundaries, and work zones.
Worker readiness	• Ensure workers are trained, certified, and briefed on safety protocols.
	• Confirm that personal protective equipment (PPE) is available and adequate.
Environmental and safety compliance	• Check for compliance with environmental and safety regulations (e.g., erosion control, waste management).
	• Verify that safety measures (e.g., signage, barriers) are in place.
Process and methodology validation	• Confirm that construction methods align with the project's quality plan.
	• Review and approve any mock-ups or sample tests.
	• Review the project's ITP and confirm understanding and capability to complete the associated tasks.

Elements of an Effective Quality Management Tool

As previously discussed, the CQM must evaluate which tools will be most effective for a particular project. Table 2.13 discusses the attributes of an effective tool.

As with all tools, a project team can get lost in analysis, forever digging deeper into the data and letting the use of the tool take away from the presence on the jobsite. Tools are essential for management of the design and construction processes—and for the achievement of a project's quality objectives. The project team must have self-awareness that the team does not rely solely on the use of dashboards, for example, to manage and understand what is actually happening on the project. The context of the actual construction is also essential to effective use of any tool.

Table 2.13 Elements of an effective quality tool.

Attribute	Meaning
Accessible	The tool and its data must be accessible to all project team members. For example, if a tool facilitates the use of inspections, those inspection templates must be available not only to the general contractor, but also to the subcontractors. It must also be accessible to other external stakeholders such as design, client, third party, and AHJ team members. Accessible also means that the tool can be accessed anywhere physically. For example, it should be not just a desktop application or website, but also an application on a mobile device such as a tablet or smartphone for use out on the project site.
Ease of Use	The tool must be easy to use. If project managers can use it easily but a superintendent cannot, it is not an effective tool.
One Platform	There is no perfect software tool. However, the quality manager should strive for the fewest programs possible. A project team member should not have to navigate between many applications to complete their quality tasks. The goal should be one software program that facilitates the proper use of all tools and tasks necessary for proper quality assurance and control (this is not always possible).
Simple	The tool must be simple to use, not require excessive clicks or cumbersome navigation to find information. Time is a real constraint on many projects, and quality tools must not be an administrative burden.
Training	Skilled personnel are a cornerstone of quality infrastructure. Ongoing training ensures workers and supervisors understand and can utilize quality tools effectively. Evaluating training programs involves measuring their relevance, frequency, and impact on worker performance. The project team members must also know *when* to use a specific tool.
Customization	The tool must include templates with lessons learned incorporated, with the ability to customize those templates with the specifics of the project.

Also, given that many of these tools facilitate data analysis, there is a significant amount of data that can be analyzed. A knowledgeable CQM will support the project team in their efforts to identify critical data to track that will lead to improvement. A strong CQM will also be able to focus on the important data, ignoring the irrelevant "noise" that will not result in improvements.

If they are not easy to use or not accessible to everyone on the team, the tools will simply not be used, and the project will not reap the results that such tools are designed to bring.

CONCLUSION

Quality management in construction depends on a synergistic relationship between advanced tools and strong infrastructure. Tools such as data analysis platforms, SPC, dashboards, and checklists provide the means to monitor and control quality effectively. Simultaneously, quality infrastructure ensures these tools are utilized to their full potential. By continually evaluating and refining both, construction professionals can deliver high-quality projects that meet or exceed expectations while optimizing resources and maintaining compliance.

QUESTIONS

1. What are three elements of an effective quality tool?
2. What are two actions the quality champion must do when it comes to evaluating what tools to use on a particular project?
3. Give an example of a project metric that could be used in SPC. Identify an associated upper and lower control limit that would trigger corrective action.
4. What is the difference between "quality tools" and "infrastructure"?

Quality Audits

AUDIT TYPES

> Identify the appropriate type of audit (e.g., first-party, second-party, and third-party) for different situations, such as personnel independence from the area being audited. (Apply)
>
> Body of Knowledge II.D.1

Audits are essential for ensuring quality, compliance, and efficiency across all project phases. They serve as crucial quality assurance tools, identifying gaps in training, processes, or experience while driving continuous improvement. Moving beyond a perception of audits as merely administrative burdens, this section explores their positive intent, emphasizing their role in fostering a culture of self-awareness. By understanding the various types, classifications, and methods of audits, organizations can proactively manage risks, ensure adherence to standards, and achieve successful project outcomes through the plan-do-check-act (PDCA) cycle.

All audits are intended to ensure planned activities are executed as prescribed, confirming that the quality system is functioning correctly. Audits are also useful QA tools to identify gaps in training, experience, or process and request support by management to promote continuous process improvement and fulfilment of the PDCA cycle. Though audits often have a negative connotation surrounding them, as they are often considered administratively difficult, the ideas presented here can help promote culture change and revitalize the positive intent of the audit with industry self-awareness.

Audits are also performed for all phases of the project life cycle. Within each audit, there is a defined relationship between those conducting the audit (the auditor) and those being audited (the auditee). Typically, the quality manager facilitates this entire process, while the auditor performs the audit efforts. The quality manager can help manage the relationship and expectations between the audit team and the auditees, whether at the organization or project level.

Audit Classifications

There are two main audit classifications: *internal* and *external*. Both are essential for maintaining quality, compliance, and efficiency. Internal audits, or first-party audits, are conducted by an organization's own personnel or hired consultants to assess adherence to internal policies, procedures, and industry standards. These audits focus on process improvements, risk management, and ensuring regulatory compliance before external scrutiny.

In contrast, external audits are performed by independent third-party entities—such as regulatory agencies, clients, or certification bodies—to verify compliance with contractual obligations, building codes, or certification standards. External audits can be second-party audits or third-party audits. If the audit is performed by a customer, it is considered a second-party audit; if it is performed by an entity that is not related to the auditee organization, it is called a third-party audit.

While internal audits help identify and correct issues proactively, external audits provide an objective assessment, often influencing project approvals, certifications, and overall credibility.

Table 2.14 Quality audit classifications.

Classification	Sub-classification	Intent	Conducted By
Internal	Internal (first-party) audit	An organization's assessment of its own processes, either voluntary for its own purposes or imposed by an external client/party	The organization's own staff, independent of the activities being audited
External	Customer (second-party/ supplier) audit	An organization's or client's assessment of its supplier's processes	A customer or contracted party on the customer's behalf
	Certifier (third-party) audit	An independent third-party assessment of an organization's processes or those of a supplier	An independent third party separate from a client

Audit Types

The various audit types include *system*, *process*, and *product* audits. Audit types determine the level at which the audit may be performed. System audits evaluate the quality system at the corporate or organizational level, whereas process audits focus on one process within the system—typically how the process is implemented at the project level. Product audits focus on the ability for specific products to meet the needs of the project as part of the supply chain. Table 2.14 illustrates a summary of audit classifications and types.

Product, Process, and System Audits

In any phase of the project life cycle, audits can be conducted on the organization's quality system and its holistic performance and capability, and also on the project level, where the system's process is directly applied and implemented.

Audits are conducted at different levels to ensure compliance, efficiency, and quality. As mentioned, these audits can be broadly categorized into product audits, process audits, and system audits. While product audits focus on verifying the final output of a specific phase, process audits evaluate the effectiveness of individual workflows, and system audits assess the overall framework governing quality and compliance.

- *Product audit*: A product audit evaluates the final deliverable of a particular phase to ensure it meets predefined specifications, standards, and client requirements. For example, in the design phase, the primary product is the design itself, which includes drawings, specifications, and engineering calculations. Auditing the design ensures that it complies with regulatory codes, project requirements, and industry best practices. This may involve checking for accuracy, completeness, constructability, and coordination across disciplines. A product audit at this stage helps identify errors or inconsistencies before construction begins, reducing costly rework and delays.
- *Process audit*: A process audit assesses whether a specific function within a project is being executed effectively and in accordance with established procedures. For example, in the construction phase, procurement and document control are critical processes that significantly impact project success. Auditing the procurement function ensures that materials and services are sourced in compliance with contractual obligations, supplier qualifications, and cost controls. Similarly, auditing the document control function verifies whether project documentation—such as drawings, contracts, and changes—is managed systematically, ensuring proper version control and accessibility. By auditing these processes, organizations can identify inefficiencies, bottlenecks, or noncompliance with internal procedures and industry standards.
- *System audit*: A system audit evaluates the effectiveness of the overall QMS in ensuring project success. Unlike process audits, which focus on individual functions, a system audit examines all interconnected elements of a quality system to determine if they collectively support compliance with an established standard, such as ISO 9001 or a client's quality requirements. This type of audit assesses policies, procedures, personnel responsibilities, training programs, and continuous improvement mechanisms. By conducting a system audit, organizations can determine whether their quality management framework is robust, consistent, and capable of meeting project objectives.

While product, process, and system audits serve different functions, they are all essential in ensuring a project meets its design intent, regulatory requirements, and quality expectations. A product audit verifies that deliverables meet specifications, a process audit evaluates whether key functions are operating efficiently, and a system audit ensures the overarching quality management framework is effective. Together, these audits provide a comprehensive approach to quality assurance, reducing risks and improving project outcomes.

Audit Methods

Combined or joint audits may be used individually or in combination with one or multiple auditors. *Combined* audits allow multiple systems to be assessed during the same audit effort, whereas *joint* audits are those audits that are performed by multiple auditing entities working as one audit team.

Audit methods include observation of activities, interviews of auditees, sampling of QC records, and in-person or remote audits.

- *Observation of activities*: This method of auditing involves directly watching work as it is performed to verify compliance with established procedures and quality standards. Observers assess whether tasks are executed correctly, following safety guidelines, and meeting project specifications. This real-time approach helps identify deviations, inefficiencies, or areas for improvement.
- *Interviews of auditees*: In this method, auditors conduct structured or informal interviews with personnel involved in the process to gain insights into their understanding of procedures, compliance with protocols, and challenges they face. These interviews help assess whether training has been effective, uncover potential gaps in knowledge, and identify inconsistencies between documented processes and actual practices.
- *Sampling quality control records*: This method involves reviewing a representative selection of QC records, such as inspection reports, test results, and nonconformance logs. By analyzing these documents, auditors can assess whether QC measures are properly documented, consistently applied, and aligned with project requirements. Sampling provides an efficient way to identify systemic issues without reviewing every record.
- *In-person/remote audits*: Another method in which audits differ is the way they are conducted; audits can be performed either on-site, where auditors physically inspect work and interact with personnel, or remotely, using digital records, video conferencing, and real-time data collection tools. In-person audits allow for direct verification of conditions, while remote audits provide flexibility and efficiency, particularly for geographically dispersed projects or when access is restricted.

  The rise of international and distributed organizations, coupled with the increase in remote work following the pandemic, has led to a widespread adoption of virtual auditing. This shift is driven by both the dispersed nature of operations and organizational cost-saving measures. To mitigate the travel and other related costs, organizations are now opting for more remote audits whenever feasible or limiting the duration of on-site audits.

Audit Techniques

Audit techniques include auditing by functional elements, auditing by technical requirements, following the root cause, and forward or backward tracing.

- *Auditing by functional elements*: This technique evaluates a process by breaking it down into its core functional elements, such as planning, execution, monitoring, and reporting. Auditors assess whether each element aligns with established standards, regulatory requirements, or company policies. This structured approach is particularly useful for verifying compliance with ISO standards, industry best practices, or contractual obligations. By analyzing functional elements independently, auditors can identify weaknesses in specific areas while ensuring the overall process remains effective and compliant. This technique is effective for auditing compliance of a process against a standard.
- *Auditing by technical requirements*: This method focuses on assessing whether a product meets its defined technical specifications, industry standards, or contractual requirements. Auditors examine design documents, test reports, material certifications, and performance data to verify compliance. This approach is especially valuable in industries where precision and reliability are critical. By auditing against technical criteria, organizations can ensure that products meet quality expectations and regulatory mandates before reaching end users. This technique is effective for auditing compliance of a product against requirements.
- *Following the root cause*: This investigative approach is used when a quality issue, defect, or failure has been identified. Auditors trace the issue back through the production, testing, and supply chain processes to determine its root cause. This may involve reviewing inspection records, interviewing personnel, and analyzing past performance data to uncover lapses in QC or procedural noncompliance. By following the incident trail, organizations can identify systemic weaknesses, implement corrective actions, and prevent recurrence. This technique is effective for auditing product quality.
- *Forward or backward tracing*: This technique involves tracing a process either forward, from raw materials to the finished product, or backward, from the final output back to its origins. Forward tracing ensures that all required steps are correctly followed in sequence, while backward tracing helps identify the source of defects or inefficiencies. This method is particularly useful in supply chain audits, manufacturing process validation, and compliance verification, as it provides a clear picture of how inputs and workflows contribute to the outcome. This technique is effective for process audits.

Determining Which Audit Type and Method to Use

The CQM identifies the needs for an audit and ensures the requirements of the audit effort are fulfilled. Knowing how to recognize proactively when an audit is required on a project or system is a honed skill. Audits must be proactive to ensure streamlined process utilization occurs at the project level, internally and with external project stakeholders.

- When an organization is onboarding a new supplier who may manufacture a product for a project and that product has not been tested in the market, an external, second-party audit may be required before that product is included in the specification.
- A client may audit a general contractor as part of the bidding process to gain confidence in their ability to perform.
- A contractor may also seek a third-party audit for certification purposes against a known standard if a client or project requires it, or if it's needed for a specific product or market.

Regardless of the audit type and need, each audit must be carried out according to a specified plan and procedure on an established interval at the organizational level. Higher-risk elements of an organization's quality program may receive more audit attention than others if they are more effective in identifying risks and associated deficiencies in project execution.

Different combinations of methods and types may be required to achieve the audit goal. These tools are outlined for use by the CQM to ensure processes and systems perform as designed and that process users are trained and understand how to utilize the process on their projects.

Audit Approaches: Compliance Versus Performance

Two approaches can be leveraged within each of the three primary audit types: compliance and performance. *Compliance* is more traditional and focused on conformance to standards, whereas *performance* dives deeper, using interview data from process users.

Compliance is indeed the primary intent of the audit, but in this sense, traditionally such audits are binary in nature: Such audits are most suitable when performing audits of products and for QC purposes. As a result of such auditing, the result is noted as either a pass or a fail. Compliance audits ensure the product meets the minimum acceptance criteria. (See Chapter 5, "Acceptance Criteria Definition and Implementation," for more information.)

Performance audits, on the other hand, review the product or process from a best practices perspective. The objective of performance audits is to measure efficiency, and the goal is not compliance only, but to exceed the minimum requirements. Such audits are therefore most suitable when processes are to be audited.

CONCLUSION

Audits play a crucial role in maintaining quality, compliance, and efficiency. Product audits ensure that deliverables, such as design documents, meet established requirements before construction begins. Process audits verify that critical functions, like procurement and document control, operate smoothly and adhere to best practices. System audits take a broader view, assessing the entire quality management framework to confirm it supports consistent project success. By implementing all three types of audits, organizations can proactively identify risks, enhance operational efficiency, and uphold the highest standards of quality throughout a project's life cycle.

Audits stand as critical pillars of QA, driving compliance and efficiency throughout project life cycles. Moving beyond mere administrative tasks, audits serve as invaluable tools for identifying gaps in training, processes, and experience, thereby fostering continuous improvement. This section has described the various types, classifications, and methodologies of audits, emphasizing their proactive role in managing risks and ensuring adherence to standards. By understanding audits' positive intent and application, organizations can cultivate a culture of self-awareness and effectively leverage the PDCA cycle to achieve successful project outcomes.

QUESTIONS

1. What various approaches can be taken when conducting an audit?
2. Give three examples where an audit may be required.
3. What is the difference between internal and external audits?
4. What are the tangible benefits of an audit?
5. What is the primary role of the quality champion regarding audits?
6. Explain the difference between auditing by functional elements and auditing by technical requirements, and give an example of when each method would be most appropriate.
7. Describe a scenario where a joint audit would be preferable to a combined audit. In your description, detail the roles of the different parties involved in each scenario.
8. Contrast the traditional compliance approach to audits with the performance approach, and discuss how adopting a performance-based audit strategy might benefit an organization in the long term.
9. Using examples from different phases of a project life cycle, illustrate how product, process, and system audits interrelate to ensure overall quality management.
10. A company is onboarding a new, untested supplier for a critical component. Detail the specific steps the quality manager should take to determine the necessary type(s) of audit, methods to be used, and the primary objectives of the audit(s).

QUALITY AUDIT PLAN

> Create and evaluate a quality audit plan that aligns with strategic objectives, key project controls, prime contractor audits, and compliance checks for areas at risk using specific, measurable, attainable, relevant, and time-bound tactics. (Create)
>
> Body of Knowledge II.D.2

A quality audit plan is a structured framework designed to assess compliance with project requirements, industry standards, and regulatory obligations. It serves as a proactive quality assurance tool to identify risks, verify adherence to key project controls, and ensure that construction activities align with strategic objectives.

An effective quality audit plan incorporates prime contractor audits, compliance checks, and targeted evaluations of high-risk areas. By using specific, measurable, attainable, relevant, and time-bound (SMART) tactics, the audit plan provides a systematic approach to monitoring quality performance, addressing deficiencies, and driving continuous improvement throughout the project life cycle.

Establishing the Audit Plan and Intent

The CQM is responsible for planning the audits. This includes identifying the need; establishing what system, process, or project will be audited; and how the necessary improvements be acted upon after the audit is completed. An example would be conducting an internal audit to self-assess that client intent is being delivered as a result of the organization's quality program implemented on each of its projects.

Audits are essential for maintaining factual information about the quality process. By virtue of their independence from process users, auditors can provide trusted information to project and organizational management and leadership to improve the quality systems, processes, and tools. (See Table 2.14 in the previous section for a summary of audit classifications and types.)

Prerequisites to the Audit

When conducting an audit, first be sure the system or process owner and associated users and stakeholders are made aware of it with substantial time available to plan. The entire team must understand the audit and what it is intended to accomplish. The CQM should be prepared to manage the potential emotional concerns regarding the audit, especially if it is internal. Many teams may feel worried that when they are under audit, their performance is being questioned—when in fact, the audit is a positive opportunity for improvement. Auditors and those involved in the project being audited should discuss these concerns for a more transparent process. Thus, it is the CQM's role to establish positive intent surrounding the audit proactively—before the audit planning begins. This also requires an established culture of improvement for the entire organization, as culture filters down to the individual projects. A suggested list of objectives and tactics is shown in Table 2.15.

While the CQM guides the audit process, the auditor and the rest of the team conducting the audit must be trained and qualified accordingly. They must also be made aware of the context of the organization, the culture, business strategy, and the *why* behind the audit, as well as the various sensitivities surrounding the people and process specifics.

A lot of information is reviewed during an audit. The audit plan should identify what information can be reviewed beforehand and what will be reviewed

Table 2.15 Audit prerequisites.

Objective/Step	Tactics
Awareness	• Send all relevant audit information to associated users and stakeholders.
	• Confirm what preliminary information is needed, what will be reviewed remotely and on-site.
	• Establish positive intent.
	• Be sensitive to the fact that a project is underway or that there are other initiatives and changes occurring in the organization. Change is indeed a constant.
Kickoff Meeting	• Kickoff meeting is scheduled with all stakeholders.
	See "The Quality Audit Handbook" for a sample agenda of an audit.
Closing Briefing and Reporting	• What did the audit reveal?
	• What improvements to the product or process are required?
	• What improvements can be made to increase the effectiveness of the audit itself?

during the audit or site/office visit. In the initial awareness communications leading up to the audit kickoff meeting and the audit itself, the CQM should discuss which methods will be used to review information and confirm the system is operating as designed, and share which tactics the stakeholders can leverage to make the audit process as effective and streamlined as possible. (Note the importance of respecting the confidentiality of shared documents.)

The audit plan should also identify if this audit is the first one of the system or if it is part of an ongoing improvement effort (such as completing an audit of the entire system annually, or of a specific process within the system each quarter). For a longer project, an audit could occur more than once.

The final prerequisite to the audit that should be outlined in the audit plan is clearly defined systems and processes. To identify areas where a process may not be performing as designed, processes must be clearly outlined and defined. This helps measure performance as well as the success of an audit in identifying conformance.

Much of this detail is covered in the audit kickoff meeting, which is why that meeting is perhaps the most important part of the process, along with quick reporting after the audit is completed.

Elements of an Audit Plan

There are several essential elements for a successful audit plan, which are outlined here.

Subject and Focus

The audit plan should specify what, in particular, is being audited. Audit plan should cross reference a standard, e.g., internal procedure or external standard, so it can be conducted objectively. Like anything on the project, not all activities should be given the same level of importance. The CQM must identify which tasks are more critical and require more attention than others, given time and attention are limited resources. This guidance by the CQM will help streamline the audit and focus on the most impactful activities. Some activities may need to be audited more often than others because they have a greater influence on the project or process quality.

Roles and Responsibilities

At the core of the quality plan—and the audit plan—is the RACI (roles and responsibilities) matrix, which outlines what each person on the team is responsible for, how they contribute to quality, and how they can help support and streamline the audit effort. (See "Roles of Quality Personnel and Internal Stakeholders" in this chapter for more information.)

Level of the Audit

There are two levels of audits: the *high level*, which assesses the performance of the organization's corporate systems; and the more on-the-ground *project level*, which assesses the performance as implemented on the project.

Given the varying length of construction projects, when auditing the process at the project level, QA and QC activities are executed only during specific phases of the project. Given that each project has different staff, materials, trade partners, design teams, clients, weather, and other elements beyond the control of the project team, the audit may be unique and not repeatable.

Scalability

The audit plan needs to scale and adapt to company policy, procedures, and the specifics of the project and its requirements.

Measure of Success

The audit plan needs to outline what a successful audit looks like. These should be tied to the leading/lagging indicators for successful projects and aligned with the organization's business plan objectives.

Identifying many issues with the process does not mean the audit was successful or unsuccessful. As the primary intent of any audit is to establish which procedures are being adhered to and which are not, the audit reveals the reasoning *why* certain procedures are done completely, partially, or not at all. This is why diving deeper is essential. If an audit only looks at the surface level—at *compliance* only—you will fail to recognize that perhaps the procedure is insufficient for the task at hand. When procedures are not executed fully, most often it is because they are too time-consuming or ineffective for the needs of the client or project, or completely unnecessary. These gaps provide opportunities for improving and streamlining the process and must be acted on and resolved immediately. See Table 2.16 for a sample audit plan.

Post-Audit Activities

After an audit is completed, the CQM must focus on communicating the results, addressing the findings, and implementing noted improvements. Also, the

Table 2.16 Sample audit plan.

What will be assessed?	How will it be assessed?	Against what standards?
Project/corporate staffing structure		
Quality assurance procedures		
Quality control procedures		
Supply chain activities		
Document effectiveness		
Task/time duration		
Training/staff understanding of process		

auditor(s) should self-evaluate their own performance to determine what was not done effectively. This process ensures that both the auditor and the organization learn from the audit to strengthen internal controls and operations for the future. The learned information should be shared and leveraged against other processes in the organization or other similar projects. Improvements can be made outside the audit environment, where they are often better received and more effectively implemented.

Part of the negative aura of auditing originates from:

- late reporting of the audit results, and
- results that were not discussed with the team during the audit.

Each recorded result should be shared with the auditees during the discussion—not after. The interview questions and audit process can be formulated to accommodate this, increasing transparency and solidifying the positive intent behind the audit, and increasing the potential for corrective actions and improvements to be made.

A Modern Perspective on Auditing

A project team's, supplier's, or organization's response or reaction to an audit is dependent on how they are informed. Many believe they are under inspection or surveillance when being audited. This feeling on the part of the project team is directly correlated to the introduction and overall handling of the audit by the lead auditor under guidance from the CQM. The surrounding culture and atmosphere of the audit depends on the positive and appropriate approach and communication style of those leading the audit, and thus its success in identifying areas for improvement. (Note that the audit itself is different from inspection and surveillance.)

Compliance

A team or organization that is being audited will be hesitant to provide honest feedback if their perspective is that of "being graded" or "evaluated for compliance."

One measure of any quality program should be its effectiveness, which includes the ability for the implemented process to save a team or organization administrative overhead while still meeting stated contractual requirements and the client's intent.

This is all based on the attitude and approach of the auditor, which stems from the CQM guiding the effort. In a typical project, the team is incredibly sensitive to their time and the outside perspective, which is why while compliance is one of the primary reasons for the audit, it cannot take the approach to a fault. Like all things, a balance is essential.

Time Requirements

Many teams already feel the burden of heavy job requirements workload on them during project execution. To also accommodate an audit—even though required—will immediately put them on the defense, impacting the audit results.

The audit leader must minimize the administrative impact on the project team. This means *not* scheduling an audit during closeout or punch list, and being sensitive to the project's schedule. This still applies at the organizational level. For example, scheduling an audit during business strategy planning season is not effective. At this time, typical business leaders or team members are likely preoccupied and may lack the time to participate in an audit interview, engage in a study, or provide candid feedback due to their need to focus on other strategic priorities.

Listening is key—not always searching for compliance for compliance's sake. Some teams may have a legitimate reason why compliance is not possible. Some teams may have a legitimate reason why they are unable to fulfill compliance to project requirements. The project requirements specification may not be applicable to the project. It may have been incorrectly cited. This is an example of a system failure. People following a system are not the cause. Internal audits can make project teams nervous, fearing that their individual performance is being assessed rather than the effectiveness of their processes. Corporate leaders can proactively address these concerns by noting the intent of the audit ahead of time, addressing the sensitivities of each team member.

Focusing on those specific failures of the process is also not the intent, though unconsciously they may be where the team put most of its attention. The audit's main purpose is to verify what is going well, rather than identifying noncompliance.[8]

Transparency

There should be no hidden agenda. Checklists used as part of the effort should be shared prior to the audit and discussed during the audit kickoff meeting. No one involved in the audit should be surprised. (A surprised reaction is a leading indicator of poor communication.)

Many view the quality system as an administrative burden, yet the audit—when executed correctly and with positive intent and total transparency—is the primary tool to expose the system's shortcomings. H. W. Chung noted that quality audits lead to continuous improvement of the corporate quality system and project quality process. Sometimes, a process or task within it may not be capable of delivering what is intended. Objectives beyond the capacity of a process are unattainable. These types of problems are not identified without an audit.

Audits also must be objective against a standard. The standards against which the program will be assessed must be shared and reviewed during the audit kickoff discussion. The checklists being used to validate the process against those standards must also be shared—like the review of inspection checklists prior to the start of the work during quality assurance to ensure those performing the work understand the client's intent.

CONCLUSION

Effective quality auditing is not merely a compliance exercise; it is a strategic imperative for project success. By establishing clear plans, fostering transparency, and maintaining a modern perspective, audits become powerful tools for continuous improvement. The CQM's role is pivotal in ensuring that audits are conducted with positive intent, addressing concerns, and leveraging findings to enhance processes. Embracing these principles transforms audits from potential stressors into catalysts for growth, ensuring projects meet objectives and exceed expectations.

QUESTIONS

1. What is the role of the CQM in establishing the intent of an audit, and why is this important before the audit planning begins?
2. Explain the differences between the three audit classifications (internal, external, certifier) outlined in Table 2.15.
3. Describe why the audit kickoff meeting is considered a crucial part of the audit process, according to the document.
4. How should an audit plan address the potential emotional concerns of the team being audited, particularly in internal audits?
5. According to the document, what are some key considerations beyond mere "compliance" that should be part of a modern perspective on conducting quality audits?

AUDIT PERFORMANCE

> Conduct a quality audit to evaluate the management system's effectiveness, review adherence to standards, and identify nonconformances through detailed observations. (Evaluate)
>
> Body of Knowledge II.D.3

The audit process is not merely the execution of an audit. It is an end-to-end audit life cycle from planning of an audit all the way through to learning dos and don'ts from audit execution (see Figure 2.8). ISO 9000 defines audits as "systematic, independent and documented process for obtaining objective evidence and evaluating it objectively to determine the extent to which the audit criteria are fulfilled."

Audit Planning

A federal organization like the U.S. Food & Drug Administration (FDA) or Internal Revenue Service (IRS) has the authority to perform surprise audits. But audits of quality systems should not be a surprise. Audits must be planned in detail to reduce disruption to ongoing day-to-day operations and make the most of it.

In many organizations, audit planning for an organization typically starts in the fourth quarter (Q4) of the current year to plan for the upcoming year. The audit plan is a description of the activities and arrangements for an audit. Planning could even start as early as the third quarter in situations where audit logistics are complex. Annual planning covers all types of audits:

1. *First-party audit*: Internal audits (system, process for conformance, compliance)
2. *Second-party audit*: Customer audits, supplier audits (award contract, qualification, maintenance)
3. *Third-party audit*: Independent registration body, for example, ISO 9001, ISO 14001, and so on (registration, surveillance)

A *system audit* is a documented process of evaluating a management system. It involves examining objective evidence to verify that the system's elements are suitable, effective, and have been developed, documented, and implemented in line with specified requirements.

Process audits ensure that activities operate within established parameters. By examining inputs, actions, and outputs, these audits verify alignment with defined requirements. Typically focused on a single process, such as concrete work, welding, or installation, process audits offer targeted assessments that differs from broader system audits. Consequently, they are generally less time-consuming than system-wide evaluations. A process audit is verification by evaluation of an operation or method against predetermined instructions or standards to measure conformance to these standards and the effectiveness of the instructions.

An *annual audit* includes the nature and type of audit, timing, and number of audit person days. For internal audits, auditors are also assigned within the organization who are trained to conduct audits. First-party audit covers a third of organizational processes each year, a third in the following year, and the final third in the third year. It is not efficient to try to cover all organizational processes every year. For supplier audits, suppliers that are critical to the organization's operation and quality are identified. Customer audits are not typically known a few quarters ahead. If available, it would be great to plug in this customer audit schedule to avoid overlap with other planned audits, i.e., first- and third-party audits. Third-party audits are predictable and planned at the appropriate time to ensure availability of the external auditors.

Once the audit is planned for the year, the audit manager responsible for the audit program obtains approval from leadership prior to execution. The CQM may also serve as an audit manager in an organization. For a large organization, due to a significant number of audits, an audit manager could be a stand-alone role reporting to leadership or the Quality Manager. The audit program may require additional resources—for example, people, travel cost, and so on. Leadership approval helps budget for the upcoming year.

Steps for Planning an Audit

As the audit year starts and the planned audit approaches, the audit manager starts planning in detail for specific audits. As the audit approaches, the audit manager, auditors, and auditee communicate to fine-tune the audit date, scope, time, location, and resource. Back-and-forth negotiations are quite normal. A formal communication is sent to the auditee organization to confirm the audit plan. The audit manager cross-references the broader annual audit plan approval from leadership to obtain confirmation from the auditee.

For supplier, customer, and third-party audits, a lot more meticulous planning is conducted to ensure the reputation and professionalism of an organization.

Plan Audit Logistics

For an internal audit, audit logistics are simple and within control, requiring a specific plan and schedule. The plan highlights the date, time, duration, process to be audited, auditors assigned, scope of the audit, access to facilities, and records (database, dashboards, etc.).

For supplier audits, the plan includes the audit location (supplier may have multiple locations), dates, start and end time each day, shifts that will be audited, total duration of the audit, product line, processes to be audited, auditors assigned, scope of the audit, access to facilities, access to personal protective equipment (PPE), and records (database, dashboards, and so on).

For external third-party audits, the plan includes the audit location (multiple-site organizational facility audited on a sampling basis), dates, start and end time each day, shifts that will be audited, total duration of the audit, business processes to be audited, auditors assigned, scope of the audit (full scope or partial), access to facilities, access to PPE, availability of leadership to cover specific requirements, and records (database, dashboards, etc.).

For all audit types, approval from the organization's leadership and auditee confirmation are required. Poorly planned and executed audits would result in the audit organization losing its credibility. The audit life cycle is shown in Figure 2.6 and described next.

Audit Life Cycle

Opening Meeting

For internal audits, a formal opening meeting may not be needed; however, it would be a good idea to avoid any misaligned expectations. For supplier audits, customer audits, and third-party audits, opening meetings are required to set expectations, state or restate the audit goals, and obtain commitment for audit performance before beginning.

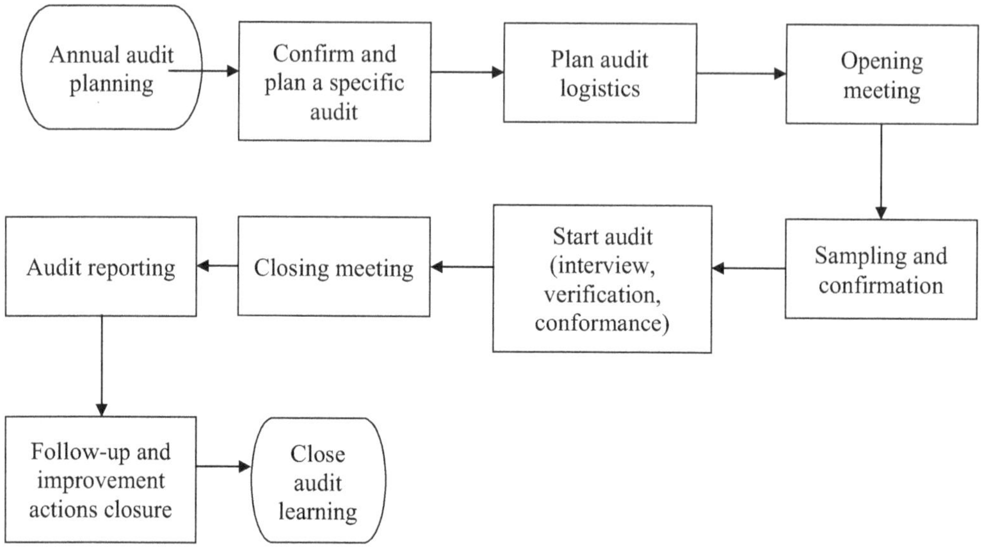

Figure 2.6 Audit life cycle.

A typical opening meeting will have items already agreed and approved during the planning. The agenda will also state that the audit is conducted on a sampling basis. Auditees will have an opportunity to clarify any questions or concerns during the opening meeting. Any unexpected changes to the audit schedule should be discussed and resolved at the opening meeting.

Audit Performed

Auditors look for conformance to business process requirements, specifications, or standards or compliance to regulatory requirements. If there is a discrepancy, they document it as a nonconformance or noncompliance. In contrast to the popular misperception, auditors do not look for nonconformance or noncompliance. Besides nonconformance, there are also other opportunities for improvement. Auditors are welcome to provide these inputs for an organization to consider.

The audit may be conducted in a combination of desk audit and on-site audit. A *desk audit*, or document review, is an audit of an organization's documents. It can be conducted at a desk since it doesn't involve interviewing people or observing activities.

When auditing a new area, function, or organization, a desk audit must be conducted before a process or system audit. This is to verify that documents meet the requirements specified in the audit criteria or standards. The document review ensures that a process or system is adequately defined before the full process or system audit.

Findings from a desk audit or document review help ensure that audit program resources are used efficiently. For instance, it would be very costly for an audit team to arrive to do a system audit, only to find that the established system was not adequate. Additionally, a desk audit or document review may be conducted periodically, or when documents (processes) are changed, to verify the adequacy of the changes.

During the desk audit, the auditor reviews the business processes for their adequacy and conformance to requirements. They also take samples from the database to evaluate the conformance to requirements. During the on-site visit, auditors meet with process owners and people who operate the process to verify conformance and effectiveness. The outcome is decided based on collective inputs from desk review, on-site audits, interviews, observation of the process, and verification of evidence.

Auditors keep the record of the auditees interviewed and collect evidence during the audit. Performing audits is a skill. Auditors do need a checklist to keep themselves focused and stay within the scope; however, a checklist can be a helpful guide to keep the audit focused. They should be ready to ask follow-up questions until they are satisfied with the response. Auditors should manage their time to ensure they cover the scope of the audit. The audit should be conducted in good faith by both auditor and auditee. The outcome of the audit should be welcomed by the auditee and addressed as an opportunity to improve business.

Closing Meeting

The auditor closes the meeting highlighting key strengths, nonconformances, and opportunities for improvements. This is an opportunity for the auditee to clarify any misunderstandings or lack of evidence that might potentially be reported as nonconformance. Typically, site or project leadership will be present at this meeting. Auditors must make use of their communication skills to ensure that the intent of the findings is well understood.

Audit Reporting

Once the closing meeting is completed, an auditor gets back to compiling all the information from the audit and prepares a formal report that identifies purpose, objective, scope, audit method, sampling, standards and requirements applied, audit findings as strengths, nonconformances, observations and opportunities for improvements. Audit nonconformances should clearly articulate the nonconformance to required specification or standards.

Audit nonconformances must include the following:

- Describe what was observed or evidenced.
- Cite what was expected from the requirement or standards, regulatory requirement.
- Articulate the gap, nonconformance, and/or noncompliance.
- Explain the risk of the gap to the organization.

Audit observation is a potential area for improvement or a minor issue that does not necessarily violate a standard or create a major risk. In other words, it is not a compliance issue but may be a preference of the auditor. Deviations from the requirements are a risk to the project and its quality. They must be linked to organizational business performance, profitability, growth, or impact to customer satisfaction. Similarly, the audit report must include the strengths of the organization to reinforce the best practices that are in place.

Follow-Up and Improvement Actions

The audit is not completed until its findings are implemented and verified for effectiveness. Auditors must review the root cause analysis of their findings and corrective action plan before taking action. They review actions taken and effectiveness either offline (desk review) or by revisiting the site for verification. For audit nonconformances that are a significant risk to the organization, a physical revisit to the audit location may be required to verify effectiveness.

Closure of the Loop: Audit Learning

The auditor or audit manager must capture lessons learned from the audit cycle and feedback to the annual audit planning and specific audit planning process. Throughout the audit cycle, there are several things that can (and will) go wrong. People availability, access issues, cancellation on short notice, leadership absence, confrontation during audit interview, closing meetings, inaction by auditee on audit finding are just a few.

CONCLUSION

In conclusion, effective audit performance is a comprehensive, cyclical process that extends far beyond mere inspection. It demands meticulous planning, clear communication, and skilled execution. From initial scheduling to thorough reporting and actionable follow-up, each phase is vital for ensuring compliance, identifying improvements, and driving organizational excellence. By embracing audits as opportunities for growth rather than punitive exercises, organizations can leverage valuable insights to strengthen their processes, mitigate risks, and ultimately, enhance their overall business performance and customer satisfaction.

QUESTIONS

1. What are the types of audit? How does audit planning rigor differ for different types of audits and why does it differ?
2. How might an auditor ensure that a remote audit is effective?
3. What are the various sources of inputs that are used to report audit performance?
4. What are the considerations for drafting a comprehensive nonconformance?
5. Why is incorporating lessons from an audit cycle important to an organization?
6. Explain how the planning and execution of internal audits differ from third-party audits, specifically addressing logistics, stakeholder involvement, and the potential impact on an organization's reputation.
7. Describe the process of creating a comprehensive nonconformance report, detailing each of the required components and why each is crucial for effective corrective action and risk mitigation.
8. Discuss the importance of the desk audit phase within the overall audit process. How does it contribute to the efficiency and effectiveness of subsequent on-site

audits, and what are the potential consequences of skipping or minimizing this phase?

9. Analyze the concept of the "audit life cycle" presented in the section. How do the stages of opening meetings, audit performance, closing meetings, reporting, and follow-up contribute to a closed-loop learning process, and what potential obstacles might disrupt this cycle?
10. Considering the shift toward virtual audits, what strategies could an audit manager implement to ensure the credibility and effectiveness of remote audits compared to traditional on-site audits, especially when addressing potential issues like time management and evidence collection?

Chapter 3
Design Phase

Design Inputs

QUALITY OBJECTIVES

> Examine the quality objectives set by the client, owner, or user from the initial scope agreement and determine how these objectives align with design inputs. Consider the delivery method and contract when reviewing design phase/ development planning. (Analyze)
>
> Body of Knowledge III.A.1

A project's quality objectives should reflect stakeholders' expectations and requirements. The project team must establish those objectives throughout the project's life cycle. Design is a critical component of any construction project, and as such the quality objectives of design must be aligned with the overall project objectives, customers' product needs, and process KPIs.

Two Types of Quality Objectives

The general requirements for quality are usually described in the contract as quality management requirements. The specific quality requirements are described for the product quality that is typically included in the technical requirements of the contract. The objectives of design quality may therefore be separated into these two basic types, general and specific.

General Quality Objectives

The general quality objectives of the project are described at a higher level. They include at least two objectives: (1) ensuring that the final design that meets the construction project's requirements is prepared by the design team to support construction without any delay, and (2) completing the final design as per the design schedule, which must be aligned with the construction schedule. The general quality requirements may describe the design objectives for some desirable aspects of the design—for example, creating a sustainable design or one that is environmentally safe. Some other examples of general quality objectives include the following:

- To implement a recognized quality management system (QMS) that is certified to a known standard (such as ISO9001 or NQA-1)
- To promote high-quality design development for an XYZ project
- To facilitate coordination between the landowners and the city
- To ensure that design plans are consistent with and promote the city's vision and civic design objectives
- To protect and enhance natural corridors and areas of environmental or community interest
- To incorporate and reinforce important natural and cultural heritage features into the community structures
- To support sustainability and safety objectives of the city (owner/client)

The higher-level (general) quality objectives are best addressed through the implementation of a QMS and its processes for quality control (QC) checking and quality assurance (QA) of design.

Specific Quality Objectives

Although the general quality objectives provide high-level guidance to the design team, specific product-based requirements are still needed for the design team to determine what should become the product-based objectives of the design.

The "product" of the design phase includes the final design drawings, calculations, and needed specifications, which must be produced through the design process and checked through design QC and QA. Whereas the process quality requirements are described in the QMS, the specific product quality requirements for the design are typically provided in the technical specifications of the contract. It should be the objective of design to also comply with the technical requirements of the product (design) that are issued for the project's various categories of construction work. Some examples of specific design quality requirements that are sometimes issued in addition to the technical design specifications include the following:

- Formal client approval is required for certain design submittals with a "critical design" designation. This is an example of higher criticality identified for some design elements.
- Design submittals are required to provide additional QA certificate from the design QA.
- Evidence of design QC checklists is required for some or all design packages.

- Evidence of design QC is required via periodic client audits.
- Design submittals with a "safety" designation are required to comply with additional safety certification from subject matter experts (SMEs).

Mutual Understanding Meeting (for Quality Objectives and Requirements)

As a best practice, mutual understanding meetings are conducted at the beginning of a construction project to help get everyone on the same page. For example, according to the US Army Corps of Engineers (USACE), there may be three (and possibly four) conferences/meetings held before the commencement of physical work. These meetings are conducted for design and build coordination.

The first meeting should be conducted at the beginning of the design phase and should include the owner's representatives as well as all design and construction stakeholders to obtain a good understanding of the design as well as construction expectations. Multiple design meetings may be conducted to ensure the design requirements are clarified completely. Such meetings are very useful to develop a mutual understanding of the quality objectives and the requirements for the design. The agenda of such meetings should include a discussion about the following quality items:

- Quality plans, including procedures to be used by the design team.
- Quality documentation and quality records (or design document control).
- Quality administration, which is the organization structure of QC, QA, and quality verification (QV) for the project. Quality verification is applicable to alternative delivery projects that have multiple layers of quality systems for the designer, design subconsultants, contractor, and subcontractors (all of whom do not always report to the owner). In such cases, the owner engages an entity that verifies both QC and QA are happening as required.
- Coordination activities to be performed (between contractor, designer, and owner). This should include establishing design working group meetings, which have a recurring agenda and frequency (e.g., weekly, biweekly, monthly, etc.).

In addition to the mutual understanding and coordination meetings, there are three more critical components of quality planning of the objectives for design: the design intent, the design quality plan, and the project delivery method.

Design Intent

ASHRAE Guideline 1–1996, the HVAC Commissioning Process, defines the term "design intent" as "a detailed explanation of the ideas, concepts, and criteria that are defined by the owner to be important." The term includes the early design decisions of the designer/architect. This is not a buzzword, but a real term with a real purpose. Following the design intent, most large-scale complex projects require the designer to issue a formal document called the "Basis of Design," which is the narrative description of what the designer will or has developed to respond to and meet the design intent.

Design Quality Plan

A *design quality plan* (DQP) is a document that should be prepared by the design firm. It acts as a guideline for the objectives of the design portion of the construction project. A DQP should outline the systematic approach, procedures, processes, and criteria used to ensure that a design meets the design intent as well as the specified quality standard. It must describe the methods used to achieve product and process design quality (procedures that address design QC and QA are described in greater detail in "Cross-Disciplinary Design Reviews" later in this chapter). A good design quality plan includes specific process definitions and procedures that describe how to achieve design QC and QA. There is a fundamental difference between the two types of design quality reviews.

1. *Product quality reviews*: These reviews are performed for compliance with the functional and nonfunctional requirements and specifications of the project. Such reviews, which are primarily for functionality and performance/usability, ensure that the design meets the technical specifications and design intent. (See "Functionality and Constructability Reviews" later in this chapter.)
2. *Process quality reviews*: These reviews are performed for compliance with the QC and QA processes of the QMS. The focus of such reviews is to ensure that the formal design QC and QA processes that are defined in the quality plan are followed by the design team. The concept is as follows: Taking care of the process quality ensures that the quality of the product (design) will take care of itself.

Both types of quality reviews are needed to create a complete design. One complements the other, and both must be implemented on every design project. Both have their advantages, but they also have their possible handicaps, which must be understood and addressed.

1. The limitation of compliance-based product QC, where the QC checker checks the design against the requirements, is that it generally tends to result in an overreliance on quality checklists. Also called "checklist syndrome," this phenomenon includes the practice of checking only the things that are on the checklist. It conditions the designers to look at nothing else. Such a mentality to "show compliance" adds false confidence to the quality reviews since the reviewer may gladly check the boxes but fail to look for things that are not on the design checklist.
2. With process quality reviews of a QMS, if the people performing the QC checking as well as independent QA auditing are unfamiliar with the subject or with the specific requirements, they may or may not check whether the technical/functional requirements have been met. By itself, this makes it an insufficient QC and QA for the design. The QMS-based

process QA may result in only a high-level assurance that creates a false sense of overconfidence.

It is only when both types of quality reviews are effectively performed together that there is a higher likelihood of producing comprehensive quality final design.

Elements of a Good Design Quality Management Plan (DQMP)

A good design quality plan includes, at a minimum, the following subelements:

- Organization chart describing roles and responsibilities for QC and QA
- Design schedule at design package level for design gates 30%, 60%, 90%, and 100%
- Design requirements analysis (for compliance purposes)
- Procedures (these must be included with a DQMP)
  - Intradisciplinary QC checking (within-the-discipline checking)
  - Interdisciplinary QC checking (across-the-disciplines checking)
  - Constructability reviews and client reviews (if necessary)
  - QA audit and QA certification of QC
  - Design deficiencies and nonconformances
  - Design changes (for field conditions and value improvements)
  - Design team meetings (agenda, frequency, and required attendees)

Note: More detailed information about these procedures is provided in "Cross-Disciplinary Design Review" later in this chapter. Also, reference Chapter 5 for field conditions and value improvement.

Project Delivery Method and Its Effect on Quality Objectives

Although the project delivery method may not change the quality objectives of the project, it does affect how those objectives may be achieved. The responsibilities for QC, QA, and QV can vary significantly in the various project delivery types. Design quality objectives must be considered carefully, right at the beginning of the project, for the type of project delivery method chosen from the project.

Following are the broad categories of project delivery methods and who is generally responsible for quality in each one (see Chapter 1 for more detail):

- *Design-bid-build (DBB)*:
  - QC and QA are the responsibility of the designer during design.
  - QC and QA are the responsibility of the contractor during construction.
  - Both are overseen (QV) by the owner or owner's representative.
- *Design-build (DB)*:
  - QC and QA are the responsibility of the DB firm.
  - Within the DB, there must be a separation of QC and QA in the design team, as well as in the construction team.
  - QV is the responsibility of the owner or its representative.

- *Construction management at risk (CMAR):*
  A CMAR project is conducted in two distinct phases:
  - In the first phase, the owner works with the performing team that has the designer on board. In this phase, QC is the responsibility of the design team, whereas QA may be performed independently by the construction management firm's representative or else by an independent quality firm.
  - In the second phase, the construction contractor is also brought on board. At this stage, the responsibility for QC and QA belongs to the contractor, and QV may be performed by the representative of the CM firm.
  - Usually, the owner and/or its representative perform an audit-based verification that QC and QA are performed effectively.
- *Progressive design-build (PDB):*
  - Even though it is issued to a DB team, PDB is a delivery method that considers the construction phase a separate agreement from the design contract.
  - In this hybrid delivery method, the DB contractor is selected before the baseline for design is defined; an integrated approach is used to deliver the project where the designer, builder, and the owner work together. The early involvement of the contractor benefits the project because the scope of design is determined collectively by the designer, contractor, and the owner. This enables a more consistent design baseline, and the subsequent construction contract therefore offers more realistic cost projections than a typical DB contract may offer.
  - The responsibility for design QC and QA belongs to the design firm, and it is verified by the owner or its representative.
  - During the construction stage, the responsibility for QC and QA belongs to the contractor firm, which is verified by the owner or their representative.
- *Integrated project delivery (IPD):*
  - IPD is one of the newer construction project delivery methods where the key parties that are involved in the design, fabrication, and construction aspects of a project are all joined together under a single contract agreement. It is different from the PDB method in terms of risk sharing and project control.
  - PDB focuses on early contractor involvement and a single contract for design and construction, while IPD involves a multiparty agreement with all stakeholders, including the owner, sharing risk and liability.
  - The responsibility for design QC and QA belongs to the design firm, but it is significantly affected by the oversight by the owner and the contractor, both of whom are part of the team right from the beginning of the project.

- *Public-private partnership*:
  - QC and QA are the responsibility of the performing company, which is mostly a DB firm.
  - QV is performed by the concessionaire or its representative.

CONCLUSION

A project's quality objectives are significantly impacted by the noted contributing factors: design intent, design quality planning, and the selected project delivery method. It is important to consider these factors in the design quality objectives because they translate forward to the remaining phases of the project life cycle and significantly impact the quality of the project in the post-design phases as well.

QUESTIONS

1. How are design quality objectives aligned with the objectives of the project?
2. How does a complex project align the conflicting objectives of multiple stakeholders?
3. How are design inputs influenced by quality objectives?
4. How are project quality objectives communicated to the design team?
5. How does the project delivery method impact the achievement of project objectives?
6. What kind of quality reviews are necessary to translate the goals and objectives into achievable results?

COORDINATION OF FUNCTIONAL AND NONFUNCTIONAL (QUALITY) REQUIREMENTS

> Explain the coordination between functional (e.g., foundation) and nonfunctional (e.g., finish) requirements and illustrate their application at various levels, such as system and component. Relate these requirements to industry standards and demonstrate their relationship to overarching quality objectives. (Apply)
>
> **Body of Knowledge III.A.2**

This section delves into the critical design phase of a construction project, focusing on the coordination of functional and nonfunctional requirements. The section elucidates the fundamental distinctions between these two types of requirements, emphasizing the mandatory nature of functional requirements, which detail what needs to be built versus the desirable yet potentially optional nonfunctional requirements, which pertain to performance and usability. Through a detailed explanation and practical examples such as bridge design, this section aims to clarify how these requirements are integrated and applied at various levels, ensuring that the design aligns with both technical specifications and overarching quality objectives.

From a construction project's perspective, the design requirements may be classified into two fundamental types: (1) functional requirements and (2) non-functional requirements.

The functional requirements are mandatory, whereas the nonfunctional requirements are desirable but may or may not be mandatory. For example, the functional requirements for the design of a bridge may be described in terms of specifications for piles, caps, bent, girders, and the deck. These are the functional elements of design that must meet the specifications of the bridge and represent its most critical features. However, the appearance of the bridge area, including the landscaping around the bridge and bridge lighting, does not affect the functional performance of the primary bridge construction and may be considered nonfunctional requirements.

Functional Requirements for Design

By definition, functional requirements are those specifications that describe the features that need to be built. These requirements therefore represent the mandatory requirements for the design project. These are typically included with the contract as the technical specifications.

Functional requirements describe what must be constructed, and thereby act as guidance standards for the designers as they prepare their final design packages (also called "construction documents") that must comply with these requirements. The functional requirements should always be stated in clear and measurable terms because they need to be inspected for compliance by the design as well as construction QC during the respective phases.

Besides the design specifications, another significant document used to supplement the functional requirements of the project is the "summary of work." This is the primary guidance document from which the designer determines the elements of the system that must be designed. The summary of work is a contractual document sent by the owner when a project is formally issued; it outlines the scope, the project deliverables, and the summary of the requirements of a project.

It must be noted that clarity in the technical specifications for design is essential for the quality of design. Wherever the design specifications are incomplete or are provided vaguely in the contract, the designer must request the owner clarify them. The specifications must be agreed to before proceeding with the design. Understanding the technical specifications (requirements) is an essential prerequisite of design, and it must be acheived through a formal activity called Requirements Management (see described later in the design chapter). Requirements management is a critical activity that must be performed at the beginning of the project; it helps identify the applicable requirements for the project and clarifies those specifications that may be vague or unclear. Also, the project requirements should be periodically revisited to ensure that any requirements that are changed are formally considered by the design team. (The processes for design changes are described in greater detail in Chapter 5.)

Examples of functional requirements for a construction project include:

- Clearance of the bridge

- Load-bearing capacity of the bridge
- Layout requirements for the building
- Material requirements (which materials may be used on the project)
- Sustainability requirements (if sustainability is prioritized for the project)

Nonfunctional Requirements for Design

The second type of requirements are nonfunctional requirements—performance or usability requirements that are desirable but not mandatory for the project. They are listed in the contract under general specifications.

In some poorly constructed specifications, nonfunctional requirements are not described clearly or are described in non-measurable terms. This frequently results in client dissatisfaction if such requirements are not met. Even if they're nonfunctional, the requirements explicitly described in contracts have a higher probability of being met by design.

Examples of nonfunctional requirements include:

- General worksite security requirements
- General safety requirements for construction
- Landscaping requirements
- Testability requirements
- Reusability requirements (including sustainability requirements)
- Maintainability requirements

Design Schedule Defines the Sequence of Design Packages

After identifying the functional as well as nonfunctional requirements, a design firm formally produces a design schedule that identifies the design packages they would prepare. The purpose is to prepare the design documents in stages, with each successive stage/phase of the design building on its previous phase. The design schedule must be aligned with the construction schedule—the elements of construction that need to be constructed first should be designed first as well. The sooner the construction contractor is hired, the more likely it is the design schedule will be aligned with the contractor's construction schedule. (Therefore, there is more consistency/alignment in a DB project than in a DBB project.)

Typically, the design is prepared in compliance with the design specifications in stages that correspond to preliminary design (30%), intermediate design (60%), and final design (90%). Also, once the 90% design is submitted to the client and formally approved, it is then issued as the 100% final design and is called the construction documents. (Stages/phases of design documents are described in "Design Review Output" later in this chapter.)

Note: Depending on the scope of the project and the preference of some clients, nonfunctional requirements are also sometimes classified as mandatory. For a successful project, the team must separate functional and nonfunctional requirements at the beginning by formally performing requirements management (RM). (See next section for details.)

Nonfunctional requirements for design may include aesthetic design requirements for the structure. Such requirements usually focus on attributes that are perceived as the client's expectations and not necessarily the client's needs. Examples may include aesthetic requirements such as creating a sleek, modern appearance, or the appearance of a '70s-era building. While they provide benchmarks, such requirements are not intended to be listed as a technical or prescriptive specifications. Other nonfunctional requirements may include those that represent the performance of the designed element, its scalability, or usability. It must be noted that some clients consider the nonfunctional requirements a mandatory requirement, though most do not.

"The appearance of a building [nonfunctional] generally involves a subjective assessment. On the other hand, technical failures [functional failures], such as leaking windows, will more likely involve an objective assessment. It is much easier to identify the aspects of a technical failure than an aesthetic one" (Atkins, 228).

From Requirements Management to Meeting Quality Objectives

Typically, through the process of requirements management (RM), the design team translates the requirements of the design into design packages that they prepare for a phased design submittal (30%, 60%, 90%, and final design).

When performed in a formal manner, RM is a best practice that offers multiple downstream benefits wherever used. This includes enabling accurate verification of deliverables, lowering the risk of miscommunication from a poorly presented scope, lowering the risk perceived by the construction contractors (bidders in cases of traditional DBB projects), and facilitating lower cost proposals.

Using Software-Based Tools to Perform Requirements Management

Requirements Management may be performed using basic spreadsheets or by using extensive Software-based tools. The scalability of whichever tool is to be used for RM should be decided by the project team after careful thought and planning. *Scalability* refers to the ability of an organization to perform well under an increasing workload. To determine which features must be included in the selected software programs, the project team should consider the complexity of the requirements and the size and scope of the project. Scalability is a critical aspect.

The primary purpose of RM is to document requirements so they may be easily communicated to all stakeholders. RM begins by enabling the conveying of the client's functional requirements (as well as performance requirements/expectations). The essentials/critical elements are listed; these may be all elements noted in the specifications as "shall," which is an imperative command indicating the noted actions are mandatory and not discretionary.

Requirements management also allows a project team to capture objective evidence that the completed work is in conformance with design reviews, QC inspections, and other significant elements that need evidence of compliance, such as construction material and suppliers' performances. Comprehensive software programs offer direct links from each task to the payment process. The scalability of the selected RM tool should be a decision made by the project team based on the criticality of requirements.

Requirements management should be implemented as a continuous process in which requirements are defined, refined, updated, and verified throughout all project phases of the construction project's life cycle. It must be performed by every project team, as it allows the team to better communicate the requirements to the performing parties, which increases the possibility of meeting them. The process helps ensure that the project's deliverables meet the needs of stakeholders. As such, the quality manager (QM) must ensure that it is performed as a formal activity that includes cross-functional collaboration between the design team and the project stakeholders.

CONCLUSION

We have so far thoroughly examined the critical interplay between functional and nonfunctional requirements within the design phase of construction projects. By establishing a clear understanding of the mandatory "what" versus the desirable "how" of design specifications, and by highlighting the role of RM in ensuring clarity and compliance, this section underscores the foundation on which successful project delivery is built. The emphasis on precise technical specifications, coupled with an understanding of stakeholder expectations, ensures that designs not only meet basic operational needs but also strive for quality and client satisfaction, ultimately leading to more robust and sustainable project outcomes.

QUESTIONS

1. What is the difference between functional and nonfunctional requirements?
2. How can requirements management be used to achieve quality objectives?
3. What are the tools that may be used to perform RM?
4. How is RM related to stakeholder engagement?
5. Why are the functional requirements for construction important to the design team?

DESIGN LESSONS LEARNED AND BEST PRACTICES

> Apply lessons learned from previous projects and industry best practices to design inputs. Understand the impact of value engineering on design. (Apply)
>
> Body of Knowledge III.A.3

The design phase of any project is critical; it lays the foundation for success or failure in subsequent stages. Applying lessons learned and industry best practices to design inputs is paramount, as is understanding the impact of value engineering. Reviewing past lessons during design and categorizing them into intradisciplinary, interdisciplinary, and constructability reviews can yield invaluable insights. These reviews highlight common failures, such as a lack of independent QC checks, poor comment incorporation, and design plans not checked against calculations, all of which underscore the need for rigorous quality management.

It may be useful to review the lessons learned during design and categorize them into various aspects of design work. There are unique lessons learned during the various aspects of design; namely, intradisciplinary design, interdisciplinary design, and constructability reviews. In addition, there are lessons to be learned by looking at the failures of the design activity in a holistic manner.

Intradisciplinary Design Review/QC Failures

Intradisciplinary checking is the checking of design within the discipline. The following list shows some of the typical intradisciplinary design review/QC failures:

1. *Design QC checking is not independent of the original designer*: This is a procedural failure and indicates a lack of conceptual clarity. The independence of the checker from the original designer must be implemented through effective quality planning and verified by QA.
2. *Comment incorporation failures*: Poor comment incorporation practices cause the deferred comments to be missed in the later design submittals, or design corrections are not verified. This represents procedural failure and is common in intradisciplinary design reviews, but it can be managed effectively and prevented.
3. *Design plans are not checked against calculations*: This is a common failure that has caused design incompatibility in many final design documents. Design calculations are the numerical basis of the design and demonstrate how engineering principles, methods, and assumptions were used to arrive at the design that is shown on the plans. Checking the calculations and cross-referencing the corresponding drawings for consistency of the calculated figures and key outputs are an essential element of QC checking. The more critical the design and the greater the visibility of the project, the more important it is to check the design plans against design calculations, especially for the final design/construction documents.
4. *Document control failures in design*: The most significantly expensive lesson for design failures is learned when the document control system used by design fails to prevent document control failures of versioning, which results in comments being made on an outdated

design. Poor design version control can lead to lost work, accidental overwriting of changes made, errors in merging the work by multiple users, and/or creation of multiple copies of the same document. All of this can result in wasted time and resources, ultimately hindering collaboration and project efficiency. Use of a proficient document control software with good version control helps prevent loss of manhours and productive labor.

Typical Interdisciplinary Design Review QC Failures

Interdisciplinary design review includes the review and QC checking of cross-disciplinary design. The following list shows some of the typical interdisciplinary design review/QC failures:

1. *Interdisciplinary design review is performed before the intradisciplinary review*: The most common design process-based failure (noted mostly in DB projects) occurs when the interdisciplinary design reviews are performed while the design is still undergoing design QC checking and resulting revisions within the original design discipline. If the within-discipline design is not completely checked yet, it should not be commented on by other disciplines because those comments could either be misread or be wrong—and therefore would need to be double-checked through repeat interdisciplinary reviews. This always adds rework and unnecessary design delays. The lesson learned is to first perform the intradisciplinary QC review and *then* perform the interdisciplinary QC review, in sequential order.
2. *Delays caused by external disciplines*: One of the most common failures of interdisciplinary reviews is the delay due to late processing by design reviewers. This can be effectively managed through coordination-based software that forces a simple but time-based workflow on team members and identifies those who are causing the design delays.
3. *No backup reviewers identified*: Many design projects are delayed because the organization and resources are not supported by a bench that would perform the IDR review in the absence of the lead staff. This is a common resource issue that can be resolved by ensuring that backups are identified.
4. *No elevation mechanism defined for conflict resolution*: Design conflicts are inevitable. A conflict resolution procedure with a description of how to elevate the issue resolution must be pre-established and communicated for successful interdisciplinary reviews.
5. *Design issues of IDR not on agenda in weekly design meetings*: In many projects, design conflicts and resolutions are not included in the agenda of the ongoing design meetings and are handled informally. This frequently results in design failures slipping through to the subsequent stages of design.

6. *Not enough time in the schedule to conduct IDR*: The most common reason for failure of interdisciplinary reviews is lack of time in the schedule. Where the design schedule does not include enough time for the IDR and its resolution, the design resulting from the interdisciplinary reviews is always deficient. Given that deficient interdisciplinary design reviews are a major cause of design defects, sufficient time should be kept in the schedule for efficient IDRs.

Constructability Review Failures

Constructability Reviews (described in detail in the following section) are the external QC reviews on the design that are performed by Construction team, and determine whether the designed structure can be built easily and efficiently). The following list shows some of the typical lessons learned due to the failures of effective Constructability Reviews.

1. *Constructability review performed late in design phase*: Constructability reviews (CRs) are described in greater detail later in this chapter. If CRs are not performed at the right time, this causes design changes and schedule delays in design deliverables. Ideally, the CR is performed at around 60% design completion. At this stage, most of the critical design elements have been defined in the design drawings, and the associated design specifications are also generally complete. If the CR is performed later, then any recommended changes due to the review may require a significant change in the design plans and related design documentation.
2. *Lack of construction experience of reviewers*: The CR should be performed by someone who has construction experience, preferably for the specific type of construction that is being designed. Lack of experience is a frequent cause of poor constructability comments. Also, the lack of field conditions assessment has resulted in siloed constructability reviews, eventually forcing more expensive design changes during the construction phase.
3. *Constructability review is more than a list of things not designed yet*: On many projects, the constructability reviewers list friendly reminders of obscure items that are a required part of the design at the next (future) submittal. Because they are already in the designers' plan to develop, including obscure items serves no purpose. Instead, thinking of the sequence of construction, the CR must be performed in the same order. For example, first perform the CR for earthwork, and then move to foundations, superstructure, and beyond.
4. *Not considering important elements of CR*: On some projects, these reviews focus on easy-to-miss items but do not review the design for testability, interoperability, compatibility, and maintainability.
5. *Roles and responsibility failures*: Whenever design teams are scattered (especially when the design is broken down into various packages that

are assigned to different design subgroups), it is extremely important to first establish the QC checking process in its entirety. Establish who is responsible for the various roles of design QC checking and how the evidence of QC will be documented, and include that information in the design quality plan.

Lack of role definition is a common failure of design quality in many projects. While it is understandable that not all projects may have a dedicated staff role for QA, it is essential that the design be checked for QC. The persons who perform the QC checking may not be dedicated staff of independent quality departments, but the roles of QC checking must be defined. And even if assigned to different persons, the evidence of their QC checking must be maintained.

Design Schedule Failures

There are design lessons learned from design schedule failures. The design failures that are attributed to poor scheduling include the following;

1. *Design schedule is not consistent with construction schedule*: For example, the design team may prepare the design for elements that are to be constructed last. Aligning the design schedule with the construction schedule is critical so early construction elements are designed first.
2. *Fast-tracked design schedule that circumvents comprehensive design QC and QA*: This is a very common failure that requires effective project management as well as a strong quality management program in place. When the QA person is independent from the production of design, they are frequently able to correct such errors through more effective QA audits.

Design Change Management Failures

This is yet another common failure of design projects and a source of lessons learned for design. Poorly established processes for design changes have failed many construction projects. Two types of failures are common with design changes:

1. *Ad hoc design changes*: "Ad hoc" means the designer is not following a consistent method and randomly adds, deletes, or moves process activities or elements from the design. Doing so only makes the design more prone to failures elsewhere. A design change must be considered a complete redesign and should undergo the same checking and review process used in the preparation of the original design. That should mean checking all drawings, schedules, calculations, and so on.
2. *Design changes made in some sections but not in the related designs*: This causes the most common design failures because correction in one element but not in the related designs outweighs the benefit the updated design may offer. All impacts of the design change on related design documents must be formally considered, especially when the design is complex.

Best Practices in Design

In addition to considering possible failures, the construction manager can identify lessons learned by applying the best practices from previous design projects:

1. *Designer qualifications*: Licensed designers have a personal stake in the quality of design.
2. *Consistency of units used*: To achieve a higher level of accuracy, the units used by the designers must be consistent across the design disciplines and the contributing design subconsultant firms. Conversions from Metric to English, and vice versa, for example, have caused a lot of known failures in the design of construction projects.
3. *Same pool of reference files*: A good design project requires good document control. Too many projects have failed because of multiple document control systems that do not coordinate with each other.
4. *Software validation of design software*: The software programs used to generate designs, especially locally developed spreadsheets, are a frequent source of design failure. These must be validated and quality audited.
5. *Comment resolution*: Formal and frequent design comment resolution meetings ensure that design comments are addressed and successive design is correct. Conversely, many projects have failed due to design changes that deferred or otherwise did not address previous comments.
6. *Cut scope, but not quality or speed*: Often in a complex project, more is required in a short time—especially when unanticipated work shows up. It is essential to not compromise the quality of the product (design) and therefore to determine when it is okay to cut scope (such as a feature that is not as significant as others so its cost outweighs the benefit). It's best to use time and resources on the primary deliverables.
7. *Using design checklists*: As Atul Gawande discusses in *The Checklist Manifesto*, the use of design checklists brings the design lessons learned from previous projects to new projects, as designers become more consistent in considering the typical design elements that should be considered for the new projects.
8. *Constructability reviews*: Frequent design review meetings with construction contractors to ensure the design is constructible prevent a laundry list of constructability comments. Bringing in the contractors helps avoid schedule delays and design changes, which result in additional costs.
9. *Recommended materials with better results in the past*: Previous design projects' lessons learned may be used to benefit the new project design team with more cost-effective designs.

10. *Stakeholders' feedback on design changes*: Another exercise that should be implemented on design projects is the formal evaluation of the design changes process. Typically, the challenges of late collaboration, effects of design QC issues, and inefficiencies in the contractors' submittals are lessons learned that every project can benefit from if implemented by the QM on the meeting agenda.
11. *Design risks*: Design risks from similar design projects in the past may be discussed, including the risks that have historically been the most significant. Use of a Pareto chart or other quality tools may help focus on which risks are most impactful and relevant to the project.
12. *Conducting formal sessions on Lessons Learned*: Industry-recognized best practices may be used to bring the various project teams together and help implement a cross-disciplinary collaboration with training that ensures all project team members are educated.
13. *Common language for the project life cycle*: Design team members must organize lessons learned in a similar manner to how the construction team members utilize and execute lessons learned during the construction of the project. Information is lost between these project phase transitions because each team discipline uses different language to establish quality objectives. Having a common language for design as well as construction phases can allow for the effective transfer of knowledge during project transitions.

Setting Up a Formal Repository for Lessons Learned During Design

Lessons learned during project and construction management are invaluable insights and experiences gained by the team throughout the project life cycle. They provide guidance and wisdom for future endeavors and can help future project managers navigate similar challenges and achieve success by preventing recurrence. These lessons learned could be of any nature, either technical or administrative. They do not have to be design-related, even if they are experienced and collected during the design phase of the project. Consider the lessons learned for all challenges that may arise with respect to design planning, communication, risk management, team collaboration, stakeholder engagement, and other factors that the team considers important.

Capturing the lessons learned can be done in several ways, but effort should be to make the process simple and esay to implement—the more complex it is, the less likely it will be used. In this respect, the best practice is to create a blank form (a template) with some easy-to-fill items that add value. It should not be a laborious, administrative process that always becomes a hurdle by itself and eventually is abandoned. For example, it is possible to automatically number the form sequentially for each time it is filled (by using software); and project-related information and the date of data entry can all be automated. The goal is to save time in data entry.

Examples of topics that add value to the lesson learned and may be included in the template include:

- Short description of failure or success
- Cause (or Causes): To save time, this could even be turned into a drop-down menu to include options such as human error/process failure/resource issue/lack of experience/lack of role clarity/lack of procedural guidance/external reason/other (explain)
- Potential risks in related areas (identify possible functions affected)
- Recommended actions

It is important to describe the lessons learned without reference to personalities. The goal of the effort should not be to assign blame, but to document what was learned and how it can be used productively in the future. In addition, there is a catch with the creation of a database for the Lessons Learned; it requires documentation in a manner that ensures clarity, consistency, and maintainability to be a useful resource, but it should not be one that could be used for litigation purposes in the future. There is a fine line between writing the lessons learned, and documenting evidence which could be used against self in case of a litigation. As such, it is necessary to consult with the senior management about the format of documenting the lessons learned.

Value Engineering (VE) and Design Quality

Value engineering (VE) is a recognized technique used to analyze and improve the value of a product, system, or process with a primary goal of reducing the life cycle cost of the project. Used in all industries due to its utility, VE is an organized application of logical and technical knowledge directed at finding and eliminating unnecessary costs. For construction projects, VE systematically analyzes already designed building features, systems, equipment, and material selections to achieve essential functions while reducing the life cycle cost of the projects.

Because the owners of construction projects everywhere are facing finite budgets and increasing demands for service and quality, using VE helps find innovative and cost-effective solutions for the construction, operation, and maintenance of the projects. By applying its techniques, innovative designs have saved taxpayers millions of dollars on publicly funded construction projects.

There are three fundamentals of VE: function, quality, and cost. Throughout VE, the team must ensure that the function is not compromised, that is, the alternate design should achieve the same function as the original design. The proposal has value when the quality of the product remains the same, but the cost is reduced. The VE method is not random but should be applied as a systematic application. Arbitrary decisions to select an alternate design cannot be termed value engineering.

Lastly, it is important to consider VE as early as possible when the impact of the decisions on life cycle costs is greatest—typically, that is the design phase of the construction project. Value engineering is a QA function. The best time for VE is during the initial planning and design stages of the project, before significant financial investments are made in terms of tooling, equipment, materials, and other resources.

Following are the five phases that comprise VE:

- *Investigation*: Identifying the functions of the project and its elements is the first logical step. By the end of this phase, the team has identified high-cost elements, functionally analyzed them, and assessed their cost/worth relationships.
- *Brainstorming*: In this phase, the power of the VE technique manifests. The team brainstorms alternatives to the current design and forces people to be creative in producing alternate solutions that do not compromise the function or the quality of the design. For DB projects and other delivery methods where the construction contractor is included in the team early, VE can benefit from the feedback of the contractor as well. (This facility is not possible when the design contract is issued individually, as with DBB contracts.)
- *Evaluation*: Alternatives to the existing design are evaluated to arrive at the best one based on the pros and cons of the possible alternatives. Formal evaluation of the cost impact of the change is made, and the best alternative is selected.
- *Development*: The best alternative is then fully developed through sketches, cost estimates, validation of test data, and other technical work to determine if the assumptions made during the study are valid.
- *Implementation*: No recommendation for cost savings is practical until it has been implemented.

Examples of VE Initiatives

Some examples of VE Initiatives are the following:

- Reducing project construction costs
- Decreasing operation and maintenance costs
- Reducing paperwork
- Simplifying procedures
- Improving project schedules
- Reducing waste in the construction project
- Increasing procurement efficiency
- Using resources more effectively
- Developing innovative solutions

CONCLUSION

Capturing lessons learned should be an ongoing effort throughout the life of a construction project. Lessons learned are a summary of the project's successes and failures and help future projects to avoid previous mistakes and common failures. According to Peter Drucker, we should learn as much from our mistakes as we do from our successes; "There is nothing so useless as doing efficiently that which should not be done at all.[3]" Our failures teach us what we should avoid in future projects.

Another best practice is to create a database of the best practices. Such a database, if located in a centralized location and accessible to the organization's future project teams, can help ensure that the information and experience of previous work is captured while the project is ongoing, and is useful to improve the processes and products in future projects.

Connecting past lessons learned to future work can start the next projects on the right foot. Retrieval of the information in the future depends on good documentation with lists, labels, naming conventions, and key search words.

Best practices examples include processes, workflows, templates, and technical as well as standardized procedures for administrative guidance, which are extremely useful for new projects and save time and effort. The "how to do it" knowledge lends the experience of the previous design team to the next that undertakes a similar design task. These can all be shared via a seminar or a virtual webinar in which the QMs are the facilitators.

Depending on the relationship, the project team may also communicate and share the lessons learned with stakeholders, other teaming partners, and design subconsultants who were also part of the team. The lessons learned during the design phase should be considered one of the sources of the overall repository for lessons learned, as more lessons are learned in the remaining phases of the project life cycle. It is the collective wisdom of such lessons from all phases that will benefit future projects the most.

Establishing a robust system for capturing and implementing lessons learned, alongside a proactive approach to VE, is paramount for design success. These practices, coupled with clearly defined roles and responsibilities, consistent documentation, and effective communication strategies, can transform design processes. By prioritizing QC, managing changes meticulously, and maintaining a centralized lessons learned repository, organizations can significantly mitigate risks and enhance project outcomes. Ultimately, integrating these best practices leads to greater efficiency, cost-effectiveness, and higher quality in design deliverables.

QUESTIONS

1. Give some examples of lessons learned in the design phase.
2. Give some examples of best practices in the design phase.
3. How is value engineering useful in the design phase, and why is it warranted?
4. What are the methods to capture the lessons learned in the design phase? Relate them with the remaining phases of the project life cycle.
5. What are the best practices for documenting in a lessons learned database?
6. Explain the significance of categorizing lessons learned into intradisciplinary, interdisciplinary, and constructability reviews, and give an example of a common failure within each category.
7. Describe the concept of VE as it applies to the design phase, and outline the five phases involved in implementing a VE initiative.

8. Discuss the challenges associated with design change management and how ad hoc design changes can negatively impact a project. What measures can be implemented to mitigate these challenges?
9. How does the timing of constructability reviews within the design phase affect the project outcome, and what factors contribute to an effective constructability review process?
10. What are the essential components of establishing and maintaining a useful lessons learned repository for design projects, and how can this repository benefit future endeavors?

Design Process

STAKEHOLDER INVOLVEMENT

> Determine key stakeholders involved in the design process for each discipline. Collaborate with the subject-matter experts from each discipline responsible for making design decisions. Analyze the impact of these design decisions on quality. (Analyze)
>
> Body of Knowledge III.B.1

Stakeholder management is pivotal during the design phase, ensuring that diverse perspectives shape project outcomes. Identifying key stakeholders at the outset—including owners, clients, and design teams—is crucial for aligning project requirements and achieving desired results. Engaging these stakeholders early on allows for a comprehensive understanding of project goals, ultimately improving the probability of client satisfaction. Effective collaboration and communication with stakeholders, especially functional leads and design discipline representatives, drive informed decisions and enhance the quality and consistency of the design work.

Stakeholders are essential for quality because they provide the input, feedback, and approval that guide the quality requirements, criteria, and performance measures of the project. They also help to identify and resolve quality issues, risks, and changes that may affect the project scope, schedule, budget, or deliverables.

Stakeholder analysis is the technique of systematically gathering and analyzing quantitative and qualitative information to determine whose interests should be taken into account throughout the project's life cycle. Stakeholder analysis should precede stakeholder engagement because it is essential to first identify the stakeholders of the project before any groundwork for effective engagement can be made. It is therefore essential that stakeholder analysis begins during the planning phase where the quality strategy is formulated. During the design phase, stakeholder analysis is further zoomed in only on those stakeholders that would make the design decisions for each design discipline.

At the beginning of the design phase, key stakeholders who will make the design decisions for each design discipline are identified. This includes identifying SMEs as well as external stakeholders who may be affected by the design and by the design changes. A formal process must be developed for stakeholder engagement, preferably in the form of a workflow that ensures enhanced collaboration where the input of all stakeholders is collected with better follow-ups and task completion.

Typically, the stakeholders of a design project include:

- Owners and their sponsors
- Client (if the designer/architect's contact is not directly with the Owner)
- Local municipalities (where applicable)
- Project's design team
- Construction contractor (the next phase customer for Designer/Architect)
- Suppliers and subcontractors (design subconsultants for design team)
- Regulators (in the case of public projects)
- Independent organizations that represent the interests of owner, client, or contractor

Engaging key stakeholders, especially customers and clients, from the initial design input stage is crucial because they define the project's requirements, features, and characteristics. Effective stakeholder management at the project's outset significantly improves the probability that the actual clients and end users will receive their desired outcomes on project completion. This activity is considered essential during the planning phase (see Chapter 2, "Project Quality Strategy Development").

During the design process stage where the day-to-day decisions are made about the design, the functional leads for different design disciplines become the representative stakeholders. In other words, at this stage, the active stakeholders are not the eventual customers and clients, but the functional leads for the various design disciplines that are involved with the design project. The process decisions can be effectively carried out by the leads as they understand the eventual goals and objectives of the project and its eventual stakeholders.

At this stage, the QM may be included in the design input process.

The goal of this closer design group interaction is to create a space to share work, receive feedback from design peers, think critically about design decisions, and ultimately improve the quality and consistency of the design work.

The design team must identify which decisions are to be made and by which internal stakeholder. Such a list must be aligned with the design inputs and other process tasks. Each collaboration point is an opportunity to review the goals of the project with the stakeholders. "It provides everyone who is to make a decision an opportunity to give and receive information"[2]. (See Chapter 2 for a more detailed description of internal stakeholders.)

Projects that have a high degree of change during the design phase should perform active engagement and participation with the project stakeholders, and include their input in the design phase. During the design process phase, the QM for this phase must ensure that the impact of design decisions on the project's overall quality is fully assessed. Collaborating with the stakeholders ensures that design choices align with the quality objectives across all the disciplines of the project and helps achieve long-term project success.

CONCLUSION

Effective stakeholder management during the design phase is paramount for project success. By diligently identifying, analyzing, and engaging key stakeholders, project teams can ensure that diverse perspectives are considered and quality objectives are met. This proactive approach, involving clear communication and structured collaboration, leads to informed decision-making and aligns design choices with overall project goals. Consequently, fostering a robust stakeholder engagement process not only enhances design quality and consistency but also significantly increases the likelihood of achieving long-term project success and stakeholder satisfaction.

QUESTIONS

1. At what phase of the project life cycle does stakeholder analysis take place?
2. Why must stakeholder analysis happen before stakeholder engagement?
3. Why are the eventual customers of the project not the stakeholders of the design decisions?
4. What makes the lead designer the internal stakeholders of the design project?
5. What are the benefits of effective stakeholder management during the design phase?
6. Why is stakeholder analysis considered a prerequisite for stakeholder engagement in the design phase?
7. How do the roles of stakeholders shift from the planning phase to the design process stage in a project?
8. Explain the significance of functional leads for design disciplines as key stakeholders during the design process.
9. Describe how active engagement with stakeholders can mitigate the risks associated with high degrees of change during the design phase.
10. What are the implications of neglecting to assess the impact of design decisions on overall project quality as it relates to stakeholder management?

APPLICATION OF QUALITY MANAGEMENT SYSTEMS IN DESIGN FEEDBACK

> Apply QMS procedures and interdisciplinary reviews to address feedback related to quality requirements in the design process. (Analyze)
>
> Body of Knowledge III.B.2

A quality management system (QMS) is a comprehensive system of formal processes that apply to the various elements and components of an organization. A QMS is built around a guidance standard that can be local or industry-specific, or it may follow an internationally accepted guidance standard (such as ISO 9000 or similar). When an organization follows a QMS, it adheres to the requirements that are specified by that guidance standard and uses it to manage quality throughout the organization.

Various quality experts, including Joseph Juran and Dr. Edward Deming, champion the implementation of a well-defined QMS to ensure product quality. Consequently, a project's quality management plan (QMP) must detail its QC and QA processes. Furthermore, the QMP should incorporate all necessary procedures, work instructions, and support tools (like checklists) that empower the design team to achieve quality by adhering to the established QMS.

A cornerstone of an effective QMS is the principle of continuous improvement. Establishing a feedback mechanism is a highly effective method for enhancing design practices. This involves gathering quality requirements and past quality experiences, and channeling the information back to the design team. This feedback serves as valuable input, offering suggestions and solutions to recurring challenges encountered in comparable projects and tasks.

Design feedback is not just about pointing out what is wrong but also appreciating what is going right. It's about nurturing growth through constructive feedback loops that designers leverage to refine their work, learn from their mistakes, and develop their skills. Design feedback loops, where established, usually lead to higher-quality and more creative future designs.

Design feedback loops offer many advantages, including alignment among the design teams and better clarity across disciplines. Feedback loops also reduce misinterpretations and design conflicts as everyone aligns with the same design concept. This reduces waste and allows the design teams to achieve better and more consistent designs in less time. A well-established design feedback loop helps align the project teams to the project's objectives.

Using a QMS for quality means utilizing QC and QA to achieve the quality objectives. Design QC results in design feedback that is used to improve the design for a superior design product. As described in "Cross-Disciplinary Design Review" later in this chapter, the design QC must be conducted within discipline (also called intradisciplinary design review) and between related disciplines (also called interdisciplinary design review). It's also important to note that whereas the two QC checking activities for design (both intradisciplinary and interdisciplinary design QC) are focused on meeting the product specifications and requirements, it is design QA audits that evaluate whether design QC is happening effectively.

Auditing is a QA activity that analyzes where QC is not achieved. QA auditing of the design identifies opportunities for improvement and forces corrective actions on the design process when it finds errors and omissions, which enables this design audit feedback process to achieve a more optimized design. "Quality Management Systems" in Chapter 2 describes a QMP, which must be developed for complex design projects. This section includes the procedures for both QC and QA.

According to Charles Nelson (2017), "The Quality Management System (QMS) is not a process that is dedicated to quality itself. It identifies the collaborative opportunities during the design process. Due to the quality management system and its procedural processes, quality becomes the outcome of performance in every aspect of the design process."

It is also essential that the design quality management plan (DQMP) describe the relationships between the various entities involved in a complex construction project. The roles of QC, QA, and in some projects independent quality firms that verify that QC and QA are implemented through the application of the QMS must be described in the DQMP, which then becomes the document that can be used as the baseline against which compliance to the QMS is audited during the design phase.

The QA audit policies, audit schedules, and audit reports formats should also be included as part of the DQMP. The frequency of revision for the procedures and evaluation of the adequacy of the QMS should also be part of the QMP. (See "Quality Objectives" earlier in this chapter for more detail about the essential elements of a DQMP.)

According to the American Society of Civil Engineers (2012), "Implementing a QMS in the design phase ensures that the designed products and services would meet the requirements and client expectations through the structured processes that it enforces. It improves efficiency and reduces errors through formal planning and management of the design effort, which is essential to achieving quality."

CONCLUSION

Integrating a QMS within the design phase is vital for achieving superior product quality and client satisfaction. By establishing robust feedback loops, conducting thorough QC and QA processes, and meticulously documenting procedures in a DQMP, organizations can significantly enhance efficiency, reduce errors, and ensure alignment with their brand vision. The structured application of QMS transforms quality from an abstract concept to a tangible outcome of every design endeavor, ultimately leading to optimized, client-centric, and successful project delivery.

QUESTIONS

1. What is a design feedback loop?
2. What is the significance of design feedback loop for design quality process?
3. How does design feedback loop act as a source of continuous improvement?
4. Describe the benefits of independent design reviews to help improve the quality of design.
5. Describe the benefits of design quality reviews (a) within the design discipline and (b) across the multiple design disciplines?
6. What are the differences between quality reviews for product quality versus quality reviews for process quality?
7. Explain the relationship between quality control (QC), quality assurance (QA), and a design quality management plan (DQMP) in the context of design feedback. How do these elements work together to improve the design process?
8. Discuss the benefits of establishing design feedback loops both within a single design discipline (intradisciplinary) and between multiple design disciplines (interdisciplinary). Give examples of how these feedback loops can prevent misinterpretations and design conflicts.
9. Explain how a QMS transforms "quality" from an abstract concept to a tangible outcome of the design process, as mentioned in the section. Use examples from the text to support your explanation.
10. How does the concept of continuous improvement relate to design feedback, and what role does it play in enhancing future designs? Detail how gathering quality requirements and past experiences contributes to this continuous improvement.

11. What is the purpose of design QA audits, and how do they differ from design QC activities? Explain how these audits identify opportunities for improvement and force corrective actions within the design process.

FUNCTIONALITY AND CONSTRUCTABILITY REVIEWS

> Participate in reviews that assess the design's functionality and constructability. Provide quality inputs that focus on various aspects, such as complexity, inspectability, testability, reliability, durability, survivability, interoperability, compatibility, maintainability, and comparative cost of ownership. (Analyze)
>
> Body of Knowledge III.B.3

In the intricate landscape of design, two crucial reviews stand paramount: functionality and constructability. While functionality reviews are internal examinations ensuring the design aligns with its intended purpose and technical specifications, constructability reviews offer an external perspective, assessing the design's buildability, cost-effectiveness, and maintainability. The former focuses on whether the design "works," validating its sufficiency and acceptability, and essentially performing product-based QC. The latter scrutinizes the design through the lens of construction personnel, ensuring a smooth, efficient, and economical building process, and bridging the gap between design intent and practical implementation.

There is a fundamental difference between these two types of design reviews; whereas the functionality review is internal and performed by the design team, the constructability review is an external review performed by the construction contractor to ensure that the design prepared by the design team is buildable, while also being cost-effective, biddable, and maintainable.

Functionality Reviews

Functionality reviews refer to whether the design is working as intended. A functionality review is a QC check that ensures that the functional (as well as nonfunctional) requirements of the project are met by the created design. Compliance checking on functional requirements is a QC check of the product to ensure that the produced/created design meets the design intent as well as the technical specifications. It is the functionality review that provides the confidence that the design is sufficient and acceptable. In other words, functionality reviews perform product-based QC of the design.

Guidance for the functionality is provided initially by the client through the scope of design; it is first developed into the formal "basis of design," which then governs the design development (DD) phase of design and lastly produces the final design (also called construction documents, or CD).

The definition of functionality is specific to the project, and only the applicable functionality must be designed. Most design projects consider only the functionality elements that are included in the technical specifications of the project and not additional functionality that has not been requested. For example, it is possible to include inspectability, reliability, interoperability, and compatibility into every product that is designed—but the designer may choose to include only those functionalities that are specifically requested through the technical specifications.

As a best practice, the following functionality elements may be considered for inclusion to make any design more comprehensive. Adding these elements enhances the quality of the project, even if they are not specifically required by the design specifications:

- *Inspectability*: The inspectability of a structure is the ease or difficulty of the postconstruction inspection of the structure for damage and wear during its useful life. This is a functional consideration that may be included in design to ensure the structure can be inspected easily during the operation and maintenance (O&M) phase. This consideration is particularly important for complex and difficult-to-reach structures such as bridges, towers, underwater piers, and tall buildings. Inspectability ensures that the designers do not simply design the structure but also consider the accessibility to parts of that structure that would need to be inspected for wear and tear in the future.
- *Reliability*: Reliability asks, "How often does the system/structure experience critical failures?" Reliability is a function of several aspects of the design and construction, all of which need to be considered as part of its functionality review for the longevity of the designed structure. These major sub-aspects of reliability may include (but are not limited to):
  - *Maintainability*: Maintainability is the ease with which a product can be kept in working order by correcting defects and repairing or replacing faulty or worn-out components. Avoid having to replace still-working parts and prevent unexpected working conditions. Maintainability maximizes a product's useful life as well as its efficiency, reliability, and safety. This concept of maintainability is equally applicable to products as well as construction.
  - *Survivability*: Survivability is a term associated with construction that is prone to be affected by disasters. Examples include structures that are constructed in extreme weather areas (locations often hit by tornadoes, earthquakes, and other natural disasters). Survivability means the design considers the use of materials as well as design considerations that would increase the survivability of the structure after a major calamity hits it.
- *Interoperability*: Interoperability may be another consideration for the functionality of design. This is generally defined as the ability of different equipment to operate in conjunction with each other. In order for the design to be optimized for cost, it is essential that interoperability

of the design be considered. One example of interoperability comes from the rail industry. The rail system is designed to enable compatible systems (such as signaling and train controls) to work with each other without a glitch. Interoperability is a significant consideration for train collision avoidance systems, which are now a standard requirement for all major rail transit construction projects.

- *Compatibility*: Compatibility is a variation of interoperability, which evaluates if the designed system conflicts with any other application or process on the new structure or system. The difference between interoperability and compatibility is that interoperability means the system or device would work fine under different environments, whereas compatibility means that different systems or devices would work fine under the same environment.
- *Complexity*: If the quality of the designed product is not enhanced by complexity, the designer must try to keep it as simple as necessary. The simplest example of complexity could be the use of a very complicated document control system when all the project needed was a simple spreadsheet. The primary goal of the designer when creating their design should be to create simple-to-operate, simple-to-maintain designs (unless the scope of work requires sophistication).
- *Durability and availability*: These two terms are functional considerations that may be mission critical for some construction, but not all. Generally, the durability of materials recommended by the designer is a common functionality for the design of most infrastructure projects.

  Availability is a term that usually relates to data accessibility and is therefore applicable to a niche in the industry for the design and construction of systems where data and its integrity are extremely important. Examples of such structures may be the design and construction of data centers or train control centers. Whenever design is significant in terms of data, the availability and durability of data should become a functional consideration for the design. There is a slight difference between these two terms:
  - *Availability*: Data availability refers to how often you can access your stored data.
  - *Durability*: Data durability refers to how safe data is from being lost.
- *Comparative cost of ownership*: The cost of ownership is a life-cycle cost that may be considered by the design team as a functionality. For example, the designer may consider designing a structure to be more sustainable and optimized for lower overall construction and postconstruction costs. For projects that value sustainability and lower cost of ownership, this may be a necessary functionality that the designer must consider.

For further information only: Life cycle cost (LCC) focuses on the total cost of owning and using an asset throughout its entire lifespan. Total cost of ownership

(TCO) considers the cost of the product from the time of purchase to the time it is delivered (i.e., when it is owned). Comparative cost of ownership (CCO) is the cost of owning and operating a specific asset, and is often used for comparing different options. All these are advanced functionalities that may or may not be considered by the designers—all dependent on the scope of the project, which guides the designer as to how comprehensively they must design.

- *Functionality of computer-aided design (CAD) files*: Design is typically produced using design software programs (such as AutoCAD, Microstation, Solidworks, or other software programs). CAD standards are considered a functionality within the CAD software. They act as a set of rules and guidelines that dictate how drawings should be created, organized, and documented. This ensures consistency across different users, as all designers on the team will utilize common style settings, layers, and line weights, thereby creating consistent drawing organization and uniformity.

Another consideration for CAD files is whether to use two-dimensional or three-dimensional design files. Each design project uses a CAD standard that is usually supplied by the client, and all designers must follow that standard while preparing the design.

There can be many functional aspects of design, only some of which have been described above. In general, the functional aspects of design are included in the scope of work and described in detail in the technical specifications of the project, which provide guidance to the designers as to how far they need to go when designing the systems and products.

What Designers Look For in Functionality Reviews

- Does the design comply with the technical specification XYZ2.2 for the design unit? (In this example, the functional review by the designer may verify that the bridge clearance is within the allowed clearance range or confirm that the overhead storage in the designed room is what the specifications require.)
- Are the correct units for the new design consistent with the measuring units' requirement? (In this example, the functional review by the designer ensures that the units used in the design are consistent with the requirement.)

Table 3.1 is an example of a functionality review checklist for a design submittal for a bridge (names and identities of the project stakeholders have been removed).

Page-Turn Review

Some design reviews for significant elements of design utilize a best practice concept famously called "page turn" in the construction industry. Typical for final design stage or for critical design elements, with the page-turn review, reviewers and stakeholders assemble for a page-by-page review of the completed design.

Table 3.1 Functionality review checklist for a design submittal.

BRIDGES, RETAINING WALLS, SOUND BARRIERS, & OTHER STRUCTURES Discipline Review Checklist
Example Construction Project -

Contract No. ABC1234
Project No. ABC4321

Submittal Name: DU9C - Roadway, Drainage, Pavement, MOT & SPM Plans
Submittal No.: PreRFC 0048
DBC Submittal Date: 11/1/2021
Contents of Submittal: Stage 3 Plans, Calculations, Level One Checklist

TECHNICAL PROVISION REQUIREMENTS						
Deliverable or Requirement	**TP Reference Section**	**General Description**	**Review Requirement**	***Review Priority Level**	**Requirement(s) Satisfied?**	**CTF Comment No.**
Unique Special Provisions (USP)	3.3	DOT provided approved USPs as Attachments which are included on SharePoint. DB Contractor may submit USPs but they must be approved by DOT in its good faith discretion.	Verify if USP is being applied and applied correctly. Review any DBC submitted USP(s). Make recommendation to CDM if acceptable or provide comments in CTF for a resubmittal	1		
Design Exceptions (DE)	3.6	Except for Design Exceptions in Sections 8.3.3 (IFA-Provided Design Exceptions) and 14.3.6 (Level One Design Exceptions), Design-Build Contractor may propose Design Exceptions and follow DOT's Design	Verify if DE is being applied and applied correctly. Review any DBC submitted DE. Make recommendation to CDM if acceptable or provide any comments in the CTF for a	1		
TP SECTION 14 STRUCTURES						
TP X.3 Design Requirements:						
Bridge Structure Requirements	14.3.1	Items 1 thru 43 defines structure requirements for ALL bridge structures	Review each bridge submittal for compliance with the TP requirements in this Section and Project Standards	1		
Sound Barrier Requirements	14.3.4	Design-Build Contractor shall comply with the Project Standards and requirements as noted in TP	Review each sound barrier for compliance with the TP requirements in this Section and Project Standards	1		
Traffic Structure Requirements	14.3.5	Design-Build Contractor shall refer to Section 11 (Traffic and Lighting) and Section 17 (ITS) for design and construction requirements as noted in TP	Review provided report info from DBC for traffic structure	1		
Level 1 Design Exceptions	14.3.6	Level One Design Exceptions have been obtained by IFA related to bridges as listed in TP	Verify if DE is being applied and if being applied correctly. Review any DBC submitted DE. Make recommendation to CDM if acceptable	1		
PROJECT STANDARDS REQUIREMENTS (Most common, Reviewers to review all Project Standards applicable to their discipline)						
State Design Manual (DM)				1		
DOT Design Standards (i.e., Standard Drawings)				1		
DOT Recurring Special Provisions and Plan Details				1		
DOT Design Memos				1		
AASHTO LRFD Bridge Design Specs				1		

This concept may be implemented for enhanced focus on quality, as the reviewers check the design against a standard (or a specification) to ensure it complies. However, the QMP must articulate if such a detailed level of review is to be implemented for ensuring the functionality of the design—it is not typical for all designs to perform "page turn" review during the design phase.

The page-turn concept makes more sense in projects where the design phase is distinctly independent from the construction phase (e.g., DBB projects). In such projects, the page-turn review becomes one of the first steps of the construction phase, as the construction contractor must understand the final design (also called construction documents) as it is issued by the designer.

Training for Consistent Functionality Reviews

The previously mentioned examples of functionality reviews have all been product-based. To be comprehensive, the reviews also require effective process-based quality. This is possible only when a process-based QMS is implemented on the project.

- For design reviews, the reviewers need to be trained on how to provide appropriate and effective comments; there should be a consistent manner of leaving reviewer comments. Part of the quality training should describe how the design review comments may be provided—for example, whether they are to be given on the design sheets or in a comment matrix and attached to the design sheets.
- Also, design quality training (which may be offered by SMEs or by the QM) should include a system for how the comments must be addressed. For example, reacting to a design comment with a response "will address" is not the correct way to handle it. Such a response is incomplete because it is only a promise and does not say *how* the

comment will be addressed. Comment resolution is effective only when the responder clearly indicates how the reviewer's comment will be incorporated. This extremely common design review problem can be addressed through good training.

- Another aspect of process-based quality is ensuring the design reviewers are qualified in their noted area of expertise. The QM can ensure that the assigned reviewers are qualified to perform the design reviews that are delegated to them as the SMEs. QA audits allow the QM to ensure that designers are not reviewing the elements of the design in isolation, or that everyone is not trying to review everything (which is neither possible nor advisable).
- Table 3.2 shows a sample comment resolution form that may be used to perform functionality reviews.

Constructability Reviews

The success of the design firm is based on its ability to develop a final design that is both "biddable" and "buildable." Whether the construction project is the traditional design-bid-build or uses the alternate delivery methods of design-build, the buildable aspect of the requirement remains the same. When the design is

Table 3.2 Comment resolution form.

Contract No. 5ZZ2

Contract # XAB 5ZZ2 Comment & Resolution Form					CODE A Accept Comment - Correct, Add to, or Clarify plans D Dismiss Comment N No Action Required C Clarify or discuss and resolve prior to next design phase R Resolve comment in next design phase				
Submittal			Package Description						
Segment		Agency	Reviewer						
Item No.	*Dwg. No. Page No.	Comment	Type	Initial Code *	Response	Final Disposition **		Reviewer Confirmed	
						Code	Date	Code	Date

not buildable, the construction project is less than successful because it may be delayed or incur cost increases. The solution is to ensure that the design of the project is buildable before construction begins. This is only possible when the design is commented on by the builder, which may identify the potential issues for construction before they are found in the field and cost time, money, and resources. Such a review of the design by a construction person is called *constructability review.*

The AASHTO Subcommittee on Construction defines constructability review as "a process that utilizes construction personnel with extensive construction knowledge early in the design stages of projects to ensure that the projects are buildable, while also being cost-effective, biddable, and maintainable" (2000). When experienced construction personnel are involved with the project from the earliest stages, their construction focus and experience can positively influence the owners, planners, designers, and material suppliers. An effective constructability review prevents significant design changes later in the design process where it is more difficult and costly to rework the design. The constructability review cycles should be included as activities in the master schedule. The design documents that must be reviewed for constructability should include design plans, calculations, any technical specifications that are prepared by the designer, and the proposed bid schedule.

Note that the constructability review is not a QC review of the design. Some authors consider it a QA review of the design, but it is not linked with QC. It is a review performed by the construction contractor to make the design more buildable. As such, it should be performed in addition to the intradisciplinary and interdisciplinary design QC reviews.

Why Constructability Reviews Are Done

The constructability review's purpose is to review the design against its ability to be constructed effectively within the proposed project site's constraints—including but not limited to geographical, environmental, and supply chain. "Overall, the goal (of constructability reviews) is to save the owner time and money by uncovering problems or potential problems that may be encountered during construction such as errors, omissions, ambiguities, and conflicts" (Pettee). These problems or flaws, if not discovered early in the design phase, may impede construction or impact any stakeholder in terms of time, cost, or quality.

Constructability reviews allow the designer to think like the contractor (builder). The reviewer looks at the design documents from the viewpoint of the contractor, the inspector, and the construction manager. The constructability review provides feedback from construction as to how the design may be built and how design changes must be made to make the design more buildable, less complex, and less costly. In addition, the construction considerations of testability, interoperability, compatibility, sustainability and maintainability should be included in the review.

Key participants in the constructability review include entities beyond the general contractor, such as subcontracting trade partners and, most importantly, product and material manufacturers. Where possible, these members of the product and material supply chain may also be involved in the earlier design stages such as finalizing the designer's intent during the schematic design (SD) as well. "They may be able to provide a design solution within the limitations of their product that has never been previously done" (Atkins and Simpson, 2008).

The use of technology (such as software-based reviews) speeds up the constructability review process and offers expedited commentary and resolution of such reviews. This is considered by many design firms to be the new way of performing design checking, as traditional methods of hard-copy designs are being replaced with centrally located design files that are checked electronically using design software. Software is only a tool to facilitate the implementation of the quality requirements; they do not change just because a software program is used to facilitate the review.

Constructability Reviews in Different Project Delivery Methods

Constructability reviews are used to ensure that the design is buildable. The goal is also to achieve improvements in cost, schedule, quality, safety, and productivity. When the designer and the construction contractor firm are independent of each other and likely disconnected (e.g., for DBB projects), the contractor is not hired until much later. In such projects, it is a challenge for design to engage with the construction contractor and obtain constructability comments at the right design stages. For such projects, constructability is sometimes deficient, and it is evidenced by more design change requests whenever the completed design is finally let out and a construction contractor is hired. With DBB projects, it is essential for the owner/client to have personnel with construction experience who may perform the constructability reviews in place of the construction contractor.

Benefits of Constructability Reviews

1. A constructability review prevents conflicts among design documents. When conflicts are minimized, the result is fewer requests for information (RFIs), fewer field orders for on-site changes, a smaller number of change orders or addendums for contractual changes, and lower administrative costs (Pettee). The proof of savings—time, money, and the weight of other impacts—is dependent on the volume, caliber, and correctness of the CR comments and how they are incorporated by the design team.
2. Good constructability reviews result in higher-quality bids because the final design offers a better understanding of the project goals and improved scopes of the projects.

When Constructability Reviews Should Be Performed

A constructability review should begin during the earliest phases of project conceptual planning and continue through design engineering, procurement, construction, operations, and project maintenance. There is limited benefit in getting contractors' comments on a preliminary design (30%), because at that stage, the design is mostly conceptual and many design decisions aren't made.

However, the 60% design completion stage is the optimal point for constructability reviews because most of the critical design elements have been completed, but the design is still not final and can be modified with relative ease. When constructability reviews are performed at a later stage, such as nearing 100% design completion, they can cause a lot of design rework. Because of the loss of time and additional cost, constructability reviews lose their value the further they are delayed past the 60% design completion. (Note that, unfortunately, the industry carries no empirical data to support this claim; it's based on experience and logic only.)

Typical Failures of Constructability Reviews

1. Comments that are made directly on the design plans are good because they show exactly where they apply—but they may be so brief that they obscure the intended message of the constructability reviewer.
2. Sometimes, the reviewers do not comment on "related designs", so a recommended change may be noted in a silo. This requires particular attention from the designers, as it is their responsibility to consider related and affected drawings due to a recommended change.
3. Delayed responses to comments or no comment closeouts are frequently the reasons for client dissatisfaction with the design.
4. Constructability review comments that are provided too close to the final design submittal (or a lot of comments offered close to the final design stage) can prevent the designer from making the necessary changes unless the additional cost of delay is accepted by the client.
5. The constructability review comments may not allocate enough time for the designer to discuss and find the resolution of the constructability issues before the issue of the final design.

CONCLUSION

It is the QM's responsibility to ensure that the impact of design on the project's overall quality and performance is considered in light of its functionality as well as constructability. This can best be achieved by including the functionality as well as constructability requirements as the essential components of the QMS for the design project.

In a well-laid-out QMS, both types of reviews are established as workflows that are forced on the design at predefined stages of design completion. As a result, the provision for these reviews is added to the schedule. Doing so ensures that such significant reviews are not missed or conveniently ignored. A well-designed QMS would also enforce nonconformances to the project whenever these reviews are not performed, thereby bringing the attention of the management to the purposeful neglect of such significant design reviews.

Both functionality and constructability reviews are indispensable for successful project outcomes. Functionality reviews ensure technical compliance and design adequacy, while constructability reviews guarantee buildability and cost-effectiveness. Integrating these reviews into the project's QMS, with clear workflows and qualified reviewers minimizes errors and enhances project delivery. Ultimately, prioritizing these reviews at the appropriate stages, particularly around 60–70% design completion for constructability, leads to optimized cost, quality, safety, and productivity, solidifying the design's transition from paper to reality.

QUESTIONS

1. What are the typical elements of functionality of product design?
2. What are some examples of nonfunctional requirements for design?
3. What are the elements of functionality in a good design quality process?
4. What is the purpose of constructability reviews?
5. When should constructability reviews be performed?
6. What are the typical failures of constructability reviews?
7. What is the primary distinction between a functionality review and a constructability review, and how do their respective focuses contribute to the overall success of a project?
8. Beyond basic technical specifications, what are three best practice functionality elements a designer might consider adding to enhance a design, and why are they valuable?
9. Why is the 60–70% design completion stage considered the optimal time to conduct a constructability review, and what are the potential consequences of delaying this review until later stages?
10. Explain how the concept of interoperability differs from compatibility in the context of design functionality, and give an example of each from the section.
11. How does the integration of functionality and constructability reviews within a project's Quality Management System (QMS) contribute to mitigating potential project risks and improving overall project delivery?

Design Outputs

QUALITY OBJECTIVES REVIEW

> Apply verification and validation to design outputs to evaluate whether quality objectives are met. (Apply)
>
> Body of Knowledge III.C.1

In the realm of design quality, meticulous validation and verification are paramount to achieving project objectives. These two distinct processes, rooted in the Latin words *validus* ("strong") and *verus* ("true"), ensure that the product both fulfills its intended purpose and adheres to specific requirements. Validation, a QA function, confirms the process's capability to deliver quality, while verification, a QC activity, checks the product against specifications. This dual approach serves as the litmus test for design success, aiming to minimize revisions and delays by identifying potential issues and enhancing overall quality.

Validation and verification were first introduced in the field of software quality, where high-profile failures forced a greater focus on ensuring that the processes and products be validated as well as verified.

The product should first be capable of meeting the intent for which it is created (validation), and then provide evidence that it is meeting the specific product requirements it is expected to deliver (verification). As such, this is a two-step process: the *validation* of the design quality process must be performed first to provide assurance the quality process will work, and second, the quality of the product is *verified* through the QC checking procedures. Together, the validation (QA) and the verification (QC) help achieve the quality objectives of the design phase. All of the design must be validated and verified during the design output stage of the design phase of the project life cycle.

For design quality, validation and verification are used as the litmus tests for design success.

Objectives of Design Quality

The typical objectives of design include identifying potential issues or conflicts in the design and enhancing the overall quality so the construction project may be performed with the least costly revisions and schedule delays.

To evaluate whether the design quality objectives are met, the performance indicators for the design phase must address both the process and product quality. Thus, the design phase must address both validation and verification. Validation is performed by QA personnel, whereas verification is performed by the QC team. (See Chapter 2 for the differences between the QA and QC roles).

Design Process Quality Objectives

Incorporating validation of the design quality objectives allows the quality manager to ensure that the design quality is aligned with the project's goals and that the designer's intent is aligned with that of the project's stakeholders. This shows that the system in place can deliver quality.

Subsequent to the validation, the verification of the design quality objectives allows the quality manager to ensure that the design process is indeed meeting the intended requirements. An example of process verification is provided below.

1. To establish and validate the design process, the best practice is to establish and implement a QMS. The QMS for design includes the design procedures, methods, formulas, principles, and theories that are to be used by the design team. Once the design QMS is implemented, it helps to achieve two things: (a) validate that the design process is capable of delivering what it is expected to deliver, achieved through the front-end activity of QA planning; and (b) at the back end of the QA, validating that the design processes worked as expected.

   The best practice method of ensuring front-end QA planning is to use design checklists. When the designers use checklists to perform their design and later their design QC, the checklists act as reminders of what should certainly be included as part of the quality checking process.

They become the front-end planning tool to help ensure the product quality requirements would be met once the design begins. Notice that all this QA work is performed at the front end, that is, before the design work actually begins.

2. In the second step, the design is performed (created) by the designers—that is, the designers create the design, which is their product. After the design (product) is created, it goes to the next step of QC checking. This is what is called the product QC and is the verification of the product. The verification of product quality is described in the next step following the description of QA.

3. Meanwhile, on the process front, after the design is performed/created, it is also possible to perform the verification of the process. Note that this QA activity is performed at the back end of the design process, and to perform the validation requires QA audits of the design process.

   Implementation of process audits is the second tool used to validate the QMS at the back end of the process. It is very important to note that *process audits* are completely different from *product audits*; the process audit determines whether or not the process established through quality planning was comprehensive. Following are some examples of process audits that may be performed to validate the quality of the processes:

- *Process audit of document control process*: A process audit on the document control process may be performed to validate if the document control process is comprehensive enough to effectively handle the documentation of the project. Examples of things to investigate in such an audit include: Does the document control process include the needed workflows? Has the document control process identified the lead people who would be required to perform the checking for the various disciplines?

- *Process audit of the nonconformance and corrective actions*: The process audit of the nonconformance system and its subsequent corrective action system validates whether or not the system is capable of delivering quality once active work begins. Such an audit may validate that the nonconformance and the corrective action both utilize the same well-established and well-linked database of project record. This in turn validates that if and when a nonconformance is issued, it would be tracked easily throughout all phases of work, all the way to closure. To perform such an audit, the effectiveness of workflows may be evaluated.

- *Process audit of design change process*: Another example of validation of the quality processes is whether the design QMS includes the formal definitions of design deficiencies and design changes. For example, when is the QA auditor supposed to issue nonconformance versus calling it just a design deficiency? Typically, in design, if a design comment/objection is knowingly deferred to the next design submittal, then it is not considered a design deficiency, nor should it be called a nonconformance. The design deficiency occurs only when a design comment is missed during the next design submittal. Even so, if the deficiency is correctable without impacting the schedule, then it may not be issued out as the nonconformance but only kept as a deficiency. (Note that a construction QM is frequently not involved with this level of design detail; these are finer points of the design QMS.)

- *Process audit of qualification of the design quality staff*: Compliance with the QMS may require that the auditors appointed to perform the process and product audits are qualified. Ideally, there should be qualification criteria for the personnel who may be asked to perform design QC. Likewise, there should be criteria to qualify the auditors who may perform such design QA audits. When comprehensive criteria are established and consistently implemented, then the likelihood of achieving the design quality objectives increases.
- *Simulated real-world conditions*: Some validation methods are used by design QMs to validate the effectiveness of the design. Examples of structural design validation fall in this category of process validation. Simulation software can be used by designers to validate that the design they created would work well under real-world conditions.

Design Product Quality Objectives

Whereas Process quality can be validated as well as verified, the product quality can only be verified; it cannot be validated. Product quality is always a Yes or No decision. The quality of product is always a binary decision that is made by those who perform the QC inspection of the product. As such, the design product quality objective is to determine whether the product passes the QC inspection or fails.

When the QC checking or review is performed on the product, it can result only in a pass or a fail; that is, if the check or review shows that the product is meeting the criteria/specification, it passes. If it does not meet the criteria/specification, it is termed a fail by the QC. As such, there is no validation of a product—only a verification. An alternate way of viewing this is to consider that if a product is to be validated, that must happen through process validation.

Product specifications have already been given to the designers in the form of the technical specifications; they are not to be established, but to be achieved. As such, the QC inspection or review of the product (design) results in only one of the two things: either the specification is met or it is not met (pass or fail). If the QC check of the design (the product of the design phase) determines the required quality is not achieved, the design must be corrected and the updated version resubmitted for QC inspection one more time. Only after the QC verifies and approves the updated design as meeting the specifications/requirements can we say that the product quality has been achieved. In other words, product quality can only be verified; it cannot be validated. And that product quality verification is a function of QC.

Meeting the process quality by itself does not guarantee product quality: One of the primary design objectives for the quality of the product is to comply with functional and nonfunctional requirements. The project quality requirements here are the technical specifications of the product against which the designers create the design. (See "Coordination of Functional and Nonfunctional Requirements" in this chapter.)

Note that by itself, following a procedure does not verify that the product requirements will be met. To have confidence that the product quality requirements are met, product specification quality verification is required. Hence, product quality can only be verified. (Review "Design Verification and Validation" later in this chapter for more information.)

Quality managers must understand validation and verification of the processes as well as of the products. In the context of a construction project, frequently the construction QM is not hired until the construction phase begins. Depending on the method of project delivery selected for the project, the construction QM may be introduced to the project life cycle during or after the design phase. Regardless of whether or not there is an independent QM identified during the design phase of the project, the design must undergo QC checking by qualified people who are capable of performing the validation and verification of the design processes, as well as the verification of the design product(s).

In ideal conditions, these roles are assigned to personnel who are independent from the production of design. However, even if the projects are of a scale that cannot allow the provision of independent personnel for such roles, the roles of QC and QA must still be identified, and the personnel who fill these roles must be noted. (See Chapter 2 for more information.) In some cases (such as in complex projects), QA roles are assigned to an independent QM that belongs to the owner or to an independent quality firm that operates on behalf of the owner. The scopes of work and role assignments in such unique cases must be carefully considered before such role assignments are finalized and approved for the project.

CONCLUSION

Mastering validation and verification is essential for achieving design quality objectives. By understanding their distinct roles and application, project teams can ensure that design outputs meet both intent and specific requirements. Validation through QA establishes the process's capability, while verification through QC confirms product compliance. This dual approach, from planning to audits, minimizes issues, optimizes resources, and ultimately drives successful project delivery. Therefore, prioritizing these quality measures is vital for ensuring robust and reliable designs in any project life cycle.

QUESTIONS

1. What are the quality objectives for the design quality control (QC) versus the quality objectives of design quality assurance (QA)?
2. What is the difference between validation and verification?
3. Why is validation a QA activity?
4. Why is verification a QC activity?
5. How are validation and verification of the quality processes achieved?
6. How is verification of product quality achieved?

7. What is the role of a quality manager (QM) in the validation and verification process during the design phase?
8. In the complex projects (such as design-build), can the responsibility for the quality of design be assigned to a construction QM from the contractor?
9. Explain the distinct roles and responsibilities of QC and QA during the design phase, particularly as they relate to validation and verification processes, respectively. Give an example of how each contributes to achieving design quality objectives.
10. The document states that "process quality can be validated as well as verified," while "product quality can only be verified." Elaborate on this distinction, and justify why product quality cannot be validated, according to the information presented.
11. Describe the concept of process audits as discussed in the context of design quality. How do these audits differ from product audits, and what specific examples are given for validating the quality of design processes?
12. Discuss the role of design checklists as a front-end planning tool in the QA process. How do they contribute to ensuring product quality requirements are met, and at what stage of the design process are they primarily utilized?
13. In the scenario of a complex project, such as a design-build, the document touches on the potential involvement of a construction QM from the contractor during the design phase. Explain the potential challenges and considerations regarding role assignments in such cases, and discuss why careful consideration is necessary before finalizing these assignments.

CROSS-DISCIPLINARY DESIGN REVIEW

> Support the design review process across multiple disciplines to identify and address gaps in design outputs using a design review checklist. Determine and address conflicts within the overarching system design, among individual component teams, and in interactions between component teams. (Analyze)
>
> **Body of Knowledge III.C.2**

Within the intricate framework of design, QC stands as a cornerstone, ensuring precision and reliability of design. This section describes design QC and is broken down into two subsections: intradisciplinary (within the discipline QC) and interdisciplinary (also called cross-disciplinary QC) design reviews.

Before a design ventures into the realm of cross-disciplinary scrutiny, it must first undergo rigorous evaluation by qualified individuals within its own field. This internal examination, mandated by quality requirements, serves as the foundation on which subsequent reviews are built. It is a meticulous process where designs are not merely glanced over but thoroughly dissected, scrutinized, and refined by experts possessing equivalent or superior qualifications to the original designer, adhering to a structured sequence designed to catch and correct any discrepancies. Both intradisciplinary and interdisciplinary design reviews are QC reviews for the design phase.

These two design QC reviews for process quality are provided below.

1. *Intradisciplinary design review*: Intradisciplinary review includes QC checking of the design within its own discipline only.
2. *Interdisciplinary design review*: Interdisciplinary review is also called cross-disciplinary review. This review is performed to look at the impact of other disciplines on the primary design. It looks for the conflicts that other design disciplines may have with the primary design and addresses them so the final design is buildable.

Intradisciplinary Design Review

The first type of QC check (review) that is performed to comply with the quality requirements is called the intradisciplinary QC review. This design review represents checking the design within a singular discipline. Before the design may be reviewed by other disciplines for a cross-disciplinary check, it must first be checked internally within the discipline by qualified checkers.

QC requires the design to be formally checked by a different person (reviewer) who must be technically qualified to review it. The concept of "equal or better" means that the checker's qualifications must be equal to or better than the original designer. At minimum, two people are required to perform an intradisciplinary design QC review—there could be more people involved in this formal review, if available, but the minimum required for a formal intradisciplinary QC review is two designers who are qualified in the discipline.

The following is a recommended sequential process for the intradisciplinary design review.

1. *Define the rules*: Before the QC checking may begin, the formal QC checking process must be defined and agreed to between the designer and the checker. The protocol for making comments on the checked design must be predefined and communicated, and the design team must be trained in the procedure for checking. Formal color codes must be assigned to the checker for agreement, disagreement, and comment resolution, and these must be included in the procedure for the intradisciplinary design review (which should be included with the design quality management plan, or DQMP). A design that does not specify the formal process for this first category of QC reviews cannot be considered formally checked.
2. When the formal checking begins, all related documentation that is needed to perform a comprehensive QC check should be provided to the checker.
3. *When to start a formal QC check*: Formal checking begins only when the design has been developed to the degree that the designer decides it is now ready. The original designer makes a judgment call as to when it is time to stop making further changes and begin formal

QC checking. This stage is also called the "design freeze," which means no more changes may be made to the design until after it has been thoroughly checked. Only after the design process is frozen (also popularly called "pencils down") can formal design QC checking begin.

4. There are four recommended steps for design QC checking within the discipline:

a. *Step #1: Checking*: As the checker reviews the completed design, they must use the same basis of design as the original designer so the design review is performed against the same design considerations.
   - The checker may agree with some of the design elements and may disagree with others; the elements of design that are in agreement may be highlighted as having been checked with no comment.
   - The disagreements of the checker with the design elements must be notated in a different color (commonly red). It is essential that the comments made by the checker are clear enough to be resolved after discussion.

b. *Step #2: Back-checking*: Once the checker has made comments on the design, these comments must first be resolved and then incorporated in the updated design. This stage of QC checking is also called "back-checking."
   - It first involves a discussion between the original designer and the checker regarding areas where the checker disagreed with the original designer. At the end of the back-checking stage, all comments made by the checker must be resolved.
   - If any comment cannot be resolved between the designer and the checker, it must be elevated to a higher-level manager, who may make the final decision.

c. *Step #3: Correction*: Once the comment resolutions have been agreed to between the original designer and their checker, they must be incorporated into the existing checked design. This step (correction) generates an updated design that represents the resolution achieved in the back-checking stage. It must then be verified before it is considered final.

d. *Step #4: Verification*: This step ensures that no inadvertent mistakes were made by the person who corrected the design as per the back-checking resolutions. This verification must ideally be performed by the checker, which ensures the closure of the entire intradisciplinary QC checking process because the checker is able to verify the resolution was actually implemented by the corrector.

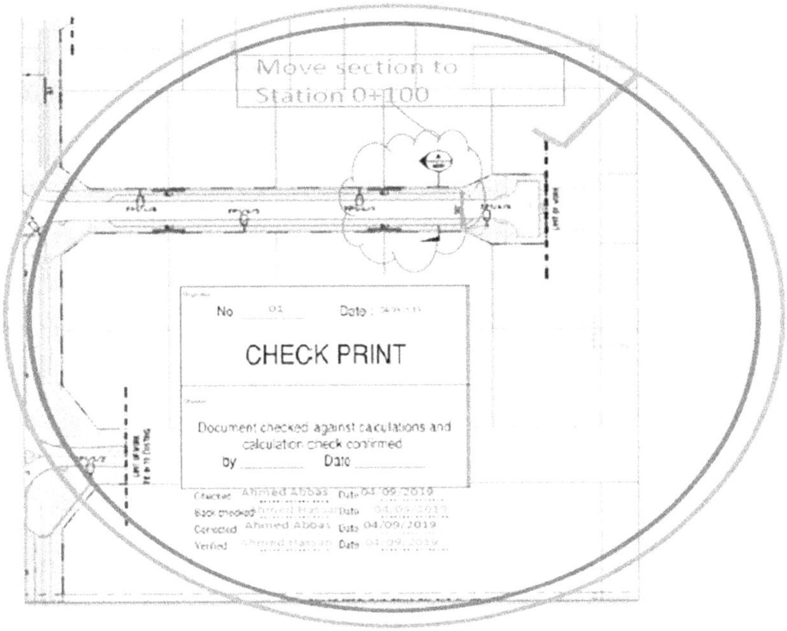

Things to Watch for During Intradisciplinary QC Reviews

- In formal QC checking, no one can check their own work; the design must be checked by a different person.
- The checker must be qualified to check the design. The recommended best practice is "equal or better," meaning the qualification of the checker must be equal to or better than the original designer.
- All related documents that were used by the original designer, including the basis of design, any design calculations, and specifications, must be provided to the checker.
- Back-checking for comment resolution must ideally be performed between the original designer and the checker.
- The person who makes corrections to the design as agreed during the back-checking resolution stage cannot verify its correctness since they cannot check their own work.
- Verification that the design has been correctly updated should ideally be made by the checker. This means the person who initially checked the design should preferably be the one who verifies that their original comments have been correctly implemented.

Essentials of a Good Intradisciplinary QC Review

- *Independent*: The QC checker must be independent of the person who originally designed it. The basic premise of formal QC checking is that it should include at least two sets of "qualified" eyes because an independent check is the most effective way of finding and correcting any potential error or omission (and checking the adequacy of the design). Independence of the checker also ensures that conflicts of interests do not arise.
- *Expert*: The formal checking must be performed by someone who is suitably trained in the same discipline as the original designer. An SME is someone who has thorough knowledge and in-depth understanding of a specific subject. Having the formal check performed by someone who is "equal or better" should be the best practice criterion for the qualification levels of the formal checkers.
- *Multidisciplinary*: Product delivery teams get asked to do a lot of different things, each of which require different skill sets. A checker who is expert in at least one area but also somewhat capable in many other technical disciplines is always the most useful type of team member for a technical project.
- *Accountable*: An accountable checker drives performance, ultimately leading to improved design because of their reliability.
- *Transparent*: Transparency fosters trust, encourages open communication, and leads to stronger relationships and better collaboration. Transparency is a necessary trait of a good QC team member.
- *Timely*: Early detection and timely delivery of work are essential for a good QC checker. Timeliness prevents potential problems and complications from moving to the next stage of design.
- *Objective*: Objective checking is more data-driven, which means that an objective person relies on quantifiable and verifiable data to perform their design, which helps with decision-making. Also, comments must be offered constructively, as that is the only way to get the acknowledgment and cooperation of the original designer.
- *Accessible*: Accessibility is another crucial aspect for the selection of good team members. An accessible person is more willing to participate fully, which fosters a more productive team environment for quality. Accessibility includes expressing comments in terms that are understood in the same way by multiple people, and are not discipline-specific or couched in jargon.
- *Balanced*: A balanced approach to design is a crucial design principle that enables the design to become acceptable to a larger variety of end users. If a checker chooses either a conservative or liberal approach, that usually ends with a lot of disagreements. The more balanced the designer and the checker are, the more likely they are to agree on the design and gain a quick approval.

Intradisciplinary QC checking of design requires personnel with specific subject-matter expertise in that discipline in order to be able to check comprehensively for the discipline-specific requirements of the design. The role of the QM in the design phase, however, is not that of QC but to act as the QA person who oversees that the "within-the discipline" QC checking of the design is performed comprehensively.

For example, QC checking for a bridge design would require a qualified SME in the field of structures to perform the intradisciplinary QC checking; this check should not be performed by an electrical engineer who is not qualified in the discipline of structures. The QA manager, however, can be a person who may not be discipline-specific in structures but understands the QC and QA processes and that the design is checked by qualified subject-matter specialists. The QM does not need to be qualified in any one subject matter, but rather needs to understand and implement QA.

CONCLUSION

The intradisciplinary design review process, with its defined steps and emphasis on qualified checkers, forms a vital safeguard in the design phase. By ensuring independence, expertise, and accountability in the review process, potential errors and omissions are identified and rectified before moving to the next stage. The principles of transparency, timeliness, objectivity, accessibility, and balance are crucial for successful reviews. This meticulous approach ensures a robust and reliable design output, fulfilling quality requirements and setting a strong foundation for subsequent interdisciplinary reviews.

QUESTIONS

1. What are the essential elements of within-discipline design QC checking?
2. How many minimum steps are involved in the typical QC checking?
3. Can the design be formally checked by the same person who generated/created the design?
4. Can the design be back-checked by the same person who created/generated the original design?
5. How are the design checker's comments addressed in the intradisciplinary checking process?
6. What are the typical failures of intradisciplinary design QC checking?
7. Explain the concept of equal or better qualifications for a checker in an intradisciplinary design review and why this principle is crucial for effective QC.
8. Describe the four recommended steps of the sequential process for intradisciplinary design review and elaborate on the purpose and significance of the back-checking step.
9. Discuss the difference between intradisciplinary and interdisciplinary design reviews, and explain why the intradisciplinary review is considered the first step in the QC process.
10. Identify and explain three of the essentials of a good intradisciplinary QC review, and elaborate on how each one contributes to a more robust review process.
11. What is meant by the term "design freeze" in the context of intradisciplinary design reviews, and why is it necessary before formal QC checking begins?

Interdisciplinary Design Review

Effective interdisciplinary design reviews are crucial for ensuring the quality and coherence of complex projects. By systematically examining design outputs across multiple disciplines, potential conflicts, gaps, and inconsistencies can be identified and addressed early in the process. This proactive approach prevents costly errors during construction, reduces requests for information from contractors, and minimizes the need for field-based changes. The primary objective of these reviews is to guarantee that design documents from different disciplines align seamlessly, ultimately leading to a more robust, efficient, and successful project outcome.

To adequately assess the impact of individual discipline design on the other disciplines, a cross-disciplinary QC review must be performed. (In the design industry, this is often called the interdisciplinary review [IDR].) It is important to note that the impact of other disciplines can effectively be considered once the design is formed within the original discipline.

Purpose of Interdisciplinary Reviews

The primary objective of the IDR is to prevent conflicts among the design documents that may be caused if the interaction of different design disciplines is not considered. The IDR addresses design conflicts and resolves coordination issues, thereby improving the quality of the design. When conflicts are minimized, it results in fewer requests for information (RFI) from the contractors during construction as well as fewer field-based design changes and change orders. It also reduces disputes, claims, or legal action.

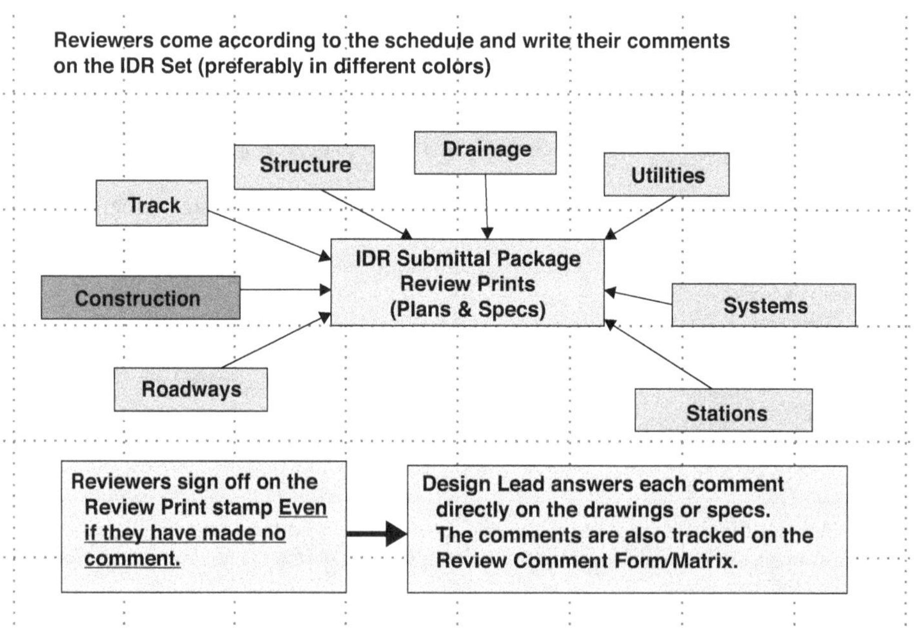

Figure 3.1 Interdisciplinary review package.

Table 3.1 Criteria for conducting an IDR.

➢ IDR is one of the BIGGEST reasons for design quality errors and failure.
➢ Poorly performed IDRs can JEOPARDIZE the success of entire design and construction.
➢ IDRs are performed to coordinate design responsibility & design details ACROSS disciplines.
➢ Conflicting comments are the MOST important in any design.
➢ The engineer of record (EOR) is responsible for accepting or rejecting design review comments.
➢ Client comments must be considered, but the final decision is made by the EOR.
➢ Formal QC MUST be performed for IDR after comment resolution of the IDR comments has taken place.

The purpose of the IDR is to identify and resolve design issues where disciplines and systems interact. Its objective is to confirm that the design is comprehensive from a systems standpoint and to mitigate risks together with the various design disciplines' coordination. IDRs (Table 3.1) remove conflicts between drawings, between drawings and specifications, and between the various specification sections. They apply to design documents such as design drawings, design specifications, engineering studies, engineering planning documents, and engineering reports. This peer QC review is performed by designers or other SMEs who have expertise in disciplines outside of the original designer's area of expertise.

When the Interdisciplinary Review Should Take Place

- Interdisciplinary reviews (as shown in Table 3.1) should be conducted at all developing stages throughout the design phase, but especially where the impact of interface design failures is high.
- On multidisciplinary projects, a formal IDR must be performed at least once during the design process, typically at the 60% design stage. As a rule of thumb, performing the IDR at the 60% completion stage offers the best value for the time and effort because some design decisions have been made, but a lot of significant decisions have not.
- IDR is *not* a substitution for communication and coordination during design. It is a formal design QC activity and a QC record.

Best Practices for Performing the Interdisciplinary Reviews

1. A formal IDR period should be strictly enforced. The review period for formal design QA after the completion of IDR must be included in the design schedule.
2. The review period for the IDR can be electronically enforced, with performance (delays) discussed during the weekly progress meetings.
3. QA after IDR: A formal QA audit must be performed after the completion of the IDR to ensure it has been done correctly.

4. Backup reviewers must be preassigned in the design QC system. Many times, the IDRs are not performed comprehensively when the primary lead person assigned for the review is not available.
5. Triggers/warnings must be sent out to the design lead electronically, indicating which reviewer is causing the delay.
6. An elevation mechanism must also be established. If there is no agreement, then the workflow should elevate the issue and get it resolved.
7. Agenda item: IDR issues and comment resolution must be a recurring agenda item of the weekly design meetings. It must also be an agenda item for the regularly scheduled meeting that takes place between the designer and contractor.

Nine Steps for Performing an Interdisciplinary QC Review Using the Traditional Method

1. *Assemble*: Assemble the set of discipline drawings and documents. It is best to compile all the submitted files into one complete set of design drawings and review one file, instead of reviewing a folder that consists of five different files.
2. *Discipline leads*: For each discipline, identify a discipline lead who must ensure the entire design package is reviewed for their discipline. They also must certify that their discipline has been reviewed by formally signing off on the review print stamp. Also, as a best practice, each discipline lead must also identify a deputy lead who will step in whenever the lead is absent.
3. *Design review comments*: Comments are issued by the discipline-specific reviewers to address the following:
    - Conformance to specifications
    - Applicable codes and standards
    - Human Factor Criteria (if applicable)
    - Safety requirements
    - Basis of the design
    - Constructability considerations
    - Quality assurance requirements
    - Holistic view of design
4. *Comment resolution*: This is the most important step in IDR. Each reviewer comment must be resolved with the discipline lead who is responsible for the comment/person who made the original comment.
5. *Elevation (if necessary)*: Wherever the comment made by the reviewer cannot be resolved with the discipline lead, it must be elevated to a senior manager/project manager/SME, who must make the final resolution.

6. *Comment matrix*: Utilize a comment matrix to keep track of the comment incorporation, particularly the comments to be addressed in a future submittal.
7. *Correction (as resolved)*: The completed review print set (with comments and responses) is then incorporated by the corrector.
8. *Verification*: Lastly, a "verifier" must sign off on the review print stamp confirming that all comments identified and responded to in the IDR have been incorporated into the updated design prints. Typically, that person should be the overall design lead.
9. *Training on how to perform the IDRs*: Training on the formal process of IDRs should be provided to the entire design team, preferably by the QM.

Eight Considerations for Performing IDRs Using the Electronic Checking Method

Traditionally, IDRs were performed using a hard-copy format. Documentation for QC reviews of IDRs consisted of hard copies of the design plan sheets and were successively reviewed by various design discipline leads for the applicable cross-disciplines for that design package. Traditional methods of quality control checking of design are increasingly being replaced by electronic and software-based QC checking methods. There are challenges and considerations with these IDRs, as follows:

1. *Software is just a tool and no more*: Several software programs offer convenient ways to perform IDRs. However, while performing the IDR using a software program, one must use the software only as a tool. It is wrong to assume the software is there to perform QC checking instead of the designer; it will not. The "quality" of design is not enhanced by the software; it is just a tool to perform the checking more efficiently. The effectiveness of the QC checking process and the experience of the designers and the reviewers after the IDR comments are made will determine the quality of the design.
2. *QC checking must remain the same*: The process of QC checking must remain exactly the same as defined for the traditional IDR with hard-copy paper prints. There should be no change in the procedure. Formal evidence of QC checking is maintained in the checkprints (a *checkprint* is the checked copy of the design, which includes the comments made by the checker as well as the resolutions of those comments to completion).
3. *Easy sharing of files*: The biggest benefit of electronic-based IDR is that it makes sharing of files with all stakeholders easier. Most software programs (Bluebeam-Revu, Autodesk-Build, Fieldwire, Adobe, and similar) allow designers to create, edit, mark up, and organize PDF files, which can then be shared with project partners.
4. *Reduction of review time*: Software-based review can cut down on QC review time, usually tabulating the comments more efficiently. However,

it is important to recognize that the use of software does not preclude the same application of QC and QA rules described in the checking procedures; the intent of QC and QA checking of the interdisciplinary design does not change when checked using software.

5. *Improved recordkeeping*: It is possible to establish a more comprehensive QC as well as QA records system in an electronic-based IDR process. However, it is important that QC processes be defined first and that users be trained on the process before they may be implemented to avoid failures.

6. *QC checks for IDRs*: The standard QC checking of IDRs should include at the minimum the following steps:

   a. *Checking:* The QC checker is the person who formally makes the comment on the drawing as part of the IDR.

   b. *Back-checking/comment resolution:* The lead designer is responsible for the resolution of the IDR comments. There are four possible resolutions:

      - *Acceptance of checker's comment*: The lead designer may accept the comment for resolution in the current submittal. They may notate the resolution next to the comment in green color.
      - *Deferring a comment to the next design submittal*: Lead designer may accept the comment but choose to implement it in the next design submittal.
      - *Rejection of a comment*: The lead designer may reject the comment but must document their reason for doing so.
      - *Clarifications*: The lead designer may decide that a clarification is required. This should be discussed at the comment resolution meeting,

   Note: The four resolutions above may be set up as the recommended options for the design lead to pick when performing the back-check/resolution. Color codes for the resolutions must be predetermined and agreed on, with training provided so the designers use the colors consistently. It is best to utilize a workflow-based QC checking process, which could be manual or software-based. A software-based IDR forces the workflow on the checking process, which manages the design process more efficiently (however, a software program by itself does not add quality to the design).

7. *Updated design (after incorporation of IDR comments)*: After the incorporation of comments as per step 6, an updated design must be developed that corresponds to the decisions that were made in the interdisciplinary QC review.

8. *Final verification*: The original checker/commentator may be brought back to close the loop and review the final drawing to ensure that the lead designer or designee did not disregard the comments without first

discussing them with the checker/original commentator. The original checker is the most appropriate person to act as the verifier that the resolution was implemented as agreed. Verification is an extremely important step, and failure to implement it has caused many failures in design IDRs.

Things to Watch for in Interdisciplinary Reviews

1. *Use tools/markup*: In most software programs, markup is done with tools and an option called "Markup" for the drawing. Using various options, the reviewer may mark up the drawing, which is then translated/ documented in the notes section (right below it).
2. *Repetitive comments*: For repetitive comments, it is best not to notate on every drawing. Instead, notate once and use "Typical," and it will notate the same comment on all the desired drawings.
3. *QC verification is* ***critical***: The original checker/commentator must review the drawing again to close the loop on QC checking. This ensures that the lead designer does not inadvertently or consciously disregard IDR comments without first discussing them with the checker/commentator.
4. *Do not flatten the design file*: Flattening involves moving the checker's markups from the PDF's annotations layer to the content layer. Doing so effectively makes them part of the drawing. If a file is flattened, markups cannot be edited without first unflattening them. Flattened files also cannot be audited to show who checked them and when. This makes the quality records incomplete, as it is no longer possible to know who changed the plans and whether they did the right thing or not.
5. *QA audit of the IDR and QA certificate*: Lastly, the IDR must be reviewed by an independent QA auditor, who typically issues a QA certificate that the entire IDR was performed correctly. A QA certificate is issued after validating that an auditable record exists for the proper execution of QC checking for the IDR. The QA certificate means that the independent QA person verified that proper design coordination has occurred between the disciplines before issuing the final design.

Typical Failures in Interdisciplinary Reviews

Wherever the IDR is not well-performed in a complex project, the resulting design quality is significantly impacted, which can translate into major cost increases and schedule delays due to the need for rework, redesign (design changes), repairs, and/or the owner being forced to accept poor-quality work.

IDR failures because of lack of effective design QC during and after the review have historically resulted in the biggest failures of design. The design conflicts between different disciplines make them "not buildable," which results in design changes after final design that always cause major cost overruns, schedule delays, and customer dissatisfaction.

Important Concerns in Interdisciplinary Reviews

- *Qualification of lead person for IDR*: The lead person responsible to oversee the QC IDRs may not be an SME in all contributing disciplines, but they should have a mature base experience that would allow them to make decisions in cases of conflicts that are elevated to the lead. The person selected for this role should have previously worked with multiple disciplines and a good multi-disciplinary background. Lack of understanding about the contributing disciplines has caused many QC IDRs to fail.
- *Using a workflow*: A proper QC IDR review also relies heavily on collaboration. It can benefit greatly from a well-planned QC workflow (shown in Figures 3.2 and 3.3) that connects otherwise independent functional disciplines and their teams. When the quality management process does not offer a good provision for the functional disciplines to work closely, the likelihood of errors, omissions, and conflicts greatly increases.

Participants in IDRs should involve all parties of the design team, including design subconsultants where applicable.

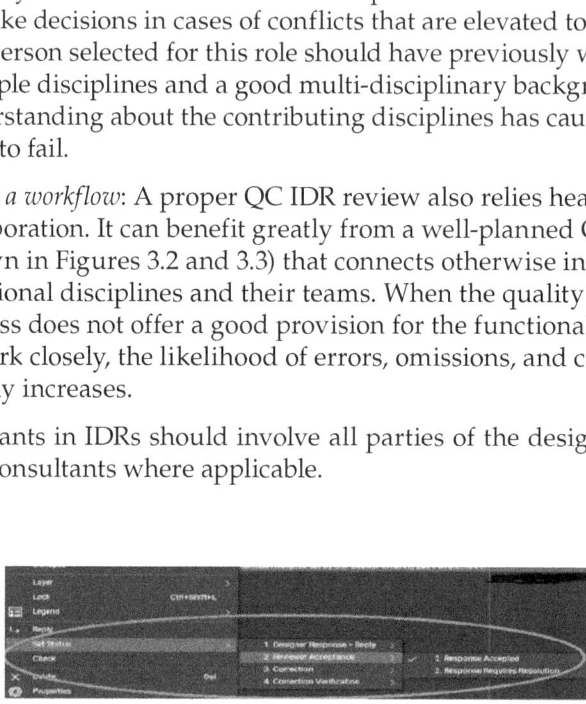

Figure 3.2 Interdisciplinary review example.

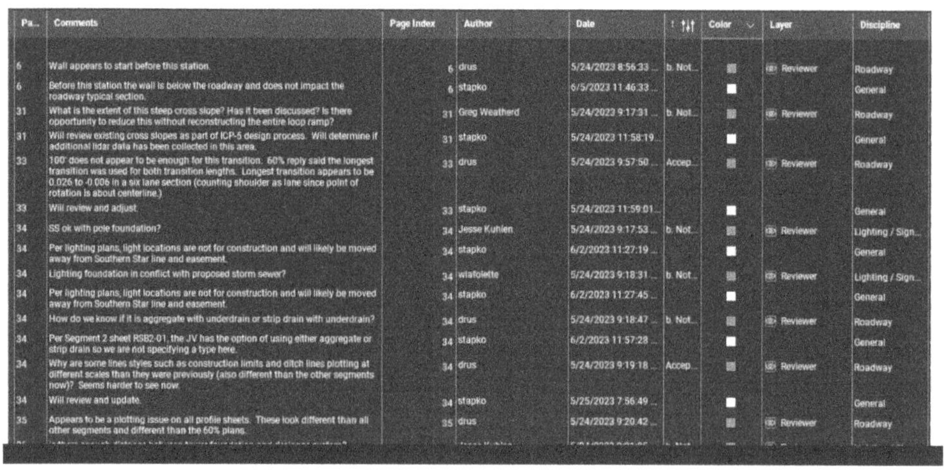

Pa...	Comments	Page Index	Author	Date	!	Color	Layer	Discipline
6	Wall appears to start before this station.	6	drus	5/24/2023 8:56:33 ...	b. Not...		Reviewer	Roadway
6	Before this station the wall is below the roadway and does not impact the roadway typical section.	6	stapko	6/5/2023 11:46:33 ...				General
31	What is the extent of this steep cross slope? Has it been discussed? Is there opportunity to reduce this without reconstructing the entire loop ramp?	31	Greg Weatherd	5/24/2023 9:17:31 ...	b. Not...		Reviewer	Roadway
31	Will review existing cross slopes as part of ICP-5 design process. Will determine if additional lidar data has been collected in this area.	31	stapko	5/24/2023 11:58:19...				General
33	100' does not appear to be enough for this transition. 60% reply said the longest transition was used for both transition lengths. Longest transition appears to be 0.026 to -0.006 in a six lane section (counting shoulder as lane since point of rotation is about centerline.)	33	drus	5/24/2023 9:57:50 ...	Accep...		Reviewer	Roadway
33	Will review and adjust.	33	stapko	5/24/2023 11:59:01...				General
34	SS ok with pole foundation?	34	Jesse Kuhlen	5/24/2023 9:17:53 ...	b. Not...		Reviewer	Lighting / Sign...
34	Per lighting plans, light locations are not for construction and will likely be moved away from Southern Star line and easement.	34	stapko	6/2/2023 11:27:19 ...				General
34	Lighting foundation in conflict with proposed storm sewer?	34	wlafolette	5/24/2023 9:18:31 ...	b. Not...		Reviewer	Lighting / Sign...
34	Per lighting plans, light locations are not for construction and will likely be moved away from Southern Star line and easement.	34	stapko	6/2/2023 11:27:45 ...				General
34	How do we know if it is aggregate with underdrain or strip drain with underdrain?	34	drus	5/24/2023 9:18:47 ...	b. Not...		Reviewer	Roadway
34	Per Segment 2 sheet RSB2-01, the JV has the option of using either aggregate or strip drain so we are not specifying a type here.	34	stapko	6/2/2023 11:57:28 ...				General
34	Why are some lines styles such as construction limits and ditch lines plotting at different scales than they were previously (also different than the other segments now)? Seems harder to see now.	34	drus	5/24/2023 9:19:18 ...	Accep...		Reviewer	Roadway
34	Will review and update.	34	stapko	5/25/2023 7:56:49 ...				General
35	Appears to be a plotting issue on all profile sheets. These look different than all other segments and different than the 60% plans.	35	drus	5/24/2023 9:20:42 ...			Reviewer	Roadway

Figure 3.3 Screenshots from a software-based interdisciplinary/cross-disciplinary review.

CONCLUSION

Interdisciplinary reviews are critical for the quality of design for the complex projects. The collaborative planning and design of multiple design disciplines requires a formal comprehensive design QC review, and is a precondition for achieving better, more responsible, and comprehensive design solutions.

Robust IDRs are essential for mitigating risks and ensuring project success. By adhering to the best practices outlined, including formal review periods, electronic tools, and meticulous comment resolution, teams can significantly enhance design quality. The effectiveness of these reviews hinges on clear communication, rigorous verification, and proactive quality assurance audits. Ultimately, investing in comprehensive IDRs translates to fewer errors, reduced costs, and a more seamless transition from design to construction, solidifying the foundation for a successful project delivery.

QUESTIONS

1. What is the purpose of interdisciplinary reviews (IDRs)?
2. At what stage of design completion should the interdisciplinary reviews take place?
3. What are the administrative prerequisites for performing a good IDR?
4. What are the steps of IDR in a traditional, nonelectronic method of design QC?
5. What are the steps of IDR in an electronic, software-based method of design QC?
6. What are the common mistakes made in performing the IDR?
7. Explain the difference between intradisciplinary and interdisciplinary reviews, and why both are important for successful project outcomes?
8. Describe three key benefits of utilizing electronic tools for IDRs compared to traditional paper-based methods, and also discuss one potential pitfall.
9. What are the engineer of record (EOR) responsibilities concerning design review comments, and how should conflicting comments be handled?
10. Discuss the significance of the 60% design stage in relation to conducting an IDR. Why is this timing often considered to offer the best value for the review process?
11. Outline the nine steps of performing IDRs using the traditional method, and highlight which of those steps is considered the most critical and why.

CRITICALITY LEVELS AND INSPECTION

> Analyze the proposed criticality levels for equipment and material. Identify appropriate inspection levels and methods (e.g., in-process inspection and third-party design review) based on criticality and budget considerations. (Analyze)
>
> Body of Knowledge III.C.3

Navigating the intricacies of project life cycles demands meticulous attention to criticality levels and inspections, especially within the design phase. This section delves into defining and analyzing these critical elements, highlighting the distinct approaches required for design outputs, procurement, and construction. Understanding the nuanced differences between criticality levels for design quality versus supplier products and construction acceptance is paramount. This analysis will guide stakeholders in identifying appropriate inspection levels and methods, balancing criticality with budget considerations to ensure project integrity and successful delivery.

To begin with, note that the criticality levels for design quality are very different from the criticality levels for the suppliers' products and services, as well as different from construction acceptance.

Although the development of the inspection and test plans (ITPs) is a collective responsibility of the designer and the construction contractor as well as the subcontractors and vendors, it is a good practice to include the designer (engineer of record) and the QM(s) with the development of the ITPs for the various phases of the project work.

Criticality Levels for Inspection, Testing, and Reviews

This section discusses the criticality levels for important phases of the project life cycle.

1. ITPs developed for procurement of equipment and materials (from suppliers and vendors)
2. ITPs developed for the construction phase of the project life cycle
3. Criticality levels for the inspections and quality reviews performed during the design phase of the project

Criticality of ITPs for the Materials Suppliers and Vendors

The criticality levels for the components and materials procured from suppliers and vendors must be defined after careful consideration of the specifications as well as other concerns, particularly with regard to the design and construction schedules developed for the project. The materials, component parts, assemblies, and services procured from external supplies frequently have high-criticality-level items that must be procured in time if the construction schedule is to be met. (This is discussed in greater detail in Chapter 4.) The items and components that may warrant more detailed and frequent inspections to ensure that their products meet the necessary standards and technical specifications should be included.

Expectations for Inspection and Testing for Material Suppliers and Vendors Usually, the technical specifications are the guidance criteria that outline the expectations for equipment and material inspection levels and methods. In well-developed bid documents, the following may be identified:

- What systems or assemblies require a mockup?
- What first work or initial inspections are needed?
- What equipment must be tested as a stand-alone unit?
- What equipment must be commissioned as a system?
- Is factory witness testing (FWT) required?

Use of Acceptable Quality Limit (AQL) in the ITPs for Material Suppliers and Vendors

The construction world does not frequently utilize the sampling plans that form the statistical basis of acceptable quality limit (AQL). The AQL is a concept that has its origins in the manufacturing world. AQLs, if implemented, determine the exact sample size and acceptance number that the quality inspectors must use when making their pass/fail decisions during in-process QC inspections of the construction materials, assemblies, equipment, and even the inspections of the completed construction of the infrastructure.

Notice that AQLs can be best applied on repetitive production, which makes the concept best applicable for the development of ITPs for the materials and component parts vendors of the construction projects. Where the AQL is implemented as the defining criteria for the ITP, the ITP may define three types of sampling plans (namely sampling plans for normal inspection, tightened inspection, and reduced inspection). The ISO 2859 standard describes the sampling schemes indexed by AQL for statistically determined samples.

In most construction projects, ITPs are still using arbitrary acceptance criteria, with some projects not even identifying or requiring statistically determined sample sizes and inspection frequencies. The move toward the use of statistically defined sampling plans is a recommended best practice, but it will only come with maturity of quality systems.

Criticality Levels for ITPs during the Construction Phase

Typically, the safety and performance criteria become the defining criteria for higher criticality in the ITPs for the construction in-process inspections. Based on their criticality for safety, the frequency of inspections and the levels of safety inspections are also identified (LetsBuild, 2024).

Usually, the development of the ITP is the assigned responsibility of the design team that is the official engineer of record (EOR) for the project. For the EOR, there is always a design engineering professional identified as the person who signs and seals the design drawings, reports, and other related construction documents issued for the project. It is customary and preferred to include the EOR in the preparation of the ITPs that are to be used in the subsequent construction phase(s).

The ITP is the official plan used by the construction team to perform QC inspections of construction activities. It includes all the critical aspects of the constructed product and defines the specific inspections and tests needed to verify quality at each stage of the construction. Development of the ITPs should be the collective responsibility of the designer (EOR) and the QMs for both the design and construction phase.

Criticality of Inspections and Review for the Design Phase

Criticality of inspection (Figure 3.4) is equally important for the design phase of the project life cycle. Given that this section is included in the design phase services of the BoK, it makes sense to also outline the criticality of design elements and the need for the right level of inspection and reviews.

The criticality of design increases as it moves from the schematic design stage through interim design and reaches the final design stage. The three stages of design are notated as follows:

1. *Preliminary design—commonly called schematic design (SD)*: This phase aims to establish the project's general scope, conceptual design, and scale. This is the concept that is sent to the client for formal approval of the overall design solution and methodology.
2. *Interim design—commonly called design development (DD)*: Typically, this is the design stage where the concepts established and approved during the SD phase are refined into more detailed and technical drawings.
3. *Final design—commonly called construction documents (CD)*: This is the final stage in the design process before construction actually begins. During this phase, the design is fully detailed and documented to ensure that it can be constructed accurately and efficiently.

Criticality Levels for In-Process Reviews of Design Submittals

The quality of in-process inspections/QC reviews for design is not governed by the sampling standards of AQL as may be possible in construction. The criticality for the design submittals is usually determined by the design team from the guidance provided in the design specifications.

Some elements of design are more critical than others. At times, the feedback from clients defines the more significant elements of design. Other times, the design team must perform its review of the requirements to determine which tasks warrant greater emphasis and enhanced quality reviews and inspections.

Additionally, different design elements carry varying levels of criticality due to their significance for the overall project. For example, the design of a tunnel in a complex construction project may include other related structures as well, but the tunnel itself may be considered the most critical design element. Likewise, in the design of a bridge, the sections of the bridge that include steel design of girders may need to be designed first due to the long lead times of the steel supplier.

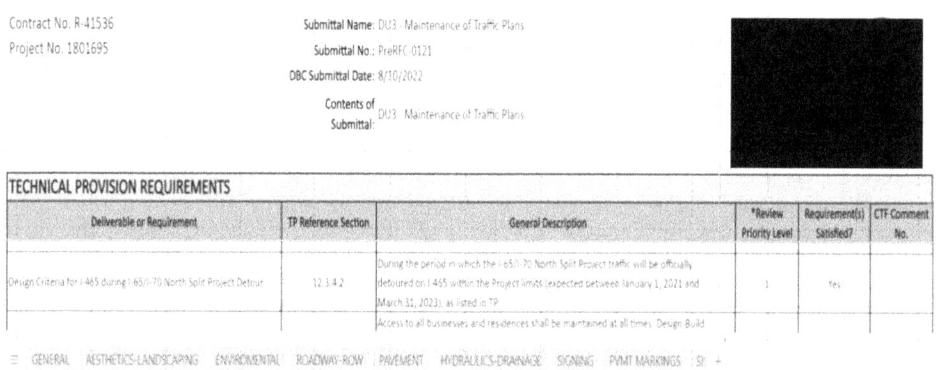

Contract No. R-41536
Project No. 1801695

Submittal Name: DU3 - Maintenance of Traffic Plans
Submittal No.: PreRFC 0121
DBC Submittal Date: 8/10/2022
Contents of Submittal: DU3 Maintenance of Traffic Plans

TECHNICAL PROVISION REQUIREMENTS

Deliverable or Requirement	TP Reference Section	General Description	*Review Priority Level	Requirement(s) Satisfied?	CTF Comment No.
Design Criteria for I-465 during I-65/I-70 North Split Project Detour	12.3.4.2	During the period in which the I-65/I-70 North Split Project traffic will be officially detoured on I-465 within the Project limits (expected between January 1, 2021 and March 31, 2023), as listed in TP	1	Yes	
		Access to all businesses and residences shall be maintained at all times. Design Build			

GENERAL AESTHETICS-LANDSCAPING ENVIRONMENTAL ROADWAY-ROW PAVEMENT HYDRAULICS-DRAINAGE SIGNING PVMT MARKINGS SI: +

Figure 3.4 Example of a design review checklist with criticality level 1 identified.

These are examples of some design elements being more critical than others based on the construction schedule.

In another example, good summer weather conditions in an elevated mountain pass may make the highest-elevation portion of the roadway the first one that should be designed. That may become the most critical design element due to the criticality for the construction schedule. The project teams must use stakeholders' meetings to determine the criticality and priority of design elements.

The design schedule should always be updated to accommodate the construction schedule; once the construction contractor is hired on the project, collaboration becomes easier. When developed with the feedback of the construction contractor, the construction schedule begins dictating the criticality of the design elements.

Notice that the criticality of design elements must be discussed and agreed to during the early design meetings. In addition to the priority, some design elements require additional reviews and approvals, such as those including regulatory and compliance aspects. Some examples of design criticality are described in the technical requirements and specifications as provided. Note that these are recommended ways to address the criticality of design submittals, but they do not represent a requirement. For example:

- Client approval is required for design submittals with a "C" designation.
- Design submittals that do not have a "C" designation do not require formal client approval. However, the design team is required to provide a QA certificate for design QA as evidence of design QC completion.
- Design submittals with an "S" designation must comply with additional safety certifications (see Spec 5XX.YY.ZZ for safety certification).
- The design teams also determine the criticality levels of the design packages based on the criticality of their delivery due dates.
- Likewise, the design elements that may cause a safety or security risk are usually identified as more critical than other design elements.
- Additional design requirements may be identified for more critical design elements.

Not all elements of design carry the same level of criticality; some design elements are more critical than others and are therefore given more significance in the critical path for the design as well as construction schedule. For example, concrete design for the airport ramp where aircraft are parked is more critical than the design of the concrete pavement on the side of the terminal building.

Higher-visibility projects with a lot of public oversight usually include higher levels of criticality for some design and construction elements, and as such warrant more advanced quality checks. An example would be a provision for performing "independent checking" for the critical elements of design and construction. If such criticality is desired, an independent organization is hired to perform design checking for the identified critical element/aspect of the design or construction. In such cases, the results achieved by the two design or two construction inspections may be correlated and accepted if within a predefined percentage of each other (say within 5%). The criticality of design elements, or the criticality of construction elements, is a topic that must be included as an important discussion item in the agenda of the ongoing design meetings.

CONCLUSION

Navigating the complexities of criticality levels and inspections across the design, procurement, and construction phases is essential for project success. By understanding the nuanced differences in criticality and tailoring inspection methods accordingly, stakeholders can ensure quality without exceeding budget constraints. This section has highlighted the importance of integrating designers and quality managers when developing ITPs and emphasized the shift toward statistically defined sampling plans. Ultimately, diligent application of these principles fosters project integrity, mitigates risks, and facilitates the delivery of high-quality outcomes.

QUESTIONS

1. What are the differences between the criticality levels for design product versus the criticality levels for construction acceptance?
2. What are the examples of higher criticality inspections?
3. How/why is a construction inspection and test plan (ITP) developed by the design team? Why is it included under Design for the BOK?
4. What are the different types of materials inspections for construction?
5. Do the critical inspections follow statistical criteria of "significance" or "acceptable quality level" for the quality control (or is it not necessary)?
6. How do the criticality levels for design quality differ from those for suppliers' products and construction acceptance, and why is understanding this distinction important?
7. Explain the role of the engineer of record (EOR) and quality managers (QMs) in developing ITPs for different project phases.
8. What is an acceptable quality limit (AQL), and why might its application be more suitable for material suppliers and vendors than for general construction inspections?
9. Discuss the relationship between the design schedule and the construction schedule in determining the criticality of design elements.
10. How does the criticality of design reviews evolve throughout the schematic, interim, and final design stages, and what factors influence these changes?

Quality Controls in Design

PROJECT CONTROLS

> Evaluate the effectiveness of project controls, including the timeline, in meeting the quality expectations of stakeholders, such as the client, owner, or user. (Evaluate)
>
> Body of Knowledge III.D.1

Project controls serve as the backbone of a successful construction project, ensuring that costs and schedules remain on track. Though often viewed in the context of the design phase, project controls are applicable throughout all stages, from initiation to closeout. By systematically gathering and analyzing project data, project controls provide a framework for monitoring, managing, and communicating financial and timeline performance. This discipline is crucial for mitigating the pervasive risks of cost overruns and schedule delays, aiming to keep stakeholder expectations aligned with actual project execution and budget constraints.

In the context of the quality of a construction project, project controls are processes for gathering and analyzing project data to keep costs and schedules on track. The definition of the project controls function is larger than the context of the design phase; it is equally applicable to all seven phases of the construction project. Typical functions of project controls include initiating, planning, monitoring and controlling, communicating, and closing out of the project costs and schedule. Through project controls, the project limits the use of funds that are designated for a specific phase and task from being used for other purposes.

Despite best efforts and intentions, many organizations find that large-scale projects miss their primary targets of cost and schedule. According to the EcoSys team, "In large-scale capital projects, 98% of projects experience cost overruns or (schedule) delays. On average, cost increases are estimated at 80% of the original value, and timelines are often delayed by 20 months or more" (2019). It is in this context that project controls are essential as they assist the project management in meeting the expectations of the stakeholders, keeping the spendings within the approved budget for all phases of the project.

It is important to recognize that project controls are a support function for project management; they are not a quality function. As such, they can help monitor and control the costs, schedules, and scopes of the various tasks of the project.

Project Controls requires the collaboration with the quality department for cost as well as schedule management. They require that the invoices that are sent by the subcontractors and suppliers be verified before they would pay them. It is in this respect that project controls cross the path of quality, whose inspectors can be a useful resource to ensure the work performed by the contractor and the subcontractors is complete and meets the quality requirements as well. Depending on the nature of the contract (project delivery method), sometimes the payments of subcontractors' and subconsultants' invoices require the affirmation of QC inspectors and/or approval by a QA supervisor/manager before they may be paid. The different methods of project delivery (particularly time and materials method) include a much closer collaboration between the project's quality staff and the project controls function.

There are further subcategories of tasks that project controls additionally perform to support the two primary objectives of cost and schedule compliance for the stakeholders' expectations of on-time and within-budget projects. However, those aspects of project controls are geared more toward supporting project management, not quality management.

The following are some of the key aspects of project controls as they pertain to quality.

Project Controls Collaboration with Quality for Cost Control

- *Invoice management*: Project controls keep the project current on invoices and progress payments (for design as well as construction phases). Completion of work is typically confirmed by QC inspectors. It is a recommended best practice for project controls to issue biweekly status reports on invoices and payments. Project controls also ensure that all costs are chargeable to the cost categories where they belong and monitor the project to ensure the costs do not exceed the budget.
- *Changes and claims records*: Decisions on requested changes by the subcontractors and subconsultants require direct input of the quality department. Also, the resolution of claims requires direct collaboration with quality that verifies the completion of corrective actions, based on which information the project management or their designee approves the payments.
- *Cost variance reporting*: Project controls monitor and report variance from the approved budget and provide reasons, which may be used by quality as feedback for continuous quality improvement. (Cost variance is an indicator of the cost-efficiency of the project). (Lessons learned in the construction phase of the project are described in greater detail in Chapter 5.)
- *Gross profit (or financials)*: Project controls also perform ongoing earned value (EV) analysis to allow the project/construction manager to assess progress, forecast future performance, and make informed decisions to keep the project on track. EV integrates the project scope with data on schedule and cost.

  Earned value is built on three metrics: planned value, earned value, and actual cost. In all these categories, the quality department (QC and QA) plays a role as they are usually asked to confirm the completion of work performed by the subcontractors, as well as design subconsultants.

- *Subcontractors' payments*: Project controls ensure that payments are approved only after quality inspectors' work completion reports are issued.
- *Force accounts*: Force account work is authorized in construction contracts when it is agreed that additional work is now needed that was not included in the original scope of work. A written "change order" affecting a change in the work is issued, which may involve either an adjustment in the contract price or an extension of the schedule.

  In these cases, work performed by the subcontractors is paid only after the actual costs for labor, materials, and equipment usage incurred in the performance of the work have been verified by the QC inspectors assigned to the project. To enable such a contract, the time and materials records of the subcontractors are verified through the quality inspections. Project controls collaborate very closely with quality for all force accounts works.

  It must also be noted that besides the time and materials method of project delivery, force accounts may also be used in those contracts where there is no unit price. Project controls would similarly depend on QC inspectors to verify the completion of work in such contracts as well.
- *Certified payrolls*: Compliance with prevailing wages is another responsibility of project controls. This is more of an administrative function that is performed by project controls and may or may not involve the quality inspectors as verifiers.

Project Controls Collaboration with Quality for Schedule Control

- *Project's completion status*: Project controls typically monitor the completion of work against the contractual requirements. Different phases of the project are reported in terms of completion of the tasks and subtasks. Project controls link the funds allocated for a phase to the completion of that phase, thereby providing the project with a very effective control over schedule as well as cost.
- *Linking deliverables with the schedule*: In some projects, the deliverables are described as CDRLs (contract data requirements list), which is a format for describing the requirements of the contract, usually in terms of specific submittals or data lists.
- *Monitoring of the schedule*: Project controls also work with the project schedulers (for design and construction) and coordinate with the owner as well. It must also be noted that they are not always a dedicated role in all projects; for smaller projects, this role is sometimes handled by the project/construction manager themselves. Regardless of whether there is a dedicated person in the role or not,

it is project controls that schedule variance from the approved project schedule. Project controls are therefore also able to identify the reasons for the delay, whether it's from the design team, the contractor, or the owner. Project controls thereby provide the management with the key performance indicator (KPI) of schedule efficiency.

- *Performance of designer (and contractor) per project schedule*: Project controls keep the designer firms as well as the contractors on track for their submittals and for completion of work, on time and on budget.

CONCLUSION

Project controls are a support function to project or construction management. They evaluate the project's conformance to the allowed cost codes and ensure that the project adheres to the discipline of using the right cost codes for all stages of work. However, they do not measure the quality of the design or construction in the traditional sense of QC or QA—that is, they are not a quality function.

Project controls enable the project to have the right accountability of the work that is performed. Also, it is important to reemphasize that they are not limited to the design phase. They are another one of those support functions that is universally applicable to all the phases of the project (namely to the contract, planning, design, procurement, construction, closeout, and O&M phases).

While project controls do not serve a quality function, their collaboration with QA and QC is vital for project success. By ensuring financial accountability and schedule adherence, project controls provide the essential framework that enables quality work to be completed within budget and on time. This support function extends across all phases of a project, facilitating transparency and accountability, and ultimately contributing to the successful delivery of project goals and stakeholder satisfaction. Thus, project controls serve as a critical element in the broader context of project management and their interaction with quality management.

QUESTIONS

1. What control processes are put in place to keep the project cost and schedule on track?
2. What are the two elements that may be controlled through project controls?
3. Why are project controls not a quality function?
4. For which project delivery methods do project controls require direct input from quality?
5. What is the role of project controls in the contract change orders and in claims?
6. How do project controls facilitate the monitoring of both cost and schedule adherence throughout all seven phases of a construction project, and why is this crucial for stakeholder expectations?
7. Explain the difference between project controls as a support function and a quality function, and then describe specific scenarios where project controls and quality departments must collaborate closely.

8. Discuss how earned value analysis, which involves planned value, earned value, and actual cost, relies on input from the quality departments (QC and QA) and why this input is necessary.
9. Describe the role of project controls in managing force accounts, detailing why QC inspectors are essential in verifying costs associated with such work.
10. How do project controls contribute to schedule control, and explain the significance of monitoring schedule variance and linking deliverables (such as CDRLs) to the project schedule?

DESIGN REVIEW OUTPUT

> Examine the design review output received at each phase (e.g., 30%, 60%, 90%) to determine its alignment with the expectations of stakeholders, such as the client, owner, or user. (Analyze)
>
> Body of Knowledge III.D.2

The design phase stands as a crucial juncture in any construction project, where concepts transform into tangible plans. Ensuring quality controls during this phase is paramount, beginning with a meticulous review of design outputs. Examining these outputs at various stages, such as 30%, 60%, and 90% completion, allows for alignment with stakeholder expectations, encompassing clients, owners, and users. Evaluating design outputs against requirements and performance criteria guarantees the final product meets the intended objectives and maintains project integrity throughout its life cycle.

At the input stage, design objectives are established after aligning with stakeholders' expectations. (This section is related to "Design Inputs: Quality Objectives," which are part of the design input.) Here, we will discuss how to evaluate the design review outputs and evaluate them against stakeholders' expectations as well as against the design's requirements and performance criteria.

Design outputs are the work products of the design phase. They are generated as a result of the design process, and they include design drawings, calculations, process specifications, materials specifications, fabrication instructions, and construction inspection (testing) procedures.

Three Gates for Design Deliverables

Design review output for construction projects is typically issued in stages/phases. It normally starts from a feasibility concept (±10%) and makes its way through the preliminary design stage (30%). The design is then developed to the intermediate design stage (60%) and finally to the final design stage (±90%). Usually for public projects, the final 100% design is the same as the 90% design with very minor changes, but with client's formal approval. (The 10% design is an arbitrary design deliverable that is less common in the design and construction industry, but it exists.)

The American Institute of Architects (AIA) divides architectural/design services into three halves and calls them schematic design (SD), design development (DD), and construction documents (CD) (Hitech BIM Services, 2021). This corresponds to 30% (schematic design), 60% (design development), and 90% (construction documents).

It may be noted that although the above-mentioned design delivery stages are common in the industry, there is currently no universally acceptable standard that specifies exactly what should be included at the 30%, 60%, and 90% design completion stages. As a result, SD, DD, and CD are still arbitrarily determined by the scope of individual projects. There may be clients who choose to have a slightly different scheme of phase completions for design and specify exactly what is expected to be completed and included for submission at each stage of the design output. Design teams should therefore get clarity on the design output delivery stages from their clients at the beginning of a project and preferably develop checklists that describe the design elements that need to be provided along with the level of detail that is expected for the various design submittals.

Note: This section is also related to "Design Monitoring" later in this chapter. Here in this section, we will describe what is typically included as design deliverables at each of the three gates. More specific information about what should be monitored at each of the three stages (SD, DD, and CD) is provided in "Design Monitoring."

Preliminary Design/Schematic Design (SD)

Thirty percent issued for review (IFR), also known as preliminary design/schematic design (SD), is the first draft of the design prepared by the design team and officially submitted to the client for approval of the overall solution methodology. (*Note*: The solutions methodology is also sometimes called the basis of design.) At this stage/gate of design deliverables, the designers (architect/engineer) ensure that they have a clear understanding of the scope of the project and will adhere to project criteria, formats, and conventions.

Typical design outputs that are generally expected to be completed by this early design stage may include the following (not an all-inclusive list):

- Basis of design (BOD)
- Conceptual design report
- Conceptual drawings using finalized alignments
- Early plan and profile sheets
- Topographic surveys (existing topography and possibly new topography)
- Geotechnical analysis of the construction area
- Environmental studies (if any were done)
- Utilities and stormwater flows

- Right-of-way plans, vicinity plans
- All design elements and their components identified
- All applicable regulatory codes and standards identified

It must be noted that not all of the above-mentioned design deliverables may be required by all projects. All construction projects are unique, and as such it is the project scope of work that determines the products and services that should make up the deliverables at the major milestones of the project.

Value engineering may begin as early as the 30% design completion stage. Typically, value analysis offers the greatest benefit when it is performed early in the product development life cycle. This is because at the early stages, not every design decision has been made, which allows for more impactful cost-saving and process optimization decisions. For those projects considering alternate methods of construction to improve the value of the constructed infrastructure, the 30% design completion stage is the most ideal stage to create alternate options in design.

Executing value engineering at this early stage of design development keeps it most economical. Performing value engineering at a later stage of design completion always adds to the cost of the value engineered solution because a lot of design decisions already made may need to be undone and/or redone.

Intermediate Design/Design Development (DD)

Sixty percent issued for procurement (IFP), also known as intermediate design/design development (DD), is generally the second QC gate for the design deliverables. The primary purpose of 60% design output is to allow the client (or the performing contractor) to begin procurement by soliciting vendors to bid on the major elements of the project, procure the materials that may be necessary for construction, and have longer lead times.

The AIA calls the 60% design stage the "design development" stage. It is an extremely important deliverable stage for the design, and as the name describes it, it takes the design from the schematic concept and develops it into the refined and final design that can be built.

Although all construction projects are unique, the following list of outputs (deliverables) may qualify as a common set of 60% design deliverables for most construction projects:

- Established reference points for surveys
- All environmental studies completed and issued
- Utility designs completed
- Interdisciplinary reviews performed, along with estimated checks to verify the design is still within the projected budget
- Constructability reviews performed and comments incorporated
- Piping and instrumentation diagrams (P&IDs) included in applicable projects
- Equipment schedules complete

Although it is not preferred, depending on the clients' wishes, the design schedules and specifications may still change at this 60% completion point. However, by then, the vendors generally have sufficient information to make accurate bids, and the procurement personnel should be able to decide who should be awarded the procurement contract while the design is still being finalized. The design outputs at the 60% completion stage generally also include sufficient guidance for the fabricators to purchase their raw materials for the critical long-duration lead items.

Using QA audits and monitoring, the QM must confirm that the design outputs align with the quality expectations set by the client and stakeholders. (Chapters 2 and 5 describe the QC and QA monitoring activities in greater detail.) This requires comparing the design output against the agreed-on scope and quality standards for the interim design/design development stage, as defined in the project scope documents.

Construction Documents (CD)

Ninety percent issued for construction (IFC), also known as construction documents (CD), is typically a design that is almost complete and is ready to be issued for construction. The design plans are expected to be complete to a degree based on which the project may be built. Granted that all construction projects are unique and may require different design deliverables, the following list of design deliverables may qualify as a common set of 90% design completion for most construction projects:

- Design schedules complete
- Final design drawings that clearly define the construction work and are ready to be issued for construction (IFC)
- Design drawings accompanied by final specifications, final design calculations, and notes
- Final cost estimate check performed and quantities determined for use by construction for planning and procurement

Technically, there should be no engineering or design changes made after this final design stage. All design comments made by different stakeholders, including the client, have been addressed. Also, there is really no difference between the 90% and 100% design output—the 90% design documents should automatically become the 100% final design documents once they are formally approved by the client. The final drawings and specifications at 90% design completion should be detailed enough for the client (or authority having jurisdiction) to offer them for bidding and construction of the project by a qualified builder (contractor).

In a well-planned quality management process, there should be no major comment or substantial change made by the client to the 90% design output. The formal approval of the client should only be an administrative step that ensures they are satisfied with the final design. An important role of the QM is to ensure that QA is effectively performed for both the product and process quality, to ensure they are aligned with the stakeholders' expectations as well as against the design's functional and performance criteria.

Alignment of Design Output with Client Expectations

It is the responsibility of the design organization to determine that the design is aligned with the project's technical requirements as well as the expectations of all stakeholders. As described in the details of the three gates above, the typical design deliverables must all be prepared in compliance with the product technical requirements and should also comply with the general alignment of the higher-level quality objectives of the stakeholders. ISO 9001 defines the following four requirements for the design and development outputs to meet the design and development goals and objectives:

1. *Outputs must meet the input requirements*: This step is crucial. The design process must verify that the output (i.e., final design) fulfills the needs and specifications that were outlined in the design input.
2. *Outputs must be adequate for the subsequent processes*: The design outputs (i.e., final design) should be adequate according to the specifications of the customer, regardless of whether that design is for products or services.
3. *Outputs must include or reference the monitoring and measuring requirements*: The design outputs must include, or refer to, the appropriate monitoring and measuring requirements as well as meet the criteria for the acceptance of design. Usually, the criteria or standard of acceptance are the technical specifications provided to the designer to design the system for. These are the specifications that are reviewed against the design by design QC. Using these specifications, design QC determines whether or not the design is meeting the project specifications.
4. *Outputs must include the characteristics of the products and services that are essential for their intended purpose*: In simple terms, this means that the resulting products or services created because of the final design should function as intended. For example, if the design specifications require the design to meet some specific safety requirement, then the design outputs should be safe and properly functioning according to those specifications and must operate as intended.

CONCLUSION

Design and development output review is essential to ensure that the product or service being designed is developed as intended by the stakeholders. Through the design process, the design input is transformed into the design output. While the design output might take many different forms, the eventual requirement that all designs must meet is the same; the final design should be able to help create a product or service that matches the final consumers' requirements and the market demand.

Rigorous evaluation of design review outputs is essential to ensure project success. By scrutinizing deliverables at 30%, 60%, and 90% completion stages, stakeholders can verify alignment with initial objectives and requirements. This process ensures the final design meets its intended purpose, fulfills all performance criteria, and complies with applicable standards and regulations. This meticulous approach to design review and output analysis ultimately safeguards the project's integrity and ensures a final product that satisfies all parties involved, from client to end user.

QUESTIONS

1. What are the typical stages/milestones for the outputs for the design phase (30%, 60%, 90%)?
2. What should be included in the schematic design (SD), design development (DD), and construction documents (CD)?
3. What is the significance of value engineering for design quality?
4. When should value engineering be initiated in a design project?
5. What are the benefits of value engineering?
6. Explain the significance of reviewing design outputs at 30%, 60%, and 90% completion stages. What specific aspects should be evaluated at each stage, and how do these reviews ensure stakeholder alignment and project integrity?
7. While the document mentions typical deliverables at each stage (30%, 60%, 90%), it also notes the lack of a universal standard. How should design teams address this ambiguity with clients at the beginning of a project? What strategies can be employed to establish clear expectations for design output delivery?
8. Value engineering is emphasized in the document. Why is the 30% design completion stage considered the most ideal time for value engineering? What are the potential consequences of performing value engineering at later stages of design completion?
9. According to ISO 9001, what are the four key requirements for design and development outputs? How do these requirements relate to ensuring the design output meets stakeholder expectations and functional/performance criteria?

DESIGN CHANGES

> Understand sources of design change communication, including requests for proposals (RFPs), change order directives, addenda, and requests for information (RFIs). Assess and track the impact of proposed design changes and determine the effectiveness of incorporating any agreed-upon changes into the contract. (Analyze)
>
> Body of Knowledge III.D.3

Design changes are an inherent part of any project, especially those involving complex infrastructure. Even after the "released for construction" (RFC) stage, adjustments are often necessary due to unforeseen issues, field feedback, or evolving requirements. These changes, known as "design changes," are crucial for ensuring the final product aligns with actual conditions and delivers optimal performance. Flexibility in the design process is essential, as acknowledged by the American Society for Civil Engineers (ASCE), allowing for necessary modifications without derailing the project's overall progress. Accepting reasonable costs for beneficial changes is also vital, promoting better outcomes and long-term value.

Design changes are sometimes made to the final design because they are necessary for various reasons, such as addressing unforeseen issues, incorporating feedback from the field, or adapting to evolving requirements. Generally, the final design is considered a very late stage to incorporate value engineering, but sometimes it is necessary. Regardless of their reason, any change that needs to be made to the final design after it has been issued as RFC must be issued by the designer as a design change.

ASCE acknowledges that not all information will be known at any one time, which is why the design process needs to be flexible enough to accommodate the inevitable changes as decisions are only made with the best information at that time. The design process should "build flexibility into the design activity plan to allow for changes and future project development, as well as associated budget and schedule revisions" (2012). This section is closely linked with Chapter 5, "Requests for Information or Changes." The description of changes from the design perspective is provided here. The description of changes from the construction perspective is provided in Chapter 5.

Two Methods for Making Design Changes

Design changes happen in two ways:

- By the designer, also called design change request (DCR)
- By the contractor, also called field change request (FCR)

Design Changes Due to Value Engineering by the Design Team

At times, a designer initiates a value-engineered design change that can achieve an equal or improved design at a reduced life cycle cost. In some cases, this is due to an identified error or a conflicting design omission that slipped through the interdisciplinary design checking.

The design change process must be formalized—that is, it must be included as part of the design quality plan. The design lead should make the changes to the plans or specifications along with all related documents. The QC checking of the new design must follow the same quality procedures implemented in the production of the original design.

Formal approval by the design lead/manager should be included in the records of the design change and maintained by the project's document control. The formal approval of the design change must be tracked, preferably through a workflow (if available). The design team's document control should also maintain a running log of all notices of design change (NDCs) that are issued by the design team along with their related documentation.

Lastly, the design change must be communicated to all stakeholders, with advisory regarding the replacement design documents (e.g., updated design plans, calculations, updated specifications, and all other documents that are changed).

Design Changes Due to Field Conditions

Sometimes the construction contractor identifies the need for a design change when they determine that the final design is not buildable as designed or encounter an unforeseen condition. Such a design change is called the field change request (FCR). A request made by the construction contractor must be processed through the project's document control system (preferably through a workflow).

- Typically, a request is sent by the contractor to the designer in the form of a request for design change. The process for formal request and initiation of design change due to a field request must also be formalized—that is, included as part of the design quality plan.
- The design as well as construction teams must be educated/trained about the design change procedure, and it must be implemented through the project QMS, and not as an independent design change process.

Typical Failures of Design Changes

The following types of failures typically occur during the design change process.

Delays Caused by Informal Design Change Process

In all projects that do not proceduralize the design change process, failures occur in the form of delayed processing and approvals. The procedures for the two types of design changes must be predefined in the design quality plan in the form of an associated procedure.

The procedures must identify the roles and responsibilities, and preferably force a workflow to the design change, which enables the project to have a more effective control over the time delays that otherwise happen due to slower processing by individuals. Design change workflows increase the efficiency, reduce errors, and improve collaboration between the different stakeholders and enhance the productivity of the design change process by streamlining the process from start to finish.

Document Control Failures

The most common failures in design changes happen due to poor document control practices on projects. A well-executed document control system that has the capacity to process the design change as a workflow can get the required approvals in a controlled time and ensure that the updated documentation reaches all stakeholders. Poorly executed or disjointed document control systems, on the other hand, become a source of failures occurring due to design changes.

Corrected/updated design must replace the previously issued design, and all related documents affected by the design change must also be updated. Many design change failures happen when the change is not implemented on all the related documents, which may either be identified late through audits or sometimes never identified until failure occurs (mostly during construction).

Another significant document control failure for design is that in many projects, the document control system of the designer is not well connected with the document control system of the contractor, and they are both not well connected to the client's document control system. This is a very common source of design failures in fast-moving design-build projects. The document control system must not only be inward-facing for the design team, but should also have the ability to interface with outside stakeholders including the contractors, owners, and any other major stakeholders who must be included in the design changes.

Projects that involve the contractor early in the design process offer benefits like enhanced collaboration and early problem resolution, and result in cost savings, improved quality, and (in general) faster project completion. Project delivery methods where the designer and builder work together from the beginning such as design-build and its variations or the collaborative project delivery through the integrated delivery method offer opportunities of common document control repository that can facilitate a direct connection between the designer and the contractor.

In other delivery methods, it becomes the responsibility of a third party to connect the designer with the contractor and ensure that document control failures do not happen. Depending on the contract and the delivery method selected, this may become the responsibility of the owner or the construction manager firm. The QMs can help establish and facilitate robust document control systems to ensure improved compliance and drive operational efficiency through a direct connection between the designer and the contractor.

Tools/Sources to Transmit Design Changes

The next sections discuss four tools that may be used to transmit the design changes: RFIs, design directives, design addenda, and change order directives.

Requests for Information (RFIs)

A request for information (RFI) is a tool for the contractor to ask the designer to comment on a design issue. An RFI may become a design change, or it may not. All RFIs should be formally documented and managed through the document control process.

Once the contractor requests designer's feedback regarding a design-centric situation, the designer must respond with their professional advice. For example, many times, RFIs are a contractor's way to request the allowance to utilize alternate materials they believe would not alter the end product, and they need the designer to corroborate their decision. In such cases, the designer needs to evaluate the request change and determine the impact on the overall design. If, after the design evaluation, the designer feels the change requested is acceptable, they may utilize the approved tools and methods to transmit the design change.

Design Directive

Sometimes during the course of their design, the designers create a parameter with which other disciplines must interface. In these cases, the design manager must issue a design directive to make the other disciplines aware of the parameter. A design directive is formal, written guidance on how the change parameter should be implemented. In complex projects, these directives can be issued by the designer to the subdisciplines within the design team.

The client or owner may also intend to issue design-related guidance to the designer team or a design directive to the designer. Because of the significance of the directives, it is essential that they be formally issued through the project's document control function, which brings it under effective controls.

The QM must ensure that the creation and development of design directives follows the best practices of consistency, clarity, and alignment with overall objectives. These directives must never be communicated verbally; they must be issued formally, preferably through the centralized document control, and identified by number and date of issue.

Design Addenda

When design changes are made that change the previously issued final design (contract construction documents) after the contract has been advertised, then those design changes must be captured in a formally issued design addendum. The addendum/addenda should note when the change is made. All design addenda should be issued formally through the central document control process so all updates are correctly documented and do not adversely affect any one discipline because of lack of communication.

Change Order Directives

A change order directive is a formal directive issued by the owner/client, and it represents formal approval to modify the original scope, schedule, or budget of a project. The design change could be one reason for getting the change order approved by the client, as it represents a change in the originally accepted terms of time, cost, and scope of work.

Design Changes and Cost

Clients and their project teams are sensitive to design changes and their associated costs. Small costs for design changes that incorporate better alternative products or account for better installation and construction methods—during the right time in the design process, or early construction administration, if it is appropriate—should be acceptable (Kelly and Male, 1993). Design changes that happen early in the design process are less expensive when compared to those that take place after the design is finally issued and introduced into full-scale production/construction.

ISO 9001 requires that for any change that is made during, or subsequent to, the design and development of products and services, the organizations must identify, review, and control the changes to ensure that there is no adverse impact on conformity to requirements (2015). Design changes typically propagate a ripple effect in the final design because a lot of related design documents must also be changed to effectively implement the change. As such, QC checking of a design change must undergo the exact same rigor of QC checking that was included in the quality reviews and approval of the original design.

When the construction quality manager is hired later in the project life cycle, it should be the responsibility of the design QM to implement quality. During the design stage, the QM must verify that once the change is approved, the design modifications are effectively incorporated into the project's contract. The responsibility to update the contract and ensure alignment with stakeholder expectations belongs to the design manager/project manager of design.

The QM may also want to analyze how the design changes are affecting the project's quality, schedule, and costs. Depending on the size and scope of the project, and the significance of various disciplines to the overall success of the project, a more comprehensive formal analysis may be performed on the design changes. Such evaluation may include the impact of the design change on three critical areas—namely quality, schedule, and costs. (See "Project Controls" for details on how to keep costs and schedules on track.) Also, records of such post-analysis may be maintained in the permanently kept records of the project as well as included in the lessons learned for future reference. (See Chapter Five, "Lessons and Improvements," for more detail.)

CONCLUSION

Effectively managing design changes is crucial for successful project delivery. Clear procedures, robust document control, and proactive communication are essential to mitigate delays and errors. Recognizing the two distinct pathways for changes—designer-initiated and field-driven—allows for tailored responses and efficient workflows. Embracing a flexible design process, as advocated by ASCE, and accepting reasonable costs for beneficial modifications ensures that projects adapt to real-world conditions and achieve optimal outcomes. By addressing potential failures and leveraging appropriate communication tools, organizations can navigate design changes successfully, safeguarding project quality, schedule, and budget.

QUESTIONS

1. What are the two methods for design change?
2. What are the common failures that occur during the design change process?
3. What tools can the designer use to communicate the design changes?
4. What are the best practices in the design change process to perform the design changes with lower cost and greater accuracy?
5. What is the best method to collaborate a design change between the designer and contractor in a traditional project that does not have them working together under one team?
6. What are the two primary methods by which design changes occur, and how do the associated processes differ in terms of initiation, documentation, and approval?
7. Describe the typical failures that can arise in the design change process, particularly focusing on document control issues, and explain how these failures can impact project outcomes.
8. Explain how ISO 9001 standards relate to the management of design changes, and discuss why a rigorous QC check is necessary for any design modification, regardless of its perceived size or impact.
9. Discuss the role of requests for information (RFIs) and design directives in the context of design changes. How does each tool facilitate communication and control within the project?
10. In project delivery methods where the designer and contractor do not have a direct, collaborative relationship from the outset, what measures can be implemented to mitigate document control failures during design changes, and who typically bears the responsibility for ensuring effective communication and coordination?

DESIGN VERIFICATION AND VALIDATION

> Understand design verification and validation and understand how the design output conforms to the design input through the use of reviews, inspections, and testing. (Understand)
>
> Body of Knowledge III.D.4

Design verification and validation are critical components of ensuring quality in any design process. This section delves into the nuances of these two distinct yet interconnected activities. Validation serves as a QA planning activity, performed before design commences, while verification is a QC process conducted during and after design. Both are essential to guarantee that design outputs conform to the design inputs and ultimately meet the client's requirements.

Difference Between Verification and Validation

Verification is the process of verifying that the product that is produced is compliant with the requirements. Validation is the process that ensures that a compliant product can be produced. Validation confirms "can we build the right product?" and verification confirms "did we build the product right?" Validation is a QA activity, whereas verification is a QC activity. It is also important to understand that not all projects have the staffing resources to formally separate the QC and QA roles. As a result, sometimes validation and verification are performed by the same staff; they just wear different hats while performing the two tasks.

Design Validation Must Be Performed First

The product that comes out of the design process is the "design." To ensure that we can build the right product, design validation must be performed before the design can actually begin. In other words, the designer must know ahead of doing it that they can design a product that would meet the client's requirements. This makes design validation a QA planning activity, not a QC activity.

Design validation should be performed before the design can commence to ensure that the resulting product (design) meets the requirements for the specified application or intended use. The simplest example of design validation is ensuring that a design software can design correctly before utilizing that software.

Design Validation Explained

Design validation means assuring that the product (design) would be capable of meeting the client's requirements and expectations (i.e., project's true needs). Generally, the design validation can be categorized under three validation activities:

1. Design software validation

2. Designer's checklists
3. Validation of the design input requirements

These three validations are described in detail below:

1. *Design software validation*: Software validation is the most significant element of design validation. Its purpose is to make sure the software is configured correctly and working as it should—so when the designer uses it, they get the correct design output.

   When initiating the design, the design team must first identify the software programs that they plan to use. The QM may verify that all software validation is either available with the designers or they prepare it before they can use it to create their designs.

   The designer must be able to answer the question: "How do I know this software is giving me the correct output?" To answer this question, especially for first-time use, the software must be validated by the designer to ensure that it is giving them the correct (or expected) result.

   The key things to consider for design software validation include the following:

   - *The application for which software is to be used*: The design software can be used to create a lot of different design applications. The project only needs the software validation for the application the designer is using it for. The software does not need to be validated for every application that it is capable of creating (nor should it be).
   - *Software version number*: Software is updated on a continuous basis by software manufacturers. If the version has changed, a new validation study is needed; otherwise, the designer may just compare the new version's results with the previous version and its previously available software validation.
   - *Software validation is the responsibility of the individual designer*: Software validation cannot be the responsibility of another person. The designer must take ownership of the software they choose to utilize for creating the design. It is not the responsibility of the software manufacturer.
   - *Design software cannot be validated for all applications under all scenarios*: It is impractical and wasteful to try to validate design software for all applications. Validation needs to be performed on a case-by-case basis, because different applications of the same software may differ from each other.
   - *A validation procedure is a necessity*: Every design organization must issue a design procedure that governs the validation of design software. A QM can assist the design team in this area as well and keep a record of all the validated software programs.

2. *Designers' checklists*: A checklist can be a valuable validation tool. A good design checklist can provide a structured, systematic way to ensure that

specific criteria or requirements are met. Design checklists specify the criteria, steps, and methods for validating design, and can help ensure the designers don't make errors and omissions. This, in turn, produces a deliverable design that meets requirements and does not need revision.

3. *Input requirements*: Designers must ensure they have the input requirements necessary to design correctly. This can be assured at the requirements management stage (previously described in greater detail in "Coordination of Functional and Nonfunctional Requirements").

Design Verification Is Performed Only After Validating the Design Process

Once the design validation has been performed, the designer knows that a good design can now be created. Next, the design must be created, and then checked and verified to ensure it meets the client's requirements. The design checking and verification must be performed both during and after the design. Design verification is primarily a QC activity.

At the tail end of this QC activity, once QA verification indicates the created design meets the requirements, a QA audit can be performed; this time around, it is performed after the design process to ensure the QC checking was indeed correctly performed—that is, proper QC processes were followed in the development of the design.

Notice that design verification results in design feedback (described in greater detail in "Application of QMS in Design Feedback"). Also, the two types of design QC checking processes are described in "Cross-Disciplinary Design Review." It is very important to understand how verification is related to validation, and vice versa.

CONCLUSION

The difference between validation (a QA activity) and verification (a QC activity) must be understood clearly, as these are two different but closely related quality processes. It is essential that the design quality procedures that are included in the DQMP describe these two concepts in greater detail for the benefit of all users.

Understanding the distinction between design verification and design validation is paramount for achieving successful project outcomes. Validation is considered a proactive QA measure because it ensures the project's feasibility/task feasibility before the design begins. Verification is a reactive QC process, confirming the completed design adheres to the established inputs. Both are essential, with validation laying the groundwork for effective verification. A thorough understanding and meticulous execution of these processes, as outlined in the DQMP, can guarantee the delivery of designs that would accurately meet the client requirements and project objectives.

QUESTIONS

1. Why is it crucial to perform design validation before the design process begins? What are the potential consequences of neglecting this step?

2. What are the various types of design validation studies?
3. Why is the validation of process quality essential before the quality of the product can be verified?
4. Explain the fundamental difference between design validation and design verification, and how their placement in the design process contributes to overall project quality. Give an example of each.
5. Discuss the significance of design software validation as a component of overall design validation. What key factors must a designer consider when validating design software, and why is individual designer responsibility essential in this process?
6. How do intradisciplinary design QC and interdisciplinary design QC relate to the broader concept of design verification? Explain how these QC checks ensure design output conforms to design input.
7. The document mentions the design quality management plan (DQMP). How does the DQMP facilitate and reinforce the understanding and application of design verification and validation principles?

Measurement and Monitoring

DESIGN MONITORING

> Examine design effectiveness and progress at various gates (e.g., schematic design (SD), design development (DD), construction documents (CD)) to align with the quality objectives and schedule. Address any deviations or issues discovered at each gate. (Analyze)
>
> Body of Knowledge III.E.1

Design monitoring stands as a crucial pillar within the design phase, ensuring both effectiveness and progress alignment with quality objectives and schedules. This section delves into the intricacies of design monitoring, exploring its role at various gates such as schematic design (SD), design development (DD), and construction documents (CD). By examining how design monitoring addresses deviations and issues, this section illuminates the critical distinction between design monitoring by a manager and QA monitoring, which focuses on evaluating the design against established quality objectives. According to ASQ's website, quality assurance (Figure 3.5) is defined as "part of quality management that is focused on providing confidence that quality requirements will be fulfilled."

This definition of QA implies that monitoring of design for effectiveness with respect to quality objectives is a function of design QA. The confidence that is provided by design QA is twofold—internally to management and externally to customers, government agencies, regulators, certifiers, and third parties.

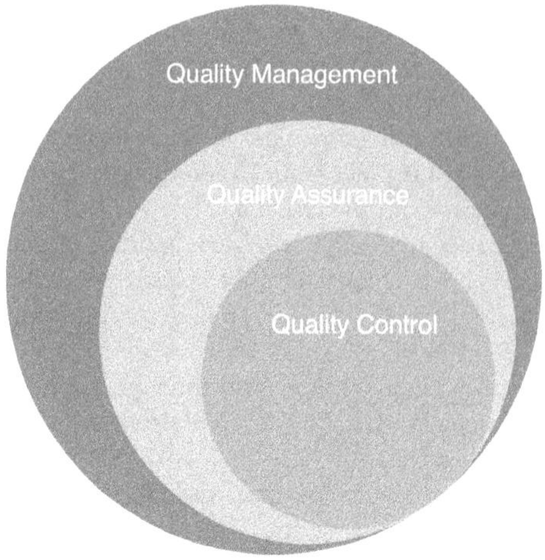

Figure 3.5 Relationship between QC, QA, and QM.

Design Monitoring Versus QA Monitoring of Design

Design monitoring is a generic term used to indicate the role of the design manager (i.e., someone who monitors the production of design). However, when design monitoring is considered in the context of evaluating the design against the quality objectives, then the purpose is not to monitor the production of design but to evaluate if the design is meeting the quality objectives. This role is to be performed by the QA manager, not by the design manager. In other words, the design manager produces the design, but the design QA manager evaluates if the design is meeting the design quality objectives.

The design QA ensures the effectiveness of the design quality against the quality objectives, as well as the progress of design against the schedule. As described in "Design Review Output" earlier in this chapter, the design of the project is developed in three successive stages, popularly called the three design gates: SD, DD, and CD. These three gates represent the three levels of design completion, namely:

- Preliminary design completion, popularly called schematic design (SD)
- Intermediate design completion, popularly called design development (DD)
- Final design completion, popularly called construction documents (CD)

The design QC processes used to perform the QC of design documents include a few formal and several informal QC checking processes. All these processes (described below) apply to all three design stages and should be implemented similarly.

- The formal quality control checking processes include the *intradisciplinary design QC checking and the interdisciplinary design QC checking*. These two QC checking processes represent the formal "process-based" QC checking (reviews) that must be implemented to achieve design quality. (See "Coordination of Functional and Nonfunctional Requirements" and "Cross-Disciplinary Design Review.")

- In addition, the *functionality reviews* represent the "product-based" QC reviews that must also be implemented to achieve the design quality. (See "Functionality and Constructability Reviews.")
- Besides these internal QC reviews for design QC, *constructability reviews* are also performed on the design to achieve design quality. The constructability reviews are also formal QC reviews but are performed by an external party, not by the design team. These external reviewers may include the construction contractors and any other independent construction-related entity such as the construction manager. (See "Functionality and Constructability Reviews.")
- Lastly, in addition to formal QC checks, there are several informal quality control checking processes performed by many design teams to improve the quality of the design. The most prominent of these is the *peer review*. (See "Peer Reviews.")

The above-described QC processes must be implemented to achieve design quality. But they need to be evaluated for their effectiveness; it is the QA function that evaluates effectiveness as well as the progress of the QC processes.

In this section, we include the information about the things that a QA auditor must consider when evaluating the effectiveness of design QC. The most commonly utilized tool for QA is auditing. The principles of quality auditing remain the same as described in Chapter Two, "Quality Audits." Following are the additional things that may be looked at during the QA audits as they pertain to the quality of the design submittals.

Quality Assurance Monitoring of Schematic Design (SD)

Quality assurance monitoring at 30% gate—with schematic review plans and concepts, also known as schematic design (SD)—is the first stage of design deliverables. Plans are issued to obtain the formal client approval of the overall solution or methodology or direction the project is going to take. Of the many options available at the beginning of the project, this design submittal enables/forces the client to select the preferred options so the design may proceed only in that direction. For example, at the SD stage, you may need to decide whether to design a steel bridge or a concrete bridge and choose the bridge type. Once that decision is made, all design development moving forward is only made with respect to the preferred option (other design options are no longer to be considered).

At the SD stage, the design is expected to include design documents that quantify the expectations of the functional performance and describe the parameters for each system that may be designed in the next stages of DD and CD.

The QA monitoring at this stage is quite basic. Design QA audits for the SD documents should verify the design has followed the QC checking procedures. There should be formal evidence that the submitted deliverables were checked in accordance with the design QC checking procedures. At this stage of design, there may be mostly intradisciplinary design QC checks that must have been performed. Some interdisciplinary comments are possible, but few constructability reviews may need to be performed because the design is not developed to the level where constructability reviews are useful.

Quality Assurance Monitoring of Design Development (DD)

Quality assurance monitoring at 60% gate—intermediate design plans, also known as design development (DD)—is the second gate for the design deliverables. At this stage, the goal is to produce a fully coordinated set of technical drawings that includes material specifications such that it is ready for procurement of vendors and suppliers. The design must be sufficiently detailed so the suppliers can be solicited for any long-lead item, or raw materials or major equipment that they may need. The suppliers should also be able to know what to provide in terms of any services that are sublet. Typically, the design at the 60% completion stage is expected to achieve this objective. A more detailed description of the typical contents of 60% design deliverables is provided in "Design Review Output."

Elements to Include in QA Monitoring at the 60% Design QC Gate

- A QA audit must verify the intradisciplinary design QC reviews were correctly performed. Evidence of the intradisciplinary QC checking may be audited by the QA to determine the effectiveness of the design QC.
- In addition to the intradisciplinary QC reviews, QA must verify that the interdisciplinary design QC reviews were performed correctly, as per procedure. This includes ensuring that comments issued by various cross-disciplines have been addressed.
- At the 60% design gate, QA monitoring must also ensure that formal records of comment incorporation by the QC are complete, particularly for all deferred QC comments. Lack of good records for QC comment resolution is typically one of the biggest failures of interdisciplinary QC design reviews, and this should be the focus of the QA audit.
- Typically, at the 60% design stage or thereabouts, the constructability reviews must also have been performed. QA monitoring at this gate must also verify the design QC did not overlook this very significant activity or delay it.

In short, the QA monitoring at the 60% completion stage must verify that QC checking for intradisciplinary, interdisciplinary, and constructability reviews were correctly performed and include formal records.

Quality Assurance Monitoring of Construction Design (CD)

For all practical purposes, QA monitoring at the 90% gate—final design, also known as construction documents (CD)—is the final design issued for construction. It is only called 90% design in publicly funded projects because there is a provision of obtaining formal approval of the client for final design deliverables, which then makes it the 100% design. (Only an administrative provision makes the 90% a final 100% design.)

Elements to Include in QA Monitoring at the 90% Final Design Stage

- The QA team should verify that design schedules have been followed by the design team and by design QC.
- Final design drawings are signed and sealed by qualified (licensed) professionals.
- Design drawings are accompanied by design specifications, calculations, and design directives (where applicable).
- QA monitoring at the final design stage may include spot-check audits of various design QC records, both for the intradisciplinary and interdisciplinary design QC check records.
- At this stage of design submittals, there should not be any major comments generated by design reviews.

CONCLUSION

Evaluation of the effectiveness of design QC should be the primary objective of the design QA monitoring. A good QA audit program provides an unbiased perspective on product as well as process quality through the independent evaluation by the QM. QA is successful when it develops and implements a good QC system that results in the design meeting and exceeding the quality requirements for that design.

In addition, QA monitoring helps determine the continual improvement strategies to implement in the QC processes. It proactively addresses potential issues in design and prevents design quality failures from occurring by utilizing the concept of lessons learned and continual improvement.

Another function of design QA monitoring is to see the trends in quality improvements and determine if the program is achieving success or failure in terms of implementing the QC program. This evaluation of effectiveness helps identify the subareas that may be producing a higher number of design errors or omissions, or most design defects or deficiencies that result in design rework or—even worse—result in more significant design changes.

Through QA monitoring of the design phase and its evaluation against the design schedule, it may be possible to identify the areas that have underutilized or overstaffed resources, or sometimes indicate which areas/disciplines of design have insufficient resources. All of these quality objectives of the design projects are best served when the evaluation of the design QC is performed by the QA manager (or their designee).

QUESTIONS

1. What is design quality (assurance) monitoring?
2. How is the QA for design different from the QA of construction?
3. What are the three stages of design completion (also called design gates)?
4. What should be monitored for quality at these design gates?
5. Why should QA monitoring be independent from QC checking?
6. What is the fundamental distinction between design monitoring performed by a design manager and QA monitoring of design, and why is this distinction critical to the overall design process?
7. Describe the significance of the three design gates (SD, DD, CD), and explain how the focus of QA monitoring shifts across these stages in relation to QC checks.
8. Explain the role of interdisciplinary QC checks in the design development (DD) stage, and why meticulous recordkeeping of comment incorporation is essential at this point according to the chapter.
9. Discuss the evaluation of effectiveness in design QC, and explain how QA audits contribute to identifying areas for continual improvement and resource management within the design process.
10. How does the chapter differentiate between formal and informal QC checking processes in design, and what examples of each are given?

PEER REVIEWS

> Identify the necessity of peer reviews (e.g., design peer reviews and third-party consultant reviews), evaluate their input, and incorporate them into the quality documentation process. (Evaluate)
>
> Body of Knowledge III.E.2

Peer reviews play a vital role in the design phase, serving as an informal yet crucial layer of evaluation. Conducted by independent experts, these reviews focus on the approach and direction of a design rather than just its compliance with requirements. Distinct from formal QC reviews, peer reviews offer a fresh perspective, potentially validating but never replacing established procedures. Their key features include the reviewer's independence from the original design team and the ability to provide candid criticism, ultimately aiming for a more comprehensive and technically sound design outcome.

Peer reviews are "informal," and are performed by an independent person who may be an internal or external reviewer but preferably is a designer who is considered an SME. Unlike a quality control review focused on the quality of the design and its compliance to the design quality requirements, peer reviews are more concerned with the approach (direction) of the design.

Following are some of the key features of peer reviews for design:

- *Not a replacement for formal QC reviews*: Peer reviews are not a replacement for the formal QC and QA reviews. They may be considered a validation of what the formal design QC and QA reviews have produced, but they are not substitutes.
- *Peer reviewer should not be part of the original design team*: The basic premise of independent review is that the reviewer should not be part of the design team that created the original design (a peer who is part of the original team may perform the formal QC checking).

  When the reviewer is independent of the influence of the original design team and is also considered an expert, they can offer a more candid criticism of the design, thereby making it more useful and comprehensive compared to a reviewer who feels limited by the influence of the original designer. The selection of independent reviewers is typically the purview of the design manager.
- *May not be subject to the same scrutiny as formal QC reviews*: Because they are usually considered informal, the peer reviews may not be subjected to the same level of scrutiny as formal QC checking. Whereas it is possible to keep a stringent standard for the peer review, the informality of the review may exist because the design reviewers do not put their engineering license/architectural license stamp over such reviews.
- *May be performed for singular discipline or cross-disciplinary*: Peer reviews may be performed within a singular discipline or may be cross-disciplinary in nature—in other words, intradisciplinary or interdisciplinary.
- *May offer a more holistic review*: The informality in a peer review may allow the peer reviewer to evaluate and holistically "review" the design the original designer developed. The idea behind a peer review is to get a fresh perspective on the design and determine independently if the direction taken by the designer for design development is in conformance with the client's expectations and the intent of the design.
- *Mostly focused on technical compliance*: Peer reviews are generally focused on technical compliance and comprehensiveness of the subject matter.
- *May be sanctioned by external or internal stakeholders*: Peer reviews are most common in the design and construction industry. They can be sanctioned by the owners and by the design firms for those aspects of design that require an additional level of confidence.
- *Documentation of peer reviews*: Although they are informal, it is still important to document the peer reviews as they are useful as a future reference. Documenting key decisions and criteria used for those decisions explains why certain decisions were made and some options were dropped.

Keeping a good record of the key design decisions, especially the ones that relate to multiple stakeholders and result in defining the direction of the future design work, is essential and should become a priority for the QM. It becomes a useful record in cases of conflicts that sometimes arise in complex projects.

Three Types of Peer Reviews

Open Peer Reviews

These are the typical peer reviews where the names of the original designers are revealed (known) and documented. Because they are not anonymous, reviewers stand behind the decisions they make and support. There are limitations to consider as well. For example, a reviewer may be more likely to give a positive rather than critical review of a colleague's work; or conversely, they may be more critical of the work of someone they do not like. Open peer reviews carry the risk of personal bias, which can be removed if using the second choice: blind peer reviews.

Blind Peer Reviews

In blind peer reviews, the identities of the reviewers are masked to reduce bias or to avoid the possibility of reprisal. This is done in those cases where very independent and honest feedback is required.

Blind peer reviews can be of further two subtypes:

- In *single-blind peer reviews*, the reviewers know the names of the original designers/authors, but their own names and identities are protected.
- In d*ouble-blind peer reviews*, reviewers do not know the names of the original designers or the other reviewers. This is done purposely so the review is completely independent of any bias.

Unless the design is extremely critical and reviewers require an extremely high level of security, blind peer reviews are usually not employed by the construction industry. But it is possible to conduct them if the situation warrants.

Third-Party Peer Reviews

Third-party peer reviews are also possible. They are often used in cases where there may be a potential conflict of interest or when a high level of expertise is required that is not available within the performing design organization.

CONCLUSION

Peer review is an important part of design development. It is designed to supplement QC processes and typically provides more specific technical expertise than QC reviews, which usually engage "equal or better" designers to perform the reviews. Peer reviews, on the other hand, include SMEs and are set up to ensure that the design meets a higher standard of quality, rigor, and validity. Peer reviews provide a means for experts to critically evaluate the work of their peers and help improve the overall quality of the design. They are widely accepted as a necessary step for ensuring the quality of the design.

A peer review serves as a valuable supplement to a formal QC review, offering unique insights and perspectives crucial for robust design development. Whether open, blind, or conducted by third parties, these reviews enhance technical compliance and comprehensiveness. By fostering critical evaluation and expert input, peer reviews ensure designs meet higher standards of quality and validity. Documenting these reviews, regardless of their informality, becomes essential for future reference, helping to solidify key decisions and maintain a record for complex projects.

QUESTIONS

1. What are peer reviews?
2. How are peer reviews different from QC reviews?
3. Are peer reviews conducted for singular discipline or are they cross-disciplinary?
4. How many types of peer reviews exist? When are they used?
5. What are the cases where independent or blind peer reviews may be warranted?
6. What distinguishes peer reviews from formal QC reviews, and why are peer reviews considered a supplement rather than a replacement for QC?
7. Explain the significance of the peer reviewer's independence from the original design team and how it contributes to a more comprehensive and candid review process.
8. Describe the differences between open, single-blind, and double-blind peer reviews, and discuss the circumstances under which each type might be most appropriate.
9. How do third-party peer reviews benefit a design project, and in what situations are they particularly valuable over internal peer reviews?
10. Despite the informal nature of peer reviews, why is it crucial to document them, and what specific information should be included in them for future reference?

QUALITY DOCUMENTATION (DURING THE DESIGN PHASE)

> Collaborate with the design review facilitator to document all quality-related activities, decisions, and changes, ensuring a comprehensive record for evaluating the design's adherence to quality standards. (Evaluate)
>
> Body of Knowledge III.E.3

In the realm of project management, the design phase stands as a critical juncture where ideas take tangible form. Measurement and monitoring during this phase are essential, but the cornerstone of quality is robust documentation. This section delves into quality documentation, exploring its role in capturing activities, decisions, and changes. Effective document control is not merely storage; it's about creating a seamless flow of information, connecting design to construction, and ensuring all stakeholders have access to the right data at the right time, throughout the project's life cycle. Document control is the most critical QA function that must be structured such that it can effectively store documents as well as records for the project for its complete life cycle (i.e., from the planning phase through design, construction, closeout, and O&M phases).

Design document control is only one element of the document control function of the project. It needs to be connected to the centralized document control of the project in a seamless manner that joins the design documents and records with those of the construction phase of the project. The objective of document control is seamless connection between these two production phases for greater overall project effectiveness, regardless of the method of project delivery. Document control must be structured in a way that reduces duplication of work and ensures there is a flow in the documentation between the key stakeholders (owner, designer, contractor), and similarly with all their subconsultants and subcontractors. Documents must be readily accessible so all stakeholders can perform the daily tasks of the project efficiently.

Quality documentation for design is a subset of the project's central document control system (described in greater detail in Chapter 6, "Impact Assessment and Compliance"). For the design phase, the quality documentation that is required to be maintained must be organized under a formally structured design document control.

The word "documentation" includes two types: front-end documents that are created to run the project and back-end records that are generated as a result of the project work. This difference between the documents and records is described in greater detail later in this section.

Design Document Control

Generally, each performing team has its own "inward-facing" document control mechanism that is centralized for them locally and enables them to create, share, and update documents as well as track the changes, comments, and approvals by the client. The project must create a centralized repository for its documents and records for the entire project life cycle. All individual document control systems of the designer, contractor, and their subcontractors should flow toward that central document control repository. Most document control systems available and in use today are digital in nature because manual document control systems can no longer manage documents for complex projects that have multiple performing teams.

The document control system for the design team must consider the following nine key aspects that should be established in their document control system:

1. The software program should be a database that is integrated (capable of integration) across the key stakeholders, which include the designer, contractor, and owner, as well as all subconsultants and subcontractors.

   Documentation does not only mean actual paper hard copies of plans. The 'living,' hard copy plans are typically considered obsolete. It's important to ensure that everybody in the project team collects the latest documents from the central source, which ensures that older versions are not distributed among the project teams and reduces the possibilities of failures from using older documents.

2. The formats in which the design documents must be stored should be pre-decided. They must also be aligned with the centralized document control of the project (which usually belongs to the owner, or to the owner's representative firm).

3. The lead persons responsible for the document control function must be identified. There should be orientation meetings between the document controllers representing the owner, designer, and contractor. Also, once the project is in progress, ongoing meetings must be scheduled between the lead document control persons of the owner, designer, and the contractor firm.
4. An experienced document control person operating through a functional QMS provides the best environment for managing the quality documents and records well.
5. Formal procedures for the development of quality documents at various levels of the project (policies, procedures, work instructions, job aids, and templates/checklists) should be established before the project begins active work. All staff must be trained in the project's document control practices.
6. The workflow of the documents must be defined at the beginning of the project and approved by the owner's centralized document control. This is extremely important for the effectiveness of document control of the project.

   When document controls are enabled through workflow-based software programs, they can enforce the protocols of formal document submittals, review periods, and responsible reviewers, and share their responses and reviews in real time to move the project forward without loss of productivity and time.
7. It is extremely important for every project to issue formal procedures that describe how the front-end documents would be stored and where. Likewise, there should be a formal procedure that describes the method to store the back-end records generated as a result of the work performed on the project.
8. Consistency and discipline are essential factors to keep the document control system effective. It is crucial to develop a habit of putting documents in their proper place as soon as they come in. Training of the project teams regarding such good practices of document control may be given by the QMs and would benefit the projects for all phases of the project life cycle.
9. The document control for design must also ensure it is capable of carrying the design files, which are usually in CAD (computer-aided design), as these files are what the designers use to create, update, and finalize their design. Usually, most design firms separate the design files repository from the final design files that are uploaded/issued to their clients in the pdf format. The document control system that a design form uses must provide clarity in terms of where the two types of files would be retained. Formal procedures about the document control, including naming conventions, must be prepared, along with training of the project staff about the document control system.

Documents Versus Records

Following are the fundamental differences between these two types of quality documentation:

1. *Quality documents*: "Documents" is the technical name of the formally developed documentation that is created at the beginning of the project to enable it to operate. As such, quality documents include all the documentation that is generated to perform the work tasked by the project.
2. *Quality records*: "Records" is the technical name of the documentation that is generated as a result of the project work. As such, a quality record is created once the quality process is completed. Records are the evidence used by QC to determine compliance with the project specifications. As such, quality records must include all documentation generated as a result of the completion of the tasks of the project, which must then be checked by QC.

Table 3.5 explains the difference between documents and records.

It is essential that the two categories of documents are clearly separated in the document control system.

Details of Document Control for a Design Project

Following are the categories of documents that must be organized by the document control at the minimum:

1. *Project documents for the design phase*: May include at the minimum the following:
   a. *Project management documents*: Project plans (e.g., project management plan, project quality plan, project risk plan, project security plan, etc.). These are all documents that provide front-end guidance to the project staff to perform the project work.

Table 3.2 Differences between documentation and records.

Document	Record
Generated as a result of Planning (Front-end Tool)	Generated as a result of Action (Back-end Tool)
Contains Information of "How to do something"	Contains information on "How a thing was done"
Cannot be used as Evidence	Can be used as Evidence
Can be Changed/Revised	Cannot be Changed (Record is Unchangeable as of the time of Issue)
Needs to be formally Approved before it may be Implemented	Approval of Records may or may not be Necessary; Approval must be against Criteria
Document may become a Record	Record cannot become a Document
Document may be in Paper form or Electronic	Record may also be in Paper form of Electronic
Duration of the Validity of Document must be defined by the Organization	Record is forever; It may need to be destroyed after a certain time, based on a Policy of Record Termination. But it is forever and cannot be changed.

b. *Specifications and standards*: The specifications that must be used to create the design need to be in a centralized location, preferably on a server with access rights given to the various members of the design team as well as outsiders (subconsultants, clients, owners). Also, the standards to which the design may need to be prepared also need to be secured in carefully structured document control folders, with a naming convention that is consistently used to identify the locations of these design documents. This folder must include a master list of all the documents that are stored in it, along with their revision numbers and dates of issue.

c. *Reference documents*: Commonly used reference documents that must be used by all teams in order to operate should be centralized by the design team's document control. Some examples may include permits, insurance documents, and most significantly the original computer-aided design (CAD) files that the designers utilize to develop/create the design plans, calculations, and subsequent shop drawings for the construction project.

2. *Project records for the design phase*: At minimum, the following documents make up the project records for the design phase:

a. *Design team submittals*: Some examples of the design team submittals include:

- The official design submittals that are issued by the design team (namely design plans for 30%, 60 %, and 90% design, specifications, and calculations that correspond to the noted stages/gates of design).
- *Contract data requirement list (CDRL)*: In some projects that are very well planned and where the scope of the design work is very clearly defined, a CDRL may be issued. The creation of CDRLs prior to the start of the project provides clarity in the proposed and expected design submittals that must be created and issued. CDRLs make the document control of design deliverables very organized and accountable.

Note: All design records must be formally issued through document control and should never be directly issued by the designers. It is a much-needed best practice to have a predefined naming convention for design record files and their revision levels, and create separately named subfolders for individual design disciplines.

b. *Contractor submittals*: This is the second category of design records, representing those that are not created by the design team but by the contractor and/or sometimes by the owner. Some examples of the contractor-facing design records that are generated by the project may include but are not limited to:

- Requests for information (RFIs)
- Field design change requests
- Shop drawings
- Catalog cuts

- Product data sheets
- Manuals of various types (e.g., maintenance manuals, safety manuals)

c. *Design document control database*: There should be a document control database created by the design's document control team. It is a best practice to have a professionally trained document control person lead this effort.

The document controller should house these design records in separate folders, according to the naming convention created and included in a standard operating procedure (SOP) for the design document control. As the design is progressively created and moves from the 30% to the 60% to the 90% design submittal stages, the record files in the database need to be updated with all related documents that the contractors send to the design team, as well as their responses. The formal status of design approvals must be notated in that database, and it should be possible to query the design work performed by the primary design firm as well as all by its design subconsultants.

The centralized document control must be tasked with creation and maintenance of project documents noted above. For the maintenance of records that are generated as a result of project work, a formal procedure for records management should be issued that addresses how project records will be stored and maintained. The procedural guidance on documents and records must indicate the process of permanent storage of the documents after the completion of the project.

Design document control database and its document controller must be able to perform the following:

a. Tracking of approved, rejected, approved as noted contractor submittals

b. Shop drawing approvals (made by the design team)

c. *Change order file management*: Design changes that are made after the issue of the final design must be carefully documented by the document control and maintained for the two types of design changes (design team changes and field-initiated design changes).

d. *Transmittals/correspondence files*: Document control should also maintain the formal transmittals as well as correspondence files, with clarity of correspondence with the different key stakeholders (contractors, owner, independent parties identified as stakeholders for more complex projects such as design-build, public–private partnerships, and a variety of other project delivery types).

e. *CAD files*: Computer-aided design (CAD) files contain both 2D and 3D designs. These are the base files that may be used by all to create models or architecture design plans. The integrity of these files is critical for the quality of all design and should follow the CAD design standards at all times.

f. *Safety and security files for design phase*: A decision must be made by the project to determine if these files may be maintained by centralized document control or by the functional department of safety and security.

CONCLUSION

Design quality documentation is a critical part of project documentation that also includes documentation for other phases of the project life cycle. In a complex project that consists of owner, designer, contractor, and a number of subcontractors and vendors, there are several document management systems that will coexist. It is essential to ensure the different repositories for the document controls of the different stakeholders are communicating with each other.

Training the staff as well as the external stakeholders avoids communication breakdowns. The best document control systems create automatic workflows based on collaboration that does not duplicate the storage and avoids repetition of data and access rights.

It is in this context that a QM can offer collaboration as a design review facilitator and enable the documentation of all quality-related activities, decisions, and changes, thereby ensuring a comprehensive record for evaluating the design's adherence to the required quality standards.

A meticulous quality documentation within the design phase is not merely an administrative task but a vital component of project success. By establishing robust document control systems, clearly defining document and record types, and ensuring seamless information flow among stakeholders, projects can mitigate risks, enhance efficiency, and maintain quality standards. The principles outlined in this section serve as a foundation for effective document management, ensuring that the project's history is well-preserved and accessible throughout its life cycle and beyond.

QUESTIONS

1. How is document control for design organized? How does it relate to the overall document control system of the project?
2. What are the key aspects of design document control?
3. What is the difference between documents and records?
4. What are some examples of project documents for the design phase?
5. What are some examples of project records for the design phase?
6. What are the procedures, processes, and workflows that must be established by design document control to effectively manage the documents for the project?
7. Where should those document control procedures, processes, workflows, and templates be included for the design phase and other users?
8. What are the administrative requirements for a good document control system for design?

Chapter 4
Procurement

Quality Requirements

> Collaborate with engineering teams to design functional and nonfunctional quality requirements for the procurement of specific products or services. Analyze if these requirements are clearly defined, complete, correct, compliant, and coordinated. (Analyze)
>
> Body of Knowledge IV.A.1

QUALITY REQUIREMENTS FOR PROCUREMENT

Organizational procurement is a company-wide process that encompasses long-term strategic sourcing and supplier relationships for acquiring goods and services. In contrast, project procurement is focused on obtaining necessary goods and services for a specific, time-limited project. The key difference lies in their scope and urgency: Project procurement addresses immediate needs with specific requirements tied to the project's goals and timeline. Procurement and quality are intrinsically linked. This chapter on quality requirements for procurement will give you an understanding of: (a) the role of quality assurance in procurement, (b) performance indicators that may be used to measure quality in procurement, (c) the collaboration required between the design and engineering teams to ascertain product requirements, and (d) the validation of those requirements.

The Role of Quality Assurance (QA) in Procurement

Quality is often the most critical aspect of the product or service being procured. As such, procurement must be linked with quality so the procured products and services consistently meet the standards, as well as the requirements and specifications, of the project. Incorporating quality into the procurement process ensures that procurement requirements are defined with quality in mind. Doing so minimizes the risk of receiving substandard products, materials, and services; reduces the costs associated with defects; and ultimately contributes to client satisfaction and overall business competitiveness. Quality assurance also plays a crucial role in regulatory compliance by ensuring the procurement processes align with industry standards and a project's legal and contractual requirements.

At the front end, the function of quality that engages with procurement is the QA team, as the effort of QA is to proactively establish the quality systems and procedures that would prevent quality issues in procured products and services. Once the QA function determines the quality plan and identifies the QC and inspection requirements for the products and services, the QC staff follow up on those quality and inspection plans to perform the QC inspections to verify whether the procurement meets requirements.

Applying the Principles of Quality Assurance to Procurement

To effectively shape the way that procurement unfolds, QA must apply the key principles that drive quality and help achieve project success. These include:

- *Clear communication*: Clear communication between the procurement team and the suppliers/vendors is crucial. QA must ensure that clear quality requirements as well as specifications are issued for the products and services to be procured from the supply chain. It is also essential that the material suppliers and vendors clearly understand those product-based requirements, as well as the expectations, in order for them to consistently produce them. This can be achieved through training (in-person or virtual) and contractual agreements that clearly outline the expectations in terms of quality, delivery, and other critical performance criteria.
- *Compliance*: Another significant aspect of quality is compliance, which requires evidence that the required features are achieved in the procured products and services. QA audits on the performance of the suppliers and vendors ensure that their products and services comply with industry standards and quality regulations.
- *Standardization*: Standardization is another QA principle that ensures consistency in the processes that produce the products and services. By adhering to established standards, suppliers and vendors also bring efficiency and cost savings to their products, which then translate down to the project.
- *Reliability*: Quality enables suppliers to offer consistency in the procurement process. By evaluating the strengths of the suppliers, as well as their weaknesses, QA determines their true capabilities, dependability, and reliability. The more established their processes are, the more consistent and reliable their products and services can be.
- *Transparency*: Another principle of quality is transparency. When QA establishes transparent sourcing processes and formal documentation

of evidence of quality, and holds the suppliers accountable for their performance (against the predefined, pre-agreed criteria and key performance indicators [KPIs]), it enables the procurement of a higher level of quality products and services throughout the project life cycle.

- *Continuous improvement*: Another fundamental principle of QA is the concept of continual improvement, which, when applied to procurement, forces the identification and elimination of inefficiencies in the procurement processes. These processes can then be improved through quality enhancements that result from corrections and corrective actions. They are also improved when proactive, preventive actions are taken to bring the products and services to the next level of quality and excellence.

Key Performance Indicators for Quality in Procurement

The project must establish some KPIs to measure and evaluate the success of quality in the procurement of products and services. Some commonly used KPIs include on-time delivery, lead times (long or short), suppliers' defect rate, customer satisfaction, and cost of poor quality (this is not an all-inclusive list).

Tracking KPIs provides useful insight into the performance of the procured products and services.

- *Material reject rate (MRR)*: The MRR is one metric used to measure the quality of the products and components procured from suppliers and vendors. MRR quantifies the percentage of the procured materials or products that are found to be defective when received and can't be used. MRR is measured as a percentage. A variation of this metric may be developed based on the different nature of products.

$$\text{MRR} = \frac{\text{Total number of rejected parts or components}}{\text{Total parts received}}$$

- *Lead time*: Lead time of the supplier/vendor is another indicator that directly impacts a project's ability to perform. A shorter, consistent lead time suggests that it's a more reliable and efficient supplier—one that manages inventory and meets customer demands well, both of which are indicators of good supplier quality. Suppliers with long lead times or inconsistent on-time performance can cause disruptions in the construction projects' schedule. The typical KPI (metric) for lead time is the measurement of the duration from order placement to delivery. Depending on the criticality of the parts, materials, and assemblies that are to be used in construction, the lead time taken by the supplier may be tracked for all materials and parts, or only for the critical materials.
- *Cost indicators*: The KPIs for the cost of the procured products, components, materials, and other services could be qualitative or quantitative in nature. The qualitative ranking of the delivery

performance could be made in comparison with the other providers of the same product or material. The qualitative cost aspects that may be evaluated can include consistency in price, or its volatility. Alternatively, the cost performance can be evaluated quantitatively by measuring the true financial impact of doing business with the suppliers; examples of quantitative cost indicators may include return on investment (ROI), cost of ownership (COI), and cost of quality (COQ).

- *Other indicators*: Other performance measures may be subjective, or they may even be turned into objective assessments in cases where more particular quality requirements are specified for the products, materials, components, or services. Examples of such KPIs may include the customer service experience, the qualifications of the supplier staff, and their understanding of the project and its requirements.

Functional and Nonfunctional Quality Requirements for Procured Materials

The concept of functional and nonfunctional requirements originated in the field of systems engineering and has been widely accepted in other industries as two types of product-based requirements usually issued by design and engineering. (*Note*: See more details in Chapter 3, "Coordination of Functional and Nonfunctional Requirements" and "Functionality and Constructability Reviews." Both of these sections must also be consulted for a more comprehensive understanding of this topic from all perspectives.

Functional Requirements

The functional requirements describe the features of the product that must be built into it. In other words, they are the mandatory requirements of the project. In general, the functional aspects of design are included in the "scope of work" and are described in detail in the technical specifications, which provide guidance to the designers as to how far they need to go when designing the systems and products.

Most design projects only consider the functionality elements that are included in the technical specifications of the project, not additional functionality that has not been requested. QC checking must be performed to ensure the functional features required in the product are included in the final design of the project. For example, it is possible to include inspectability, reliability, interoperability, and compatibility in every product designed, but the designer may choose to include only those functionalities specifically requested through the technical specifications. As a best practice, the following are the additional functionality elements that may or may not be considered for inclusion to make any design more comprehensive:

- *Inspectability*: The inspectability of a structure is defined as the ease or difficulty of the postconstruction inspection of the structure for damage and wear during its useful life.

- *Reliability*: According to the ASQ website, *reliability* is the probability of a product performing its intended function under stated conditions without failure for a given period of time. A quality test is considered reliable when it gives the same repeated results under the same conditions. Reliability is a function of several aspects of design and construction, all of which need to be considered as part of its functionality review. Among its many sub-aspects, two of the most significant sub-aspects of reliability are maintainability and survivability.

- *Interoperability*: Interoperability may be another consideration for the functionality of design. In the context of the construction industry, this is defined as the ability of different equipment to operate in conjunction with each other. In order for the overall design to be optimized for cost, it is essential that the interoperability of the design be considered as a functional requirement in the design of the infrastructure.
- *Compatibility*: Compatibility (a variation of interoperability) evaluates if the designed system conflicts with any other application or process on the new structure or system. Interoperability means the system or device would work fine under different environments, whereas compatibility means that different systems or devices would work fine under the same environment.
- *Complexity*: If the quality of the designed product is not enhanced through its complexity, the designer must try to keep it as simple as necessary. Technically speaking, the complexity of a design should not be a functionality criterion—but in a way, it becomes so. For example, a document control system that is developed to handle a million-dollar infrastructure project can also easily handle a smaller-scale project; but the additional complexity that is good for the larger project may be too complex for a smaller, simpler project.
- *Durability and availability*: These two terms are functional considerations that may be mission critical for some construction, but not all. Generally, for most construction projects, the durability of the constructed infrastructure refers to the long-term performance of that structure, while availability pertains to the accessible and timely supply of all its resources. In some niche design projects, where the design is very data-centric, the availability may refer to how the data is accessed, whereas durability would refer to the guarantee that data would not be lost. Note that these are niche sub-categories of functional requirements, which may not apply to all construction projects.
- *Comparative cost of ownership*: The cost of ownership is a life cycle cost that may be considered by the design team as a functionality. For example, the designer may consider designing a structure to be more sustainable and optimized for lower overall construction and postconstruction costs. For projects that value sustainability and lower

cost of ownership, this may be another functionality the designer must consider.

Obviously, the list of functional requirements is specific to the individual project and would change from one project to the next. But at their core, the functional requirements are for all those elements of the construction project that are mandatory to build.

Nonfunctional Requirements

The second type of requirements is called nonfunctional requirements, and they represent the performance or usability requirements. These requirements are desirable but may not be mandatory. For most construction projects, such requirements are described in the contract under "General Specifications." In the context of construction and its components and materials, some examples of nonfunctional requirements may include the following:

- General requirements for worksite security
- Requirements for landscaping
- Requirements for testability, reusability, and/or maintainability

Many times, these requirements are not considered mandatory but desirable for the user experience. If so, then the project may be considered "substantially complete" even if some of the nonfunctional requirements are still on the punch list. (Note that it is not possible to close out a project if the mandatory/functional requirements have not been met and remain on the project's punch list.)

Role of Procurement in the Collaboration Effort

When the project is at the procurement phase, the role of procurement is to act as the collaborator between the design/engineering group and the construction contractor; this collaboration ensures the designed materials and components are not only acceptable per the engineering design but also readily available and cost-effective for the project. In cases where the designed materials, components, or supplier parts or services are not available, or are not cost-effective, alternate options are sometimes suggested by the contractor. In such a case, the request for change is submitted by the contractor; it must first be approved formally by the design/engineering through the design change requests process before it may be obtained by procurement.

Validation of Functional and Nonfunctional Quality Requirements

Validation is a QA function for design/engineering, where the design QA ensures the design can meet its specifications or expectations. This topic has been described in greater detail in Chapter 3, "Design Validation and Verification," as design validation primarily happens in the design phase. This topic is reintroduced here to emphasize that, when in the procurement phase, it is essential for procurement to coordinate with design/engineering and validate that the requirements are met with the prescribed design of the products, components, or materials. Subsequently, procurement can include these requirements in the RFP that it may issue to obtain proposals and bids from prospective vendors and suppliers.

CONCLUSION

Procurement must not only ensure from design/engineering that the validation requirements have been met for both the functional and nonfunctional requirements, but also ensure that they are clearly defined, complete, correct, and compliant with the technical specifications of the project. To achieve that, procurement must collaborate with the designer as well as the design QA manager. This collaboration is critical because it ensures that the two phases of design and procurement are not considered in silo, but that they are connected and cross-referenced so the project procures the right components and materials for the construction phase.

QUESTIONS

1. How is the procurement in an organization different from the procurement for a project?
2. What is the role of quality assurance (QA) in procurement?
3. What are the principles of QA that may be applied to procurement?
4. What are some key performance indicators (KPIs) for QA in procurement?
5. What is the difference between functional and nonfunctional requirements?
6. How are the project specifications and requirements translated into the inspection and test plan (ITP) for quality control (QC) inspection of the procured materials?
7. What is the role of procurement in the validation of the project requirements?
8. Explain the difference between organizational procurement and project procurement, and why this distinction is important for understanding quality requirements in each context.
9. Describe the role of QA in the procurement process, specifically outlining how it interacts with procurement both proactively and reactively to ensure product and service quality.
10. Discuss how the principle of "transparency" in QA can impact procurement outcomes. Give an example of a transparent practice and its benefits.
11. Contrast functional and nonfunctional requirements in the context of procurement, and explain how each type of requirement is validated.
12. How does the material reject rate (MRR) serve as a KPI in quality procurement, and what factors might influence the determination of acceptable MRR levels for a project?

TRADE PARTNER REQUIREMENT ALIGNMENT

> Develop a strategy, incorporating methodologies such as first article inspection (FAI) and factory acceptance testing (FAT), to determine whether the trade partner's inspection and testing plan aligns with established quality requirements. Incorporate a rigorous review process for the inspection and test plans to ensure comprehensive evaluation. Additionally, integrate specific quality requirements for subcontractors and vendors, focusing on their scope of work, historical performance, compliance with standards, and the thoroughness of their documentation and plans. (Create)
>
> Body of Knowledge IV.A.2

In the construction industry, the term *trade partner* refers to specialized contractors who work on a collaborative, long-term basis with the general contractor. In the construction industry, the term can also refer to specialized contractors who work on a collaborative, long-term basis with a general contractor. The use of the term *partner* signifies a broader relationship between the two organizations and implies a collaborative and shared responsibility with a mutual goal. This is a change from previous years, when the terms *supplier* or *subcontractor* were used in the industry for all firms that the primary contractor would hire to provide component parts, assemblies, or materials, or professional or field services. The formerly hierarchical relationship is now considered in a more collaborative manner.

For a construction project, a trade partner could be classified as either a supplier or a vendor. Although frequently used interchangeably, a supplier is defined as the organization that provides raw materials or components for the construction project, whereas a "vendor" is an organization that provides finished products or services. These two terminologies are frequently used for trade partners and are often used interchangeably. (*Note*: In this section, the term "supplier" may be used interchangeably for "trade partner.")

The difference between trade partners and subcontractors is more than just semantics. Trade partners are onboarded early in the project life cycle to participate in the development of the execution strategies, schedules, and target costs; subcontractors, suppliers, and vendors are hired at a later stage, mostly on an as-needed basis and usually after the execution plan commences (Coleman, 2016).

The alignment of trade partner quality requirements means aligning the quality requirements that are prescribed for the suppliers and vendors of the project. Before describing how to make this alignment happen, it makes sense to first introduce the quality system requirements that a project must put in place for its trade partners.

Quality Plans for Trade Partners (Supplier/Vendors)

Supplier quality is the ability of the supplier to deliver goods or services that satisfy the client's needs. A supplier/vendor quality plan is a document prepared by the supplier to comply with the requirements of the primary client. Ideally, the supplier quality plan should be a part of the supplier's comprehensive QMS that has been established to monitor the performance of the supplier and keep track of their product quality, cost, compliance, and other critical metrics. Having an established QMS that is evaluated and formally approved by the client ensures that process-based quality is implemented at the supplier that includes process steps and sampling plans, and other features to ensure that the suppliers' products meet the client's requirements. Using a structured process that monitors, evaluates, and continually improves the performance of the supplier, such as a QMS, ensures that the goods and services received by the project consistently meet required quality standards as well as offer consistency in the supply chain.

Supplier quality management must begin early in the design of the product as an essential component of the supplier selection process. It should continue through the entire life cycle of that product and for the duration of the relationship with that particular supplier. A good supplier quality management system has a process for taking input requirements and effectively converting them to deliverable outputs that are deemed valuable by the client. Supplier quality management (in the context of the trade partners' quality alignment) is a strategic initiative that engages the project's important suppliers, vendors, and service providers; aligns their quality requirements to those of the project; and then oversees them as they execute their quality system.

Aligning Quality Requirements for Trade Partners

Supplier quality assurance (SQA) is the process that ensures the quality of products or services that a project acquires from its trade partners is aligned with the quality standards and specifications of the project. The absence of alignment results in fragmented execution as the team pulls in different directions. When the quality of the trade partners is aligned with the quality strategy of the project, the trade partners become an extension of the primary team. For the project to meet its quality standards, it is important that its trade partners (suppliers and vendors) from whom the project procures its component parts, assemblies, and materials comply with the project's quality standards.

Achieving supplier quality alignment starts with planning. First, the selection criteria for suppliers are developed, along with the definition of KPIs of their quality. Following that, the "quality of product requirements" is defined for the suppliers and the vendors by following the quality standards for compliance. Also, for complete alignment, the document and records management systems of the trade partners must also be consistent with that of the project. For longer-duration construction projects and especially those that utilize many supplier-based parts, assemblies, and materials in their construction installations, it is best to establish a "certified supplier quality program." This may include the minimum criteria of performance excellence for qualification as a certified supplier as well as performance indicators that are the most significant for the project (which may vary from project to project).

Common Vendor and Supplier Selection Criteria

Supplier selection involves choosing the most suitable suppliers for the project. The project team must establish formal criteria for selecting vendors and suppliers in advance. This ensures that procurement relies on quantifiable insights and adheres to a standardized, strategic, repeatable, and objective process. Formal selection criteria not only bring consistency but also offer more benefits than simply securing the lowest price vendor. Following this approach enhances quality, reduces costs, lowers risks, and fosters stronger, long-term relationships with suppliers, contributing to both short- and long-term success. Consider some of these key criteria when selecting a supplier or vendor.

- Supplier's previous experience with the product/service to be procured
- Supplier's past performance (if available) with the product/service to be procured
- Relative level of sophistication of the supplier's quality system, including meeting the regulatory requirements or any mandated quality system registration (e.g. ISO 9001 or the applicable industry standard)
- Supplier's ability to meet the current and potential capacity requirements on the desired delivery schedule of the project
- Financial stability of the supplier/vendor
- Availability of technical support and the technical depth of their team in developing and optimizing their processes
- Total cost of dealing with the supplier, which may include the material cost, the cost of communication/technology, the cost of inventory, and document control
- Supplier's track record for business performance and continual improvement

Historical Performance and KPIs of Trade Partners

The historical performance of the trade partners (suppliers and vendors, as well as any other service providers) must be monitored and audited against predefined KPIs. It is widely recognized that performance improves when people know they will be evaluated. The KPIs must be predefined and the frequency of evaluation agreed with the trade partners. The most common KPIs for trade partners include quality, delivery, and cost.

Quality Performance

The goal of quality performance must be zero effects. This is a philosophy and quality approach that emphasizes "doing things right the first time." It aims to eliminate errors, omissions, and defects in all aspects of production and service, such that quality is improved, costs are reduced, and clients are satisfied. A commonly used industry metric for supplier and vendor quality is the MRR (Material

Reject Rate), also called "scrap rate." This metric measures the percentage of materials or parts that are unusable or rejected during a production process, often due to quality issues or defects.

Delivery Performance

The goal of delivery performance of the trade partners should include the reduction of cycle times (delivery times), short lead times, the ability to offer quick turnaround during emergency needs, and backup options in case there are supply chain problems. Many supplier evaluation programs evaluate this metric on a qualitative basis (e.g., on a scale of 1 through 5, or similar).

Cost Performance

Generally, the cost performance of trade partners is measured with respect to two factors: relative direct cost in comparison to other suppliers/vendors and relative cost trend in comparison to any applicable industry benchmarks as identified by the organization.

Other Criteria of Trade Partners' Performance

Several other criteria of performance may be included in the evaluation of historical performance of the trade partners, which include but are not limited to:

- *Technical quality of deliverables:* This is used in cases where the supplier/subcontractor also provides technical reports along with the parts and services. The quality/comprehensiveness of services may also be ranked qualitatively (on a scale).
- *Timeliness (delivery)*: This is usually measured in terms of the preidentified lead times and the on-time delivery performance and the reliability of the suppliers and vendors.
- *Invoicing*: The accuracy of the invoices and correct charging of preapproved rates are a significant time-saver and performance-based criteria for trade partner services.
- *Change in orders and performance of contracts department*: In cases of changes, the promptness of the revised change orders is always helpful to the project performance.
- *Dependability/reliability*: The reliability of the trade partners is a function of the quality of their processes, as well as their commitment to the project and its superior performance.
- *Communication*: The most well-aligned trade partners are connected to the prime contractor through a standardized communication network, which offers quick responses.

There are several different ways that the trade partners may be evaluated. An example template (Figure 4.1) is provided for evaluating the historical performance of trade partners. The project team must determine which aspects of the trade partners' performance are the most critical, and scale the evaluation criteria up or down based on the significance and complexity of their work.

Subcontractor/Supplier Name:

Project Number: **Project Name:**

Work Package Manager: **Procurement Representative Name:**

Approximate Value of Work: **Period of Performance:**

Type of Work/SOW:							
ELEMENT	**EVALUATION CRITERIA**	**Exceeds Expectations**	**Meets Expectations**	**Needs Improvement**	**Serious Problem**	**Not Applicable**	**COMMENTS**
TECHNICAL QUALITY	Completeness, staff competency, accuracy, precision, relevance	☐	☐	☐	☐	☐	
Deliverables	Clarity, requested level of detail, data presentation format, submitted per schedule	☐	☐	☐	☐	☐	
TIMELINESS	On-time delivery, agreed-upon cycle time in turning around requests	☐	☐	☐	☐	☐	
COST	Actual cost of service (direct or indirect)	☐	☐	☐	☐	☐	
INVOICING	Clarity, traceability, accuracy	☐	☐	☐	☐	☐	
AGREEMENT/ CHANGE ORDERS	Conditions, negotiations finalized, unilateral CO's issued, claims submitted	☐	☐	☐	☐	☐	
DEPENDABILITY/ COMMITMENT	Consistency, credibility,	☐	☐	☐	☐	☐	
COOPERATIVENESS	Taking responsibility, responsiveness, flexibility,	☐	☐	☐	☐	☐	
COMMUNICATION	Keeping informed on progress/problems/procedural changes	☐	☐	☐	☐	☐	
OVERALL ASSESSMENT OF PERFORMANCE		☐	☐	☐	☐		

Issues/Suggestions:

How can the company improve service to Parsons?

EVALUATOR'S NAME: **CONTACT NO.:** **DATE:**

PROJECT MANAGER'S APPROVAL: ______________________________ **DATE:**

Figure 4.1 Template for the performance assessment of trade partners.

Document Control Alignment of Major Trade Partners

For large complex projects, it is also useful to bring consistency and standardization into the document control processes of the key suppliers and vendors. Utilizing the same document control numbering system and software allows for a seamless connection between the documents and records. Many large projects utilize cloud-based platforms and a centralized document control that offers access rights to all trade partners. This ensures that the document management of the entire project follows the same version control, access management, and real-time collaboration as that between the primary stakeholders of the project.

Comprehensive Review of Inspection and Test Plans

Inspection and test plans (ITPs) are essential tools the construction QC team uses to ensure the work complies with quality requirements. (The ITPs that must be used to verify the compliance of the work performed in the construction phase are described in greater detail in Chapter 2, "Quality Control.")

Alignment of the trade partners' ITPs ensures verifying that they implement similar plans for the inspection and testing of their products and services as the project's primary QC and ITPs (which have been put in place by the primary construction contractor). The quality manager (QM) must ensure the supplier's ITPs are included in their quality plan, and that these plans conform to the requirements of quality for the products and services produced by the trade partner.

The role of procurement is not to develop ITPs for the trade partners; it is the responsibility of the trade partners to do that, and their ITPs must be included in their quality plans, which must be formally approved by the project's QA function. Procurement must only coordinate with QA and ensure that comprehensive evaluation of the trade partners' quality performance includes a thorough evaluation of their ITP as well. (See Chapter 2, "Quality Control" to gain more understanding about the inspection and test plans and how they must be prepared for inspection QC.)

Alignment of the Suppliers and Vendors with Established Quality Requirements

In addition to the primary construction work performed on the construction site that needs to undergo QC, there are additional QC activities that may need to be performed at the fabrication facilities of the material providers or at the trade partners' facilities.

Depending on where they apply, two of the most relevant QC tests performed at the manufacturing/fabrication facilities of the suppliers and vendors include the factory acceptance test (FAT) and the first article inspection (FAI).

Factory Acceptance Test (FAT)

A factory acceptance test (FAT) is a test or a series of tests that are performed on equipment (usually assemblies or parts that must be installed as part of the construction) (DXP Marketing, 2023). As the name indicates, a FAT is performed on-site at the supplier's facility (factory) to ensure product quality as well as the process that manufactures those parts and equipment. These tests verify if the equipment that is being manufactured through that process meets the client's requirements and product specification, and will be ready for installation when the shipment is received at the construction site.

FATs help buyers make informed decisions and choose suppliers and vendors that offer better/more consistent processes that can consistently meet quality and compliance requirements. FAT tests usually also highlight areas that may need improvement, and the bottlenecks or inefficiencies can then be converted into opportunities for continuous improvement for the supplier's process.

The tests that may be included for the FAT should be selected carefully to provide information about the quality of the critical equipment and whether parts can be consistently produced to the same level of quality as tested and approved. FAT requirements are often decided by the design specifications.

Many of the tests on equipment and components include an evaluation of the prototype parts and the processes that manufacture those parts. The inspector who visits the supplier/vendor facility to perform the FAT may use their discretion to select the elements and the tests. Usually, the FAT does not expand to the quality or management system of the supplier; it remains limited to product quality.

Some examples of elements that are generally included in the FAT are the following:

- Checking of the raw materials of the supplier for quality and quantity
- Assessment of the functionality of machines/equipment that are used to produce the critical components and assemblies manufactured for the client
- Audit of the production records for similar products (either for the same client or a similar product developed for a different client)
- Evaluation of the QC and QA processes and records of the supplier through formal samples of production records. It is best to sample those records that show product deficiencies in particular. Subsequently, corrective and preventive actions that have been put in place by the supplier must also be evaluated to ensure they are satisfactory and reliable.
- Verification of safety practices
- Usage of safety equipment
- General inspection of the workplace for cleanliness and other facilities
- Qualification of the staff responsible for manufacturing/assembling the product
- Evaluation of the production schedule for critical equipment and its actual performance against the planned baseline manufacturing schedule of those parts

All the above should be the scope of the selected tests, as they evaluate the supplier's ability to consistently meet the technical specifications of the critical products they are manufacturing for the client.

A checklist for FAT should enable the quality manager (or their delegate) to verify, at a minimum, the following items (note that this list is not all inclusive and may be extended if project-specific needs indicate that other elements must be added):

- Proof of conformance to specifications (e.g., certificates of compliance)
- Test plans for the critical equipment that is produced

- Standard work procedures and ensuring the availability of reference standards
- Sample results of production (of the part/assembly)
- Equipment studies (for the equipment on which the part is being manufactured)
- Processes qualifications (e.g., welding qualifications)
- Calibration certificates for the equipment (where applicable)
- Control system that is being used by the fabricator
- As-built technical drawings
- Equipment data sheets (and their organization)
- Materials data sheets
- Standard maintenance procedures
- Training materials for the production staff
- Ergonomic assessment (if applicable)

First Article Inspection (FAI)

As described by Partida (2020), first article inspection (FAI) is another QC check that is usually performed at the trade partners' facility. This test validates their production process and verifies whether a new or modified production process is capable of producing conforming parts that meet the manufacturing specifications detailed in technical or engineering drawings. FAI occurs when a qualified/authorized inspector takes one or several parts from the first production run and then compares them to the client's specifications to verify that they conform. The FAI does not necessarily include checking the first parts coming off the production line. It is acceptable to take parts from within the initial batch, but one should make an effort to select the parts randomly.

When Is First Article Inspection Needed?

The first production run could be for either a new or modified production process and need an FAI. Also, when a production lapse of some years has occurred, an FAI should be considered part of the requirement for restarting manufacturing. Similarly, if any alterations occur in the materials, sourcing, process, tools, or location, there should be accompanying FAIs to show that the changes did not cause adverse consequences for the product's functionality or other quality attributes. FAIs are common requirements for subcontractors (suppliers or vendors) working on federal construction projects.

Method for Performing an FAI

- The first step for conducting an FAI involves creating an ITP for the part. The ITP should include the (product) requirements for the inspection.

- It is best to include the reference to the associated specifications and relevant tolerances that are allowed for the part.
- After comparison with the specifications, the inspector must prepare the first article inspection report (FAIR). This report summarizes whether a part passed or failed the examination from the QC inspector.
- Also, it is a good practice to use a checklist to perform the FAI to help achieve thoroughness in testing.
- Dimensional characteristics of the parts (which include size, shape, and feature location) are usual elements that must be measured during the FAI. These features are normally inspected using calibrated tools.
- Typically, measuring tools such as vernier calipers, go/no go gauges, and more sophisticated coordinate measuring machines (CMMs) are used to measure the dimensional features of the parts and to compare the produced parts with their specifications. The goal is to see if they differ. If differences exist and are outside of the allowable tolerances, correction or corrective action may be necessary. After that, one must repeat the FAI to ensure conformance.

CONCLUSION

Establishing robust trade partner relationships and ensuring quality alignment is paramount for project success. By defining clear requirements, implementing rigorous inspection and testing, and fostering collaborative communication, projects can mitigate risks and ensure the delivery of high-quality products and services. The integration of tools like FAI and FAT further strengthens this alignment, providing tangible measures of compliance and performance. Ultimately, a well-managed trade partner ecosystem translates to a resilient supply.

QUESTIONS

1. Explain the distinction between a "supplier" and a "vendor" in the context of a construction project, and discuss why the term "trade partner" suggests a shift in the traditional relationship.
2. Describe the significance of a supplier quality management system, and detail why it is important for it to be evaluated and formally approved by the client.
3. What two key QC tests may be performed at a trade partner's manufacturing/fabrication facility, and how do they contribute to ensuring alignment with project quality standards?
4. Discuss the role of inspection and test plans (ITPs) in ensuring quality alignment with trade partners. Whose responsibility is it to develop the ITPs, and what is procurement's role in the process?
5. Explain why historical performance and key performance indicators (KPIs) are crucial for evaluating trade partners. Give two examples of KPIs related to quality and one example related to cost performance.

Resource Management

PRE-ACTIVITY MEETINGS

> Schedule, implement, and oversee pre-activity meetings (e.g., pre-assembly, pre-inspection, quality kick-off) to review the project's requirements with the relevant stakeholders. (Evaluate)
>
> Body of Knowledge IV.B.1

From the perspective of the construction contractor, pre-activity meetings are conducted during the construction phase and are not necessarily considered a procurement phase activity. The current certification BoK for the CCQM exam includes pre-activity in the procurement phase, most likely because the BoK includes pre-assembly meetings also as part of the pre-activity meetings. Therefore, it's important to outline the difference between pre-assembly meetings and pre-construction meetings.

In projects that include major pre-construction work (e.g., installation of assemblies in the construction of a train control room), the completed systems that must be installed during the construction phase may include component assemblies and other custom-designed mechanical or electrical units that are pre-built at the suppliers' fabrication facilities and shipped over to the final construction sites for installation. Pre-activity meetings for such projects must be conducted during the pre-construction phase and then conducted again separately once the assembly reaches the construction site during the construction phase. This is one example of a construction activity that may be considered an overlap between the procurement and construction phases. (Note: The BoK generically refers to pre-assembly meetings that are performed for such projects as pre-activity meetings.)

The description in the next section includes details about what the pre-activity meetings are and how they must be conducted to facilitate construction readiness, regardless of whether they are conducted during the construction phase or in the pre-construction (procurement) phase of the project. Improvements in safety, quality, scheduling, and customer satisfaction are among the key topics that must be included on the agenda for pre-activity meetings.

What Is a Pre-Activity Meeting and Why Is It Important?

A pre-activity meeting is an *internal meeting* held by the project team before the work is executed. The purpose of this meeting is to map out potential issues and familiarize the team with the expectations of the work before it begins. Usually, during the construction stage, the most common application of a pre-activity meeting is the pre-construction meeting. During this meeting, all team members are alerted to their responsibilities, and the work procedures are explained to everyone. Proper communication and scheduling then happen, which sets the foundation for the construction initiation.

Pre-activity/pre-construction meetings are important for many reasons. The following are key benefits:

- *Make introductions*: Even a small building project involves a lot of stakeholders. This is a chance for everyone to get to know key stakeholders, breaking down siloes between procurement, safety, quality, scheduling and management, and so forth.
- *Review drawings, designs, and submittals for product data*: At the start of the meeting, all drawings, designs, and submittals for product data should exist in a state of approval. However, if there's a problem, it's best to catch it early on. The pre-activity meeting offers the chance to submit and review that everything is ready one final time before construction begins.
- *Create a roadmap*: A pre-activity meeting gives the team a roadmap of how to perform the task at hand. Typically, a project begins with high-level introductions about the project, the project owner, the roles and responsibilities of the project team, the chains of command and the communication channels, the document reviews, and all other aspects prior to the pre-activity meeting. The pre-activity meeting is performed at the task level, not at a high level, to ensure the construction team's readiness to start the construction work. This meeting is the last opportunity to help avoid any misalignment before construction begins.

Who Runs a Pre-Activity Meeting and Who Attends It?

Deciding who leads the meeting depends on whether the meeting is conducted during construction or the pre-construction phase. Typically, construction crews consider the pre-activity meeting a production meeting for the construction staff. As such, pre-activity meetings are typically headed by the construction superintendent responsible for the construction work. Pre-activity meetings must include significant contributions by the construction quality team, including the CQM.

Deciding who attends the pre-activity meetings depends on the scope of work (SOW) and the definable feature of work (DFOW) for the project. Because the meeting is at the task level, the DFOW should become the basis of the selection of meeting attendees. Some of the typical roles that may be included in the pre-activity meetings are:

- General contractor's project manager/superintendent
- General contractor's quality manager (e.g., a CQM)
- General contractor's safety manager
- Trade partner's superintendent/foreman (as identified by the DFOW)
- Other stakeholders directly involved with the task at hand, including the owner's representatives for quality, safety, and other administrative workers

The main reason is to have all stakeholders collectively map out the best way for the pre-construction tasks to proceed.

Pre-Activity Meeting Agenda

A pre-activity meeting agenda should cover the key aspects of the work activity that the team will undertake. High-level plans and strategic direction of the team are not discussed during the meeting. This is a tactical meeting aimed at ensuring the team is ready to undertake the construction (or the pre-construction) activity. Depending upon the scope, stage of work, and complexity of the task at hand, the agenda may include some or all of the following:

- *Introductions:* Introductions of the team that is involved with the task (activity) at hand.
- *Contract review:* The portion of the contract that applies to the work activity.
- *Final design documents:* The applicable construction documents for the activity at hand must be available to the pre-activity team. These plans must be at the "approved" stage.
- *Contractor submittals:* If the construction activity has additional contractor submittals such as shop drawings, specifications, product data, mock-ups, samples, and all the applicable RFIs related to the activity together with the designer's approvals, they must all be available to the team before the activity is initiated.
- *Safety requirements:* All general and specific safety requirements for the activity at hand must be made available to the team during the pre-activity meeting.
- *Permits:* All required permits that may be applicable to the work activity must be included, discussed, and confirmed at the pre-activity meeting.
- *Sequence of work:* The work plan for construction, along with the sequence, schedule, and material plan. This should include confirmation of the availability of materials that are to be used in the work activity.
- *Material plans:* Information about the materials that are to be used, and confirmation of their availability once the work activity begins. Also, instructions on how to manage any remaining materials after the completion of the work.
- *Inspection and test plans:* All QC inspection and testing that is related to the work activity must be on the agenda and discussed during this meeting. It is extremely important to discuss the inspection and test plan (ITP) with the personnel who will be responsible for performing QC and QA inspections.
- *General quality requirements:* Besides QC inspection requirements that are included in the ITP, the pre-activity meeting must also discuss any

higher-level QA requirements that may be applicable to the work (e.g., any applicable requirements that pertain to the QMS). Instructions for "as-builts" must also be included along with the quality requirements.

- *Site conditions:* The review of current site conditions is another essential part of the agenda. Also, any special instructions, such as site policies and procedures (e.g., authorization for site visits, safety, sustainability, or other important areas) must be included in the meeting and discussed.
- *Project schedule:* The schedule of project delivery, especially as it applies to the work activity, must be included in the agenda. This should focus on key milestones for work activity, as well as major deliverables.
- *Administrative aspects of the work activity:* Key administrative tasks that must be included in the agenda may include the resources required for that work activity. This should include ensuring that all procedures that apply to the work activity are available to the team.
- *Roles and responsibilities:* The pre-activity meeting, whether it is in the construction phase or the pre-construction phase, must discuss the roles as well as the responsibilities of all the people that are going to work on the activity.
- *Applicable document management process:* During the pre-activity meeting, the team should also discuss the internal and external document management requirements of the project. This discussion may include access rights and assurance that all team members have access to the documents needed to perform the task at hand.
- *Visual aids:* The pre-activity meeting checklist is a visual aid that can help achieve the intended goals of the meeting. When teams consist of multi-functional staff, it is always helpful to present by using visual aids like a whiteboard so attendees can take notes, make lists, and stay engaged with the critical points of the presentation.
- *Reporting requirements:* Reporting requirements for the work team (both internal and external) must be included, as well as who must receive the results of the pre-activity meeting.

A preconstruction meeting agenda helps the team plan for the meeting. It also serves as a guide during the meeting, making it easier to keep on topic. Thanks to the preconstruction meeting agenda, the attendees know what to expect. The pre-activity meeting should foster discussion and conversation among the participants as they discuss key agenda points. The meetings should also have formal minutes prepared and distributed for recordkeeping (note that artificial intelligence software can perform record-keeping and note-taking).

CONCLUSION

Pre-activity meetings lead to increased overall project efficiency. They minimize risks and work-related incidents and, as a result, can often lead to increased profits. The goal is to complete the project on time and on budget while reducing risks and costly delays. In addition, involving the project team at the start of the project helps identify any pitfalls and risks, and allows the team to mitigate and remove those risks to help ensure the work is built correctly the first time.

QUESTIONS

1. Beyond simply reviewing drawings and schedules, what is a key benefit of pre-activity meetings in terms of team dynamics and relationships?
2. What key roles provide significant input during these meetings, and why is their contribution important?
3. While a pre-activity meeting checklist is crucial, what specific aspect of the checklist needs to be tailored to each project or client, and why is this customization necessary?
4. What is the distinction between understanding one's individual role and the more comprehensive understanding that should be achieved through the pre-activity meeting?
5. When does a pre-activity meeting come within the purview of procurement?

MATERIAL CONTROL AND RESOURCE IDENTIFICATION

> Incorporate material control, material traceability, and vendor requirements into the project prior to commencement of work. Evaluate the resources needed for various inspections (e.g., self-inspection and certification, third-party inspection), including those conducted at the source and as mandated by regulations. (Evaluate)
>
> **Body of Knowledge IV.B.2**

Construction material management stands as a cornerstone of successful building projects, weaving together intricate threads of planning, procurement, logistics, and quality control. It is the linchpin that ensures material resources are not just available but available at the right time, in the right quantity, and at the right price, transforming design blueprints into tangible structures. This critical function encompasses a vast scope, potentially impacting every facet of an organization or a project and directly influencing project efficiency, budgetary adherence, and ultimately its timely completion.

Materials management on projects is different from traditional materials management for organizations in that projects can neither afford to have surplus nor undersupply of the materials. Ordering excess materials can result in significant challenges of storage, holding, potential obsolescence or spoilage (in case of the materials that may be affected by environmental factors and therefore have a limited productive life). Overages in materials management can directly increase a project's hard costs because the unused, leftover, or damaged materials represent tied-up cash, usually forcing significant discounts or write-offs. Material shortages, on the other hand, can result in construction delays and work stoppages as the construction crews wait for critical components or materials to arrive at the construction site.

Successful construction projects require on-time delivery of quality materials and careful management of the flow of materials from initial sourcing to final installation. Precise logistics, efficient communication, and effective supplier quality management are necessary to ensure that quality materials and equipment arrive at the construction sites when they are needed, preventing construction delays and cost overruns. Effective materials procurement and control/management are also critical for maintaining the cost-effectiveness of the construction project as well as to ensure its timely completion. It prevents project delays that could occur due to material shortages, avoids the financial waste that could occur from overages and spoilage, and also ensures that the materials used meet quality standards, all of which in turn helps maintain a safe, consistent, organized, and dependable work environment.

Material Control Requirements

The most significant material control problem in construction projects is inaccurate forecasting, which leads to oversupply or undersupply of materials. Inaccurate forecasting happens because of several failures in materials control, not one. To address this most common material control problem, contractors must first establish a comprehensive materials management and control process (on their end), and then work with trusted supply chains that include reliable suppliers and vendors. Material controls are effective if both partners (contractor as well as supplier) are in sync; focusing only on one does not address the material controls problem. In the absence of a good materials management and control process, many suppliers and vendors apply ad hoc controls that do not provide consistency and cause failures in the long run. (Note: In the construction industry, such suppliers and vendors are frequently called trade partners; see the "Trade Partners" section earlier in this chapter.)

The suppliers and vendors that are selected to support a construction project with the materials needed and component parts should be responsive, responsible, organized, and capable of delivering the right materials with the right quality, on-time, and at the most optimal cost. Having such material suppliers is the only way to ensure that no disruption in the construction schedule occurs due to material shortages. The selection of the trade partners is performed with the collaboration of the procurement and the design functions, along with direct contribution by quality; it should not be the role of a singular department to select the material suppliers.

Five Objectives of Materials Control

The following are the five primary objectives of material control management:

1. *Right material*: The first objective of material control management is to ensure that suitable materials are available for production. It involves identifying the materials required for production and ensuring they are of the correct quality, specification, and quantity. By ensuring the right materials are available, the contractor firms can minimize the risk of production delays and ensure customer satisfaction.

2. *Right time*: The second objective is to ensure that the right materials are available at the right time. It involves managing the movement of materials within the warehouse, reducing lead times, and improving the efficiency of delivery processes.
3. *Right amount*: The third objective is to ensure that the right number of materials is available for production. It involves determining the optimal inventory level to maintain and implementing processes to manage the movement of materials within the warehouse. By ensuring that the right amount of material is available, contractor firms can minimize the risk of stock shortages, reduce the cost of storage and handling, and increase efficiency.
4. *Right price*: The fourth objective is to ensure that materials are purchased at the right price. This involves negotiating with suppliers (and vendors) to obtain the best possible prices and implementing cost-saving measures, such as reducing waste, reducing lead times, and improving the efficiency of delivery processes.
5. *Right sources*: The fifth objective is to ensure that materials are sourced from the right sources. This involves identifying reliable suppliers, developing partnerships with suppliers, and ensuring that materials are purchased from approved suppliers only. By sourcing materials from the right sources, using an established procedure for conformance to the project and referencing code requirements, contractor firms can reduce the risk of defective materials, minimize the risk of production delays, and ensure customer satisfaction.

Next, we will describe the material control requirements for contractor firms and those that must be required from suppliers/vendors.

Material Control Requirements (Implemented by the Contractor)

The prerequisite to effective materials control is a well-defined and well-implemented procurement process used by the contractor. At the minimum, the following material control requirements must be implemented by the contractor's firm (the more complex a project is, the greater the need to implement these minimum controls):

- *Defined procurement approval process:* Approval authorization limits ($ limits of procurement orders) must be pre-defined. If the material value of the orders requires higher-level approval, this must be known to those who need the materials (namely, field staff). Also, the method through which orders must be placed needs to be documented. (Most organizations and projects utilize some construction management software that has a procurement module to handle all procurement orders; manual systems are outdated). Training must be provided for staff who may be involved with placing orders for materials by following the defined process and using the software. It is always best to have at least one formally defined procedure that describes the procurement process, and identifies the method, and the order authorization limits, as well as other necessary details of procurement. All those concerned with order placements, as well as the field staff who typically need those materials, must be familiar with the procedures. A CQM can assist in this process by collaborating and leading the entire effort.

- *Inventory management of the contractor firm*: The list of materials required by construction, along with the status of their inventory with the contractor, must be known and available. The status of existing materials inventory must be current. The trigger points (reorder points) specifying when needed materials are to be ordered must be well-established and known to the responsible construction staff. For a good, seamless process, connect the inventory of the suppliers' materials with that of the contractors'. (Some firms do this to facilitate better on-time delivery of the materials.)
- *QC inspection process for inspection of incoming materials*: The process to verify the quality of the incoming materials must also be formally defined, along with the responsibilities of who will perform these quality inspections. Typically, construction QC inspectors perform the incoming materials inspections—although for smaller-sized projects and teams, this responsibility is sometimes given to the construction staff. The CQM must engage with the construction crew and ensure that the responsibilities of QC and QA for incoming materials are defined, known, and implemented.
- *Accurate construction schedule:* An accurate construction schedule is vital for timely project delivery, which in turn is dependent on the timely delivery of materials and parts to the construction sites. Resource allocation for a construction project and all its workflows is similarly dependent on the accuracy of the project schedule. Construction projects that do not have accurate schedules, or that lack enforceable work plans or quality plans, run the risk of delays in the procurement of construction materials as well. Delays in the procurement of materials and parts typically result in work delays, cost overruns, and potentially further subsequent failures.
- *Accurate materials forecasting by the contractor firm:* The most significant material control requirement for construction projects involves the contractor firm having an accurate forecasting system, which is integrated with their material storage/inventory. Accurately forecasting what materials are required by the project and the time when they are needed enables the procurement department to place the correct orders of materials and parts with suppliers at the right times. Also, the decisions, such as which materials must be ordered from which (possibly preferred) suppliers, as well as lead times for the delivery of those materials, are all part of good forecasting. For most construction organizations, bad forecasting constitutes the biggest failure of materials procurement.

Thus far, we have described the essential prerequisites of materials control for the construction contractor. In addition to these internal requirements that the contractor must meet internally, it is essential that similar control requirements be applied to suppliers/vendors so procurement of materials is not negatively impacted. (The section below, as well as "Trade Partner Requirements" and "Vendor Performance Evaluation" in this chapter, includes further details on this topic.)

Material Control Requirements (Implemented by Material Suppliers/Vendors)

Like the contractors' material procurement system, there are control requirements that must be implemented by the materials suppliers and part vendors. The best practice is to include any procurement-based material control requirements in the contract terms within the contracts made with their suppliers. Depending upon the criticality of the material suppliers, some or all of the following material controls may be implemented with the suppliers and vendors:

- *Inventory management system of the supplier/vendor*: If the material supplier or the part vendor is supplying materials or components that are critical for construction, especially items that have a longer lead time, then the contractor may require a formally managed procurement system to be put in place. A well-developed and well-implemented inventory management system that controls inventory and can provide current information about the materials and their availability is very useful for materials control.

  It is possible for the prime general contractor firm to specify the requirements that a supplier/vendor must have in their inventory management system. The decision to enforce specifications and requirements must be dependent on the criticality of the parts, components, and materials that are being procured. Examples of such additional requirements that may be imposed include the following (not all may be applicable for all construction environments):

  - *Specifications*: Project specifications define the required type, grade, quality, standards (e.g., ASTM, ISO, BS), and performance characteristics of every material that is being procured. This may need to be provided to the supplier/vendor by the contractors' procurement department.
  - *Bill of materials (BoM)/material take-off (MTO)*: Engineers and estimators quantify the exact amounts of specified materials that are needed based on the design drawings. This informs procurement what amounts need to be procured.
  - *Procurement strategy*: A procurement plan may be developed that details how and when materials will be sourced, ordered, and delivered to align with the construction schedule. This should identify the long lead-time items, as they are critical.
  - *Submittal schedule*: A schedule may be created that outlines when suppliers/vendors must submit their material data (shop drawings,

product data, samples, certifications) for review and approval by the design team or owner's representative before ordering or fabrication. This must be coordinated with the contractors' quality department. A CQM can coordinate this with the procurement department of the contractor firm.

- *Storage and handling plan*: A plan may be made by the contractor firm for receiving, inspecting, storing, and protecting materials on-site (or at designated off-site locations) to prevent damage, theft, or degradation before they are incorporated into the work. This plan must consider space requirements, security, and environmental controls (e.g., climate control for sensitive items).

- *Coordinated inventory management*: Another best practice in procurement of parts and materials from suppliers/vendors is to align the inventories of the suppliers with those of the contractor. (This makes business sense only if the criticality of the suppliers' materials and parts is very high.) For projects that are complex and include multiple stakeholders, it may be cost-effective to utilize a common software for inventory management; such a software can then be used to coordinate the inventory management between the suppliers/vendors and the prime contractor firm.
- *Formally defined order/reorder points*: For critical materials that represent high levels of material supply risks, suppliers may be contractually required to formally prepare quality plans that include the details of the inventories as well as the reorder system. It is possible to create sophisticated inventory controls that can track materials stock levels and calculate the reorder points as well. All these control requirements must be implemented, subject to the scalability of the project and its significance. The decision to bring sophistication into the system should be justifiable through a business case for it.
- *Lead procurement resources of the supplier/vendor:* The names and organization structure of the supplier/vendor must be known, along with their contact information. For critical items, it is best to collect further information about the procurement process and the backup staff in case of critical supplies.

It is noted that the need for parts and materials in most construction projects consists of limited quantities, and ensuring their continuous availability is generally possible through simple collaboration between the contractor and all suppliers. Where good supplier management processes are established, the risk of stockout is significantly reduced. (See also "Trade Partners Requirements Alignment" in this chapter for more information.)

Good coordination, establishing clear procedures for purchasing, receiving, and issuing materials, rigorous quality control for outgoing materials, and accurate record-keeping are other best practices that can be implemented as control requirements for the procurement of parts, assemblies, and materials. All the above-mentioned requirements may be placed upon the suppliers and vendors based on the scalability of the project and the business need for a higher level of coordination between the primary contractor and the suppliers/vendors.

Material Traceability Requirements

Material traceability refers to the ability to track the journey of raw materials throughout the supply chain, from their origin to the final product. The most significant aspects of traceability of materials include:

- *Real-time information or manually managed inventories:* The leading technologies of facility and materials management include many software-based solutions that can record the movement of inventory in real time. However, such technologies are mostly utilized in the manufacturing industries, where their utility makes them cost-effective. For a materials supplier or a parts vendor that only supplies to the construction industry, the scalability of real-time tracking of the material inventory may or may not be economically viable. Materials suppliers must optimize the capability of their traceability systems based on the criticality of the materials and parts they supply, as well the significance of the need for real-time information.
- *Technologies for traceability of the materials:* Technologies that are currently available to track the movement of materials include barcode scanners and radio frequency identification (RFID), which uses radio waves to identify objects from a distance without making any physical contact or requiring a line of sight. Other technologies that are available include laser markings that allow permanent, durable identification on materials and items.
- *Push or pull systems:* The traceability requirements for construction materials may utilize either a push or a pull system. Push systems are used by materials suppliers who produce materials in response to their customers' forecasted demands, even if those customers have not yet placed orders. Conversely, materials suppliers can utilize pull systems where the traceability of the suppliers' materials is based on actual placed customer orders, not on the basis of the contractors' demand for the materials.
- *Integrated record management system*: Most supply chains utilize construction management software, which have multiple modules that connect the various aspects of construction; inventory management, traceability, and QC inspections of shippable materials and component parts can all be recorded through a centralized software source that can be secured against data and cybersecurity breaches. Utilizing such solutions can improve the resource performance of the materials suppliers and vendors, but they all must be used based on the economic justification of their use for the construction project.
- *Lot traceability versus piece traceability*: In some high-risk construction projects, it is critical to know exactly which materials and parts are

utilized in the construction of the building or infrastructure. In such cases, lot traceability and sometimes even piece traceability are required. Lot traceability identifies the products by their production batch (which is also called a lot). It allows for quick identification of possible issues and, if necessary, efficient recall for a group of items that may have been used during construction. In contrast, a piece traceability system tracks and documents each individual item. For piece traceability, each piece/item must utilize a unique identifier, thereby providing the highest level of detail, precision, and traceability. Obviously, it becomes the most expensive and complex form of traceability. All construction projects are different, and the extent to which traceability of the lots or parts is warranted must be a decision that is rationally made by the CQM, together with the procurement department of the construction contractor.

Traceability requirements represent the various options that a construction contractor firm has when they engage with material suppliers and parts manufacturers. If a well-managed supplier selection and evaluation system is put in place, it helps turn the procurement from an ad hoc activity into a smooth, organized, process. It subsequently results in more efficient construction, mitigates construction delays, and offers reduced costs and other benefits. (More specific guidance on how to set up a vendor selection and evaluation system is provided in "Vendor Performance Evaluation" later in this chapter).

Material Inspection Requirements for Suppliers/Vendors

The material inspection process is an essential part of any construction project. It involves the testing and evaluation of materials to ensure that they meet the necessary quality standards and are safe for use in the construction of buildings and infrastructure. Various types of QC and QA inspections can be implemented with material suppliers and part vendors. Some of these may include self-inspections that must be performed by the suppliers/vendors; those suppliers may be required to provide quality certifications along with their shipments. Other inspections may be external to the suppliers/vendors and require the contractors' CQM to perform those inspections upon the suppliers/vendors.

Depending on the nature of the materials, as well on their criticality, some of the inspections that may be implemented on the materials and parts include the following:

- *In-process material inspections at the supplier facility:* For some construction projects, materials and component assemblies are manufactured at fabrication plants. It is common to include typical in-process QC inspections as requirements for material suppliers and parts and assembly vendors. Most of these quality inspections are the internal QC inspections that are to be performed by the material supplier on their production. Alternately, these in-process inspections could also be performed by independent testing organizations by utilizing external QC inspectors, or they could be performed by the quality representatives of the construction contractor.

It is a best practice to require a material supplier to include in their quality plan the details of all in-process QC inspections they routinely perform on their production of the materials.

Two of the most applicable and well-known QC tests that are performed at the manufacturing/fabrication facilities of the suppliers and vendors are tests that are usually performed by the construction contractor. These tests include the factory acceptance test (FAT) and the first article inspection (FAI).

The FAT is performed on-site at the supplier's facility (factory) as an external test and is performed by the construction contractor's quality representative. The purpose of a FAT is to ensure product quality and verify the process that manufactures those parts, materials, and/or equipment that are shipped to the contractor's construction site.

The FAI, on the other hand, focuses on the capability of the "first production run" at the manufacturer's facility. It may be performed either as an internal test (i.e., performed by the material supplier/part manufacturer themselves) or as an external test (by the contractor's quality representative). (See more detail in "Trade Partners Requirements Alignment" in this chapter.)

- *Outgoing (pre-shipment) material inspections at the supplier facility:* These are the QC inspections that are typically provided by QC staff in fabrication plants. Depending on the nature of the materials, these inspections could include the measurement of product characteristics that define the quality of the product and materials. These could be discrete or continuous measurements, and the materials suppliers may be required to provide material certifications for the quality of the shipped materials and parts.

  The results of QC tests performed on outgoing shipments by the supplier should be formally documented in detail. These tests may generate the following certifications and reports:
  - *Mill test reports (MTRs) or mill certificates for metals:* A mill test report is a QA document issued by the manufacturer of the material/metal. It verifies the chemical and physical properties of that material and certifies the compliance of the manufactured material with industry standards (like ASME or ASTM).
  - *Certificates of conformance (CoC) or compliance:* A CoC is an official document issued by an official quality representative of the material supplier, or their approved representative. The CoC officially attests that the product meets the specific quality standards as well as other contractual requirements that it is expected to meet.
  - *Batch records (for concrete, grout, coatings, etc.):* A batch is a defined quantity of items or materials that are produced together in a single run under uniform conditions. It is a record of the production/

manufacturing of the materials that moved through each stage of the production process as one group. Batch records are typically required for projects that need to include the complete manufacturing/production history of a specific product batch. Typically, such information is most useful for construction projects where the project team wants more detailed information about the entire batch of the material. If the batch is used in multiple locations at the construction site, a possible failure in the material would affect all those sites. Batch records allow the contractor to mitigate the possible failure and offer a better chance of quick identification and correction of the defects.

- *Chain-of-custody documentation (for certified wood or sustainable materials):* This is yet another type of quality record for the materials, and it is useful when materials received from suppliers have a possibility of deterioration or lack of integrity for any reason. Chain of custody documentation, if provided, chronologically tracks the possession, control, transfer, analysis, and disposition of that material. This requirement is useful when the integrity of the materials is of utmost importance.

All the above-mentioned test results are not always needed for all of the materials and component parts manufactured at a supplier's facilities—some may apply, and others may not be logical. It is important for the CQM to work with the procurement department, rationally determine which internal/external tests and QC inspections should be applied to a project, and then require only those from the materials suppliers.

- *Material handling inspections at the supplier facility:* Material handling inspections are aimed at verifying that materials and products sent to construction sites will be free of handling defects. When defect-free materials and products are shipped from fabrication facilities, the likelihood of receiving good materials and products at the construction sites significantly increases. Material handling inspections may be required for those cases where poor handling could cause defects during shipping. Pictures of the shipped material, along with visual quality reports, may be required.
- *Incoming material inspections at the construction facility:* Another commonly implemented QC inspection for material inspectors and vendors is an incoming inspection performed at the construction site. The primary purpose of such an inspection is to ensure that the material arrives at the construction site free of any defect—that no loss of quality has happened as a result of shipping from the fabrication house to the construction site.

All these QC inspection requirements must be carefully considered by the quality manager and their staff, and communicated to procurement. The principle of common-sense logic must be followed to determine which tests and inspections make the most sense for all procured materials, as well as for the component parts procured for the construction sites. (Chapter 5, "Laboratory Accreditation," must be studied with this topic to gain a more comprehensive understanding of the quality requirements for testing laboratories, because they are also procured as a service (vendor) organization to assist the construction project.)

CONCLUSION

In summary, effective material control and resource identification are indispensable for project success. By meticulously integrating traceability, establishing stringent vendor requirements, and thoroughly evaluating resource needs for inspections from the outset, organizations can navigate the complexities of construction with greater precision. This proactive approach ensures that materials are not just available, but aligned with the project's specifications, timelines, and budget. Consequently, these best practices minimize disruptions, uphold quality standards, and enhance overall efficiency, proving that diligent material management is indeed the backbone of successful project delivery.

QUESTIONS

1. What are the five primary objectives of material control management as outlined in the chapter?
2. Explain the difference between material identification and material traceability, and give an example of a scenario where piece traceability would be necessary.
3. Describe at least three documents or plans that are developed to incorporate material control, traceability, and vendor requirements into a project prior to commencement of work.
4. Identify and briefly explain critical challenges encountered in materials control management.

RESOURCE MOBILIZATION AND MANAGEMENT

> Integrate resource qualification and training, along with warehouse quality and material management into the resource management process. Assess the need for specific resources tailored to each inspection type, ensuring that these resources are available and properly allocated, and aligned with the established inspection and test plan. (Evaluate)
>
> Body of Knowledge IV.B.3

In the realm of construction project management, the seamless integration of resource mobilization and management stands as a cornerstone of success. Resource mobilization and management involve the processes of providing and removing resources, materials, and personnel to and from project worksites. One of the objectives is to ensure that personnel, materials, and equipment are not only available, but also qualified, trained, and managed to meet the quality objectives of the project. From initial site preparations to ongoing material control, the project needs a strategic approach to resource management. Effective resource management, driven by a comprehensive resource management plan aligned with the project's quality plan, is crucial for project efficiency, safety, and profitability. Efficient mobilization strategies not only save time and resources but also contribute to the overall project success.

The CQM can support project management in resourcing the project by ensuring that all resources—personnel, materials, and equipment—are suitable for their intended purpose and are managed in a way that supports the quality objectives of the project. This can be achieved by embedding quality considerations into the standard resource management life cycle (planning, acquisition, allocation, management, and control).

Because the CCQM BoK defines a construction project in terms of the project life cycle, resource management for the project must also be considered in terms of all the phases of the project—not just for construction. The concept of resourcing the project with qualified personnel, equipment, and materials remains applicable throughout all seven phases of a construction project's lifecycle.

The resource management process must include planning the sequence of operations and managing logistics. This is where procurement usually gets involved, to ensure that necessary resources are available across the supply chain. Managing resources throughout all phases of the project means qualified and trained people, who are positioned accordingly, are supported with the proper equipment and materials.

The first step in resource management and mobilization is the development of a resource management plan, which is described next. If the construction project is issued to multiple performing organizations (such as in a DBB project), then each production phase of the project must develop its own resource management plan.

Resource Management Plan

A resource management plan is a strategic document that outlines how a project's required resources, including people, equipment, materials, storage, data, and financial resources are to be identified, acquired, allocated, and monitored throughout the project lifecycle. The goal of a resource management plan is to ensure that resources are used efficiently and effectively to meet the project's quality objectives (which include product quality objectives, prevention of delays and waste, and maximization of project profitability and client satisfaction).

Before the characteristics of an effective resource management plan may be considered, an essential prerequisite is the context in which the plan may be written. The resource management plan must be described in the context of the project delivery method and the phase for which the plan is prepared. It is critical to recognize the significance of this prerequisite for creating an effective resource management plan; the plan is only useful if it offers strategic guidance for resourcing that is specific to the project phase and the project delivery method. Also, the alignment of the resource plan to the strategic quality objectives of the project allows the resource plans of the different phases of the project to be consistent throughout the lifecycle of the project.

The essential elements of a resource management plan for a project may include five stages:

1. *Resource identification:* In the first stage of resource management, the project team defines the resources it needs to deliver its objectives. A comprehensive list of all the resources that the project requires must be created. If the construction project is delivered in independent phases, then the resources needed for the successful completion of each phase must be identified in resource management plans that are developed independently for each phase. Resources that must be identified may include the following:
   - *Human resources:* The plan must define the roles that must be filled and subsequently determine the responsibilities that each of those roles must fulfill.
   - *Materials and component parts:* The plan must identify the raw materials and component parts needed to deliver the end products (of the project, or the phase).
   - *Storage and warehouse resources required:* The plan must also include the resources needed for the storage of the materials and component parts.
   - *Resourcing the receiving inspections:* The plan must also include the reference to the project's receiving quality. The CQM can assist the construction manager in the development of resources for effective incoming inspections to take place.
   - *Equipment and machinery needed:* The plan should identify the equipment the project (or phase) may need. It must ensure that the correct tools, machinery, and other equipment are available when needed, they are not over-utilized or under-utilized, and the project stays on schedule and within budget.
   - *Time resources/allowed schedule:* The schedule of the project and its deliverables should be described in the plan. The best practice is to develop the project schedule at varying levels of detail. Typically, the levels of the project schedules and the time resources that are needed for the project (or phase) are described in increasing level of detail from level 1 (for high-level stakeholders) to level 4 (the execution-level schedule with greater detail).
   - *Information and data resources needed:* The plan should describe the information and data resources that are needed by the project (or phase).
   - *Financial resources/budget:* The plan must assess the current financial status, as well as future goals, and create a strategic roadmap to allocate money effectively. This part of resource planning generates the budget for the project (or phase).
   - *Identification of external resources:* The plan must identify areas needing external resourcing (such as third-party inspection services).

Depending on the size and scope of the project, a good resource plan may require additional elements that need resourcing. But these are the most essential elements for resource identification at the first stage of the project. The CQM can assist the project manager in identifying needed resources.

2. *Resource acquisition:* For the second phase of resource management, the processes for acquiring items identified as needed resources must be included in the resource plan. In other words, the resource plan should clearly describe which resources would be purchased, rented, or hired, and how the mobilization of those resources will take place for the project. The action items during this phase/stage of procurement include negotiations with providers and vendors, and establishing contracts that subsequently acquire the right resources at the right time and cost.
3. *Resource allocation:* In the third stage of resource management, an allocation strategy must be prepared. Allocation means assigning specific resources to project tasks and deliverables. A resource calendar, which can allocate the resources and present it to all stakeholders, is a useful tool.
4. *Resource mobilization:* Once an allocation has been made, it is time to mobilize the team according to the allocation. Where the mobilization of staff and resources follows a well-thought-out mobilization plan, the project avoids conflicts, delays, and frustrations. In reality, all projects follow a resourcing curve that is bell-shaped and has project resources are allocated over time, peaking in the middle and tapering off at the beginning and end. Thinking through such details ensures that the right staffing and resourcing are done at the right time, which makes it most economic as well as efficient.
5. *Resource monitoring and control:* After the resource plan has identified, acquired, and allocated the resources, the next stage is resource utilization and management. Depending on the size and scope of the project and its criticality, this stage is where resource loading and resource leveling may be required. For simpler projects, this may not be needed.

   A resource management plan should guide the project by providing a holistic view of all moving pieces in each project. It should include the five sequential stages of resourcing, which address the essential subelements of resourcing such as budget, staffing, workload requirements, and technology requirements. Following a logical, methodical process of resourcing allows a project to identify, acquire, allocate, mobilize, and monitor the resources that help run the project in the most efficient manner.

Resource Identification Stage

The first step in resource management is identifying the resources needed for the project and creating a comprehensive list. If the construction project is delivered in independent phases, then the resources needed for the successful completion of each phase must be identified in the resource management plan that is developed for each phase. Resources must be identified for the following categories.

1. *Human resources:* First, define the roles that must be filled and subsequently determine the roles and responsibilities. Clearly define who is responsible for managing and utilizing specific resources and any critical tasks that must be included in the resource plan.

2. *Materials and component parts:* Next, determine what raw materials and component parts are needed to deliver the outcome or results of the project. One concept that is very useful in creating the list of materials and component parts is the creation of a bill of materials (BOM), which is a comprehensive, hierarchical list that breaks down the final product into its individual parts, sub-assemblies, and raw materials.

3. *Storage and warehouse requirements:* Resources must also be identified for the storage of the materials and component parts. The quality manager may assist the project manager/construction manager with the resourcing requirements for site logistics (e.g., temperature/humidity control, stacking limits, shelf-life limits, protection from UV/weather, or segregation needs). Shelf-life monitoring and protocols such as First-In-First-Out (FIFO) or Last-In-First-Out (LIFO) may need to be implemented.

   It is in this aspect of resourcing that a CQM may assist the construction superintendent. A CQM may review the manufacturers' recommendations and project specifications to define any specific storage requirements for the materials (e.g., temperature/humidity control, stacking limits, shelf-life limits, protection from UV/weather, segregation needs). This information is usually fed into the site logistics and warehouse planning process.

4. *Resourcing the receiving inspections:* The quality manager may also support the resourcing of the receiving inspection of the parts and materials that are received for a construction project. Typically, the resourcing needs for inspection of the materials include detailed receiving inspection procedures and checklists that outline what needs to be verified upon on material delivery (e.g., quantity, identity vs. PO, visual condition/damage, correct labeling, expiry dates, or required certifications like MTRs/CoCs). The CQM can assist the construction manager in developing resources for effective incoming inspections. Also, another related requirement must be the provision of segregated materials based on incoming quality inspections.

5. *Equipment and machinery needed:* The next resource that must be identified is the equipment the project (or phase) may need. The resource plan must ensure that the correct tools, machinery, and other equipment are available when needed, without being over-utilized or under-utilized, to keep the project on schedule and within budget.

6. *Time resources/allowed schedule:* The best practice is to develop project schedules and the accompanying deliverables at varying levels of detail. Typically, the levels of the project schedules and the time resources needed for the project (or the phase) are described from level 1 (for high-level stakeholders) to level 4 (execution-level schedule with greater detail).

7. *Information and data resources needed:* The next resource that should be described in the resource plan should be the information and data resources needed by the project (or phase). This involves assessing the current and future needs for information and data, and accordingly making decisions about how data and information will be stored, accessed, and used across the project or the organization.
8. *Financial resources/budget:* The resource plan must also assess the current financial status, as well as future goals, and create a strategic roadmap to allocate money effectively to achieve the project goals. This part of resource planning generates the budget for the project (or phase).
9. *Identification of external resources:* The resourcing of third-party inspection services is one example of external resourcing. Typically, the CQM works with the lead designer and procurement staff to ensure that contracts are awarded to qualified suppliers and vendors in a timely manner. This resourcing effort includes tasks such as clearly defined vendor/supplier qualifications, reporting formats, and calibration standards that must be implemented by the external resources.

Depending on the size and scope of the project, a good resource plan may require additional elements, but these nine elements are essential for resource identification at the first stage of the project. The CQM can assist the project manager in identifying the resources needed. Also, by analyzing project specifications, codes, standards, and risk assessments, the CQM may help define the minimum qualification requirements for personnel performing specific quality-affecting tasks. The CQM may also provide input for job descriptions and subcontracts. Qualification requirements, especially for jobs that affect quality directly, may be provided to HR and to procurement, and included in job descriptions, pre-qualification criteria for subcontractors, and the tender documents.

Resource Acquisition Stage

This is the stage of resource management where the project or the organization obtains the necessary resources identified in the first stage. Now, contracts take place, and the hiring gets underway. Whereas this stage is mostly a function of project management, the CQM can assist the project manager or to the HR by providing verification of the resources being acquired.

- *Key staffing:* One of the most critical aspects of resource acquisition is to ensure that the key positions identified for the completion of the project are filled by qualified staff. Projects in the technical fields frequently fail to recruit niche experts, sometimes even when they are identified as essential staff. In other projects, understaffing causes performance and delivery issues. All of this must be addressed through effective resource management.
- *Verification of staff qualifications:* It is important to verify the qualifications of the acquired human resources (e.g., quality inspectors, quality engineers, or other personnel who may

affect quality, both internal and external to the project). This might involve reviewing resumes, checking the validity of their professional references, ensuring the qualifications of the selected staff are sufficient, confirming their credentials with certification issuing bodies, or reviewing their performance on past projects. This resource acquisition activity must be performed formally for key positions that are pre-identified for the project or the organization.

- *Qualification testing:* Perform qualification testing for the quality and workmanship of the inspection staff (the QC people). This is a technical task that cannot be performed effectively by HR. Verifying the skill qualifications for critical skills that are needed on the project (for example, the workmanship and qualification for welding or electrical circuits) is something that the CQM may coordinate or oversee. In other words, the CQM can act as an extension of project management when it relates to the qualification testing of the candidates who may be formally approved before they may perform the quality inspection work. The CQM may also analyze project specifications, codes, or standards, and ensure that the resources, human as well as equipment, technology, data, schedules, and other specifications, are being met by the staff that is considered by resource acquisition.

Resource Allocation Stage

Once the resources are acquired, the next stage in resource management is the allocation of those resources to the tasks that must be performed for the delivery of the project. One core principle for projects and organizations is, "Resources are always limited." This is a core principle because even if a resource exists in larger quantities now, it is not infinite and will run out eventually. It is therefore essential for every project team to make wise choices and consider all resources to be limited. Resource allocation means strategic decisions are made so those limited resources are used efficiently and effectively by balancing the competing needs and priorities of the project.

- *Multidisciplinary staff:* Depending on the nature of the tasks on the project, resource allocation may vary from one project to the next. For base-level tasks that do not require a very high level of niche qualifications, project teams generally hire a multi-disciplinary staff that can perform general construction (GC) work. Typically, the technical specifications of a project include information about the definable features of work (DFOW), which are specific, manageable work packages that have distinct trade or discipline requirements. It is logical to utilize the DFOW as a guide and specify those trades in the resourcing for the project. Furthermore, the control requirements included with the DFOW may be used for resourcing as well as to determine the permitting requirements for the project. (See more detail about the role of DFOW in Chapter 5, "Required Permits.")

- *RACI chart and qualification matrix:* Two tools most used for resource management include the RACI and qualification matrix. A RACI matrix identifies the roles of all internal and external stakeholders and project team members with respect to four areas: responsible, accountable, consulted, and informed. The qualification matrix, on the other hand, is a project-specific matrix developed to show the critical tasks of the project. It lists the personnel qualified to perform those tasks. The CQM collaborates with construction superintendents and the construction manager to ensure that the qualification matrix only lists personnel who are formally qualified for the positions and are assigned to perform specific quality-critical tasks.
- *Mobilization plan:* A mobilization plan is a structured approach outlining the resources needed to prepare for a project before the work begins. It primarily includes the plan for the mobilization of human resource allocation that assures that site management (i.e., project manager, production superintendent/supervisor) has the internal personnel resources (QC inspectors, trained site engineers, and foremen) assigned to the construction activities.
- *Integration with project schedule:* Effective resource allocation requires that the inspection durations, frequencies, and required sequences (especially hold/witness points from the ITP) must be included in the project's construction schedule. Resourcing at this level means turning the QC inspection activities into scheduled tasks that are linked to construction progress. To achieve this goal, the CQM typically works with the project scheduler and uses the project look-ahead schedules (e.g., three-week, six-week) to anticipate upcoming inspection workloads and confirm resource readiness.
- *Budget allocation:* Although it is typically considered a function of project controls, budget allocation is also a resourcing activity. The cost estimates for third-party services, equipment rental/calibration, specialized training, and travel (for source inspections) are typically provided to project controls for inclusion and tracking within the project budget. A CQM's role is typically limited in terms of budget allocation, but they may be included for some projects.

Resource allocation and mobilization is the stage of resource management which makes a project operational. After this stage is complete, the project is usually considered in operation. The next stage of the project is ongoing project monitoring and control, which has different resource requirements than initial stages.

Resource Mobilization Stage

Mobilization is another stage that must be considered in the construction phase of the project lifecycle. Mobilization begins in the preconstruction phase, if the construction project is issued in independent phases (e.g., in a DBB project, mobilization plans must be developed independently for the design phase and the construction phase). Mobilization of the construction staff involves detailed assessments of project requirements, resource availability including required qualifications and training, and other potential challenges.

A construction project mobilization plan should include the description of activities such as setting up site facilities and temporary utilities, transporting equipment and personnel, and securing the necessary permits and documents. Key steps in mobilization for the construction phase should include forming a mobilization team, which takes on the mobilization of the construction team as a mini project by itself. Tasks may include scheduling the site preparation, selection of office spaces, establishing communication, technology, and other administrative tasks like permitting and licensing of the new construction spaces. A well-structured plan outlines a timeline for these activities, assigns responsibilities, and helps prevent delays and budget overruns. CQMs are usually not directly responsible for these tasks, but they could be included in the task teams responsible for the mobilization.

Each construction project is unique. Resource mobilization for the construction phase must consider project-specific factors, such as location, size, and complexity, to tailor the mobilization plans that align with the specific needs of the construction phase.

Whether the construction site requires establishing new office spaces or renting the office spaces from those available in the area, the mobilization of the construction team has a host of tasks to attend that must be documented in the mobilization plan for the construction project.

Resource Monitoring and Control Stage

At the earlier stages of resource management (namely, resource identification, acquisition, and allocation), the effort is to establish the project and make it operational. However, an operational project is dynamic and subject to constant change in conditions. The construction industry has a reputation for having one of the highest employee turnover rates across all industrial sectors. Because most construction projects are long term, resource management remains an ongoing subject for a construction project; it is never a one-time acquisition and allocation that can solve the resource challenges of the project. Resource monitoring and control are ongoing processes.

Once resources are acquired and a project is operational, the focus of resource management shifts from execution to monitoring and control. The need to hire employees in the critical areas of the construction operations due to employee turnover may require optimizing available resources, re-allocation, and other monitoring and control phase resourcing. Some of the techniques used during this phase of resource management may include the following:

- *Resource leveling:* Resource leveling is a project management technique that adjusts the task schedule and balances it for the use of labor, equipment, and materials. The technique adjusts the schedule and other variables that may be flexible with the aim of achieving efficiency and preventing overallocation of resources. Resource leveling is most applicable if the construction work consists of repetitive work. If, however, the work is significantly different in different phases, which happens more often, then resource leveling is not useful.

- *Audits and surveillance:* During the resource monitoring and control phase, the CQM may assist the project management by performing periodic audits of the project resources in all aspects of work (human resources, materials, equipment, data and information, etc.). The goal of such audits is to ensure the project continues to perform with the right resources, including the assigned key staff personnel with the required, current qualifications.
- *Ongoing training needs:* Another task that may be considered in the domain of resource monitoring is ongoing training on the project. The need for continual training and updates is continuous for every project, more so for a project that has high employee turnover. New staff may need quality training on various subjects. It is once again in this respect that the CQM may provide or coordinate the training of the current staff for all essential areas of knowledge (e.g., understanding the project quality objectives, project quality plan, or training needed for specific inspection procedures, defect reporting, or new material installation techniques). To develop training materials, the CQM may develop or source quality-specific training modules, checklists, or work instructions.
- *Addressing resourcing nonconformances:* If quality issues arise that are traceable to deficiencies in skills and employee training, then the CQM may provide retraining, requalification, or removal of the individual from that specific task.

CONCLUSION

Effective resource management is pivotal in steering any organization toward its goals. This process involves a strategic approach to managing various essential resources. By integrating qualification, training, and material management, and aligning resource allocation, project teams can achieve not only compliance but also optimized efficiency. The emphasis on planning, availability, and communication ensures that the right resources—human, financial, or material—are deployed at the precise moments required. This holistic approach fosters quality, safety, and ultimately, the profitability necessary for project success.

QUESTIONS

1. What are the five stages of resource mobilization?
2. What are the resources that must be identified during the resource identification stage?
3. What are the major sub-elements of mobilization during the resource acquisition stage?
4. How is resource allocation different from resource acquisition?
5. What are the essentials of a resource mobilization plan?
6. What are the ongoing resource challenges of a construction project?

Risk Management

Risk management in construction quality management is crucial for ensuring project success, safety, and compliance with regulatory requirements. Construction projects often face numerous uncertainties and challenges due to their complexity, multiple stakeholders, and ever-changing site conditions. Therefore, it is vital to identify, assess, and mitigate risks systematically to mitigate quality issues, prevent delays, and reduce cost overruns.

The uncertainty inherent in construction projects, combined with the complex interplay of multiple disciplines, makes effective risk management a critical component of successful project delivery. Risk management is the systematic process of identifying, analyzing, evaluating, treating, and monitoring risks to minimize their impact on project outcomes. It is an ongoing activity that requires continuous monitoring and adaptation as the project progresses. In construction quality management, risk management is not just about preventing failures or defects; it's also about ensuring that the project meets its quality objectives in a safe, efficient, and compliant manner.

RISK IDENTIFICATION AND ANALYSIS

> Analyze various risk identification and assessment tools, including failure mode and effects analysis (FMEA), fault tree analysis (FTA), and strengths-weaknesses-opportunities-threats (SWOT) analysis. Examine risks based on quantitative and qualitative data related to potential impact on the project. (Analyze)
>
> Body of Knowledge IV.C.1

Risk identification and analysis form the foundation of risk management. In construction, identifying and analyzing risks involves understanding the project's complexity, stakeholder expectations, and market conditions. A systematic approach to risk identification and analysis helps prevent procurement-related disruptions that could lead to project delays, cost overruns, and quality failures. There are several tools, such as FMEA, FTA, SWOT, and others that can be effectively utilized for risk identification and analysis in construction procurement. These tools help a CQM systematically identify risks, assess their impact and likelihood, and develop appropriate mitigation strategies.

Failure Mode and Effects Analysis (FMEA)

Failure mode and effects analysis (FMEA) is a systematic, proactive tool used to identify potential failure modes within a process, assess their effects, and prioritize them based on their severity, occurrence, and detectability. Originally developed in the 1940s by the U.S. military and widely used in automotive and manufacturing industries, FMEA has become a vital tool in construction project management. It helps in identifying potential risks, evaluating their impact on the project, and developing strategies to mitigate those risks. This approach ensures that the project proceeds smoothly with minimal disruptions related to procurement issues such as material shortages, late deliveries, or substandard quality.

Steps for Conducting FMEA

FMEA involves several systematic steps, each designed to ensure comprehensive identification, assessment, and mitigation of risks. The following are the recommended steps, adapted for use in construction procurement:

1. *Define the Scope and Assemble the FMEA Team*
   - *Define the scope*: Clearly define the boundaries of the FMEA analysis. For construction procurement, this may include specific procurement activities, such as supplier selection, material delivery, or contract management.
   - *Assemble the team*: Form a cross-functional team with representatives from procurement, project management, quality, and any other relevant disciplines. Diverse perspectives will ensure that all potential risks are identified and evaluated.
2. *Identify Failure Modes and Their Effects*
   - *List potential failure modes*: Brainstorm and list all potential failure modes in the procurement process. A failure mode is any way in which a procurement process could fail to meet requirements. Examples include late delivery, incorrect quantities, code violation, or receipt of non-compliant materials.
   - *Determine the effects of each failure mode*: For each failure mode, describe its potential effects on the project. For example, if a critical material like steel is delayed, it could delay the structural work and, subsequently, the entire project schedule.
3. *Assess Severity (S), Occurrence (O), and Detectability (D)*
   - *Severity (S)*: Rate the severity of each failure mode's effect on a scale of 1 to 10, with 10 being the most severe. Severity measures the potential impact of the failure mode on the project.

     Example: In a high-rise construction project, a delayed delivery of concrete (critical for foundation work) would score a high severity rating due to its potential to halt progress.
   - *Occurrence (O)*: Assess the likelihood of each failure mode occurring. Rate the probability on a scale of 1 to 10, with 10 being the most likely. Historical data, supplier performance records, and expert judgment can be used to determine occurrence ratings.

     Example: If a supplier has a history of late deliveries, the occurrence rating for late deliveries could be 8 or 9.
   - *Detectability (D)*: Evaluate how easily the failure mode can be detected before it impacts the project. Rate detectability on a scale

of 1 to 10, with 10 being the least detectable (high risk) and 1 being the most detectable (low risk).

Example: If there is no automated system to track material delivery schedules, the detectability rating would be high, indicating a high risk.

4. *Calculate the Risk Priority Number (RPN)*

   The risk priority number (RPN) is calculated by multiplying the S, O, and D scores for each failure mode:

   $$\text{RPN} = \text{S} \times \text{O} \times \text{D}$$

   RPN values range from 1 to 1000, with higher values indicating higher risk levels. Failure modes with high RPN values should be prioritized for mitigation.

5. *Develop Mitigation Strategies*

   Based on the RPN values, determine appropriate mitigation strategies for the highest-priority risks. Mitigation strategies could include selecting alternative suppliers, increasing inventory levels for critical materials, or implementing stricter supplier evaluation criteria.

6. *Implement and Monitor Mitigation Actions*

   Assign responsibility for each mitigation action and implement the strategies. Continuously monitor the effectiveness of these actions and update the FMEA as necessary.

7. *Review and Update the FMEA*

   FMEA is not a one-time activity. Regularly review and update the FMEA to reflect any changes in the procurement process or project scope. This ensures that new risks are identified and managed proactively.

FMEA Template for Construction Procurement

Table 4.1 provides a structured approach to conducting FMEA in construction procurement. This template can be customized based on the specific needs and requirements of the project.

FMEA is a powerful tool for proactively managing risks in construction. Continuous review and updating of the FMEA will ensure that it remains relevant and effective throughout the project life cycle.

Fault Tree Analysis (FTA)

FTA is a top-down, deductive failure analysis tool used to identify the root causes of undesirable events and understand the relationships between these causes in complex systems. It is widely used in construction to analyze and manage risks associated with procurement failures, such as delayed deliveries, noncompliant materials, or contract disputes. FTA helps project managers and procurement teams visualize the cause-and-effect relationships of various procurement-related risks and develop targeted strategies to mitigate these risks.

Table 4.1 Example FMEA template for construction procurement.

Process Step	Potential Failure Mode	Potential Effects	Severity (S)	Potential Causes	Occurrence (O)	Current Controls	Detectability (D)	RPN ($S \times O \times D$)	Recommended Action	Action Taken	Revised RPN
Material Delivery	Late delivery of critical materials	Delays in project schedule	9	Supplier capacity issues	8	Weekly status check-in calls	5	360	Identify backup supplier	Alternative supplier contracted	180
Material Quality	Receipt of noncompliant materials	Rework and additional costs	8	Lack of supplier QA/QC	7	Inspection on receipt	6	336	Increase frequency of supplier audits	Quarterly audits scheduled	168
Supplier Payment	Payment delays affecting supplier performance	Reduced supplier reliability	6	Internal processing delays	5	None	9	270	Implement automated payment tracking	Automated tracking implemented	135

Steps for Conducting FTA in Construction

Conducting FTA involves several systematic steps that ensure a comprehensive understanding of how various procurement risks can lead to critical failures in construction projects. The following steps outline the recommended approach for using FTA in construction procurement:

1. *Define the Top-Level Undesired Event*
   - *Establish the scope*: Begin by defining the top-level undesired event in the procurement process. This is the ultimate failure that could disrupt the project's objectives, such as delays in material delivery, receipt of defective materials, or failure to meet regulatory compliance requirements.
   - *Specify the context*: Clearly describe the context and boundaries of the FTA. For example, the scope may include the procurement process for a specific critical component or service in the construction project.

     Example: In a chemical plant construction project, the top-level undesired event might be "failure to procure compliant safety valves." This event could lead to delays in construction or, worse, safety violations.

2. *Identify Intermediate Events and Contributing Factors*
   - *Decompose the top-level event*: Break down the top-level event into intermediate events or contributing factors that could cause the undesired event. These could include failures in procurement processes, issues with suppliers, or logistical problems.
   - *Use Boolean logic*: Represent the relationship between these intermediate events using Boolean logic gates such as AND, OR, and NOT. This helps in visualizing how multiple factors combine to cause the top-level undesired event.
     - *AND gate*: The top-level event occurs only if all intermediate events below it occur. For example, delays in both material production and shipping could lead to overall delays in material delivery.
     - *OR gate*: The top-level event occurs if any of the intermediate events occur. For example, either inadequate supplier qualification processes or lack of supplier quality oversight could result in non-compliant materials being delivered. These failures could stem from poor supplier evaluation criteria, insufficient incoming inspection, or unverified material certifications.

3. *Identify Basic Events*
   - *Determine the root causes*: Identify the basic events or root causes of each intermediate event. These could be factors such as inadequate supplier capacity, lack of supplier quality controls, or insufficient communication between procurement and project management teams.

Example: In a high-rise building project, basic events contributing to the late delivery of structural steel could include delays in raw material procurement by the supplier, transportation issues, and customs clearance problems.

4. *Construct the Fault Tree Diagram*
    - *Create a visual representation* (Figure 4.2): Construct a fault tree diagram using symbols to represent the top-level event, intermediate events, and basic events. Use arrows to show the flow of causation from basic events to the top-level event.

      Example: For a metro rail project, a fault tree diagram can visually depict how issues such as delays in the delivery of rail tracks, nonconforming dimensions of the rails, or lack of adequate storage facilities can all lead to "failure to complete track installation on schedule."
5. *Assign Probabilities to Basic Events*
    - *Quantify the likelihood of basic events*: Assign probabilities or frequencies to each basic event based on historical data, supplier performance metrics, or expert judgment. These probabilities indicate the likelihood of each basic event occurring.
    - *Calculate the probability of intermediate events*: Use the probabilities of the basic events and the logic gates to calculate the probabilities of the intermediate events. This helps in understanding which root causes have the highest likelihood of leading to the top-level undesired event.
6. *Analyze the Fault Tree and Identify Critical Paths*
    - *Determine the probability of the top-level event*: Calculate the overall probability of the top-level undesired event using the probabilities of the intermediate events. This helps prioritize which root causes to address.
    - *Identify critical paths*: Determine the critical paths, which are the sequences of events most likely to cause the top-level failure. Addressing these critical paths can significantly reduce the probability of a top-level event.

      Example: In a residential housing development, if the top-level event is "failure to receive occupancy certification," critical paths may include incomplete electrical inspections, noncompliance with fire safety regulations, or delays in installing utility connections.
7. *Develop and Implement Risk Mitigation Strategies*
    - *Create mitigation plans*: Develop targeted risk mitigation strategies for the high-priority basic events. Strategies could include selecting alternative suppliers, increasing inventory levels, or improving communication channels with suppliers.
    - *Implement and monitor*: Assign responsibility for implementing the mitigation strategies and monitoring their effectiveness over

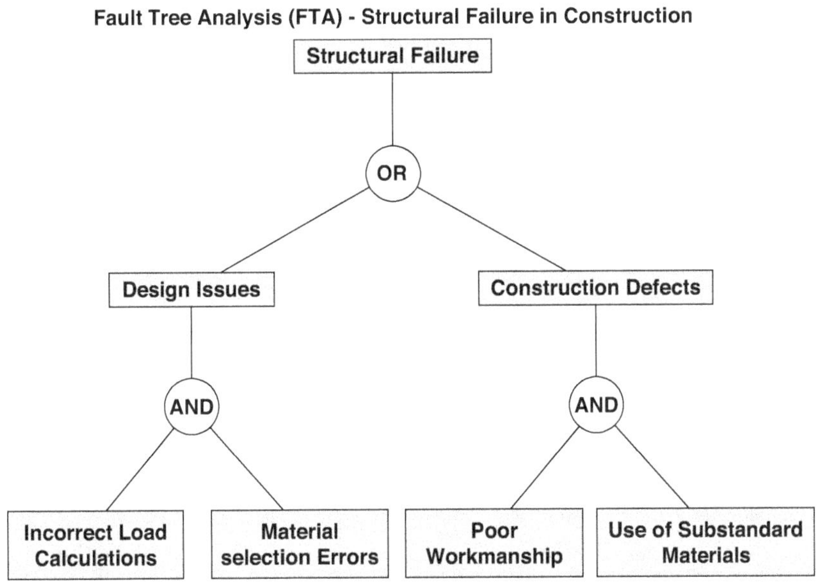

Figure 4.2 A simple fault tree analysis of structural failure in construction.

time. Update the fault tree as new risks emerge or as mitigation strategies are applied.

8. *Review and Update the FTA*
    - *Regularly review the fault tree*: The fault tree should be reviewed and updated regularly to reflect any changes in the procurement process or project scope. This ensures that new risks are identified and managed proactively.
    - *Communicate changes*: Any changes to the fault tree and associated mitigation strategies should be communicated to all relevant stakeholders, including procurement, project management, and QA teams.

Fault tree analysis is a powerful tool for understanding the complex relationships between various procurement risks and their potential impact on construction projects.

SWOT Analysis

Strengths, weaknesses, opportunities, and threats (SWOT) analysis is a strategic planning tool used to identify and analyze internal and external factors that can impact a project or organization. It is a structured approach to evaluate an organization's strengths, weaknesses, opportunities, and threats to create strategies for improvement. In construction procurement, SWOT can help project teams identify factors that could influence the successful acquisition of materials, subcontractors, or services, allowing for proactive decision-making.

Steps for Conducting a SWOT Analysis in Construction Procurement

1. *Identify Strengths*
   - *Definition*: Strengths are internal attributes or capabilities that provide an advantage in meeting project objectives. In construction procurement, strengths might include a robust supplier network, strong relationships with subcontractors, or an experienced procurement team.

     Example: A large residential developer might have long-standing relationships with reputable suppliers, ensuring consistent access to high-quality materials at competitive prices.
   - *Strategy*: Leverage these strengths to negotiate better contract terms, maintain material quality, and avoid supply chain disruptions.
2. *Identify Weaknesses*
   - *Definition*: Weaknesses are internal attributes that place the project at a disadvantage. These may include limited supplier options, inadequate inventory management, or lack of a well-defined procurement process.

     Example: A commercial builder lacking digital tools for procurement management may face inefficiencies in tracking orders and managing supplier performance, leading to project delays.
   - *Strategy*: Implement a procurement management software to streamline order tracking, enhance communication with suppliers, and reduce manual errors.
3. *Identify Opportunities*
   - *Definition*: Opportunities are external factors that the organization can capitalize on to achieve project objectives. These can include emerging technologies, favorable market trends, or government incentives.

     Example: A public transit construction project could benefit from new green building incentives offered by the government, making it feasible to procure sustainable materials at reduced costs.
   - *Strategy*: Align procurement strategies to take advantage of these opportunities by investing in sustainable material options and applying for incentives.
4. *Identify Threats*
   - *Definition*: Threats are external factors that pose risks to achieving project objectives. Examples in procurement include volatile material prices, supplier bankruptcies, or changes in regulatory requirements.

     Example: A power plant construction project may face the threat of material shortages due to geopolitical tensions affecting the supply of critical components.

Table 4.2 Example of risk analysis techniques.

Technique	Application	Example
FMEA	Assessing risk of material shortages	Identifying failure modes like late deliveries, incorrect quantities
FTA	Analyzing causes of procurement failures	Identifying root causes of elevator component delays
SWOT	Evaluating internal and external factors	Analyzing regulatory threats and supplier strengths
Monte Carlo Simulation	Quantifying the impact of cost fluctuations	Predicting budget variance due to steel price changes
Decision Tree Analysis	Assessing the consequences of procurement decisions	Evaluating outcomes of choosing alternative material suppliers

- *Strategy*: Establish alternative suppliers and secure long-term contracts to hedge against material price volatility and supply chain disruptions.

Quantitative and Qualitative Risk Analysis

Risk analysis in construction should consider both quantitative and qualitative factors. Quantitative analysis involves the use of numerical data and statistical methods to assess the probability and impact of risks. Techniques such as Monte Carlo simulation and decision tree analysis (see Table 4.2) can quantify the potential impact of supply chain disruptions or material cost fluctuations.

Qualitative risk analysis, on the other hand, relies on expert judgment and experience to evaluate risks based on predefined criteria. It involves categorizing risks as high, medium, or low based on their potential impact on the project. For instance, qualitative analysis can be used to assess the risk of a supplier's bankruptcy based on financial health indicators and historical performance.

CONCLUSION

Effective risk management in construction quality management is essential for ensuring project success, mitigating potential delays, and maintaining compliance with industry standards. Utilizing tools such as FMEA, FTA, and SWOT analysis provides a structured approach to identifying, analyzing, and addressing risks proactively. By integrating both quantitative and qualitative data, construction teams can better assess the likelihood and impact of risks, prioritize mitigation efforts, and ensure that quality objectives are achieved efficiently and safely throughout the project life cycle.

This comprehensive approach ensures projects stay on schedule, within budget, and meet quality objectives. Continuous monitoring and adaptation are vital for managing the ever-evolving complexities of construction, ensuring project success, safety, and compliance with industry standards, ultimately leading to successful project delivery and stakeholder satisfaction.

QUESTIONS

1. What are the key differences between failure mode and effects analysis (FMEA), fault tree analysis (FTA), and SWOT analysis when identifying and analyzing risks in construction projects?
2. How can the use of quantitative and qualitative data improve risk assessment in construction quality management?
3. Explain how the risk priority number (RPN) is calculated in FMEA, and discuss how a high RPN value should influence risk mitigation strategies in construction procurement.
4. Describe the difference between AND and OR gates in FTA, and give a construction procurement example for each type of gate and its implication on risk management.
5. How can the use of historical data and expert judgment be balanced when assigning probabilities to basic events in FTA? What are the potential drawbacks of relying too heavily on one over the other?
6. Discuss how SWOT analysis can be applied to identify both internal and external risks in construction procurement. Give an example of how a "weakness" can also be a driver of a "threat."
7. Compare and contrast quantitative and qualitative risk analysis methods, and explain why integrating both approaches is crucial for effective risk management in construction quality management.

RISK TREATMENT

> Evaluate risk mitigation strategies such as risk transfer, retention, acceptance, avoidance. (Evaluate)
>
> Body of Knowledge IV.C.2

Risk treatment is a pivotal component of the risk management process in construction quality management, focusing on how identified risks are managed to ensure minimal impact on project objectives. This phase builds on the insights gained during risk identification and analysis to devise strategies that either reduce the likelihood of risks occurring or mitigate their potential effects (see Table 4.3). In construction projects, where uncertainties can stem from supply chain disruptions, material defects, or adverse weather conditions, selecting the appropriate risk treatment strategy is essential for maintaining schedule integrity, cost control, and quality standards.

The process of risk treatment involves evaluating each risk based on its severity, probability, and potential impact to determine the most suitable management approach. The four primary strategies (also called risk treatments)—

Table 4.3 Strategies for risk treatment.

Strategy	Description	Example
Risk Transfer (Shift Responsibility to Another)	Shifting the impact of a risk to a third party, usually through insurance, warranties, or contractual agreements. This strategy ensures that external parties bear the risk of potential issues like material defects or delivery delays.	*Healthcare Construction*: During the construction of a hospital, procurement contracts can include clauses that make suppliers responsible for defects found in medical equipment during installation and testing phases. This ensures that the hospital does not bear the cost of replacing faulty equipment.
Risk Mitigation (Ensure Retention)	Accepting a risk that cannot be eliminated, nor transferred, and choosing to mitigate it. The project team develops a contingency plan to manage the risk if it materializes. This plan must reduce the impact of the risk. Used when the risk is minor, not major. This strategy is not recommended for major risks (all major risks must preferably be eliminated or transferred).	*Residential Construction*: If there is a risk of minor delays in receiving finishing materials like flooring, the project team may choose to retain this risk and plan for minor schedule adjustments rather than incurring additional costs for expedited shipping.
Risk Acceptance (Accept the Risk and Monitor)	Acknowledging a risk without taking proactive measures to reduce its likelihood or impact, often because the impact is negligible or no feasible mitigation strategies are available. The team monitors the risk and reacts if it occurs.	*Infrastructure Projects*: For a remote highway construction project, the team may accept the risk of supply chain disruptions due to adverse weather conditions, as these events are unpredictable, and mitigation strategies are not cost-effective.
Risk Avoidance (Eliminate the Risk)	Changing the project plan to eliminate the risk entirely. Typically used for high-impact risks that could have severe consequences on project outcomes. This strategy involves changing suppliers, altering project scope, or avoiding certain activities.	*Commercial Construction*: In a commercial high-rise project, if a supplier has a history of providing substandard steel beams, the procurement team may avoid using that supplier altogether, even if it means paying a higher price for materials from another, more reliable vendor.

risk transfer, risk mitigation, risk acceptance, and risk avoidance—offer a comprehensive framework for addressing various challenges in construction procurement. Each strategy must be tailored to the specific context of the risk, balancing project constraints such as budget, timelines, and stakeholder expectations.

The following provides an overview of the primary risk treatment strategies, supplemented with practical examples from different sectors within the construction industry.

Steps for Implementing Risk Treatment in Construction

The following steps outline a structured approach to risk treatment in construction procurement:

1. *Risk Evaluation and Prioritization*
   - Review the identified risks and prioritize them based on their severity, probability, and potential impact.
   - Utilize the scoring system from tools like FMEA to determine which risks require immediate attention and which can be managed with lower-priority strategies.
2. *Selection of Risk Treatment Strategy*
   - Determine the most appropriate risk treatment strategy for each prioritized risk.
   - Consider factors such as project constraints, budget, and stakeholder preferences when choosing between risk transfer, retention, acceptance, or avoidance.
3. *Development of Mitigation Plans*
   - Create detailed mitigation plans for risks that will be transferred, retained, or avoided.
   - For risks that will be accepted, ensure that contingency plans are in place to manage their impact if they materialize.
4. *Implementation of Risk Treatment Plans*
   - Implement the chosen risk treatment strategies and integrate them into the project management plan and procurement process.
   - Update procurement contracts, create contingency reserves, or revise project schedules and scope as necessary.
5. *Monitoring and Reviewing Risks*
   - Continuously monitor identified risks and the effectiveness of the treatment strategies.
   - Update the risk treatment plans as new information becomes available or as project conditions change.

Risk Treatment Plan Template

Table 4.4 shows an example using simple template for documenting risk treatment plans in construction procurement.

Effective risk treatment is a cornerstone of successful construction quality management, providing the means to mitigate potential challenges and ensure project continuity. By systematically evaluating and prioritizing risks, selecting

Table 4.4 Risk treatment plan example.

Risk ID	Risk Description	Treatment Strategy	Action Plan	Responsible Party	Monitoring and Review
1	Delayed delivery of steel reinforcement	Risk Transfer	Include penalty clauses in the contract with the supplier for late deliveries. Ensure backup suppliers are prequalified and ready to supply.	Procurement Manager	Review weekly supplier performance reports and logistics data.
2	Supply chain disruptions due to weather	Risk Acceptance	Acknowledge the risk and prepare for possible delays in project schedule. Communicate with stakeholders and update project timeline if delays occur.	Project Manager	Monitor weather forecasts and update stakeholders accordingly.
3	Use of substandard electrical components	Risk Avoidance	Avoid using suppliers with a history of noncompliance. Source components from suppliers with third-party quality certifications (e.g., ISO 9001).	Quality Assurance Lead	Review supplier quality records and third-party audit results.
4	Price volatility for raw materials	Risk Mitigation (Retention)	Allocate contingency budget for potential cost increases. Track market trends and purchase bulk materials when prices are low.	Financial Controller	Review quarterly budget and adjust contingency reserves.

appropriate treatment strategies, and implementing robust mitigation plans, project teams can significantly reduce the adverse effects of uncertainties. The continuous monitoring and revision of risk treatment plans further enhance their effectiveness, ensuring that risks are managed dynamically as project conditions evolve. By integrating these strategies into the overall project management framework, construction teams can achieve their objectives while maintaining high standards of quality and efficiency.

QUESTIONS

1. What are the four primary risk treatment strategies, and how do you determine which is most appropriate for a specific risk in a construction project?
2. How can continuous monitoring and regular updates to risk treatment plans enhance their effectiveness in addressing evolving risks in construction projects?
3. Explain the interplay between risk probability, severity, and potential impact in selecting the most appropriate risk treatment strategy. Describe a hypothetical scenario in a construction project where two risks have similar potential impacts but different probabilities, and explain how you would treat each risk differently.
4. The document outlines five steps for implementing risk treatment in construction procurement. Discuss how each step builds on the previous one and how neglecting any of these steps could compromise the effectiveness of the overall risk management process.
5. Analyze the Risk Treatment Plan example (Table 4.4). Describe how you would modify this template to incorporate more specific financial considerations for each risk, particularly for risk retention strategies, and why this modification is important.
6. Contrast the risk acceptance and risk mitigation/ retention strategies, detailing the specific circumstances under which you would favor one over the other. Give an example of a nuanced situation in a construction project where the distinction between these strategies is not clear-cut.
7. The document mentions integrating risk treatment strategies into the project management plan and procurement process. Explain how cross-departmental collaboration (e.g., between procurement, project management, and quality assurance) is essential in effectively implementing and monitoring these strategies. Give an example of what might happen if these departments did not collaborate effectively.

RISK MONITORING

Use risk monitoring tools (e.g., risk-based audit programs) and perform surveillance audits to regularly evaluate identified risks and monitor key factors (e.g., cost, time, safety, supplier performance), and the criticality of material and equipment. (Evaluate)

Body of Knowledge IV.C.3

Risk monitoring is a critical component of quality management that ensures risks are managed effectively throughout the construction project's lifecycle. It involves systematically tracking identified risks, reassessing them as conditions evolve, and evaluating the effectiveness of implemented risk responses. This process not only maintains alignment with project goals but also enables teams to remain proactive in addressing new risks as they emerge.

In construction projects, where uncertainties are inevitable, risk monitoring tools such as risk-based audits, surveillance audits, performance dashboards, and continuous risk reviews provide critical insights. These tools help teams evaluate key factors such as cost, time, safety, supplier performance, and material criticality. By integrating these techniques into their workflows, CQMs can optimize resource allocation, make data-driven decisions, and enhance the likelihood of successful project outcomes.

Risk Monitoring Tools and Techniques

Effective risk management is an ongoing process, not a one-time event. In the dynamic environment of a construction project, new risks emerge, existing risks change in severity, and mitigation plans must be continuously evaluated. This requires a robust system of risk monitoring, utilizing specific tools and techniques to ensure project success. Several tools and techniques are commonly used to monitor risks in construction procurement (see Table 4.5). These tools enable project teams to track risk status, measure performance, and identify new risks that could affect project delivery. Below are key tools and techniques used in risk monitoring, along with practical applications in various construction sectors.

Table 4.5 Risk monitoring tools and applications.

Tool/Technique	Description	Application in Construction
Risk-Based Audit Programs	Prioritize audit activities based on the risk levels of various project elements. These programs ensure compliance with quality requirements and safety standards.	Risk-based audits are used in high-risk areas such as structural steel fabrication or electrical installations to verify adherence to specifications and identify potential issues early.
Surveillance Audits	Periodic assessments that verify compliance with quality management processes and evaluate the effectiveness of risk controls.	Conducted in residential construction to monitor ongoing work quality, adherence to safety protocols, and material conformity during various project stages.
Performance Metrics and Dashboards	Real-time data tracking of project performance through metrics and visual dashboards that highlight deviations and trends.	Used in commercial construction projects to monitor schedule adherence, budget control, and safety incidents, providing early warnings of potential risks.
Continuous Risk Review	Regular project status updates and risk register reviews that assess the status of risk treatment actions and identify new risks.	Implemented in large infrastructure projects to evaluate the effectiveness of risk responses and plan for new mitigation actions during monthly project reviews.

Risk-Based Audit Programs

Risk-based audit programs involve prioritizing audit activities based on the risk associated with different project elements. This approach helps project teams focus their audit resources on areas that pose the highest risk to project success, ensuring that potential issues are identified and addressed proactively.

Example: In a transportation infrastructure project, a risk-based audit program might prioritize audits on structural elements, such as bridges or tunnels, where the impact of defects could be severe. The audits would focus on verifying adherence to design specifications, material quality, and safety standards.

Implementation Steps

1. *Identify high-risk areas*: Review the risk register and prioritize project elements with high-risk ratings (e.g., critical path activities, structural components, or complex installations).
2. *Plan audit schedule:* Develop a schedule that focuses audit activities on high-risk areas during critical phases of the project.
3. *Conduct audits*: Perform detailed assessments of the high-risk elements to ensure compliance with quality, safety, and design requirements.
4. *Document findings*: Record any nonconformances or deviations and assess their potential impact on project objectives.
5. *Follow up on corrective actions:* Implement corrective actions for identified issues and track their resolution in subsequent audits.

Surveillance Audits

Surveillance audits are periodic assessments that ensure quality management processes are followed and risks are effectively controlled. They are conducted at regular intervals to monitor key project parameters, such as cost, time, safety, supplier performance, and the criticality of materials and equipment.

Example: In a high-rise commercial building project, surveillance audits can be conducted every month to monitor the quality of concrete pouring, electrical installations, and fire safety systems. Any deviations identified during these audits are addressed immediately to prevent rework and schedule delays.

Steps for Conducting Surveillance Audits

1. *Develop an audit checklist*: Create a checklist that includes key project parameters and risk areas to be assessed.
2. *Schedule regular audits:* Determine the frequency of surveillance audits based on the project's risk profile and criticality of the work.
3. *Perform on-site audits:* Assess ongoing activities to ensure they meet quality standards, safety requirements, and project specifications.

4. *Review audit findings:* Share audit findings with the project team, and develop corrective actions for any identified issues.
5. *Update risk register:* Incorporate new findings into the risk register, and revise risk treatment plans as needed.

Performance Metrics and Dashboards

Performance metrics and dashboards provide real-time visibility into project performance. Key performance indicators (KPIs) such as schedule adherence, cost control, and safety incidents are tracked to identify deviations from expected outcomes. Visual dashboards present this information in an easily interpretable format, enabling project managers to quickly identify trends and anticipate potential risks.

Example: In an industrial plant construction project, dashboards can track the progress of critical activities such as equipment installation and testing. If a delay in installation is detected, project managers can take corrective actions to avoid schedule slippage.

Continuous Risk Review

Continuous risk review involves regularly scheduled meetings, such as weekly or monthly project status updates, where the project team revisits the risk register, evaluates the status of risk treatment actions, and decides on additional actions if required. This approach ensures that the project remains agile and responsive to emerging risks.

Example: In a residential housing development, weekly project status can be used to review risks such as material shortages or subcontractor delays meetings (see Table 4.6). The project team can then update the risk register with new mitigation plans or decide on immediate actions to prevent further delays.

Table 4.6 Risk meeting template.

Continuous Risk Review Meeting Template
Project Name: [Insert Project Name]
Meeting Date: [Insert Date]
Agenda:
1. Review of Current Risk Register
2. Status of Risk Treatment Actions
3. Identification of New Risks
4. Planning of Additional Risk Responses
5. Action Items and Responsibilities
Attendees: [List attendees and their roles]

CONCLUSION

Risk monitoring is essential for maintaining effective risk management throughout a construction project's life cycle. By leveraging tools such as risk-based audit programs, surveillance audits, performance dashboards, and continuous risk reviews, project teams can proactively identify and address risks. These tools ensure that project objectives are met by enabling real-time performance tracking, facilitating timely decision-making, and adapting risk responses to evolving project conditions. Continuous monitoring not only mitigates risks but also builds resilience within the project framework, ultimately contributing to successful and efficient project delivery.

QUESTIONS

1. How can performance metrics and dashboards assist project managers in monitoring risks and proactively identifying potential project issues?
2. Explain the role of continuous risk reviews in maintaining effective risk management throughout a project's life cycle.
3. What are the benefits and challenges of integrating real-time performance dashboards into the risk monitoring process in construction projects?
4. Explain how risk-based audit programs contribute to proactive risk management in construction projects, and give a detailed example of their implementation in a high-risk area.
5. Compare and contrast surveillance audits with continuous risk reviews in terms of their frequency, focus, and impact on project risk management.
6. Discuss the advantages of utilizing performance metrics and dashboards for risk monitoring, and describe potential challenges a project manager might face when integrating these tools.
7. Describe the steps involved in implementing a continuous risk review process, including the key components of a risk meeting template and why each component is important.
8. How do the various risk monitoring tools (risk-based audit programs, surveillance audits, performance metrics, continuous risk reviews) collectively contribute to building resilience within a construction project framework and ensure successful project delivery?

Quality Reporting

REVIEW AND INTERPRETATION OF QUALITY DOCUMENTS

> Evaluate inspection and test reports, quality records, and manufacturers' data books to ensure traceability. Assess conformance with product specifications and quality criteria. (Evaluate)
>
> Body of Knowledge IV.D.1

Quality documentation applies to all the phases of the project lifecycle. The following section addresses the typical requirements of documentation that must be considered for the project. Documentation management for the design phase is described differently than the process for the construction phase. (See "Quality Documentation During Design" in this chapter.)

Regardless of how the project is contracted out (whether as a DBB, DB, or any other variation of project delivery), the way the document management of the designer, contractor, and the suppliers/vendors must be organized is intrinsically linked with the centralized documentation management for that project.

Quality Documents Versus Quality Records

Following are the fundamental differences between these two types of quality documentation:

1. *Quality documents*: "Documents" is the technical name of the formally developed documentation created at the beginning of the project to enable it to operate. As such, quality documents include all documentation generated to perform the work tasked by the project.
2. *Quality records*: "Records" is the technical name of the documentation that is generated as a result of the project work. As such, a quality record is created once the quality process is completed. Records are the evidence used by QC to determine compliance with the project specifications. As such, quality records must include all documentation generated while completing project tasks, which must then be checked by QC.

Table 4.7 shows the difference between documents and records.

Table 4.7 Differences between documentation and records.

Document	Record
Generated as a result of planning (front-end tool)	Generated as a result of action (back-end tool)
Contains information on "how to do something"	Contains information on "how a thing was done"
Cannot be used as evidence	Can be used as evidence
Can be changed/ revised	Cannot be changed (record is unchangeable as of the time of issue)
Needs to be formally approved before it may be implemented	Approval of records may or may not be necessary; approval must be against criteria
Document may become a record	Record cannot become a document
Document may be in paper form or electronic	Record may also be in paper form or electronic
Duration of the validity of document must be defined by the organization	Record is forever; it may need to be destroyed after a certain time, based on a policy of record termination. But it is forever and cannot be changed.

Quality Reports and their Frequency

Quality reporting stands as the cornerstone of effective project management, particularly in the demanding realm of construction. It meticulously documents the status, progress, and outcomes of quality management endeavors, providing critical insights into how project work impacts overall quality objectives. Throughout the project life cycle, quality reports serve as vital tools, especially during the design and construction phases, where the project's tangible outputs are measured against specific criteria. The frequency and detail of these reports must be carefully calibrated, balancing compliance needs with operational efficiency, to ensure they add value without becoming a bureaucratic burden.

A quality report is a document that details the status, progress, and results of quality management activities within a project or organization. The main purpose of quality reporting is to look at the various aspects of the project work and report them with respect to their impact on quality. A quality report may be generated to summarize QC or QA activities (or both).

Quality reports should be generated throughout the life cycle of the projects to describe the performance of the project against the quality objectives, but they are of particular importance during the two production phases of the project life cycle (design and construction). While all phases of a construction project are important, the design and construction phases are generally considered the most significant. The design phase lays the foundation for the project by translating the client's vision into detailed plans and specifications. The construction phase then brings these plans to life, transforming them into a physical reality (usually into a structure). As such, these are the two phases of the project life cycle that "produce" an end product that can and should be measured against the product-specific quality criteria. The metrics for quality that may be included in the quality reports for the different phases of the project life cycle are different. Also, ongoing quality reporting is different from special-purpose quality reporting, such as those based on quality audits.

The frequency of quality reporting is a topic of interest to all stakeholders of a construction project. The optimal reporting frequency in any project depends on its size, complexity, and the needs of its stakeholders specific to the different phases of the project.

Quality reporting is crucial for showing compliance with the requirements as well as for identifying inefficiencies, optimizing the processes, and demonstrating the effectiveness of quality in the project activities. However, it is not in anyone's interest when the projects spend more time on the reporting than on the actual delivery of the project work. It is essential that the frequency and duration of the quality reports be discussed proactively among the stakeholders; reports must appropriately provide the best value to all stakeholders without turning into a non-value-added and bureaucratic deliverable.

Recommended Quality Reports for the Design and Construction Phases

The two production phases (design and construction) are usually considered more significant primarily because these two phases generate the products of a construction project that can be reviewed/inspected/tested for product quality. Quality reporting for these phases may include:

1. *Design Quality Reports*: These typically include the quality reporting on the design audits that are performed by QA on the design packages delivered during the reporting period. Design quality reports may be prepared to provide the status of the intradisciplinary and interdisciplinary QC checking performed on design submittals. Also, the status of constructability reviews and the resolution of comments may generate useful information about design errors and omissions. Addressing these improves the quality of the design and ensures that the construction phase will include fewer design changes. A CQM, in consultation with project stakeholders, should decide which should be the ongoing quality reports during the design phase, along with the frequency of reporting (weekly, monthly, or quarterly).
2. *Construction Quality Reports:* The significance of quality reporting during the construction phase cannot be overstated. There are generally two types of ongoing quality reports that may be prepared during this phase: QC and QA reports. In addition, there may be specific product-based quality reports that are also generated, sometimes to approve specific materials and products, when requested.

    a) *Construction QC Reports:*
    - *Daily inspection report:* The most significant QC report is the daily inspection report, which is produced to document the quality of the constructed work. Typically, the daily inspection reports are prepared by QC inspectors who inspect the construction (i.e., constructed product) and qualify those inspections as either pass or fail.
    - *QC reports from suppliers/vendors/other service providers*: For all those functions and products that are the responsibility of the suppliers, vendors, or other service providers (e.g., testing labs), the same requirements for QC must be flowed down as are implemented for the quality of the work performed by the primary construction contractor. More specific guidance on vendor management is provided in "Vendor Performance Evaluation" later in this chapter.

    It is important to understand that all QC inspections are those where the constructed product is subject to a pass/fail decision. The quality reports that evaluate the performance of QC with analysis and trends over time are the QA reports (discussed below).

b) *Construction QA Reports*:

The QA reports for construction include all those reports that evaluate the performance of the project's QC. The fundamental difference between QC and the QA reports is that the latter evaluate the effectiveness of quality, or the lack of it.

Depending on the preference of the project, and as determined based on scalability, the projects must also determine the frequency as well as the level of detail that they require in their QA reports. It must be noted that these reports offer analysis and evaluation of the QC results, whereas QC reports only provide the status of compliance (pass or fail), with no further analysis, breakdown or trending. The analysis may include statistical evaluation and trending, which provides significantly more valuable information than just a compliance-based pass/fail report from QC checking. QA reporting typically requires additional skills in data management, and the quality managers (e.g., the CQM) are expected to offer these additional skills to interpret the data over weekly and monthly trends, and provide the additional information to the management.

c) *Product-based Quality Reports for approvals of Materials*:

The third type of quality reports are those that may or may not be ongoing; such reports are generally prepared on request. For example, a quality report may be prepared to describe "product-based quality" of construction materials or other products. Likewise, the quality approval of "materials" may fall under a separate category because these quality inspections are usually not routine. Quality reports for these task-based requests may be prepared by quality personnel when requested. Such reports are product-specific and may require a different protocol for the collection of test data, testing methods, and for their review and approval. Depending on the product approval protocol, material quality reports may require additional proof of testing, such as testing from independent testing laboratories, and may include statistical studies on the results that are obtained.

Quality reports for product-specific testing are usually more advanced in nature and typically include additional information such as material specifications, product data sheets, test conditions, test methods, acceptance criteria, sample sizes, and sometimes the qualifications of the personnel who may be authorized to perform such advanced quality tests.

Characteristics of a Good Quality Report

There are several features that distinguish a good quality report:

- *Based on pre-established criteria*: Quality reports are more authentic when they are generated as an output of a pre-validated quality system. An individual's (quality) report may be subjective and include the bias of human interpretation and judgment, influenced by their individual perspective. It is best to first establish the reporting process, including standardized evaluation criteria for the quality reporting (which defines the various levels of quality compliance and excellence), and then let the process self-generate quality reports based on the system's analysis of the QC data in the system. Doing so removes individual biases and human subjectivity from quality reports, and introduces transparency into the QA analysis, which is based on the methods and assumptions used in the analysis.

- *Reports from a "quality" database*: The quality reports generated from a well-designed and well-maintained quality database of project records can be more verifiable and accurate than those prepared by individuals. The QA should strive to establish a quality database into which the QC inspection data is fed, and then it may self-generate analysis and evaluation quality reports. Such quality reports offer increased data reliability and reduced errors if the database is based on the accepted principles of best quality requirements and specifications. Also, data entry errors and omissions are removed through the data validation rules that such systems can implement, thereby eliminating the errors and omissions of individual QC inspectors.
- *Use of dashboards*: Another method of quality reporting includes the development of "dashboards" that can offer the advantages of improved decision-making, better communication to management and other stakeholders, better data transparency, and increased efficiency if they are established correctly and maintained. There are, however, known potential drawbacks, such as data overloads, misinterpretation, and misaligned metrics that can be generated out of ill-maintained dashboards. Besides the high maintenance cost of the dashboards, which may include software-based licensing rights in some cases, there is the drawback of technical challenges (coding expertise, limited resources, and integration limitations). As such, what could be a useful system for quality reporting can quickly turn into a burden if the dashboard is not supported by ongoing management commitment or becomes unjustified due to lack of use.
- *Essentials of quality reporting*: The essentials of good quality reporting also include the fundamentals of reporting, namely identifying the following: (1) the contact person(s) responsible for preparing and issuing the quality reports (names, titles), (2) the backup person(s) responsible for the quality reporting, (3) the frequency of the reports, and (4) the target audience for the quality reports (this should include the list of recipients).
- *Summaries of the quality deficiencies versus nonconformances*: There is a difference between quality deficiencies that are quality fails (and are easily correctable) versus quality nonconformances that cannot be corrected without impacting the schedule. This information is

very useful for discussing the status of open issues during ongoing status meetings so they may be removed where possible or mitigated. Depending on the scale of reporting, some QA reports break down the deficiencies and nonconformances into functional disciplines that provide more useful information to management. Other reports break down this information with respect to physical areas where the nonconformances are identified. Such quality reports represent a very detailed level of reporting.

- *Stop work orders*: Further details in quality reporting may include information on "stop work" orders, if they are issued.
- *Status of design changes*: For construction projects, quality reports may also include summaries of performed design changes. The design changes could be based on field conditions that change or else could result from the designers' value engineering (a common practice during design and construction projects).
- *Status of requests for information and submittals*: Another aspect of QA reporting for construction projects includes the status of the requests for information and the contractors' submittals.

Evaluating the Quality of the Quality Reports

"Evaluation of the quality inspections and testing" is a QA activity. Evaluating a test report or a quality record for traceability means that someone acting as QA is ensuring that the quality inspection results contain accurate, complete, and easily traceable information about the inspected process and/or product. For a quality inspection and testing process to achieve this goal, a number of steps can be taken:

1. *Data entry validation*: If the data entry for the inspection and testing is performed electronically, the easiest solution is to include data validation rules in the testing and inspection report. This prevents data input errors and ensures that wrong information cannot be entered because the test (QC) inspector would be notified immediately of an error in data entry. For manual data entry systems, this is not possible, so wrong data or information can be logged in as the test record.
2. *Training in accuracy and traceability*: It is important to train quality inspectors so they understand the need for data accuracy and traceability. Training for quality test and inspection reports to contain accurate, complete, and easily traceable information about the inspected process and product requires the following essentials:
    - *Clear definition of testing criteria:* The clear and consistent definition of testing criteria should be included in the ITP (a formal document made available to the persons performing QC testing).
    - *Consistent testing methods:* The testing methods for the inspections and tests must be consistent. The concept of verification and

validation (described in Chapter 4, "Design Verification and Validation") requires that testing methods be standardized so that the results are reliable, reproducible, and comparable across different tests and operators. This requires the inspectors to follow a systematic testing process and standardize their testing methods.

- *Use of correct equipment and tools:* Using the correct testing equipment is necessary for quality inspections to obtain reliable and accurate results. This is an essential prerequisite for test result accuracy.
- *Evidence of compliance through the test results:* Depending on how the inspection and test form is designed, compliance (or lack of it) can be shown readily on the form or inspection report.

3. *Verification of the QC inspection and test reports*: The QC inspection and test reports should be reviewed to ensure the correctness of information as well as to confirm that the inspections are conducted properly and the work that is produced and checked meets the required standards. Verification of QC has traditionally been performed on the construction projects in several different ways:

   a. Many public sector projects in the industry still operate with the concept of a "resident engineer," who not only represents the public agency (owner) and oversees the construction but also ensures quality and other aspects of the project work while managing the contract. Traditionally, daily inspection reports (inspection and test reports collected by the inspectors) are signed off by the resident engineer to assure they contain accurate, complete, and easily traceable information about the inspected process and product.

   b. In private construction industry projects, the oversight role of the resident engineer is typically split into responsibilities of construction management and of quality of work. Responsibilities of construction management are usually handled by field supervisor(s) and their staff, who ensure the construction work is completed as scheduled. On the other hand, the responsibilities for quality of work—inspections and testing—are handled by QC (field) inspectors who collect information about the "quality characteristics" of the construction work performed. After this, the inspection and test results collected/documented by the QC inspectors are reviewed and evaluated by a different person performing the role of QA.

The two types of arrangements for QA, as noted above, represent slightly different methods of project delivery that are prevalent in the public and private construction sectors. No matter who performs the QA role, the commonality here is "the need to ensure the reliability of the inspection results." It is the reliability of the inspection and testing that becomes the basis of the quality of work performed on the project.

CONCLUSION

Quality reporting is frequently the aspect of a quality system that is not well-planned. As a result, quality reports are neither traceable nor consistent and cannot provide a high level of confidence. It is important to plan the quality reporting structures and use best practices in setting up good-quality reporting structures to obtain more accurate results for the quality systems.

Meticulous quality reporting is paramount to project success. It ensures traceability and compliance, and provides valuable insights into project performance. By prioritizing clear, consistent, and well-structured reporting, stakeholders can identify inefficiencies, optimize processes, and demonstrate the effectiveness of quality management efforts. Investing in robust quality reporting systems and practices prevents bureaucratic burdens while enhancing data reliability and transparency. Ultimately, a well-planned and well-executed quality reporting strategy elevates project outcomes and fosters stakeholder confidence in the deliverables.

QUESTIONS

1. What are the key differences between quality control (QC) reports and quality assurance (QA) reports during the construction phase, and why is understanding these differences crucial for effective project management?
2. Discuss the factors that should be considered when determining the optimal frequency of quality reporting in a construction project, and how stakeholders can balance the need for compliance with operational efficiency?
3. Explain the significance of using pre-established criteria and a quality database for generating quality reports, and how these methods contribute to the authenticity and accuracy of the reports compared to individual reporting.
4. Describe the challenges and potential drawbacks associated with implementing quality reporting dashboards, and what measures should be taken to ensure their successful and beneficial use in a project.
5. How does the concept of "traceability" apply to quality documents such as inspection and test reports, and why is it important for quality assurance to verify this traceability?

DOCUMENTATION AND COMMUNICATION OF QUALITY INSIGHTS

> Evaluate and document key insights from the quality management process to inform future actions. Communicate quality-related insights and determine the appropriate timing, escalation, and method to alert stakeholders about significant quality issues. (Evaluate)
>
> Body of Knowledge IV.D.2

The documentation of quality insights should be completed as part of a lessons learned initiative. It is essential for all projects and should be done formally and continuously throughout the entire project life cycle (not just the procurement phase). Quality insights are the specific aspect of lessons learned that look at the impact of project decisions on quality—what went right for quality, what went wrong, and why. Such a summary of the project's successes and failures with respect to quality can help the future phases of existing projects as well as future projects.

Types Of Quality Insights

It is a best practice to categorize quality insights into three broad categories: quality compliance, performance excellence, and improvement. These broad categories can be subcategorized in several different ways. Depending on the preference of the project team, there are two ways that the quality insights could be documented:

1. *Quality insights in project phases*: Document quality insights under project phases, such as quality insights during precontract, planning, design, procurement, construction, closeout, and operations maintenance phases, and preferably a separate one for quality in past projects.
2. *Quality insights in functional areas*: Document insights for technology, project management, communication, business processes, resourcing, and so on.

Tools for Collecting Quality Insights

Several tools can be used to collect quality insights. Quality managers can choose to utilize any or all of them, depending on resources and the scale, size, complexity, and criticality of a project. The tools include:

- *Brainstorming sessions for quality*: A brainstorming session should focus on generating a wide range of ideas and insights from current experiences and past projects, with the goal of identifying areas for quality improvement and preventing future quality mistakes. It is key for such sessions to be open, supportive, and focused on quantity of ideas rather than immediate evaluation.
- *Cause-and-effect sessions on quality issues*: The team must understand the difference between correction and corrective action; if a quality lapse is an error (an incident that can be corrected), then a corrective action may not be needed. However, if the quality manager determines the identified quality problem to be systemic or more complex than a simple error, root cause analyses may be required. A process that identifies the root cause(s) of the commonly occurring quality problem becomes a source of quality insights for the future.
- *Surveys and questionnaires on quality issues*: The quality of a survey or questionnaire is of prime importance and helps collect accurate and reliable quality insights. A good questionnaire or survey is characterized by clear, unbiased questions; a logical sequence; and a focus on gathering relevant and reliable data. It is possible to develop good surveys that provide useful quality insight with the least amount of time required by the responder.

- *Dedicated case studies on quality issues*: Sometimes projects have the budgets and resources to perform dedicated case studies on investigating quality issues and opportunities for quality improvements.
- *Focus groups on quality issues*: Focus group discussions are facilitated discussions held with a small group of people who have specialist knowledge of a given subject. In this case, the use of subject-matter experts (SMEs) to discuss the quality issue(s) and collect their interest in a particular quality topic can be explored. Focus groups offer a "qualitative" method of research, where the limited group of invited individuals discuss a specific topic (e.g., quality issues) and allow for in-depth exploration of experiences, opinions, and needs of the SMEs for the project.
- *Interviews (audits) for quality issues*: Interviews that are specific to quality issues, or are conducted as part of QA audits, are another source of quality insights. It is important to plan these audits so their results can be readily used as inputs into quality insights/quality lessons learned for future projects.
- *Photographic and video evidence*: Audio and video evidence of quality issues, as well as pictures, are helpful for future training and learning sessions.

The Sequence of Documentation and Communication

The first order of business is to create a repository for quality insights (lessons learned). The best practice advice for a lessons learned repository is that it should be simple, easily searchable, and easy to document. Ease of use is the most frequent reason for the success or failure of these repositories. Creating a user-friendly system that requires less time to document the quality insights leads to increased use and is more successful than a complex or tedious system that requires a lot of effort for documentation.

It is essential to define quality insights through a formal policy or procedural guidance that describes what is to be documented and how. A procedure or policy narrative that describes the communication of the quality insights and the process that must be followed allows consistent collection of good data and fact-based insight that helps everyone. Not defining the format in which insight may be collected results in otherwise useful information and data that is in random order and cannot be easily used to develop reports and analyzed for trends or other analyses.

Another typical holdup is the legality of what can and should be used as a lesson learned. The description of the format as well as the protocol must include guidance that personal names and critical information about a legal issue, such as a claim, must be made with careful consideration by authorized people on the team. If possible, describe quality problems and solutions without making them person-specific. If this care is not taken, otherwise-useful information could be used negatively in legal cases. This can be prevented through a formal procedural guidance that tells what should be documented and how.

A lesson learned or quality insight should not be used to assign blame. Instead, when the issue is looked at in a positive manner, it can serve as a source of learning and future training. The communication protocol must also include guidance on the following important aspects:

1. *Methods of communication*: Decide if communication should be verbal or written and choose where comments should be posted.
2. *Escalation*: Escalation means assigning the right person to document the quality insight. Procedural guidance on the severity of quality issues can inform a team member when to utilize the escalation—for example, if the quality issue is controversial, or has turned into (or can turn into) a potential claim, then its documentation should be escalated to the authorized people. All other matters may be recorded by all users in the easy-to-use repository for lessons learned.
3. *Impact on quality goals and objectives*: The impact of the quality issue on the project's quality goals and objectives is an important aspect of quality insights. This requires an understanding of the goals of the project and all its stakeholders, and allows the documentation of quality insights to serve the cause of stakeholders' satisfaction through comprehensive corrective actions and prevention of recurrence.

CONCLUSION

Quality managers must understand and ensure the implementation of quality management in all phases of the project. This sets up a comprehensive system that can be used to collect information and data about possible quality failures as well as actual failures that can happen during all the phases of precontract, planning, design, procurement, construction, closeout, and beyond. The quality data collected through the application of QC and QA during all these phases allow the quality managers to use this wealth of information constructively for future phases (and projects).

By maintaining open lines of communication, quality managers can ensure that stakeholders are informed promptly of quality issues. This proactive approach helps in addressing problems before they escalate and allows for a preventive mindset and culture with regards to the project and the organization. It can play a critical role in maintaining the integrity and quality of the project, ensuring successful outcomes.

Establishing a robust system for documenting and communicating quality insights is paramount for project success. By employing the suggested tools, maintaining a user-friendly repository, and adhering to clear procedural guidance, organizations can leverage lessons learned to drive continuous improvement. The proactive engagement of stakeholders, coupled with the systematic collection and analysis of quality data across all project phases, fosters a culture of quality and prevention. Ultimately, these efforts not only mitigate risks and prevent future failures but also significantly enhance project integrity and ensure successful outcomes. (Detailed information about lessons learned can also be found in Chapter 5, "Lessons and Improvements," and Chapter 8, "Learning and Best Practices Analysis.")

QUESTIONS

1. How can the quality insights be documented for current and future guidance of projects?
2. What are the tools to collect quality insights and lessons learned?
3. What are the features of a good document repository of quality insights?
4. What procedural guidance should be issued to ensure the insight is collected in a consistent format that is readily usable for reporting and QA analysis?
5. What should be the procedure for escalation of responsibility to document controversial quality issues?
6. Explain the distinction between "correction" and "corrective action" in the context of quality issues, and describe how this distinction influences the documentation of quality insights.
7. Compare and contrast the two methods for categorizing quality insights outlined in the chapter (by project phases vs. by functional areas). What are the potential benefits and drawbacks of each approach?
8. Describe the key elements of a user-friendly repository for quality insights, and discuss why "ease of use" is considered the most frequent reason for the success or failure of such a system.
9. What are the legal considerations involved in documenting quality insights, particularly regarding personal information and potential claims? How can procedural guidance address these considerations?
10. Beyond simply documenting quality insights, how does the chapter suggest these insights should be actively communicated and utilized to impact project goals and objectives, and what role does stakeholder engagement play in this process?

VENDOR PERFORMANCE EVALUATION

> Analyze vendor performance periodically using quality key performance indicators (KPIs) and use supplier performance incentives (e.g., awards, more wallet share, premium). (Analyze)
>
> Body of Knowledge IV.D.3

In the realm of procurement, ensuring quality goes beyond initial selection; it necessitates a robust vendor performance evaluation system. This section delves into the critical aspects of establishing and managing such a system, which is designed to measure the success of vendor and supplier relationships. Through standardized scorecards and meticulous analysis of KPIs, organizations can effectively monitor contractual obligations and desired outcomes. This evaluation process not only documents vendor performance over time but also influences future contracting decisions, fostering continuous improvement and streamlined information sharing across internal departments, and ultimately driving quality and efficiency.

Vendor performance evaluation measures the success of an existing or new supplier relationship. It assesses the vendor's quality of service and evaluates how well they meet contractual obligations and performance expectations. A vendor performance evaluation system, also referred to as "vendor scorecards" or "vendor report cards," is a standardized way to capture a record of a vendor's performance to monitor whether a contract's desired outcomes are being met.

Implementing a vendor performance system enables a project team to document vendor performance over time and to support vendors in improving how they deliver goods or services. It influences which vendors your company chooses to contract with based on previous positive or negative experiences. It facilitates information sharing across internal departments. (See "Trade Partner Requirement Alignment" earlier in this chapter for more details.)

Creating the Vendor Performance System

Following are the recommended steps in creating a vendor performance evaluation system.

1. *Define the objectives of the vendor performance evaluation system:*
   The first step is to define the objectives of the vendor performance evaluation system. To achieve consensus in a multi-organization complex project, the objectives of the vendor performance system can only be finalized after they have been agreed to by the stakeholders of the project. If the project is performed by a single organization, it is relatively easy to achieve consensus on the objectives, since all project members belong to the same organization and thereby follow the same objectives.
2. *Design the vendor performance evaluation system*:
   A well-designed vendor performance evaluation system involves clearly defined evaluation criteria. The best practice is to utilize weighted criteria for evaluation because all factors that determine vendor performance are not weighted equally. A weighted criteria system assigns different weights to the performance criteria based on the preferences of the project's stakeholders. This makes the gathering of information and the analysis of data more effective, as the emphasis is placed on the elements of vendor performance that are considered most important. The vendor performance evaluation system needs to be secured, streamlined, and organized so it can manage and track the vendor performance effectively against the preferred criteria of performance.
3. *Vendor performance may be evaluated against the following criteria of evaluation* (the list is not all inclusive):
   - Availability of team
   - Cost control
   - Fulfillment of key performance criteria
   - Project and contract management
   - Quality of work performed

- Regulatory compliance
- Responsiveness and professionalism
- Schedule/timelines of performance

Surveys and Questionnaires to Collect Vendor Performance Appraisal from References

It is also recommended to obtain vendor's performance appraisals from references (direct customers of those vendors). This may include supervisors, procurement agents, or other personnel in the field, such as field engineers or QC technicians. References who connect directly with the vendors are useful, as they may provide specific information about the vendors, such as a vendor's knowledge, their willingness to provide service, their problem resolution, and overall satisfaction with the vendors' performance. Some examples of questions include:

- How would you rate the vendor's overall performance? (Choose from Excellent, Good, Fair, Poor)
- Would you recommend this vendor? (Choose from Yes, No, Conditional)
- Does the vendor's staff have sufficient knowledge and expertise in the subject matter?'
- Does the vendor identify and implement new areas for improvement?
- Does the vendor have a strong response to problem resolution?
- How does the vendor rate across the following performance categories? (Choose from Above Standard, Standard, Below Standard)
  - Vendor requirements met as to quality of goods/services
  - Vendor requirements met as to timeliness of delivery or performance
  - Vendor requirements met as to price/budget
- Business relations (professionalism, responsiveness, change management)

Evaluation Can Be Performed on a Qualitative or Quantitative Basis, or Both

The evaluation can be on a qualitative or quantitative basis, or it can be a combination of both.

- For example, using a Likert scale, the vendor performance may be ranked on a 5-point or 7-point scale; this would mean the evaluation question may include options such as Strongly Agree, Agree, Undecided, Disagree, Strongly Disagree.
- Alternatively, the performance may be rated on a scale of 1 through 5, with 1 showing the lowest score and 5 the highest score achieved in any category of performance.

- In some evaluations, the performance is measured in terms of adjectives such as "Needs Improvement," "Successful Performance," or "Exceptional Performance."
- There are many variations in which the vendor evaluation can be made. It may be the preference of the quality manager to determine which scale to use, and how to structure the questions.

Key Performance Indicators (KPIs)

Elements that are commonly considered the key performance indicators (KPIs) include quality, reliability, and delivery performance of the vendors. Other performance criteria may be considered, depending on the preference of the clients or project stakeholders. KPIs are the metrics by which vendors are measured. Setting KPIs matters because they outline expectations for third parties and help businesses define the relationship at the negotiation stage. They are crucial elements of vendor management best practices. KPIs are mutually agreed-upon criteria for performance. Once put in place, they help to keep the vendors accountable when it comes to fulfilling their obligations. ("Trade Partner Requirement Alignment" earlier in this chapter describes the alignment of the trade partners' criteria for better performance and recommends performance indicators that can be used to obtain consistent performance from all vendors and material suppliers.)

Benefits of Vendor Management KPIs

KPIs offer a robust program to help improve vendor performance by enabling more consistent performance and reduced costs. KPIs are also a critical move toward lowering business risks. Their goal is to improve the business to world-class through better vendor management, but, depending on the starting point, organizations (and projects) evolve through the following stages.

Best practices in selecting vendor KPIs include:

- They must be established for each critical element of the product and/or service and should be realistic and achievable.
- They should be both quantitative and qualitative and relevant to the importance of the vendor relationship.
- They should be based on a minimum acceptable required level of performance.
- If they attract penalties, there should be incentives for the vendor:
  - Punishment may not be the best way to improve performance. Penalties without incentives will not enhance either service delivery or the relationship with a vendor.
  - Penalties may keep a vendor on track, but incentives can be used to encourage innovation and continuous improvement.
  - Incentives such as a fixed price with award fee, cost-plus with award fee, or cost-plus with incentive fee (cash awards) are useful.

- KPIs should be reviewed regularly and realigned to the project's quality objectives as required.

CONCLUSION

Vendor performance management promotes long-term relationships with suppliers. As per point 4 of Deming's 14 Points for Quality management, the long terms relationships of all involved organizations and prevent vendor selection on price alone. Vendor management is an area within procurement that must be addressed with careful planning. Proven quality management methods that have been known for more than 50 years but remain only partially deployed for vendor management need to be deployed to achieve improved vendor performance, create supply advantages, reduce costs, and achieve quality improvements in all aspects of quality of the vendors and suppliers. ("Trade Partner Requirement Alignment" earlier in this chapter includes information about the specific quality requirements for subcontractors and vendors. The content may be reviewed for additional information related to the quality aspect of vendor performance.)

QUESTIONS

1. What are the key benefits of implementing a weighted criteria system in vendor performance evaluation, and why is it considered best practice compared to an unweighted system?
2. Explain the difference between qualitative and quantitative methods of vendor performance evaluation, and give an example of each as discussed in the text.
3. How do key performance indicators (KPIs) contribute to effective vendor management, and what are some best practices for selecting and utilizing them?
4. Why does the chapter emphasize obtaining vendor performance appraisal from references in addition to internal evaluation, and what specific information can these references provide?
5. According to the conclusion, how can implementing proven quality management methods in vendor management contribute to long-term relationships and overall quality improvement?

Chapter 5

Construction

Quality Plan Deployment

The project's technical specification prescribes the requirements that services or products must conform to (Ashford, 1989). The specifications usually contain all contractual QA and QC requirements, from evaluating materials ("Material Evaluation") to acceptance criteria ("Acceptance Criteria and Implementation"), to necessary coordination meetings ("Meetings and Participation").

Construction documents (CDs) are the comprehensive, legally binding set of architectural drawings and specifications created by architects and design engineers to detail a building's design and materials for a contractor to follow during construction. As the project enters the "construction documents" phase, all related documents required to construct the building are assembled and released for bidding. Also, a specification writer who specializes in selecting materials, outlining acceptance criteria, and translating the client's intent into written form produces specifications in parallel with the drawings, with associated revisions resulting from reviews at each phase of the design. (See Chapter 3, "Design Review Output," for more information about the final design.) It is important to recognize that the design of the building/structure that must be constructed by the contractor is the responsibility of the designer or engineer of record.

CONSTRUCTION PLANS AND SPECIFICATIONS

> Assess plans and specifications (e.g., permanent and temporary work) in relation to the project's quality plan to meet established quality standards. (Evaluate).
>
> Body of Knowledge V.A.1

The construction contract will incorporate a set of design drawings and specifications, each complementing one another and serving as the primary design deliverable for the project. The specifications do not supersede the drawings, nor do the drawings outweigh the intent of the specifications. Chapter 3, "Design Output" and "Design Monitoring Gates," describe how the design is created. The

CDs (construction documents) are the final design documents (also sometimes called construction plans). These are the final, highly detailed graphical drawings created by architects and design engineers to guide the construction of a building. These construction plans are complemented by design specifications that use written text to detail the project's quality standards, materials, and performance requirements. Atkins and Simpson (2008) note that specifications and drawings are intended to work together to compile the contract documents. Specifications set the quality standard for components and materials, while drawings illustrate the quantity, layout, and location of those elements. The specification is incomplete without the drawings, and the drawings are incomplete without the specification.

Quality of Construction Drawings and Specifications

The project drawings and specifications must be reviewed prior to construction to ensure they are constructible. Prior to evaluating the specifications, the CQM must validate that the final design (CDs), as issued by the design engineer or architect, has not undergone any further change. The quality of the drawings and specifications will greatly impact the ability of the construction team to perform (Atkins and Simpson, 2008).

If the final design (CDs) has been changed, the differences should be addressed via the formal means outlined in the design change process. The CQM should confirm the construction team has the latest set of final design drawings that have been issued formally by the designer. Potential design changes must also be evaluated against procurement and other risks and managed accordingly. (See "Risk-Based Inspection and Testing Strategy" later in this chapter.)

Design Phase Interpretation

Many construction projects delivered using the DBB method have a limitation: the construction contractor's input was not included in the final design documents issued by the designer because the contractor was not hired when the design was completed. In all such cases, when a pre-completed design is given to the contractor to use as the construction plans and specifications, it becomes essential for the contractor to perform the design phase interpretation from the contractors' perspective.

To do this, prior to the beginning of construction, the final drawings (CDs) and specifications should be evaluated for constructability as well as for their validity and applicability to the specific project, geography, and supply chain. To constructively evaluate and attain team alignment on intent and quality objectives, have a specification review meeting. If there is a need for modifications to the final design, it is best for the contractor firm to issue a request for information (RFI) or a field design change request (FCR).

Evaluating the Specifications

The understanding and interpretation of the specifications on the part of the construction team is the most important element determining the success of a project because, eventually, the specifications and drawings are the vehicle that translates the strategic vision into actionable steps and tangible construction deliverables.

(To make the quality strategy specific to the project, see Chapter 2, "Project Quality Strategy Development"). The quality strategy itself acts as a framework that can be implemented and extended into a quality plan for any project, incorporating the corporate requirements as well as the project specifics and stakeholder concerns. Table 5.1 provides a strategic approach to reviewing the construction drawings and specifications for pertinent quality information.

Table 5.1 Evaluating quality requirements from the specifications and drawings.

Strategy Action	Objective	Examples
1. Review the project's specifications.	Identify the technical requirements. Confirm tolerances and allowable deviations.	• Material requirements • Workmanship standards • Performance requirements
	Identify code requirements and reference best practices/ guidelines.	• Local/municipal • National • Trade best practices
	Organize the project by critical element (e.g., definable features of work, or DFOW). This may be individual specification sections, combinations of multiple sections, or multiple systems within one section. [See V.B.1 Risk-Based Inspection and Testing Strategy for additional tactics on how to do this.]	• Concrete • Finishes • Building envelope • Elevators • Building automation • Fire alarm
	Outline submittal and shop drawing requirements for each component and system.	Create a submission log to track and validate. [See V.A.3 Construction Submittals for more information.]
2. Review the project's drawings.	Review each discipline and confirm coordination between each set.	• Confirm dimensions. • Outline scope responsibilities between sets for critical details. • Confirm documents are consistent with associated specifications.
	Identify other key components and systems, and understand scope interdependencies.	• Mechanical • Electrical • Plumbing • Low voltage/communications • Structural
3. Outline the inspection and testing requirements by component and system.	Identify specific required *inspections* and milestones.	• Mock-ups • Deliveries • Benchmark/first work • Structural steel connections • Waterproofing installation • Code/municipal expectations
	Identify specific required *tests* and milestones for finished systems and components.	• Commissioning • Soil compaction • Concrete strength • Code/municipal expectations

(continued)

Table 5.1 *(continued)*

Strategy Action	Objective	Examples
4. Incorporate actions from 1 to 3 into project quality plan.	Input requirements into quality tool/template at the corresponding phase of construction to translate the appropriate information to all internal and external stakeholders and those installing the work.	• Kickoff • Coordination • Preconstruction • Preinstallation • Inspection checklists • Testing documents
	Leverage the responsibilities matrix from II.C.2 Roles of Quality Personnel and Stakeholders to confirm responsibilities. It is the quality manager's job to ensure each person assigned to those respective tasks is capable of and trained for doing so and is guiding that process accordingly.	• Closeout/punch list requirements • Tools/equipment/calibration
5. Implement the documented plan.	Confirm how this information from the drawings and specifications will be communicated to the team and how often.	Identify how each task will be assessed for understanding and completion.

It is essential to evaluate how the specification sections rely on one another and work together. The end goal is understanding and interpreting the specified standard of quality outlined in the contract documents (drawings, specifications, industry standards, and best practices) as set forth by the design team and the client. Constant validation and review of these standards throughout the project is essential, especially as change starts being incorporated.

These reviews should not occur in isolation or by any one individual or company. There needs to be a team perspective in interpreting the intent from the final design plans and specifications. That interpretation of the specifications and its correlation to the client's intent must be communicated among all project team members.

The technical requirements, once outlined, must be reviewed with the owner and the design team to confirm the overall interpretation of the construction team.

The following can serve as a checklist of questions for the construction contractor to ask the owner and designer during the construction team's review of the specifications and drawings.

- Confirm with the client and designer: is our interpretation of the specifications correct?
- Is this the intended level of quality to be achieved?
- Do the specifications align with local building codes and municipal interpretations?
- Are the requirements aligned with the project's geography?
- Are the specifications compatible with the overall specifics of the project?
- Do the acceptance requirements in the specification meet the client's intent?

- Are the specified standards clear and attainable?
- Do the specifications meet the client's intent and minimum expectations?
- Confirm that these actions and deliverables on the part of the construction team are what the design team and client want to review in detail.
- What planning needs to occur?
- What meetings are required? What is the agenda or intent of each? Who needs to be there (see also "Meetings and Participation")?
- What plans need to be documented and understood by other project team members?
- What validation needs to occur?

Asking these questions as part of the specification review helps the construction contractor ensure that the construction quality plan will be developed with the advice of multiple disciplines, where everyone understands the context so they can achieve the desired quality proactively (and not through costly design changes at later stages of construction). These points of clarification can be reviewed continually during design review and stakeholder meetings, where the construction contractor meets with the designer as well as the owner and other stakeholders.

Design Changes in Drawings and Specifications

Often, changes in the final design and specifications are necessary. This happens a lot more on projects where there is no overlap between the designer and the contractor (e.g., DBB projects). Questions about the differences between details in the drawings and statements in the specifications are inevitable during the construction phase. "Requests for Information or Changes" later in this chapter outlines the standard methods and procedures for incorporating changes in the specifications and drawings per the project quality plan's design change process. (See also Chapter 3, "Design Changes," to understand the process that a designer or the architect firm uses to implement field-based design changes.)

Technology

Modern software capabilities allow specifications and drawings to be easily accessible while installing the work. This helps reduce conflicts resulting from changes. If a change comes at the start of a critical path activity, the field installation teams may find a freshly printed drawing is obsolete. The CQM may ensure that the role of technology is to expedite the change process and ensure that quality is not compromised.

CONCLUSION

Achieving the established quality standards for a project requires a comprehensive and collaborative approach to evaluating plans and specifications. The integration of drawings and specifications, informed by the client's intent and aligned with project-specific requirements, ensures clarity and consistency in execution. Regular review, validation, and communication among team members, coupled with proactive planning, mitigate potential conflicts and set the foundation for success. While modern technologies can enhance efficiency, the critical thinking and collective expertise of the project team remain indispensable in interpreting and implementing the quality standards defined in the contract documents. By fostering alignment and accountability across all phases, the team can confidently deliver a project that meets or exceeds the desired outcomes.

QUESTIONS

1. What is the intent of the specification?
2. What governs: the specification or the construction drawing?
3. What must be done with the drawings and specifications prior to the start of construction?
4. Name two types of requirements that would come from the specifications.
5. Who should be involved in a review of the specifications?
6. Explain the relationship between construction drawings and specifications, and why it is crucial that they are reviewed together prior to construction. Give an example of what might happen if they are not reviewed for coordination.
7. Describe the strategic approach to reviewing construction drawings and specifications as outlined in Table 5.1 of the document. How does this approach ensure pertinent quality information is captured and implemented?
8. What are some of the key questions that should be asked of the owner and designer during the construction team's review of the specifications and drawings? Why is it important to confirm the team's interpretation with these stakeholders?
9. Discuss the role of technology in accessing and managing construction plans and specifications. While technology offers benefits, what potential pitfalls should be avoided, and what remains indispensable in ensuring quality?
10. Explain the concept of "design phase interpretation" as it relates to construction specifications. Why might specifications need to be evaluated for validity and applicability prior to construction, and what measures can be taken to ensure team alignment on intent and quality objectives?

REQUIRED PERMITS

> Understand the importance of securing all essential permits before project commencement. (Understand)
>
> Body of Knowledge V.A.2

Understanding why certain permits are required is essential to a project's overall success. This is applicable during both the design and construction phases. Additionally, the planning effort put forth during the early schematic design phases can help streamline construction mobilization while establishing relationships with key stakeholders in the local municipality. A city's specific permitting and testing expectations, interpretations of national codes, and additional or more rigorous requirements must be translated into the final design deliverable and shared with other construction stakeholders to ensure project success from all perspectives—quality and safety alike.

Codes establish the minimum quality standard for a building and its primary components (Kubal, 1994). These minimum standards, whether noted in the contract documents or not, must be met to attain turnover milestones. The architect or engineer of record must be knowledgeable of the local permits required, based on the type of building and where it is being built, while construction team members must be equally knowledgeable of which activities are required to keep permits active and close them out.

Understanding Required Permits

There are two main sources for identifying the required permits:

1. The project's specifications (contract documents)
2. The local building/municipal/code office (authorities having jurisdiction, or AHJ). (See "Regulatory Body or Authority Having Jurisdiction [AHJ]" later in this chapter for more information.)

A review of the project specifications (see "Construction Plans and Specifications" earlier in this chapter) will help the project team understand the following:

- What trades are involved in the project?
  - Understanding what trades are involved will help identify required building permits, as well as the associated inspection and testing requirements. (See "Risk-Based Inspection and Testing Strategy" later in this chapter.) Usually, the different trades that are included in the technical specifications are described such that they may be considered as definable features of work (DFOW) with unique control requirements. These control requirements can serve as the basis for quality control inspections and are included in the inspection and test plans (ITP) when such plans are developed. (Note that the ITPs are the most significant component of the construction quality plan.) Such QC inspections and tests must be accounted for in the project schedule and shared with the subcontracting trade partners.
- What are the complicated or specialty components/systems/elements of the project?
  - Certain systems, such as elevators and fire alarms, will require separate permits, tests, and inspections involving multiple trades.
  - Hazardous materials may also require special permits or certifications.

While a review of the specification establishes a foundational list for the required permits, coordinating early with the Authority Having Jurisdiction (AHJ) and associated stakeholders allows you to incorporate specific interpretations of building codes into the project's execution plans. Specific test and inspection nuances may be required or expected in one municipality and may be very different in another, even in the same geographic region. Navigating these variations requires early communication and planning to understand the expectations for permitting and interpretations of specific codes that may not be outlined in the specifications.

Types of Permits

Most AHJs require a general building or construction permit to be applied for and issued for any project to commence, regardless of the project scope. Depending on jurisdiction, the different types of general building permits may be issued based on project type:

- New construction
- Addition of structure to existing construction
- Alteration

This permit will act as the overall framework for all work to be performed, list code-required inspections and tests, and may also list additional permits required for specialty or temporary work during the execution of the construction project.

Some common specialty permits required for permanent work in construction projects include:

- Electrical
- Plumbing
- Elevator/escalator
- Concrete
- Roofing
- Roadway signage
- Mechanical/HVAC/boiler
- Fire protection (sprinkler/fire alarm)
- Swimming pool

Some common permits required for temporary work during a project include permits for:

- Roadway/sidewalk closure
- Environmental (e.g., asbestos abatement)
- Demolition
- Crane
- Scaffold/shed

Additionally, permits for temporary and permanent utilities through a separate utility company may also be required both during a project and for the completed work. (See "Regulatory Body or Authority Having Jurisdiction [AHJ]" for interactions with utility companies.)

These utility permits could include:

- Electrical
- Water
- Sewer
- Gas
- Telephone/data

Best Practices in Permitting

A CQM—and other project leaders—can ensure success by being proactive and understanding the permitting process early. The key is integrating the necessary permits with all other project planning tasks.

1. *Understand the permitting requirements early.*
    a. Review the specifications and outline what permits are required.
    b. Inquire with municipal stakeholders what local regulations are in place, what applicable building codes are, and the associated permits required.
2. *Establish relationships with permitting authorities and code officials.* Planning is streamlined when there is trust among team members, particularly with external stakeholders. When the project team reaches out proactively to understand code officials' concerns and incorporate them into the project planning documents, officials are more willing to participate in essential project meetings. This allows the team to understand the inspection schedule and requirements clearly.
    a. Initiate early conversations with municipal stakeholders.
    b. Set up a recurring coordination session where requirements can be understood and then integrated into the broader project planning work sessions (see "Meetings and Participation" later in this chapter).
3. *Integrate the permitting process and associated expectations into the project planning and meeting cycle.* When the permits and associated actions are incorporated into project planning documents, the entire team and all stakeholders can be trained on their roles and responsibilities in ensuring the permitting requirements are met. This also creates an open, transparent environment where potential risks in adhering to the permitting process can be brought up early and resolved before being escalated. These requirements can be incorporated into a checklist to help manage expectations and planning.

   a. Identify permit holders and their responsibilities to maintain active permits and ultimately close them out (see Chapter 2, "Roles of Quality Personnel and Internal Stakeholders"). Such permit holders may be the client, a subcontracting trade partner, or the general contractor.

4. *Integrate permits with the audit process.* This allows issues to be resolved before delays are incurred. To integrate permits in the audit process, the following steps should be considered:
   a. Confirm that the necessary documentation is in place (see Chapter 2, "Quality Audits") and that the required inspections are occurring commensurate with the work.
   b. Create a compliance tracking system or permit checklist to be referenced as part of other key meetings and planning sessions.
   c. Add specific verifications per each permit to the project's auditing plan.
5. *Integrate permits with the construction schedule.* Ties between permitting activities and construction activities within the scheduling software should be made to ensure those activities that require a permit do not proceed without the necessary permits being secured. Permits are necessary to ensure the project does not experience delays.
   a. Review the schedule and the specific permitting activities early with building officials to confirm all permits and associated tasks are accounted for.
   b. Incorporate the appropriate timelines for the municipal review process, confirming what the turnaround time for securing a permit may be for the specific project.
   c. For each of those permits, tie these activities to the associated inspections.
6. *Understand infrastructure requirements.* There may be required plans to present to the local code department as part of the permitting process before permits can be secured. These tasks must be coordinated during the design phase to ensure successful mobilization. Some examples of infrastructure requirements for permitting include the following (this is not an all-inclusive list):
   a. *Site/logistics plans*: This helps local authorities to understand utility connections, building layout, setbacks, landscaping, and general impact on the community and surrounding areas. This also illustrates site access and temporary road requirements for construction activities and future occupants of the finished building.
   b. *Fire/emergency response plans*: Where will the fire marshal and emergency response teams access the building? This is applicable both during construction and after certificate of occupancy.

c. *Soils/geotechnical reviews*: Confirming the soil's composition, bearing capacity, and stability helps confirm the suitability of the structural design.

d. *Environmental impacts*: For larger projects, local habitats may be impacted. Construction activities can negatively impact the nearby ecosystem, and a mitigation plan may be required for certain jurisdictions prior to permitting. This includes drainage requirements and an understanding of the stormwater systems to ensure there are no runoff problems, soil erosion, or potential for flooding.

e. *Land use/zoning*: Based on the intended use of the finished building, compliance with local zoning laws must be validated.

f. *Utility plans*: Plans confirming water, gas, sewage, and electrical utility connections and capacity are required for both temporary (during construction) and permanent connections prior to permitting.

7. *Understand the project closeout process.* If a permit is missed or not closed out correctly, the project can be shut down, creating cost and schedule issues—as well as trust and credibility problems for the team. Include details on what activities are needed to close each specific permit.

Benefits of Aligning with the Permitting Process

Understanding and following the appropriate procedures surrounding permits provides many benefits to the current project as well as future business endeavors. Permits are not just for the construction process but for the entire life cycle of the building. When the CQM and the project team understand the project's permit procedures, the following benefits can be expected:

1. *Establishes trust and credibility for project teams doing business in that municipality.* Relationships with code officials, working well with the city from the beginning of the permitting process, translate all the way through to the inspection and turnover process, and future operations with the client and city relationships.
2. *Ensures compliance with local regulations.* Not complying with the permitting process can have financial and legal impacts. Fines are incurred when projects do not follow the processes set forth by local jurisdictions. This also shows the design and construction teams are not skipping important steps and are willing to work within the processes and systems that ensure the project will comply with regulations.
3. *Ensures that environmental risks and potential impacts are mitigated.* Construction can be a volatile process when not managed well. It impacts the regional environment well beyond the project site in the form of runoff, material transportation, drainage, and destruction of local habitats. Environmental permits ensure the project will not damage local ecosystems and that sustainability is a focus for the project.

4. *Promotes a safe work atmosphere.* Permits also ensure the project will provide a safe environment for workers and for the future occupants. When workers feel safe, their work quality increases, as does labor productivity, streamlining the entire process. Permits also reduce risks of critical system failures, such as fire (electrical) or water damage from plumbing and building envelope systems.

CONCLUSION

Ensuring a thorough understanding and proactive management of required permits is paramount to project success. By integrating permitting processes into all phases, from planning to auditing and scheduling, the project team fosters trust with stakeholders, ensures regulatory compliance, and mitigates environmental risks. This meticulous approach not only prevents costly delays and legal issues but also promotes a safe work environment. Ultimately, aligning with the permitting process establishes credibility, streamlines operations, and sets the foundation for successful project delivery and future business endeavors.

QUESTIONS

1. What is the primary intent of the building code?
2. How can the project team be proactive in identifying what permits are required?
3. How can a CCQM streamline the permitting process?
4. Name two benefits of aligning with the permitting process.
5. How do a project's infrastructure requirements impact the permitting process?
6. Explain the dual sources for identifying required permits on a construction project and discuss how discrepancies between these sources should be addressed to ensure project compliance.
7. Describe the difference between general building permits and specialty permits, providing examples of each for both permanent and temporary work, and explain why this distinction is important for project planning.
8. Detail the six tactics a CCQM can employ to be proactive in the permitting process and explain how each tactic contributes to the overall success of the project.
9. Discuss the benefits of aligning with the permitting process, and explain how these benefits extend beyond immediate project completion to impact future business endeavors.
10. Analyze how infrastructure requirements impact the permitting process, providing specific examples of infrastructure plans that may need to be presented to local code departments, and explain why early coordination on these plans is crucial.

CONSTRUCTION SUBMITTALS

> Review various types of construction submittals (e.g., action and deferred/delegated), taking into consideration the interdependency across trades and the project's quality plan and standards. (Apply)
>
> **Body of Knowledge V.A.3**

Construction submittals are the linchpin of a successful project, acting as the critical bridge between design intent and physical realization. They are more than mere paperwork; they are the tangible evidence that materials, products, and methods align with contract requirements and quality standards. This process, from initial design review to final approval, ensures that all stakeholders are on the same page, fostering collaboration and minimizing costly errors. Effective management of submittals is paramount, transforming them from a potential bottleneck into a streamlined pathway.

A construction submittal is any written or physical information sent by contractors for approval by the design team, confirming that items such as materials, products, or equipment planned for construction meet the contract requirements. The primary purpose of construction/contractor submittals is to ensure that all products and materials meet quality standards and project requirements before they are procured or installed. As such, this is a contact between the contractor and the designer and acts as the critical checkpoint before construction begins.

The contractor submittal process begins with a careful review and extraction of the design documents. Whether that includes the drawings, specifications, or a combination of both, the required submittals will need to be determined by mining this design information. A CQM or their designee (e.g., an office engineer) will need to start by developing a submittal log to organize the necessary submittals into a matrix to see which submittals need to be prepared for review by the owner and the designer (architect or engineer of record).

In the past, this submittal log was generated manually by the CQM or project manager, who reviewed the plans and specifications and input the information into a table format (usually a spreadsheet). However, now CQMs may use software designed to review and extract the design-required submittals. Having submittal logs generated automatically is a welcome tool, but it is important to ensure that the information created is accurate and complete. This type of software uses keyword recognition to extract the pertinent information and organize it in order from the design documentation. This can be complicated because each specification is produced by different specification writers, based on the division or scope of work. Dependence on any software assumes that the technical specifications are all completely accurate with no extraneous information, which is usually not true; so human oversight is required to determine what is logical and rational and avoid duplication of effort.

Submittals are a vital tool in the quality process of any construction project and therefore cannot be viewed or treated as a paper-pushing exercise. Designers, architects, and engineers put careful detail into the design that needs to be represented in the individual product, material, or equipment scope submittals.

Examples of Contractor Submittals

Submittals can include many items to reflect how the contractor is planning to construct the design intent, such as product data sheets, samples, mock-ups, manuals, and more. Before work starts on site, submittals need to be presented to the owner and their design representatives to ensure the plan matches the design intent. This is the first physical QC of the scope of work before any physical work, and it helps ensure the contractors and designers are on the same page.

Before construction begins, various types of submittals are typically required. These can include shop drawings and product data, samples, mock-ups, engineering calculations, deferred submittals, vendor information, warranties and manuals, and as-built drawings. During the project closeout phase, additional submittals such as product warranties and as-built drawings are often necessary. The list provided is not exhaustive and may vary for each contractor or vendor involved in the project:

1. *Shop drawings*: Shop drawings, or component design drawings, are detailed component drawings for elements of the design that need to be manufactured off-site (or on-site too, such as reinforcement installation drawings for a slab on grade) as a necessary product to complete the overall design of the project. These are usually custom-fabricated or specially built equipment that need to be designed separately by a subject matter expert (SME) designer and verified to ensure they will incorporate into the fit and function of the rest of the project.

   Shop drawings are usually produced by the contractor, or by their supplier, manufacturer, subcontractor, or fabricator. An example of shop drawings is the structural steel elements of a large structure. Each component of the structure may need to have its measurements and connections detailed beyond what is shown in the structural design drawings. These drawings may include the interactions and calculations to verify that they will meet or exceed the strength requirements for the building loads. They may also reflect how many of the items will need to be produced. Other examples of shop drawings may include drawings created for elevators, structural steel, trusses, precast concrete, windows, appliances, cabinets, air handling units, and millwork.

   It is important to note that the shop drawing submittals are sent over to the EOR to obtain their "general concurrence" with the shop drawing. Approval indicates that the EOR considers it consistent with the design intent of the designer. (However, approval does not indicate that the EOR is responsible for the shop drawing.)

2. *Product data*: Product data refers to documentation or "cut sheets" provided by a manufacturer for standard products planned to be used in the project's construction. These documents contain standard product details like sizes, performance data, preparation, and installation instructions. This is also where manufacturers display laboratory testing documentation for any product certifications. For example, a manufacturer typically manufactures windows in a factory using standard products and configurations that are then tested and rated for performance. The designer may select windows based on the visual design, but confirmation that they meet the performance requirements for the design is vital, too.

3. *Samples*: Samples are physical examples of materials to be installed. The design team wants to verify that the materials match the intended quality, color, texture, or structural requirements. Common examples include flooring, tile, and paint.

4. *Mock-ups*: A construction mock-up submittal is a crucial part of the construction process where a model or prototype of a particular element or assembly is created and presented for approval before full-scale production and installation. This involves the following:
   - A mock-up is either a small-scale model or a full-size replica of a construction element. This could be anything from a section of a wall, a window assembly, or a specialized fixture.
   - The primary goal of a mock-up is to verify the design, materials, and workmanship. It ensures that the final construction will meet the aesthetic and functional requirements specified in the project plans.

   Imagine a project involving the installation of a complex glass façade. Before proceeding with the entire façade installation, a mock-up of a section of the glass wall is constructed. This mock-up includes the glass panels, framing, sealants, and any other relevant components. The design team and stakeholders review the mock-up, checking for visual appeal, structural integrity, and overall quality. Any issues or improvements identified during this review are addressed, ensuring that the final installation will be successful.

   A mock-up is an invaluable tool in construction, providing a clear and tangible preview of the project's components and setting a standard for quality and craftsmanship.

   a. *Types of mock-ups*

      - *Off-site mock-ups*: These are created at a separate location, usually in a controlled environment like a workshop or a factory, and are then transported to the site for review.
      - *First-in-place mock-ups*: These are built directly at the project site, integrating with the actual construction environment. They are particularly useful for assessing how the mock-up interacts with existing site conditions.

   b. *Process of a mock-up submittal*

      i. *Preparation*: The contractor prepares the mock-up using the exact materials and construction methods intended for the final product.

      ii. *Review*: The mock-up is presented to the project stakeholders, including the design team, general contractor, and sometimes the client, for a thorough review.

      iii. *Documentation*: Detailed photographs and documentation of the mock-up are created to capture its specifications and quality.

      iv. *Feedback*: Stakeholders provide feedback, and any necessary adjustments are noted. This is the phase where potential issues can be identified and corrected before mass production.

      v. *Approval*: Once the mock-up meets all the specified requirements and the stakeholders are satisfied, it receives formal approval. The approved mock-up then serves as a benchmark for the actual construction work.

c. *Importance of mock-ups*
   - *Quality control*: QC ensures that the final product will meet the desired quality and performance standards.
   - *Cost efficiency*: A mock-up identifies and rectifies potential issues early on, reducing the risk of costly changes later.
   - *Stakeholder confidence*: They build confidence among stakeholders by providing a tangible example of what the final construction will look like.

5. *Engineering calculations*: Construction engineering calculations are an essential part of the submittal process in construction projects. These calculations provide detailed mathematical analyses and engineering data that support the design and construction of various project elements. Here's a breakdown of their role and importance:
   - Engineering calculations are detailed mathematical analyses that verify the structural integrity, safety, and functionality of various construction elements. These can include structural calculations, energy calculations, load calculations, and more.
   - The primary goal is to ensure that the proposed design meets all relevant codes, standards, and project specifications. They help in identifying potential issues and ensuring the construction will be safe and durable.

   Imagine a project involving the construction of a multistory office building. The structural engineer prepares calculations to determine the load-bearing capacity of the building's columns and beams. These calculations are submitted to the design team for review. If the design team finds any issues, the engineer revises the calculations and resubmits them for approval. Once approved, these calculations guide the construction process, ensuring that the building is structurally sound and safe.

   Engineering calculations are a critical component of the submittal process, providing the necessary data to ensure that construction projects are built to the highest standards of safety and quality.

   a. *Types of engineering calculations:*
      - *Structural calculations*: These determine the strength and stability of structural elements like beams, columns, and foundations.
      - *Load calculations*: These assess the loads that a structure must support, including dead loads, live loads, wind loads, and seismic loads.
      - *Energy calculations*: These are used in designing HVAC systems and other energy-related components to ensure efficiency and compliance with energy codes.
      - *Thermal calculations*: These help in designing insulation and other thermal control measures to maintain energy efficiency.

b. *Typical roles involved in the submittal process for engineering calculations:*
    - *Preparation*: Engineers prepare detailed calculations based on the project's design and specifications.
    - *Submission*: These calculations are included in the submittal package and submitted to the design team for review.
    - *Review*: The design team reviews the calculations to ensure they meet all requirements and standards.
    - *Feedback*: Any discrepancies or issues identified during the review are addressed, and the calculations may be revised and resubmitted.
    - *Approval*: Once the calculations are approved, they serve as a basis for construction work, ensuring that the project is built to the required standards.

c. *Importance of engineering calculations:*
    - *Safety*: Ensures that the structure will be safe and capable of withstanding various loads and stresses.
    - *Compliance*: Verifies that the design complies with all relevant codes and regulations.
    - *Quality control*: Helps in maintaining high standards of construction quality and performance.
    - *Risk mitigation*: Identifies potential issues early on, reducing the risk of costly errors or rework during construction.

6. *Deferred submittals*: Deferred submittals are also an important part of the construction submittal process, especially for the complex projects. They refer to elements or components of the project that are not included in the initial set of drawings and plans but are submitted for review and approval at a later stage. Deferred submittals are parts of the project that are not included in the initial design documents and are submitted later. These can include specialized components or unique elements that require additional design or approval. Here's a detailed look at deferred submittals.

    a. *Reasons for deferred submittals:*
        - *Specialization*: Some elements require specialized design expertise that is best handled by a specific contractor or designer. Examples include fire protection systems, prefabricated elements, and custom interior finishes.
        - *Owner preferences*: Sometimes, the project owner may want a particular contractor to design a specific component, such as a custom bar or kitchen equipment in a restaurant.
        - *Complexity*: Certain elements may be too complex to design and detail upfront and are better addressed later in the project timeline.

b. *Process of deferred submittals:*
   - *Initial design*: The project begins with a set of initial design documents.
   - *Identification*: Elements that need to be deferred are identified and listed on the title or cover sheet of the original plan.
   - *Pre-submittal conference*: A meeting is scheduled to discuss the deferred submittals and define a schedule for their submission.
   - *Submission*: The deferred submittal documents are prepared and submitted for review.
   - *Review and approval*: The documents are reviewed by the architect/engineer of record and then submitted to the building official for approval.
   - *Integration*: Once approved, the deferred submittals are integrated into the project.

c. *Challenges and best practices for deferred submittals:*
   - *Coordination*: Managing deferred submittals requires careful coordination among various stakeholders to ensure timely submission and approval.
   - *Documentation*: Clear documentation and communication are essential to avoid misunderstandings and delays.
   - *Budgeting*: Planning for deferred submittals in the project budget is crucial to avoid unexpected costs.

d. *Typical deferred submittals:*
   - *Fire protection systems*: Sprinkler systems and fire alarms that require specialized design and approval.
   - *HVAC systems*: Heating, ventilation, and air conditioning systems that need detailed engineering calculations and design.
   - *Custom elements*: Unique architectural features, custom furniture, or specialized equipment.

   Deferred submittals help ensure that all aspects of the project are thoroughly reviewed and approved, contributing to the overall quality and success of the construction project.

7. *Vendor information*: Particularly on large commercial or public projects, the project owner often has limited visibility of the parties that the general contractor or their subs hire to perform specialized portions of the work. At times, the owner may request a list of all subs and suppliers providing work and materials on a project, along with contact information and a description of their scope of work.
8. *Warranties and manuals*: Contractors and suppliers are often required to provide documentation about the terms of warranty coverage for products, materials, or equipment used in the project. Operation or maintenance manuals provide detailed instructions on how to operate

and maintain various components or systems in the building. If these are reviewed during the construction submittal process, they might also be asked to provide actual executed warranties for the project during the closeout submittal process.

9. *As-built drawings*: As-built drawings, simply called "as-builts," are the final diagrams submitted by a contractor on completion of their work. As-built drawings are often prepared by the original design team, which is sometimes also called the architectural & engineering (A&E) firm. The contractual title of the design firm is the engineer of record (EOR) or architect of record (AOR).

They depend on the "redlining" made on the construction drawings by the contractors' field staff to indicate as-built conditions. There may be situations where as-built drawings are not always prepared by the contractor, as the contractor may submit the redline drawings to the original designer so the designer prepares the as-builts.

Important Items to Include in a Submittal

A submittal cover sheet should be formatted to include a summary of what is in the package, including revisions and status of the submittal. An area should be reserved on the cover sheet for comments and approval notes.

Typical information needed on submittal cover sheet:

- *Project name*
- *Spec section*: The specification number to which the submittal or package corresponds
- *Submittal number*: A unique identifier for the submittal
- *Revision or version number*: For use when versions of a submittal are created in response to a revision request
- *Package number*: If the submittal is part of a package, include the number of the package
- *Submittal manager*: Name and contact information for the person responsible for document management—typically a general contractor, project manager, or project engineer
- *Submission date*: The date the submittal was sent
- *Received date*: The date the submittal was received
- *Approver(s)*: Name and contact information for the people responsible for review and approval
- *Response*: Typically preset fields for approved, approved as noted, revise and resubmit, and rejected
- *Notes/remarks*: Blank space for the reviewer to provide details about changes or revisions requested

Steps in a Contractor Submittal

Although there is no one-size-fits-all tool or solution for the submittal process for every type of project, the outline of the process remains the same.

The basic prerequisite of a good contractor submittal process is the provision of an effective submittal process. The best practice for contractor submittals is to utilize the project's document control system as the mechanism (tool) through which the submittals are processed. Benefits of using project's document control for contractor submittals include the following.

- It makes the submittal process inclusive, more organized, and provides a more consistent method of record management of the submittals.
- It also offers the security of controlled access and accountability for the data and submittals, making known who are responsible for processing submittals.
- Formal document controls also enable better tracking of submittals; a workflow tied to a project schedule can enforce due dates for review as well as offer better progress on those reviews.
- A good contractor submittal system also offers the ability to report its effectiveness through predefined reporting mechanisms that generate such reports based on actual data. Doing this forces effectiveness on the contractor submittal process, mitigates delays, and prevents possible lack of processing.
- Establishing a submittal process that offers access rights to stakeholders (designer/EOR, contractor, owner, and any other as directed) forces organization and consistency to the submittals.
- Also, any delays due to long, drawn-out processing by the designer or by other stakeholders are effectively prevented as the document workflow can identify where the bottleneck is in real time. This avoids delays that may be negatively impacting the project in terms of the schedule as well as budget.

When contractors start to review the design information to produce contractor submittals, they may be able to identify gaps in information with the expertise as the SME of the scope of work. These gaps are then formulated into questions to the owner and design team to clarify expectations and confirm the plan moving forward. Those questions are documented as requests for information, or RFIs. Once the RFIs are answered, the contractor may be ready to prepare the submittal.

There are four basic steps that a contractor submittal goes through:

1. The contractor prepares the submittal documents. Following must be included in the submittal:
   - *Scope of the submittal*: Describe whether the submittal includes shop drawings, product data with specifications, material samples, mock-ups, test reports, calculations, warranties, or installation guides.
   - *Applicable criteria*: By carefully reviewing the project specifications, the contractor must list the reference to the applicable criteria

that must govern the submittal. This may require attention to the technical requirements (usually the requirements are listed as Division 1, 2, or 3 specifications). The applicable criteria that should govern the submittal must be included with the submittal.

- *Applicable codes, standards, and performance criteria recommended by the design*: If the specifications for an item include codes and standards, or any specific equipment manufacturers and models that may have been specified by the design team, that information should also be included along with the submittal.
- *Project plans and specifications*: The contractor must include the applicable design plans and calculations that may have been issued by the designer.
- *Exceptions*: The submittal must also reference the exact exceptions that the contractor may be requesting (if any). This means the contractor's decision to deviate from the design or from the minimum standards along with the reason.

2. Once the general contractor has thoroughly reviewed the submittal to ensure it is comprehensive and meets the contract requirements, it should be forwarded to the design team for them to begin to review the submittal. Depending on the nature of the project delivery method, the contractor may need to send the submittal to the project owner, who in turn may connect with the design team. For projects where the designer and the contractor are working together and/or overlap, this connection may be direct.
3. The design team (which could include the architect, engineers, and a construction manager or owner's representative) reviews the submittals after the general contractor's initial review to ensure the information aligns with the design requirements.
4. The design team ("design team" here means EOR or AOR) then responds to the submittal. Their response can be issued in one of four ways:
   - *Approved*: The submittal has been approved with no exceptions.
   - *Approved as noted*: The submittal has been approved contingent upon adherence to the comments made by the EOR.
   - *Revise and resubmit*: The submittal has not been approved and needs to be resubmitted; this is actually a "rejection" of the submittal in its current form. If the contractor disagrees with the rejection, this warrants further discussion and formal resolution before the contractor may proceed.
   - *Not reviewed*: The EOR has received the submittal for record purposes only but does not review or comment on it.

   Each of these responses gives guidance to the general contractor on what direction to take next, either greenlighting the work to proceed or asking for more information before proceeding.

   Important note: Just because a designer or team approves a submittal does not absolve the contractor from errors and defects caused by the approved scope of work. The engineer's approval is typically issued only for a partial review of the design features; it is not a complete check on the methods of assembly, methods of installation, or construction. Approval cannot be construed as allowance for any departure from the contract. As such, due diligence is required from the contractor to ensure compatibility throughout the project, as well as to coordinate the submittals and work of multiplate trade partners.

The design team usually sets the process for reviewing submittals, as detailed in the construction contracts. The general contractor may also gather submittals from subcontractors and send them to the design team. Typically, the submittal process starts long before any work is happening in the field to ensure time to allow for changes that may be based on procurement needs. The submittal process generally continues all the way to the end of the project and into closeout. It can vary slightly, depending on the project and the preferences of the owner and design team. The submittal review stages can also differ, but it should follow the above-mentioned four basic steps (more or less). For example, an owner might want to review all interior design submittals closely but leave the electrical submittals to the general contractor's discretion. Note that if any exceptions are implemented on a project, they must be documented formally as part of the contract so as to disclose who is responsible for making important decisions—because deviating from the final design can easily translate into significant cost and schedule impacts.

CONCLUSION

The CQM's role in contractor submittals is usually limited; in smaller projects, the CQM may be tasked to keep a submittal log and act as a liaison between the contractor and the design team. In larger projects, this responsibility is usually given to an office engineer, who is the person responsible for providing administrative support to the construction project. A CQM is frequently expected to audit the submittal process, as well as periodically perform checks on sample submittals to make sure the process is effective. Auditing helps determine if this process is followed consistently.

Mastering the construction submittal process is crucial for project success. From meticulous logging and review to final approval and documentation, each step ensures quality and adherence to design intent. By understanding the various types of submittals—shop drawings, product data, samples, mock-ups, and more—and the importance of clear communication and coordination, project teams can streamline workflows, mitigate risks, and deliver projects that meet or exceed expectations. Effective submittal management transforms potential obstacles into opportunities for collaboration and excellence.

QUESTIONS

1. Describe the significance of the submittal log in the construction submittal process, and explain the potential challenges associated with submittal logs generated both manually and automatically.

2. Compare and contrast the purpose and content of shop drawings versus product data submittals, providing examples of when each type would be required during a construction project.
3. Explain the process and importance of mock-up submittals, detailing the differences between off-site and first-in-place mock-ups and how they contribute to QC and stakeholder confidence.
4. Discuss the role of engineering calculations as part of the submittal process, and elaborate on how their review and approval ensure project safety, compliance, and quality standards.
5. Outline the four basic phases a submittal goes through, emphasizing the responsibilities of the general contractor and the design team at each stage, and discuss why a designer's approval of a submittal does not absolve the contractor from errors and defects.

MATERIAL EVALUATION

> Evaluate materials to verify their conformity to project specifications. (Evaluate)
>
> Body of Knowledge V.A.4

Ensuring the success of a construction project hinges on meticulous evaluation and management of materials. Beyond design and procurement, the physical materials form the project's substance, directly impacting the fulfillment of the owner's expectations. A seemingly minor oversight in material conformity, delivery, or storage can lead to substantial setbacks. Therefore, a robust material evaluation process, encompassing comprehensive lists, timely delivery schedules, proper storage solutions, and rigorous site acceptance criteria, is paramount to guarantee project quality and timeliness. A project can do some things right, but if the materials are incomplete, incorrect, or late, it may fail; a project can do some things wrong, but if the materials are complete, correct, and on time, it can succeed. This statement emphasizes the importance of evaluating and confirming the materials for a construction project, as materials make up the substance of the project.

A CQM can organize the efforts to ensure the materials match the design and the approved submittal documents. This can be done using several tools (listed next).

Comprehensive Materials List

The first step in controlling the flow of building materials is to know what materials the project is going to need. A well-prepared materials list can significantly streamline the construction process, helping to avoid delays and budget overruns. This list is a critical foundation for any project, ensuring smooth execution and efficient resource allocation.

Key Attributes of a Comprehensive Materials List

The key attributes of a materials list should include the following:

1. *Itemization by specification section and scope of work*
   - *Clarity and organization*: Break down the materials list into manageable sections, making it easier to identify and track specific items.
   - *Enhanced project planning*: Align material requirements with the project's specific goals and objectives.
2. *Supplier and installer information*
   - *Efficient procurement*: Streamline the sourcing process by providing clear supplier details.
   - *Effective coordination*: Ensure seamless collaboration between suppliers and installers.
3. *Quantities, dimensions, and types*
   - *Accurate material procurement*: Prevent over- or under-ordering of materials.
   - *Cost-effective planning*: Optimize material usage and reduce waste.

Using Project Management Software for the Comprehensive Materials List

Creating a list of building materials requires thorough planning, detailed categorization, and wise economic considerations. Utilizing project management software to create a list is a recommended best practice. There are many project management (and better-suited construction management) software programs that help achieve more productivity and effectiveness in the creation and maintenance of building material lists. Using any good software generally improves cost accuracy and project efficiency, as it can automate orders as well as provide tracking, prevent material shortages, and optimize material delivery and storage. Such software programs can also provide real-time data for better budgeting and scheduling, and centralize the materials information for all stakeholders. Other benefits also include reduction of waste, and timely and precise order quantities. Following are some of the characteristics of a centralized construction management software for creating a comprehensive materials list:

- *Centralized database*: Stores all material information in one accessible location.
- *Real-time tracking*: Monitors material availability, usage, and potential shortages.
- *Automated calculations*: Quickly calculates material quantities and costs.
- *Integrated scheduling*: Links material requirements to project timelines.
- *Collaboration tools*: Facilitates communication between team members, suppliers, and installers.

- *Document management*: Stores relevant material specifications, purchase orders, and delivery schedules.
- *Reporting capabilities*: Generates detailed reports on material usage, costs, and procurement status.

How to Create a Comprehensive Materials List

Following are the considerations to create a comprehensive materials list:

- *Involve the entire team*: Collaborate with architects, engineers, contractors, and suppliers to ensure accuracy and completeness.
- *Utilize building information modeling (BIM)*: Leverage BIM to generate detailed material takeoffs and visualizations.
- *Consider material lead times*: Account for delivery times to avoid delays in the project schedule.
- *Regularly review and update*: Monitor material availability and adjust the list as needed.
- *Implement QC measures*: Ensure that materials meet specifications and industry standards.

By creating a thorough materials list and leveraging project management software, you can significantly improve project efficiency, reduce costs, and enhance overall project success.

Integrating Material Procurement with Project Delivery Schedule

Once a comprehensive materials list has been established, the next crucial step is to integrate it with the project's delivery schedule. This requires carefully analyzing the delivery schedules, lead times, and specific requirements for early material delivery.

Following are the key things to do to ensure that the materials in the material list are linked with the receiving schedule and the construction delivery schedule:

- *Critical path analysis*: Identify the critical path of project delivery and the materials necessary to maintain the project's delivery timeline. This requires the project team to develop a comprehensive schedule that considers task dependencies, sequence of tasks, and task durations.
- *Buffer time*: Account for potential delays in the supply chain or unforeseen circumstances by incorporating buffer time into the delivery schedule.
- *Regular communication*: Maintain open communication with suppliers to monitor delivery status and address any issues promptly.

- *Early material delivery for mock-ups*: For early material deliveries in cases where mock-ups and performance testing are planned, it is best to do the following:
  - *Identify mock-up requirements*: Determine which materials are needed for early mock-ups or prototypes.
  - *Prioritize procurement*: Ensure that these materials are prioritized in the procurement process.
- *Coordinate with the construction team*: Work closely with the construction team to schedule and execute mock-ups efficiently.

Using Software to Track Materials

There are several benefits to tracking materials by using a software program instead of manual methods. Material tracking performed by utilizing web-based software programs can offer advanced monitoring of ordered materials from procurement orders through the delivery of those materials by using real-time data. These systems can offer multiple benefits for improved efficiency and productivity of the construction projects. Following are some of these benefits:

- *Real-time visibility*: Monitor the status of material orders, deliveries, and inventory levels.
- *Automated alerts*: Receive timely notifications for potential delays or shortages.
- *Improved forecasting*: Use historical data to forecast future material needs.
- *Cost control*: Track material costs and identify opportunities for savings.
- *Enhanced collaboration*: Facilitate communication among project teams, suppliers, and subcontractors.
- *Flexible delivery schedules*: Negotiate flexible delivery schedules with suppliers to accommodate potential changes in the project timeline.
- *Contingency planning*: Develop a contingency plan for material shortages or delays, including alternative suppliers or expedited shipping options.
- *Quality control*: Implement rigorous QC procedures to build confidence that materials will meet specifications and standards.

By aligning the materials list with the project schedule and utilizing material tracking software, it is possible to minimize delays, optimize resource allocation, and ensure the project's successful completion.

Plan for Material Storage

Once the materials list and delivery schedule is identified, the next critical step is to plan for material storage. Consider factors such as storage space, environmental

Table 5.2 On-site vs. off-site pros and cons.

	Pro	Con
On-site storage	– Quick access to materials – Reduced transportation costs – Greater control over material security and conditions	– Requires significant space and resources – Potential for theft or damage – Increased security measures may be necessary
Off-site storage	– Lessens the burden of on-site storage – Reduced risk of theft and damage – Can be more cost-effective for long-term storage	– Increased transportation costs – Potential for delays in material delivery – Less control over storage conditions

Table 5.3 Pros and cons of RFID versus bar code scanners.

Pros	Cons
RFID (Radio Frequency Identification)	**RFID (Radio Frequency Identification)**
Real-time tracking of materials	Higher investment costs for RFID technology
Remote monitoring of materials location and condition	Requires specialized infrastructure and software
Faster inventory checks and audits	
Improved accuracy and reduced human error	
Bar Code Scanners	**Bar Code Scanners**
Lower initial cost	Requires line-of-sight scanning
Widely available and easy to use	Slower and less accurate than RFID
Scanners are suitable for basic inventory tracking	

conditions, and security. See Table 5.2 for the pros and cons of on-site versus off-site storage.

Material Tracking Through Technology

To ensure efficient material management, another best practice is to consider implementing technology solutions like RFID tags or barcode scanners. See Table 5.3 for the benefits and drawbacks of using these approaches.

Choosing the Right Technology for Materials Inventory Management

The best choice depends on the specific needs of your project, including:

- *Material value*: High-value materials may benefit from the enhanced security and tracking capabilities of RFID.
- *Inventory frequency*: Frequent inventory checks may favor RFID's speed and accuracy.

- *Storage conditions*: Harsh environments may require durable RFID tags.
- *Budget constraints*: Barcode scanners offer a more cost-effective solution for basic tracking needs.

By carefully considering these factors and implementing effective material storage and tracking strategies, you can optimize resource utilization, minimize waste, and ensure the timely delivery of materials to the construction site.

Site Acceptance Criteria for Materials Inventory Management

Site acceptance criteria are a set of standards and requirements that materials must meet on delivery to the project site. These criteria ensure that the materials are of the correct quantity, quality, and type, and are suitable for use in the project.

1. *Material specifications*
   - *Quality standards*: Ensure materials meet specific quality standards (e.g., ASTM, ISO).
   - *Material properties*: Verify material properties like strength, durability, and fire resistance.
   - *Manufacturer certifications*: Check for necessary certifications and warranties.
2. *Quantity verification*
   - *Delivery documentation*: Verify the quantity against the delivery documentation (e.g., packing slips and invoices).
   - *Physical inspection*: Conduct a physical count or measurement to confirm the quantity.
3. *Condition inspection*
   - *Damage assessment*: Inspect for any damage during transportation or storage.
   - *Expiration dates*: Check for expired materials, especially for consumables like adhesives or sealants.
   - *Storage conditions*: Ensure proper storage conditions have been maintained.

Roles and Responsibilities for Site Management of Materials

- *Project manager*: Oversees the entire process, ensuring compliance with project specifications and standards.
- *Site supervisor*: Coordinates with the delivery team and conducts initial inspections.

- *QC inspector*: Conducts detailed inspections, tests, and certifications.
- *Receiving foreperson*: Verifies delivery documentation and quantity.
- *Material supplier*: Provides accurate documentation and ensures material quality.

Best Practices for Site Acceptance

- *Clear documentation*: Maintain detailed records of material deliveries, inspections, and acceptance.
- *Regular communication*: Establish effective communication channels among all parties involved.
- *Training and certification*: Ensure personnel are trained to recognize material defects and perform quality checks.
- *QC plans*: Implement a robust QC plan to minimize defects and rework.
- *Contingency planning*: Have a plan in place for handling material discrepancies or delays.

By establishing clear site acceptance criteria and assigning responsibilities, you can ensure that only suitable materials are used on the project, reducing the risk of defects, delays, and cost overruns.

Material Protection Plan

A material protection plan is a strategy to safeguard materials and equipment from damage, theft, and deterioration throughout the construction project. This plan is essential to minimize losses, maintain project quality, and ensure timely completion.

Key components of a material protection plan include:

1. *Risk assessment*
    - Identify potential risks to materials, such as weather conditions, theft, vandalism, and accidental damage.
    - Assess the severity of each risk and prioritize mitigation strategies.
2. *Storage and handling procedures*
    - *Proper storage*: Store materials in designated areas, protected from weather and environmental factors.
    - *Secure storage*: Implement security measures like fencing, locks, and surveillance systems.
    - *Safe handling*: Train workers on proper handling techniques to avoid damage during transportation and installation.

3. *Inspection and maintenance*
    - *Regular inspections*: Conduct regular inspections to identify and address potential issues.
    - *Maintenance procedures*: Establish procedures for cleaning, repairing, and replacing damaged materials.
4. *Defect tracking and trade damage reporting*
    - *Defect tracking system*: Implement a system to track and document material defects, including their cause and resolution.
    - *Trade damage reporting*: Establish a process for reporting and addressing damage caused by subcontractors or trades-people.

Role of the Software in Material Protection

The use of project management software can significantly enhance the effectiveness of a material protection plan. Key features include:

- *Centralized database*: Store information on material specifications, quantities, and locations.
- *Real-time tracking*: Monitor material usage and inventory levels.
- *Automated alerts*: Receive notifications for potential issues, such as low stock or damage.
- *Document management*: Store and access relevant documentation, such as inspection reports and maintenance records.
- *Reporting capabilities*: Generate reports on material usage, costs, and damage incidents.

By implementing a thorough material protection plan and using project management software, you can safeguard your project's valuable resources and minimize the risk of costly delays and rework.

First-in-Place Compatibility Evaluation for the Materials

A first-in-place compatibility evaluation is a critical step in the construction process to ensure that materials and components installed early in the project are compatible with subsequent installations. This is particularly important for materials that may interact with other materials or systems, such as:

- *Structural elements*: Ensuring that foundations, walls, and floors are treated and placed correctly to receive any elements that rely on their placement accuracy.
- *Plumbing and electrical systems*: Verifying that pipes, conduits, and wiring are installed correctly, can accommodate future connections, and are in the surface plane.
- *HVAC Systems*: Checking that ductwork, piping, and equipment are properly sized and aligned.

- *Finish materials*: Confirming that paint, flooring, and other finishes are compatible with underlying substrates and each other.

Key Considerations for First-in-Place Compatibility Evaluations

- *Material specifications*: Verify that materials meet project specifications and industry standards.
- *Installation standards*: Ensure the materials are installed according to the manufacturer's instructions and industry best practices.
- *Tolerance checks*: Measure dimensions and alignments to ensure they are within acceptable tolerances.
- *Field verification*: Conduct visual inspections and field tests to identify potential issues.
- *Documentation*: Document the results of the evaluation, including any deviations from specifications or standards for future reference and reporting.

Benefits of First-in-Place Compatibility Evaluations

- *Early detection of issues*: Identifying and addressing potential problems early in the construction process can save time and money.
- *Improved quality*: Ensuring compatibility between materials and systems can lead to a higher-quality finished product.
- *Reduced risk of rework*: By addressing issues early, you can avoid costly rework and delays.
- *Enhanced project coordination*: Effective communication and collaboration between trades can help prevent conflicts and streamline the construction process.

By conducting thorough first-in-place compatibility evaluations, you can significantly improve the overall quality and efficiency of your project.

Material Inspection and Testing Plan

A material inspection and testing plan is a detailed document outlining the procedures and standards to be followed to ensure that materials used in a construction project meet specified quality requirements. This plan is crucial to maintaining project quality, safety, and compliance with building codes and regulations.

Key Components of a Material Testing and Inspection Plan

1. *Material specifications*
    - Clearly define the required material properties, such as strength, durability, and fire resistance.
    - Reference relevant industry standards and codes.

2. *Testing procedures*
    - *Laboratory testing*: Outline the specific tests to be conducted in a laboratory setting, such as compressive strength, tensile strength, and chemical analysis.
    - *Field testing*: Describe the field tests to be performed on-site, such as moisture content, density, and visual inspections.
3. *Inspection procedures*
    - *Visual inspections*: Detail the visual checks to be made for defects, damage, and compliance with specifications.
    - *Dimensional inspections*: Specify the tolerances for dimensions and measurements.
    - *Material certification review*: Verify that materials are accompanied by valid certifications from manufacturers.
4. *Quality control tools*
    - *Checklists*: Create checklists to guide inspectors through the inspection process.
    - *Data sheets*: Use data sheets to record test results, inspection findings, and corrective actions.
    - *Statistical process control (SPC)*: Employ SPC techniques to monitor and control material quality.
5. *Documentation and recordkeeping*
    - Maintain detailed records of all testing and inspection activities, including test results, inspection reports, and corrective action plans.

Benefits of a Material Testing and Inspection Plan

- *Enhanced quality*: Ensures that materials meet specified quality standards.
- *Reduced risk of defects*: Identifies and addresses potential issues early in the construction process.
- *Improved project performance*: Minimizes delays and rework caused by material failures.
- *Compliance with codes and regulations*: Verifies that materials comply with applicable building codes and standards.
- *Cost savings*: Reduces the long-term costs associated with material failures and repairs.

By implementing a thorough material testing and inspection plan, you can significantly improve the overall quality and durability of your construction project.

Installed Material Protection Plan

An installed material protection plan is a strategy to safeguard materials and equipment that have been installed on a construction site. This plan is important to protect the investment in materials and ensure the project's successful completion without unnecessary delays. The components of an Installed Material Protection Plan include:

1. *Risk assessment*
   - Identify potential risks to installed materials, such as weather damage, accidental damage, and theft.
   - Assess the severity of each risk and prioritize mitigation strategies.
2. *Protective measures*
   - *Weather protection*: Implement measures to protect materials from rain, snow, and extreme temperatures, such as tarps, covers, and temporary enclosures.
   - *Impact protection*: Use protective barriers or padding to shield materials from accidental damage.
   - *Security measures*: Implement security measures like fencing, lighting, and surveillance systems to deter theft and vandalism.
3. *Inspection and maintenance*
   - *Regular inspections*: Conduct regular inspections to identify and address potential issues.
   - *Maintenance procedures*: Establish procedures for cleaning, repairing, and replacing damaged materials.
   - *Documentation*: Maintain detailed records of inspections, maintenance activities, and any damage or repairs.
4. *Subcontractor coordination*
   - *Clear communication*: Establish clear communication channels with subcontractors to coordinate material protection efforts.
   - *Shared responsibility*: Ensure that subcontractors are responsible for protecting materials within their scope of work.
5. *Insurance coverage*
   - *Review insurance policies*: Ensure that adequate insurance coverage is in place to protect installed materials from loss or damage.
   - *Consider additional coverage*: Evaluate the need for additional insurance coverage, such as installation floater insurance, to protect materials during the installation process.

By implementing an installed material protection plan, you can safeguard your project's investment, minimize delays, and ensure a successful outcome.

Material Warranty Process

A material warranty is vital for project closeout to happen without interruptions. Each of the technical specification divisions will call out the required warranty information. By beginning with the end in mind, the CQM will organize the material coordination efforts to ensure an intact warranty is delivered to the owner at project closeout. This process is covered thoroughly in Chapter 6, "Final Turnover and Closeout Documentation," and "Warranties."

CONCLUSION

Comprehensive material evaluation is fundamental to project success. From establishing detailed lists and ensuring timely delivery to proper storage and rigorous site acceptance, every step is crucial. Implementing material tracking technologies, crafting protection plans, and conducting thorough inspections guarantee quality and mitigate risks. By prioritizing first-in-place compatibility and meticulous testing, projects can avoid costly errors. Ultimately, a well-executed material evaluation process ensures project efficiency, compliance, and the delivery of a high-quality finished product that meets all expectations.

QUESTIONS

1. Describe the potential consequences of neglecting to align a comprehensive materials list with the project's timeline, and explain how material tracking software can mitigate these risks.
2. Compare and contrast the advantages and disadvantages of on-site versus off-site material storage, and discuss the factors that should influence the decision between using RFID tags or barcode scanners for material tracking.
3. Explain the significance of "first-in-place compatibility evaluation," and give three examples of project elements where this evaluation is particularly critical, detailing the potential issues that could arise without it.
4. Outline the key components of a robust material protection plan, and discuss how project management software can enhance the effectiveness of this plan in minimizing losses and ensuring project quality.
5. Detail the roles and responsibilities of different stakeholders involved in site acceptance criteria, and discuss the best practices that should be implemented to ensure only suitable materials are used on the project, reducing the risk of defects and delays.

REQUESTS FOR INFORMATION OR CHANGES

> Participate in the process to track requests for information (RFIs), field change requests (FCRs), and design change requests (DCRs) to ensure the project remains on track and that objectives adapt when necessary. (Apply)
>
> Body of Knowledge V.A.5

Change is inevitable, which is why the project's quality plan must take the form of a flexible framework to accommodate forthcoming changes (as discussed in Chapter 2, "Project Quality Strategy Development"). Change is normal, thus the quality system must be a dynamic framework. However, the amount of change should also be managed, as excessive change can be detrimental to a project's success. Change must therefore be controlled (Ashford, 1989).

Note that change is described here in the context of design changes that are made to the final design (which is issued to construction by the EOR at the end of the design stage). This change is different from contractual changes that are focused on alterations to the scope, cost, or schedule of the contract (which must be addressed through a different process for scope and contract change).

Also, given that design changes are essentially a design activity, this section describes how the contractor must connect with the designer and utilize a request for information (RFI) to formally collect the designer's decision about the proposed change to the intended final design. The best practice is to utilize the contractor submittal process (described earlier in "Construction Submittals") for all RFIs so there is a complete record of contractor requests and subsequent designer dispositions. The design of a constructed infrastructure remains the responsibility of the EOR, and it cannot be changed without the explicit approval of the designer. A good construction quality plan must include formal procedures for RFIs and the construction submittal process, and it must also be consistent with the design change process that is specified by the designer in their design quality plan. It is the responsibility of the CQM to ensure this is done.

There are four critical tools available for managing change See Chapter 3, "Design Changes" for more information (where this concept has been described from the perspective of the designer/ architect):

- Request for information (RFI)
- Design change request (DCR)
- Field change request (FCR)
- Change log

Tools for Incorporating Changes

The RFI, DCR, and FCR serve as a means of formally incorporating changes into the design documents (Table 5.4). The change log is a log of all the proposed design changes, and the disposition that is made on all those proposals by the designer/ EOR. The change log should keep a record of the RFIs, DCRs, and FCRs, and it should be accessible to the contractor, designer, and other applicable stakeholders (such as the owner).

An RFI is a tool that facilitates the communication between the contractors and designers in the form of a documented process. It also ensures that all stakeholders get a clear and common understanding of the design changes that are made on the final design, whether by the designer's own value analysis improvement (DCR) or because of the field-based recommendations of constructability (FCR). It minimizes confusion and errors and stays as a record of the major decisions that are made on the project. When RFIs are not used appropriately by either party, there may be schedule and cost risks for every member of the project team. One study found that the average RFI response costs construction teams $1,080, and that the total potential loss across an entire project could be as high as $859,000

Table 5.4 RFI, DCR, and FCR definitions and uses.

What	Definition	When to Use	Relation to Other Tools
Request for information (RFI)	A formal question from one party to another, asking for clarification on inconsistencies, ambiguity, or missing information.	When a formal change needs to be incorporated into the contractual design documents; or when additional information is needed to complete the work; or when clarification is needed.	• An RFI can trigger a DCR or FCR. • If an RFI leads to a change in the original design, a DCR would formally incorporate the change into the contractual design documents. • An RFI may trigger an FCR to allow the contractor to proceed given the new information during construction.
Design change request (DCR)	• A formal request for proposing modifications or alterations to the original project design. • Most DCRs represent designer's own value analysis and improvement in the previously issued final design.	Correcting errors in the original design, accommodating for unforeseen site conditions, necessary changes initiated by the client, or regulatory or code-related changes. DCRs can be used during either the design or construction phase.	• A DCR may follow an RFI if the RFI response requires modification to the original design. • A DCR may also follow an FCR if a constructability or other field issue requires a major change to the design.
Field Change Request (FCR)	Similar to a DCR, the FCR is also a formal request for change; however, its use is limited to the construction phase. In other words, the changes that pertain to the constructability of the design are usually issued as FCR.	Unforeseen site conditions (e.g., poor soil conditions) or constructability challenges that were not anticipated during the design phase, safety concerns, productivity improvements.	• An FCR may trigger a DCR if proposed changes cannot be accommodated with minor field adjustments. • However, if the proposed changes can be incorporated with minor field alterations, then a DCR or FCR may not be required.

(Medina, 2024). RFIs, while a necessary tool, can create administrative overhead when not used effectively. (See Chapter 3, "Design Changes" for additional information on RFIs.)

Below are best practices for improving the overall RFI process, and thus more effective use of RFIs:

1. An established written procedure should be developed and put in place, while defining what the RFI will be used for within the context of the project (Medina, 2024).
2. When RFIs are submitted, a potential solution should also be provided as part of the request. This decreases the review time required by the design team and owner.
3. When planning efforts are conducted early, RFIs can be asked long before that portion of the work begins, reducing the schedule pressures for all involved.
4. Must ask one question per RFI. Multiple questions will complicate responses and increase response time.
5. Review RFIs with the design team and owner—using the change log (see Table 5.4)—prior to formally submitting the RFI. This provides context and may also decrease response times. It is best to include status of RFIs as a standard discussion item in the weekly/biweekly meeting agenda.
6. Note what is stated in the specifications regarding the protocols for RFIs, DCRs, and FCRs. The process that must be used for RFIs, DCRs, and FCRs is usually described in the technical specifications.

Regarding DCRs and FCRs, the fundamental difference is that DCRs are initiated by the designer/EOR. They are an improvement of the final design that was previously issued by the designer. DCRs are usually the result of value analysis performed by the design team, which initiates the need for the change in design. The FCRs, on the other hand, are initiated by the contractor and usually represent a constructability issue. When the construction contractor believes the final design as issued by the designer cannot be constructed, they may issue an RFI to obtain clarification or an FCR to propose the design changes due to constructability issues in the original design.

RFIs are the tool that the contractors must use to request information from the designer/EOR; some RFIs may proceed to become a DCR or FCR, but others may be simply answered by the designer and do not generate either a DCR or an FCR. DCRs can also be initiated by any stakeholder if they notice errors or required improvements to the design or for technical issues or code compliance concerns.

Managing and Tracking Change Effectively

The change log is the primary document outlining all changes to the project, including incorporated and forthcoming changes (an example is shown in Table 5.5). It documents how the change will be reconciled—as an RFI for example, or an official, contractually acceptable change to the contract documents. Each item on the change log must progress from "potential" to "incorporated," where potential may mean that some part of the project *may* change but is still being discussed. Incorporated means the change has been incorporated into the design documents and is formally approved.

Table 5.5 Example change log item.

Change Log Attributes	Example
Change Title	Additional elevator lighting
Description	Lighting levels at the elevator doors on the second floor are below the code-required 10 foot-candles. The L14 fixture in the lighting schedule is unable to provide adequate illumination. The light can be swapped with an L16 fixture if the client and designer approve of the color. L14 would be included in attic stock.
Date Added	7/10/2024
Purpose	Local city code required 10 foot-candles at all elevator entries.
Source	Elevator inspector
Rough Order of Magnitude (ROM) Cost	None if approved by 8/15/2024
Schedule Impact	None if approved by 8/15/2024
Contract Adjustment Vehicle	RFI# 17
Status	Pending
Next Action	Mechanical trade partner to submit shop drawings showing proposed shutoff valve location.

These three sources of change—RFIs, FCRs, and DCRs—must be reviewed regularly in an open, transparent meeting using the change log as part of the meeting agenda and minutes. For example, a change log should be used on every project and be a topic of the owner, architect and contractor (OAC) meeting and other key meetings in the meeting flow (see "Meetings and Participation" later in this chapter for effective project meeting flows).

A design change may be so extensive that it generates the need for modifying the contract; in such cases, a potential change order to the contract is issued, and the design change along with the change log must be included as part of the change order submitted by the contractor. As such, the design change may be considered a component of the change order. When reviewing change orders or a change of any kind to the project (cost or not), it is best to do so in an open, transparent way where key stakeholders can participate (Atkins and Simpson, 2008). Sending the change log and associated meeting agenda prior to the meeting between all participants allows them to review the proposed changes and allows for an effective conversation during the actual meeting.

The change log guides the discussion during the meeting between stakeholders. Changes that are most time-sensitive should be reviewed first. For example, RFIs should be listed in order of importance according to the project schedule and when a response is needed. Accordingly, forthcoming contractual changes being incorporated into the design deliverable should also be discussed so that when they are released for pricing or the team is being directed to proceed, the changes are clearly identified. (The usual practice is to circle or outline a modification on a drawing with a revision cloud, commonly termed "bubbled.")

CONCLUSION

The documentation associated with change can be burdensome, as is ensuring those installing the work are made aware of change in a timely manner. When a change occurs, it should not be a surprise. Surprise is a lagging indicator of poor communication. It is what delays necessary changes from being incorporated into the projects; some changes may be necessary in achieving the project's quality objectives.

RFIs, DCRs, FCRs, and change logs are all available tools in most modern project management software. This allows for change to easily be incorporated and accessible to everyone on the project team, pushing the latest information out to the field. "Construction is better served when the project team has the flexibility to revise the original project design as construction proceeds. These changes help to produce a facility that is consistent with project objectives" (American Society of Civil Engineers, 2012).

The CQM's involvement in the RFI, FCR, and DCR processes is essential to maintaining project quality and compliance. By staying actively engaged in tracking and reviewing these changes, the CQM ensures that construction aligns with contract documents and project objectives. Additionally, participation in these processes allows for more effective auditing, as familiarity with the change management system enables proactive oversight rather than reactive corrections. Most importantly, failing to properly manage and document changes can lead to costly rework, schedule delays, and diminished project performance. A well-structured and transparent approach to RFIs, FCRs, and DCRs—supported by a comprehensive change log—ensures that the design change is not a disruption but a controlled and necessary element of a construction project's success.

QUESTIONS

1. What is a request for information (RFI)?
2. What is the difference between a design change request (DCR) and a field change request (FCR)?
3. What are the critical elements of a change log?
4. How can a change log be used to manage change and keep stakeholders informed?
5. Who can initiate an FCR?
6. Explain how an RFI can potentially lead to both a DCR and an FCR, and describe a hypothetical scenario illustrating each outcome.
7. Describe the potential financial risks associated with improperly managed RFIs, and outline the steps a project team can take to mitigate these risks.
8. Discuss the role of the change log in maintaining transparency and effective communication regarding project changes, and explain how it should be utilized during OAC meetings.
9. Compare and contrast the initiation and scope of DCRs and FCRs, and identify a situation where an FCR might necessitate a subsequent DCR.
10. Analyze the statement "surprise is a lagging indicator of poor communication" in the context of change management, and discuss how the use of technology can help prevent such surprises in a project.

Process Management

RISK-BASED INSPECTION AND TESTING STRATEGY

> Develop a risk-based inspection and testing strategy by identifying risks. Establish priorities for inspections, including modifications based on changes in project requirements (e.g., scope modifications), and testing in line with specifications (e.g., concrete testing) and accepted standards or regulations (e.g., welding certifications). (Create)
>
> Body of Knowledge V.B.1

Risk management involves identification, assessment, and control (Griffith, 2011). Risks to be identified, assessed, and controlled include not just those that may impact quality but also safety and environmental risks. Each of these management systems—quality, environment, and safety—support one another when risk presents itself. For risk to be quantified properly for a project, the project's quality plan must include QA and QC activities that accommodate environmental and safety risks as well as quality risks, given their impact on delivering the client's intent.

All projects and organizations must consider risk-based thinking as they establish their processes for quality management. This thinking moves beyond the traditional corrective action mindset and involves risk analysis as a proactive means to achieve quality by preventing the likelihood of negative outcomes. This section on construction quality process management is closely linked with Chapter 4, "Risk Management," where more details are provided on how to perform operational risk management. For the construction phase, this section describes how the CQM may use risk-based thinking to develop the inspection and testing strategy for the project. For the operational details on risk management that includes qualitative and quantitative assessments, risk mitigation, and monitoring, refer to Chapter 4.

Risk is traditionally transferred down the supply chain from the general contractor to subcontracting trade partners. Open communication and understanding by all stakeholders bring knowledge to the forefront, allowing the team to properly plan for known risks while leaving flexibility in the quality plan to incorporate changes based on newly discovered risks. Quality seeks to *allocate* and *mitigate* risk, rather than merely *transfer* it contractually (Kubal, 1994).

Risk Identification

The first step in risk management is the identification of risk. A *risk awareness matrix* is a tool used for identifying risks and developing mitigation strategies. A sample matrix is shown in Table 5.6. The risk awareness matrix is somewhat similar to the failure mode and effects analysis, where the probability of the risk

Table 5.6 Sample risk awareness matrix.

Risk				Resolution		
Description	*Category*	*Probability*	*Impact*	*Plan*	*Responsible*	*Date*
Permitting will take longer than three months.	Municipal	Medium				
Cold weather impacting concrete pours	Concrete	High				
Poor soils or groundwater present	Site work	Low				

and its potential impact are used to prioritize what risks should be given more attention than others and planned accordingly.

- *Description*: A description of the identified risk.
- *Category*: This helps group risks by category to identify stakeholders and plans for resolution. Example categories include municipal, project team staffing, contract, schedule, procurement, estimate, design, and construction.
- *Probability*: The likelihood of the risk coming to fruition.
- *Impact*: If the risk materializes, this column articulates the potential impacts on the project.

The *risk awareness matrix* (Table 5.5) helps create awareness of the risks and provide context to all team members. When plans for mitigating risk are developed early, it is much easier to execute them later. High-impact risks create cost issues and schedule delays, and if those risks are properly forecasted, implementation is much more effective. Also, it is important to note that if the project scope is modified, the risk awareness matrix should be able to identify (and manage) the risks that arise from the scope change.

In *How Big Things Get Done*, the authors discuss "reference-class forecasting" where project teams can use data from previous projects that articulate the cost impacts from unforeseen surprises (Gardner and Flyvbjerg, 2023). All organizations have available data to help determine the risk baseline from previous projects. Lessons learned from past projects can also be used to add risk items to the risk awareness matrix for assessment.

The Risk Planning Meeting

A *risk planning meeting*—or discussion—is a focused discussion to identify the risks of the project. The recommended frequency of risk meetings is not universal; frequency should be based on the criticality of the project, the criticality of the risks, and how quickly the risk landscape of the project is expected to change. Based on these primary factors, the project team must make the strategic decision about how often risk meetings must be called to discuss, identify, and later to evaluate the status of the project risks. Once that strategic decision is made, the frequency of these meetings is determined.

It is a best practice to call in all stakeholders of the project to a common strategic risk meeting at the beginning of the project to initiate the entire risk management process. However, not all stakeholders may be present at this meeting, as some (including the contractor) may not yet be hired. Also, realistically, any risk matrix that is developed during an early stakeholder meeting may be able to identify only the broad, high-level risks for the project. To make risk management effective, it is essential that such high-level identification be considered only as the baseline of the project risks that are considered significant by all the stakeholders.

The primary goal of the high-level stakeholders' risk meeting should be to arrive at strategic decisions regarding overall project risk. For example, the decisions should include the identification of lead personnel belonging to all the stakeholders who should be made responsible for the project risks, the frequency of meetings and levels of risk analyses, and stakeholders' consensus on the operational methods to use to break down the risk management further into its operational phases.

The next goal of the project should be to break the high-level risks down with respect to the different project phases, and then initiate operational risk management for those phases. It is only at the operational risk management level that the project teams can address the actual risks through more detailed qualitative and quantitative assessments. Risk mitigation strategies are less effective when the risks are considered at a high level; the closer the risk assessment is to the operational day-to-day level, the more effective the risk management strategy is as the mitigation can address the root causes of those risks. The risk awareness matrix (Table 5.5) can be used as the agenda template for the risk identification meeting, where the completed/filled matrix may become the initial identification of the project risks.

Inspection and testing is a QC function. Inspection and testing plans (ITPs) are best prepared by a collaboration of the project's QA staff and the design team's staff, as well as select QC personnel from the construction phase of the project. Other contributors to the identification of risk for inspection and testing may be the local authorities having jurisdiction (AHJ) and the supply chain. Diverse perspectives are helpful to not only identify but also assess and manage the project risks as a team.

Inspection and Testing Strategy

Before the project may begin inspection and testing, it needs to have a clear and agreed-on scope and acceptance criteria for quality testing. This helps avoid ambiguity, confusion, and misunderstandings. (See "Acceptance Criteria Definition and Implementation" later in this chapter for more information.)

As described in Chapter 2, "Project Quality Strategy Development," the means by which a project is organized will improve the effectiveness of the quality plan and ability to plan and execute tests and inspections. "The core idea is to organize the project efficiently to manage quality assurance and quality control to mitigate risk" (Dupslaff, 2024).

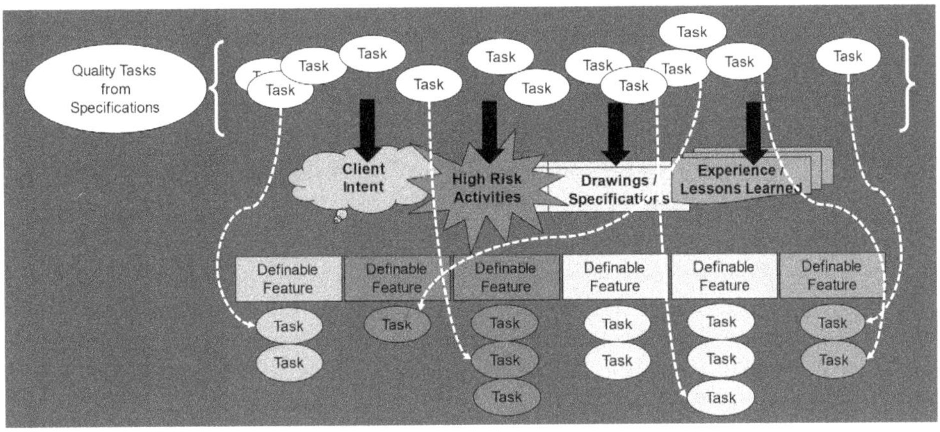

Figure 5.1 Organizing specification tasks by DFOW.
Source: Dupslaff, 2024.

One common concept that is universally understood by the construction contractors is the concept of *definable features of work (DFOW)*. The DFOW are generally created after construction mobilization and represent separate, distinct tasks within the project that would require specific and unique quality control and inspections. As such, the DFOW can be used as the basis of inspection and testing strategy for a construction project. It is frequently true that the DFOW are identified in terms of different trades or disciplines and can effectively break the project down into manageable work that can be inspected and tested for quality.

Figure 5.1 provides an illustration of how project tasks may be organized with respect to the project's technical specifications. At a minimum, each section of the project's specifications can be considered as a DFOW (it is also possible to have multiple DFOWs under one specification). For example, electrical, mechanical, HVAC, plumbing, masonry, concrete work, and so on are all DFOWs. It is important for the CQM to realize that before deciding on the DFOW, the project must define its features narrowly enough so that each feature can be inspected or tested with adequate quality controls. However, it is equally important not to define them so narrowly that it may result in creating a lot of unnecessary inspection and testing work that offers little value. "Whether the organization leverages definable features of work or something else, the means for organizing a project should be consistent across the project life cycle" (Dupslaff, 2024).

The second source of items or tasks that must be included in the ITP includes the critical items identified in the risk awareness matrix (shown in Figure 5.1). This matrix identifies the "high-risk activities" of the project and usually includes physical elements that the risk meetings of the stakeholders identify as the most critical risk factors. It is a good practice to assign QA and QC activities to monitor and measure these risk factors as well.

Depending on the project team's risk tolerance, the team must decide which risks they must consider first. The risk awareness matrix from Table 5.5 is created as a result of the project risk meetings and includes the risks that are considered by the stakeholders to carry the highest probability of occurrence.

- When they use these tools to establish their inspection and testing strategy, the project team uses what is termed "risk-based thinking." After risk identification, the project team must begin developing mitigation plans for these identified risks. Risk mitigation must be performed based on a priority that is determined by using qualitative and quantitative risk management.
- The risk-based inspection and test strategy, which helps prepare the ITPs, can establish priorities for inspections based on the significance of identified risks and their impact. A change to the project scope necessitates extended evaluation of risk for the modified scope and may result in additional QC testing. For operational details, see Chapter 4, "Risk Management."

Translation of Risk Assessment and Control Throughout

There are many quality tools at the CQM's disposal to perform risk analysis. These tools translate the known risks from the risk awareness matrix into tangible actions to be followed by the responsible project team members, which are identified using the roles and responsibilities matrix (RACI) from Chapter 2, "Roles of Quality Personnel and Internal Stakeholders." Identified risks need to be discussed during key project meetings, such as preconstruction and preinstallation meetings (see "Meetings and Participation" later in this chapter), to translate the necessary plans from Figure 5.1 through subcontracting trade partner onboarding and planning.

These meetings should be seen more as working sessions where the risks and their associated mitigation strategies—what inspections and tests will be required to eliminate risks—are articulated to those performing and installing the work. This is where QC activities are identified, evaluating the acceptance criteria against the means by which the work will be validated and controlled (see "Acceptance Criteria and Implementation" later in this chapter).

Not all risks are known at the start of a project. Even after initial scoping, buyout, discovery, and due diligence, decisions can only be made based on the information available at the time. As the project progresses, new risks will emerge, requiring swift mitigation strategies. Upfront planning helps accelerate project completion and reduce overall risk exposure. However, risks and their severity evolve throughout the project. Low-impact risks early on may escalate if left unaddressed, while high-impact risks may diminish once they are no longer relevant. As change occurs—through change orders or unforeseen conditions—new risks will arise, necessitating continuous updates to the risk awareness matrix, along with revised mitigation plans that must be reviewed and implemented. (See "Requests for Information or Changes.")

Testing in Line with Specifications (and Beyond)

For most traditional design-bid-build construction projects, the design plans and specifications are considered complete and final at the time of the mobilization of the contractor. While the specifications are essential—and by contract outline the testing and inspection requirements—they alone do not contain all inspections and tests that may need to be completed based on the specific project risks. In addition, the risk awareness matrix helps identify the remaining risks that are collectively identified by the project team and its stakeholders.

Stakeholders may have experience or lessons learned from similar previous projects that could be applied. For example, the specification may not require pretesting prior to the start of functional testing as part of the commissioning process, but the lead project manager may require pretesting to reduce the risk of extending the project schedule due to major commissioning issues arising later in the project.

In addition, the preconstruction phase is where stakeholder involvement has the greatest impact. Risk planning and mitigation require the involvement of all internal and external stakeholders, each with a unique blend of experience and perspective that will prove useful for the project.

CONCLUSION

A proactive, risk-based inspection and testing strategy is essential for ensuring project success. By systematically identifying risks, prioritizing inspections, and aligning testing procedures with specifications and regulations, project teams can better manage uncertainties. The risk awareness matrix serves as a cornerstone for ongoing risk assessment, evolving as new challenges arise throughout the project life cycle.

Effective risk management is not a one-time activity but a continuous process that requires collaboration among all stakeholders. Open communication, knowledge sharing, and data-driven decision-making—leveraging lessons learned from past projects—help refine mitigation strategies and improve outcomes.

Ultimately, while project specifications define baseline requirements, real-world risks demand a flexible approach. Risk planning meetings, stakeholder engagement, and adaptive QC measures ensure that potential issues are addressed before they impact cost, schedule, or performance. By integrating risk awareness into every phase of a project, teams can reduce uncertainties, enhance efficiency, and deliver results that align with the client's intent.

Establishing a robust risk-based inspection and testing strategy is vital for project success. By meticulously identifying, assessing, and mitigating risks, projects can navigate uncertainties and maintain adherence to quality standards. The risk awareness matrix and stakeholder meetings serve as crucial tools for proactive planning and continuous improvement. Embracing flexibility and adaptability throughout the project life cycle ensures that evolving risks are managed effectively. Ultimately, this strategic approach not only minimizes potential disruptions but also reinforces the project's ability to meet and exceed the client's intent and deliver exceptional results.

QUESTIONS

1. What is the goal of quality when it comes to risk?
2. What is a primary tool for outlining a project's risk?
3. What are potential sources and examples of risk?

4. What role do the specifications play in risk management?
5. Explain the difference between "transferring" and "allocating/mitigating" risk in the context of project quality management, as discussed in the chapter. Why is the latter preferred?
6. How does the risk awareness matrix contribute to a project's risk management strategy, and what key components should it include? Describe how a scope modification would impact this matrix.
7. Discuss the importance of reference-class forecasting and using data from previous projects in risk assessment. How can this information be incorporated into the risk planning process?
8. The chapter emphasizes stakeholder involvement in risk planning. Why is it crucial to have diverse perspectives during risk identification and assessment? Give examples of stakeholders and their potential contributions.
9. Explain how definable features of work (DFOW) can be utilized to translate identified risks into tangible QA and QC activities. How does this approach connect the risk awareness matrix to practical project execution?

QUALITY MANAGEMENT PROCESSES

> Identify inspection processes for various stages, including source, initial, intermediate, and continuous. Prioritize program inspections and tests based on their criticality, respective project stages, and potential impact on overall project quality, ensuring alignment with the project's evolving needs and scope changes. Identify the need for procedures for specialized tasks, such as welding, heat treatment, coating, and soldering. (Evaluate)
>
> **Body of Knowledge V.B.2**

Effective quality management is paramount to the success of any project, particularly during the construction phase. This section delves into the critical processes necessary to ensure quality at every stage, from initial sourcing to final handover. By identifying and prioritizing inspections, implementing specialized procedures, and adapting to evolving project needs, teams can mitigate risks and deliver superior results. This discussion emphasizes the importance of aligning inspection efforts with project risks, client expectations, and lessons learned, thereby fostering a culture of accountability and quality from the outset.

As discussed in Chapter 2, "Project Quality Strategy Development," a project generally proceeds through the following phases:

1. Precontract
2. Planning
3. Design
4. Procurement

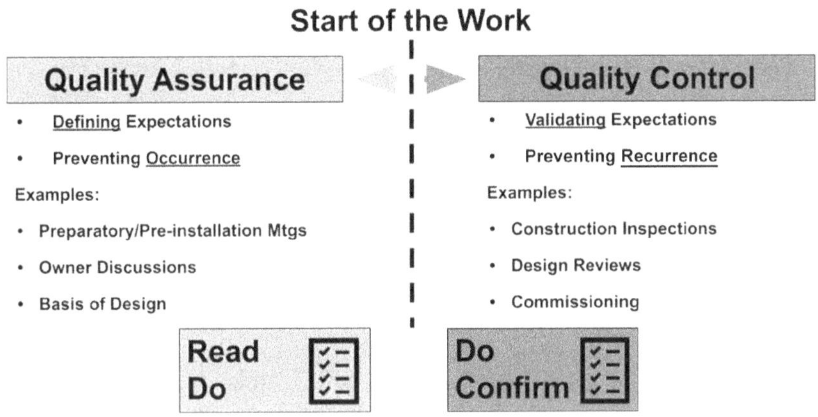

Figure 5.2 Checklists, quality assurance, and quality control.
Source: Dupslaff, 2023.

5. Construction
6. Project turnover and closeout
7. Transition to owner operation
8. Operations and maintenance

At each phase, there are physical inspections of the work. Inspections are conducted using a checklist containing critical checks to confirm at that stage. Different checklist types must be used at different phases.

Figure 5.2 is a continuation of the discussion in Chapter 2, "Quality Assurance." The READ DO checklist and DO CONFIRM checklist types are discussed in Atul Gawande's "The Checklist Manifesto: How to Get Things Right." A READ DO checklist is where the checklist user *reads* the checklist, then *does* the associated actions. A recipe for baking a cake is a great example of a READ DO checklist. A DO CONFIRM checklist is what is most often used during QC, where the work is *done* first, then inspected to *confirm* it is installed correctly.

While true "inspection checklists" are not used in this manner in QA—*before* the work starts—meeting agendas used for planning during these early phases are also "checklists for conversations." While this section focuses on inspections in QC—*after* the work starts—it is essential that these principles also be applied to the tools utilized during QA. (See Chapter 2, "Quality Assurance," for more details on planning and quality tasks *before* the work starts.)

Prioritizing the Inspections and Phases

Prioritization of inspections should be based on their criticality, on respective project stages, and on the potential impact of the inspections over project quality. This general guidance must be considered by the CQM as they plan the quality inspection and testing for QC to implement.

Inspections contain *explicit* knowledge—documented checks in the checklist being used to validate the work. Each bit of explicit knowledge must fit somewhere within the project's phases. The CQM must evaluate the project's risks, objectives, the client's intent, and other key project elements, and guide the project team in placing that explicit knowledge in the right area. This allows that knowledge to be known to the right person at the right time in the project's life cycle to mitigate the risk accordingly.

Each project phase has inspections that are conducted to ensure the work is being done correctly. For example, in design, there are inspections that designers will conduct on the drawings to ensure they captured the technical requirements appropriately. The estimators will also inspect the proposal prior to sending it to the client to confirm all the appropriate budget items are accounted for. These inspections commonly use a checklist that contains explicit information to check at that time.

The same occurs during the construction phase. There are specific inspections to be conducted at those times to validate the progress of the work based on the stated requirements (see "Compliance Monitoring" later in this chapter). Prioritizing both which inspections are conducted and what information is contained within the checklists for those inspections is critical to the success of the project's QC efforts.

Inspections and Risk

Prioritizing program inspections is all about understanding the associated risks (see "Risk-Based Inspection and Testing Strategy" later in this chapter for more information on a risk portfolio). The project team must decide collaboratively on the risk and probability of the work being installed incorrectly. The result of this discussion is the understanding of what level of inspection should be performed for each critical portion of the work (for each DFOW, for example). The level of risk will also help dictate the necessary inspection frequency.

Inspections and Lessons Learned

The incorporation of lessons learned—or tacit knowledge—into the inspection tools is critical to preventing recurring issues from prior projects or materials used previously. (See "Lessons and Improvements" later in this chapter.) Lessons learned will be company-specific, whether at the design level, the general contracting level, or the trade partner or tiered trade partners level.

Inspections and the Client

In addition to the project's risks, the client may also require that special attention be given to certain elements of the project that are of particular importance to them. The critical question that the CQM must ask continuously throughout the entire project is: "What are the things that the client is particularly interested in us getting right?" These elements of the project are part of the achievement of the project's quality goals and the client's intent. Even if all contractual elements of the project are delivered successfully, the client may not feel quality has been achieved—or their *intent delivered*—if the important elements they care about are not given special focus and attention. These client "hot buttons" should be the focus of each critical meeting. (See "Meetings and Participation" for more on how to incorporate essential content into the right stakeholders at the right time.)

Inspections and Timing

By taking into account the inspection and test requirements of the specifications—along with those based on the project's risks and the client's intent—the project team will have the proper context for the successful execution of QC via the ITP. Deming noted that quality cannot be inspected into a product (1982). This very much applies to construction. Once the work is installed, it is too late to prevent defects. Installers must know what they need to know to be successful, which comes from planning during QA and discussing the details of what will be on the specific inspection *before* the work begins. This is why the impact on quality is greater earlier in the project. Inspections themselves are necessary, but they are not the only tool to be used for ensuring the work is installed correctly.

(1) The checklist here is a checklist for a discussion.

(2) Work in place is also used for WIP, or "work in progress," follow-up (as noted by the U.S. Army Corps of Engineers and the Three Phases of Quality), or ongoing and continuous inspections.

(3) Benchmark also used for first of a kind, first work, or initial inspection.

Table 5.7 illustrates the various "inspection" or QC "categories" that the necessary inspection information may fall into. The CQM must evaluate the project's specifications, client needs, and risks to ensure the correct content is in one of those buckets in Table 5.7.

- Estimate checks (also include any checks prior to the start of work). Checklists for estimators and also design reviews (design items are not shown in Table 5.7 but are also critical to construction QA activities).
- Factory/source inspections
- Delivery inspections
- Benchmark inspections
- Work in place inspections

The ITP must clearly outline these inspection buckets for each DFOW, and the project team must understand *what* will be inspected at each stage. This ITP should be developed during planning (see Chapter 2, "Project Quality Strategy Development") and executed *after* construction mobilization. For example, depending on the risk of a product or material, it may be necessary to have the fabrication plant do source or factory inspections prior to delivery. This ensures the product meets the project's standards when it leaves the plant.

Table 5.7 Inspection categories.

	Preconstruction		Construction Quality Control	
Inspection Checklist	Estimate check	***Start of the Work***	• Factory inspection • Delivery inspection • Benchmark inspection (3)	Work in place inspection (2)
Meeting (1)	Budget reviews		Preconstruction (scope-specific)	Preinstallation (DFOW or trade-specific)

(1) The checklist here is a checklist for a discussion.

(2) Work in place is also used for WIP, or "work in progress," follow-up (as noted by the U.S. Army Corps of Engineers and the Three Phases of Quality), or ongoing and continuous inspections.

(3) Benchmark also used for first of a kind, first work, or initial inspection.

Different Types of QC Inspections for Construction Phase

The QC inspections in construction can be classified into different types: source, benchmark, and work in place/continuous inspections.

Source Inspections

To identify the need for source inspections, the project team must know what is being produced off-site. Then the team must identify the risks associated with each fabricator.

- Has the team worked with the specific supplier before?
- Is the product custom?
- Is the product on the critical path of the job?
- What level of testing/inspection is required for the product?
- What is the client's past experience? Have they had problems with this type of product in the past?
- Is a QC team member needed at the plant full-time?
- Do the plant and its manufacturing processes need to be audited as part of prequalification?

Once the team understands the risk, an appropriate source inspection program can be developed.

Benchmark Inspections

A benchmark inspection should be identified for each DFOW. (Recall that the project's DFOWs must be identified during the early planning stages—again, see Chapter 2.) The benchmark inspection is intended to inspect the first portion of work to verify that it is correct, meets expectations and requirements, is within specified tolerances, and so on. If the benchmark inspection identifies issues, then adjustments to the work need to be made. If the benchmark inspection yields no issues, then the work continues and WIP or follow-up inspections are conducted to ensure the work is going according to plan.

Work in Place/Work in Progress (WIP)/Follow-Up/ Continuous Inspections

A work in place (WIP) inspection is an inspection of the work at some point after the benchmark inspection and prior to any functional testing. These are typically formal inspections conducted throughout the installation of an area of work or DFOW, but they can also be informal as project team members are walking the site.

Formally documented WIP inspections must utilize a checklist as required in the specifications or as prerequisites for testing. The ITP must integrate inspection activities to ensure components and systems are ready for functional testing.

Formal WIP inspections will also be required to maintain active permits throughout the project's duration (see "Required Permits" later in this chapter).

Tests Versus Inspections

While Table 5.6 outlined a framework for inspections, the same must be done for tests. Note that specific inspections may need to be completed prior to tests, such as system readiness inspections as a prerequisite for commissioning functional testing. Inspections are for static work and systems, while tests are for dynamic functions of systems and components. (Source: Dupslaff: 2025)

Specialized Procedures for Unique Construction Tasks

Specialized procedures—such as welding, heat treatments, soldering, or other activity that requires certain qualifications for personnel executing the work—must also be identified in the ITP. The project team must first understand what work involves specialized tasks, which come primarily from the product submittals (see "Construction Submittals"). If specialized procedures are required, the CCQM must outline:

- The steps involved
- The inspection points
- What the specifications call for regarding qualifications
- Whether separate submittals are required to document specialized procedures and have them approved prior to starting the work

Specific sections that also discuss requirements for specialized procedures include:

- Construction Plans and Specifications
- Destructive and Nondestructive Tests
- Compliance Monitoring
- Acceptance Criteria Definition and Implementation

Using Technology

Much of the checking, validating, inspecting, and testing requirements—the project's QC activities—come from the specification. Modern project management software allows users to search for keywords such as "test," "inspect," "verify," or other such synonyms. These are triggers for QC activities that need to be incorporated into the ITP, the baseline required tests and inspections for the project.

In addition to the inspection effort, technology allows the easy creation of nonconformance reports (NCRs) for noncompliant work and can reduce the associated administrative overhead (see "Compliance Monitoring" later in this chapter).

Inspection Responsibility

The project's responsibility matrix (see Chapter 2, "Resource Assessment and Mobilization") must outline who is responsible for conducting and overseeing what inspections and testing activities, and ensuring they are qualified to do so. Inspections and tests will be shared responsibilities between:

- General contractor
- Design team
- Third party
- Agency

This shared responsibility means that trust must be part of the project team's culture. When a party indicates their inspection is completed, the CQM must validate initially that they did in fact do what they stated they would do. One individual on the team will not be able to validate every inspection on the project. The ITP must outline responsibility and accountability regarding testing and inspections. When those responsible for installing the work are aware of the acceptance criteria, it is much easier for a culture of accountability to ensure the work is compliant with the project's expectations.

Inspections and Changes

As discussed in "Requests for Information or Changes," change is inevitable—and when it occurs, the ITP must also be updated accordingly. The benefit of using modern project management software is the ability to keep the quality plan and associated ITP "living" and able to incorporate client and scope changes. Gone are the days when a drawing would be printed for installation. Change on some projects is so prevalent (unnecessarily so) that as soon as a drawing is printed, it is obsolete. Digital ITPs allow all project crews to have access to the latest drawings with marked changes (RFIs or pending design changes).

Flexibility is a necessary requirement even when planning and scheduling inspections. The ITP, like the quality plan itself, must therefore be treated as a flexible framework that can accommodate sudden change.

Table 5.8 Common definable features of work (DFOWs).

Common *Definable Features of Work*		
Building automation system	Structural steel	Chilled water system
Hot water system	Building envelope	Foundations
Site utilities	Roofing	Elevators
Precast	Concrete	Overhead MEP
IDF rooms	Electrical rooms	Framing
Interior finishes	Emergency generators	Lighting/lighting controls

Inspection Criticality

Table 5.8 showcases some of the more common critical project elements that will incorporate all inspection types, from source to WIP.

Knowing which inspections are more critical than others requires the evaluation of the project's goals, risks, and critical physical elements. It is impossible to inspect everything on the project. The CQM must help the project team focus on what inspections need to be conducted over others, what systems and components are more critical, and to what level of detail—which requires alignment with the project's risk profile.

Regarding the level of detail on each inspection, the inspection checklist (or tool) itself can only contain so much information. Not only do the most critical elements of the project need to be inspected, but the level of detail within each inspection must be balanced with the needs and risks of the project. The checklists used for inspections should only contain the top 10 essential items that will cause the greatest impact if missed or installed incorrectly. If a team adds too much detail to an inspection, the critical items may get missed amidst everything else.

Inspections are required and can have a significant impact on quality so long as they are not conducted too late in the process. Critical inspections must be part of the project schedule, as failed inspections are often lagging indicators of poor planning.

CONCLUSION

Effective quality management hinges on meticulous planning, strategic inspection processes, and continuous adaptation. By aligning inspection priorities with project risks, client needs, and lessons learned, teams can ensure compliance and exceed expectations. Balancing inspection depth with efficiency, and understanding that prevention is superior to detection, are critical principles. Ultimately, robust quality management, driven by collaboration and a clear understanding of project goals, leads to successful outcomes and strengthens client relationships, solidifying the foundation for future projects.

QUESTIONS

1. What are the two main types of checklists, and how can they be used?
2. Why is the timing of an inspection important?
3. What are the different types of inspections?

4. Why must an ITP be flexible?
5. How can a CQM understand what level of detail is required regarding what inspections to conduct as well as what each of those inspections should check?
6. Explain the distinction between "READ DO" and "DO CONFIRM" checklists, and discuss why "DO CONFIRM" checklists are more commonly used during QC in construction projects.
7. Describe how a project's risk portfolio influences the prioritization and frequency of program inspections, and how definable features of work (DFOW) play a role in this process.
8. Discuss the significance of incorporating both explicit knowledge (documented checks in checklists) and tacit knowledge (lessons learned) into inspection tools, and give examples of how each can be applied.
9. Explain why the inspection and testing plan (ITP) must be considered a "flexible framework" and how it accommodates project changes, referencing the role of modern project management software.
10. Elaborate on the various categories of inspections outlined in Table 5.6 (estimate checks, factory inspections, delivery inspections, benchmark inspections, work in place inspections) and explain how the CQM ensures the correct content is allocated to each category.

QUALITY CONTROL AND MONITORING

> Develop a system to address nonconformances, manage nonconformance reports (NCRs) and corrective action requests (CARs), and implement corrective and preventive action (CAPA). Design quality protocols and establish criteria for stop-work notices. Establish a robust framework for maintaining quality records. Design a strategy for conducting site quality audits and create a mechanism for generating and distributing quality reports. (Create)
>
> **Body of Knowledge V.B.3**

This section dives deep into the critical aspects of QC and monitoring within construction process management. Effective QA lays the groundwork, but it is through rigorous QC that project plans are truly executed. Recognizing that even with meticulous planning, nonconforming work is inevitable, this section outlines a comprehensive system to address such issues. By establishing protocols for managing nonconformance reports (NCRs) and corrective action requests (CARs), a project can minimize defects and maintain adherence to contractual standards, reduce the cost of poor quality, and thus ensure overall project success. Nonconforming work is described in great detail in "Compliance Monitoring" later in this section. In addition, readers are advised to review Chapter 6, "Turnover and Closeout," for a comprehensive understanding of the topic.

As previously discussed in Chapter 2, "Quality Assurance" and "Quality Control," *quality assurance* is where the quality tasks and associated efforts are organized. A project's requirements are extracted from the contract, and tracking mechanisms are established, as well as the tools required to deliver the quality objectives and other goals. *Quality control* is where the work planned in QA gets executed—the implementation of plans and tasks set forth as part of QA.

While planning during QA is critical, regardless of how much planning is conducted, defective work will be discovered (Ashford, 1989). This is not to say the project team should reduce its focus on planning; rather, there will always be some portion of the work that is nonconforming, though the quantity of such work is greatly reduced when planning is given sufficient attention and effort.

Nonconformances and Corrective Actions

A *nonconformance* is a completed element of the work that is unacceptable (Hart, 1994). Nonconforming work can be identified by any qualified member of the project team who has the training and expertise to determine nonconformance in accordance with the contract documents. For example, a member of the design team may identify nonconforming work during a site walk or document it in a field report, and the client may note a similar concern during a key project meeting. Nonconformance can occur in both product and material during production *prior to* installation and *after* final installation. The project's QA system must outline the required tasks during both production and installation that will mitigate and identify nonconforming products and work.

A *nonconformance report* (NCR) (Table 5.9) is a formal document that not only identifies but also records incidents of nonconformance.

All project stakeholders must have a clear understanding of what is considered a nonconformance. Without a universal definition for the project, nonconformance can become subjective rather than being subject to the standard outlined in the contract documents. Nonconformances include work that does not conform to required standards and deviations from plans or installation drawings.

Table 5.9 Basic format of a nonconformance report.

Sample Nonconformance Report
Nonconformance ID/#
Citing party:
Description (Describe objectively: hat, Where, When, Why important, Who impacted, How much/many):
Location:
Contract drawing reference:
Responsible for action:
Nature of impacts (schedule, cost, quality, safety, environment, sustainability, others):
Corrective action (Prevent recurrence):
Preventive action (Prevent occurrence):

A *corrective action request* (CAR) is a formal request to address a nonconformity. NCRs, such as the example of format shown in Table 5.9, include associated CARs. CARs also include the associated root cause and methods for the prevention of future similar nonconformances. CARs are utilized for process-related concerns as well as physical material and equipment nonconformances that may result in safety, budget, schedule, or quality problems on the project.

A *corrective and preventive action plan* (CAPA) is a structured plan that incorporates both NCRs and CARs into one cohesive system. CAPAs are also formal responses to process issues identified during audits, as well as installation and equipment-related issues.

Managing NCRs and CARs Together as a System via a CAPA

A CAPA brings NCRs and CARs together as part of the QC system to identify nonconforming work, establish a means for correcting it, and prevent future issues. A CAPA brings formality to NCRs and CARs. Within a CAPA, CARs are linked to NCRs but are also separate given their distinctive purpose as part of QC. The NCR documents the nonconformance, while the CAR is the associated required correction in response to the NCR. An NCR identifies the problem, while a CAR resolves it.

The general framework for NCRs and CARs within a CAPA is as follows:

1. *Identification*: An NCR is issued by an inspector, owner, design team member, or other qualified project stakeholder to document a nonconformance.
2. *Root cause analysis*: The root cause is identified.
3. *Corrective action*: The corrective action is implemented by the responsible party that fully resolves the NCR based on the root cause.
4. *Verification*: The corrective action is reviewed by the party who identified the problem or by another qualified project stakeholder and confirms if the corrective action was effective.
5. *Closure*: The CAR is closed when all responsible parties confirm the correction is acceptable.

The NCR and CAR are more localized and focused on the project level, while the CAPA is a strategic, broad approach to systems as a whole. The differences are shown in Table 5.10.

The project's specifications often identify specific steps for NCRs and CARs within the context of the project, along with timelines for responses, as well as with potential contractual consequences. NCRs cannot remain open for extended periods of time, and CARs must be implemented as soon as responsibly possible to prevent greater schedule, budget, safety, or quality implications. As work continues to proceed in the field, an NCR of lesser impact becomes a larger issue; as an example, when drywall is installed, covering up a preexisting nonconforming condition, suddenly it will require more work to resolve the issue.

Table 5.10 CAR and NCRs as a system.

	Corrective and Preventive Action Plan (CAPA)	
	Nonconformance Report (NCR)	**Corrective Action Request (CAR)**
Purpose	Identify nonconforming elements of the work	Identify and track corrective actions
Focus	Issue identification (what went wrong?)	Resolution and future prevention of issues (how can it be fixed?)
Scope	Individual issues	Root causes and systemic problems
Timing	Immediately on nonconformance identification	After NCR is issued and root causes are identified
Output	Record of the problem	Formal plan to correct the issue

Implementation of a CAPA

Any modern project management software can implement the CAPA, NCR, and CAR framework. Deficiency logs (such as Procore's observations of the deficiency type) serve as collections of NCRs with associated CARs. Based on the specifics of the project, the CQM must create a CAPA system that not only defines what nonconforming work is, but also captures them in NCR and associated CARs. These NCRs and CARs need to be tracked and discussed regularly in key project meetings (see "Meetings and Participation") as well as part of the project's key reporting metrics (see "Reporting and Analysis").

CAPA is effective when the project team's culture allows for open issues to be discussed without blame or concern for reprimand. The same is also required at the corporate level so the CAPA can be used to share lessons learned and associated root causes with other projects to prevent quality issues across multiple projects in a portfolio (see "Lessons and Improvements"). These issues should also be reported along the entire associated supply chain for the equipment, material, or assembly.

Designing Protocols for Stopping Nonconforming Work

NCRs often are issued *after* the work is installed; however, work that is nonconforming can be identified as it is installed. A stop-work order can be issued if the work is clearly incorrect to prevent further cost, schedule, safety, and quality implications. However, formally stopping work can be a sensitive action, given that, again, nonconforming work can be subjective based on each individual's internal motivations as part of the project.

What qualifies as "nonconforming work" must be clear to all project stakeholders and formally documented as part of the project's acceptance criteria (see "Acceptance Criteria Definition and Implementation" later in this chapter). And a clear protocol for stopping nonconforming work needs to be outlined as part of the project's quality plan; it must identify which project team members have the authority to do so (see Chapter 2, "Roles of Quality Personnel and Internal Stakeholders").

Documentation and Framework for Keeping Quality Records

The purpose of documenting nonconformances—beyond legal, auditing, and contractual obligations—is to promote a culture of awareness and transparency on the project team. A critical element of QC is keeping track of the process failures that allow defects and nonconformances to occur. Documentation helps create discipline for checking work (SOURCE needed). This culture is essential to being able to identify root causes that can be used to prevent similar incidents on future projects.

Documentation is also the source of explicit knowledge. Records from the CAPA framework across multiple projects are essential because the root causes contained within the reports and other project data are used by all involved organizations to mitigate future risks and incorporate lessons learned (see "Lessons and Improvements"). The established framework should include a management system that tracks root causes to positively impact future projects (Hart, 1994). "Reporting and Analysis" later in this chapter provides further detail on setting up an integrated reporting system. However, proper reporting can only be accomplished once the results are documented.

Auditing and Reporting of Nonconforming Work

Each project and organization will have separate requirements and needs related to the documentation and reporting of nonconformances across projects. Prior to creating any reporting framework, the organization must identify what data it needs and how it will be used, after which the associated projects within its portfolio can be informed and the framework put into place.

For example, within the CAPA, there are various opportunities where a log of nonconformance reports and associated CARs can be included for communication to broader audiences for information only, and also for execution of the CAR. Any key project meeting with the appropriate stakeholders could serve as a vehicle for communicating NCRs and CARs at a corporate level (see "Meetings and Participation"). The team must identify from the beginning what information they want to track (Marsden, 2019). Alignment of the corporate and project-level quality objectives provides the strategic direction of reporting—as well as the QA and QC activities themselves.

QA and QC activities—as well as the associated documentation—also have cost, whether as part of the project's budget or as part of the organization's overhead. A cost of quality analysis should be conducted to ensure the returns for tracking NCRs help promote the profitability of the organization (see Chapter 8, "Cost of Quality Analysis" for more information). If the cost of the CAPA (prevention costs) across multiple projects is more than the prevented (failure costs), the project team's attention may need to be diverted elsewhere, or the wrong NCRs are being tracked.

Knowing what NCRs need to be part of the reporting framework will help teams use these quality budgets effectively. To prevent these NCRs, the CAPA needs to outline what NCRs should be captured. What may be considered a minor NCR at the project level may seem insignificant; however, if the associated material or equipment is used across multiple projects—and the same NCR is present elsewhere—it can have a significant risk impact for the organization, not just the project. Use of a Pareto chart can help a project team evaluate what nonconformances need to be reported; not all NCRs have the same importance or impact. Leading and lagging indicators can help with this evaluation (see "Reporting and Analysis").

The auditing process (see Chapter 2, "Quality Audits") can identify if the CAPA documentation and reporting framework is effective.

- Are the root cause analysis tools revealing the correct solution?
- Are future incidents in fact being prevented?
- Are estimating and design teams aware of incidents originating prior to the construction phase?
- Are the incidents the result of training issues, performance issues, or process failures?

In addition, what is critical is how the specific NCR is handled as part of the CAR.

- *Remove and replace*: Neither the client nor the design team need to approve. The construction team tears the nonconforming work out and redoes it.
- *Rework*: The work is redone but not completely removed. Neither client nor design team approval is required.
- *Repair to an acceptable standard*: The work is repaired to "good enough," likely not meeting all requirements, but fit for use. Client and design team approval is required.
- *Use as is*: The client agrees to accept as is. Design team and client approval are required. Often results in schedule and cost at the contractor's expense.

CONCLUSION

Robust QC and monitoring are indispensable to successful construction project management. By implementing a comprehensive system encompassing NCRs, CARs, and a CAPA framework, projects can effectively address nonconformances and prevent future occurrences. The emphasis on clear communication, documentation, and a blame-free culture fosters continuous improvement. Ultimately, a meticulous approach to QC not only ensures adherence to contract standards but also contributes to project profitability and long-term organizational learning, setting the stage for consistent delivery of high-quality outcomes.

QUESTIONS

1. Explain the distinction between QA and QC, and how they relate to the nonconformance report (NCR) and corrective action request (CAR) processes.

2. Describe the five steps of the general framework for NCRs and CARs within a corrective and preventive action (CAPA) system, and discuss the significance of each step.
3. Discuss the challenges and considerations involved in implementing a stop-work order in a construction project, particularly in relation to defining "nonconforming work."
4. How do the documentation and reporting of nonconformances contribute to both project-level quality improvement and broader organizational learning across multiple projects?
5. Explain the concept of "cost of quality" in the context of managing NCRs and CARs. Why is it important to conduct a cost of quality analysis, and what are the potential consequences of neglecting it?

DESTRUCTIVE AND NONDESTRUCTIVE TESTS

> Determine the appropriate scenarios for using destructive and nondestructive test methods and verify those performing the tests are accredited or certified. (Evaluate)
>
> Body of Knowledge V.C

The primary quality tool for planning and implementing QC testing during the construction phase is the inspection and testing plan (ITP). The ITP is an essential component of the project's construction quality plan; it is detailed map of the inspection and testing that is required at different stages of construction. Two main approaches to testing exist: destructive testing (DT) and nondestructive testing (NDT). While both aim to assess the integrity and reliability of materials, they differ significantly in methodology, applications, and impact on the tested item. The destructive and nondestructive testing included in the ITPs are described in detail next.

Note that the development of ITPs is a collective responsibility of the designer and the construction contractor. More details on how the ITPs are typically developed with the assistance of the CQM in collaboration with the designer is provided in other sections of this handbook (see Chapter 2, "Risk-Based Inspection and Test Strategy," as well as Chapter 3, "Criticality Levels and Inspections").

Destructive and nondestructive tests are necessary to ensure the quality and safety of a newly constructed building. The CQM must be aware of the project's critical materials, components, and assemblies and be familiar with available testing methods to confirm adherence to the stated requirements.

In a destructive test, materials are tested to their failure point, evaluating their durability, strength, and other elements of their performance. Destructive tests typically validate samples rather than an entire assembly and are best used when seeking to understand the in-depth parameters of performance. A nondestructive test evaluates the performance of materials or a system without damage. Where destructive tests focus more on the individual elements of an assembly, a nondestructive test can assess the performance of an entire system.

Both types of tests may be conducted at the project site or off-site in a certified laboratory environment. Not only does the test procedure need to adhere to specified requirements, but the facility itself must also be accredited according to the specifications.

Evaluating Testing Requirements

The development of ITPs is the collective responsibility of the designer and the construction contractor. Typically, the designer/EOR determines what material, component, assembly, or system may be used that is consistent with the design requirements listed in the technical specifications. It is here that the construction contractor's field construction team and the CQM collaborate with the design team. Together, they finalize the ITP schedule for the destructive and nondestructive testing that the project would require for QC inspections.

Understanding the project's critical elements, such as DFOW (or other related terminology), the CQM can determine what resources are required to achieve the QC tasks outlined in the associated specifications. Table 5.11 notes typical DFOWs, though the list is not all-inclusive.

While the specifications are the foundation for a project's testing requirements and will outline whether destructive or nondestructive tests are needed, local and international codes and regulations may also need to be consulted.

The project's roles and responsibilities matrix (see Chapter 2, "Roles of Quality Personnel and Internal Stakeholders") should articulate who will oversee and confirm that the test is planned, and that those involved in conducting and evaluating the test are trained and qualified to do so.

The ITP outlines what DFOWs are subject to testing as well as the type of testing required. For those projects where the designer and the contractor work independent of each other (e.g., DBB projects), the ITPs may be developed during the construction phase. In such cases, the proposed ITP must be submitted to the

Table 5.11 Example definable features of work.

Concrete	Soils
Rebar	Structural steel
Finishes	Masonry
Building envelope	Elevators
Prefabricated assemblies	Plumbing systems

designer of record by using the project's submittal process, where the construction team's interpretation of the specification is confirmed. The ITPs also help to align third-party inspector and local AHJ expectations, once they are approved by the required project stakeholders.

Destructive Testing Methods

Table 5.12 provides examples of common destructive tests. Note that this list is not all-encompassing.

Nondestructive Testing Methods

Table 5.13 provides examples of common destructive tests. Note that this list is not all-encompassing.

Once the testing requirements are identified, the project team should review the ITP in a work session attended by all project stakeholders—internal and external—to confirm all requirements are accounted for (see "Meetings and Participation"). This should be done prior to material or equipment procurement.

In *Construction Quality Management for Contractors* (2020), the U.S. Army Corps of Engineers (USACE) provides essential questions to confirm during the early planning processes, which ensure that all testing needs are understood by the necessary stakeholders:

- Has all required testing been identified?
- Are test reporting requirements understood?
- Have laboratory facilities and testing equipment been verified as acceptable?

Table 5.12 Destructive test examples.

Test	Purpose
Compression test	Evaluates material compressive strength (concrete, masonry)
Tensile test	Evaluates tensile strength and elongation (steel)
Bend test	Evaluates the ability of building materials to resist bending (slabs, beams)
Shear test	Determines material shear strength
Impact test	Evaluates toughness and resistance
Fatigue test	Evaluates endurance limit under cyclic loads
Creep test	Evaluates the deformation of materials under sustained high temperature and stress over time
Hardness test (Vickers, Rockwell, Brinell)	Evaluates material resistance to surface damage
Bond test	Evaluates bond adhesive strength between two materials

Table 5.13 Nondestructive test examples.

Test	Purpose
Ultrasonic test	Detects internal material flaws using high-frequency sound waves
Radiographic test	Identifies cracks and voids using x-ray or gamma rays
Magnetic particle test	Identifies surface (or near-surface) damage and discontinuities for ferromagnetic materials
Rebound hammer test	Correlates surface hardness to compressive strength (concrete)
Ground penetrating radar test	Locates voids, utilities, or other obstructions in concrete and soil
Thermal imaging	Identifies temperature variations, which helps evaluate moisture infiltration or other material defects
Electrical resistivity test	Evaluates moisture content and corrosion potential in masonry and concrete
Visual inspection	Conduct inspection without the use of complicated equipment, leveraging the naked eye or drones
Penetrant test	Uses dyes to identify surface cracks

- Are laboratory personnel qualified?
- Has the calibration of equipment been verified as accurate?
- Is there a procedure for documenting corrective steps?"

Finally, it is imperative that the ITP outline *who* will be present at each stage of the test, whether as part of planning, reporting, or attending the actual test itself.

Submittal/Procurement And Other Testing Requirements

The submittal or shop drawing data should outline the applicable standards against which the material or equipment was tested. Certain tests may be required for warranty purposes (such as most roofing materials), or laboratory tests required prior to shipping to the project site and being installed. Those team members reviewing the submittals—including the CQM—must be able to evaluate what the passing standards are and be able to confirm that the material indeed meets the intent of the specifications.

Reporting And Tracking Test Status

All test reports should note the specification they are to confirm to, the sampling methods and procedures used and signed by the qualified supervising authority (Hart, 1994). Also, the reporting methods must confirm that all test results are being submitted according to the specified submittal procedures, that the reports are complete and accurate, and that any failed tests are retested accordingly, referencing the root causes of previously failed tests. (See "Reporting and Analysis" later in this chapter for more details.)

CONCLUSION

Understanding the nuances of destructive and nondestructive testing is crucial for ensuring project quality and safety. Implementing a robust ITP, coupled with meticulous adherence to specifications and regulatory requirements, is paramount. By clearly defining roles, ensuring personnel qualifications, and meticulously documenting test procedures and results, projects can achieve the desired quality standards. Proactive engagement and thorough review of ITPs with all stakeholders further solidify testing expectations and foster successful project outcomes.

QUESTIONS

1. Explain the importance of integrating the inspection and testing plan (ITP) with other project plans during the planning phase, and discuss potential consequences if this integration is not effectively carried out.
2. Compare and contrast destructive and nondestructive testing, detailing the scenarios where each method would be most appropriate, and justify your choices with examples from the document.
3. Describe the role of the roles and responsibilities matrix in ensuring that testing is conducted by qualified personnel, and outline the steps that should be taken if unqualified personnel are identified.
4. Discuss the significance of USACE's questions regarding testing during the early planning process, and explain how addressing each of these questions contributes to the overall QC of a project.
5. Elaborate on the requirements for test reports, including the consequences of failing to meet these requirements, and explain the process that should be followed when a test fails.

Compliance Management

This section of the BOK includes three subsections that address the elements of construction that relate to compliance management. Compliance management is the process of monitoring and assessing the systems to ensure that they comply with the industry and security standards, as well as with their own corporate and regulatory policies and requirements. Compliance management is more applicable to those industries and projects that operate in regulated environments; it ensures that the project teams understand the requirements that are relevant to their business, industry, and jurisdiction, and ensure that their staff comply with those requirements. For the construction phase of the project, the areas where compliance management apply the most include the use of accredited laboratories for QC inspections, as well as any other mandatory compliance requirements that apply.

LABORATORY ACCREDITATION

> Evaluate laboratory accreditation and conduct periodic reviews or surveillance audits based on project quality requirements. (Evaluate)
>
> Body of Knowledge V.D.1

Laboratory accreditation stands as a cornerstone of QA, particularly in industries like construction where reliable test results are paramount to safety and durability. It is a formal recognition of a laboratory's competence to perform specific measurements, assuring stakeholders of the validity and accuracy of the data produced. Through rigorous evaluation and periodic surveillance, accreditation bodies ensure laboratories adhere to international standards, maintain effective management systems, and consistently demonstrate technical competence. This process instills confidence in the marketplace, reduces risk, and guarantees that laboratories provide trustworthy data for informed decision-making. Accreditation applies to various types of laboratories, including calibration and environmental testing laboratories.

Value of Accreditation for Laboratories

Laboratory accreditation is an attestation of a laboratory's competence to perform measurements based on a defined scope of accreditation. Accreditation is accomplished through a conformity assessment of the laboratory by a competent assessor or team of assessors. The assessors evaluate the laboratory's management system and conformity assessment methods. Validity and appropriateness of methods, suitability of equipment, competence of personnel, and QC procedures are a few of the areas of focus. As part of the assessment, the laboratory will demonstrate technical competence through method witnessing as a technical competence evaluation. Ultimately, accreditation is an independent demonstration that the systems and processes of the laboratory conform to specified requirements, are effectively implemented, and produce valid results.

Accredited laboratories can produce valid data using technically competent staff, valid test methods, traceable measurements, and suitable measuring equipment. Confidence in the data generated is further enhanced through the independence and impartiality of the laboratory. Conformity assessments and subsequent accreditation of the laboratory demonstrate competence, reduce risk, ensure consistency, and promote confidence in the products and services subject to the measurements performed. Formally recognizing a laboratory's competence to perform measurements promotes confidence in the validity of results generated by the laboratory. Laboratory accreditation is an internationally recognized means for evaluating a laboratory's management system and technical operations. Accreditation ensures that laboratories are competent to produce reliable, accurate results to promote confidence in the marketplace.

Laboratory Accreditation Authorities for Construction Materials Testing

(Please note that American standards are identified in this section as examples. You may find an equivalent national standard from your geography.) There are several laboratory accreditation authorities for construction materials testing typically listed in the contract for the establishment of accreditation of construction materials testing laboratories. The laboratory's scope of accreditation typically includes the test methods required by the contract's quality requirements and by ASTM standards as appropriate to the testing field as listed below:

- Laboratories engaged in testing of construction materials are required to meet ASTM E329.
  - Testing of concrete and concrete aggregates is required to meet ASTM D3666.
  - Testing of bituminous paving materials is required to meet ASTM D366.
  - Testing of soil and rock as used in the engineering design and construction is required to meet ASTM 3740.
  - Engaging in nondestructive testing (NDT) is required to meet ASTM E543.
- Under the guidance of the National Institute of Standards and Technology (NIST), a National Voluntary Laboratory Accreditation Program (NVLAP):
  - The Construction Materials Testing laboratory accreditation program (LAP) began in 1978 as the Concrete LAP. The program gradually expanded beyond concrete testing to include other testing areas such as admixtures, aggregates, cement, geotextiles, road and paving materials, soil and rock, and steel materials.
- American Association of State Highway and Transportation Officials (AASHTO):
  - The American Association of State Highway and Transportation Officials (AASHTO) is a standards-setting body that publishes specifications, test protocols, and guidelines used in highway design and construction throughout the United States.
  - The AASHTO Accreditation Program (AAP) formally recognizes the competency of testing laboratories to perform specific tests on construction materials.
- Under the guidance of the American National Standards Institute (ANSI), the ANSI National Accreditation Board (ANAB):
  - The goal of the ANAB laboratory accreditation program is to provide third-party assurance of the quality of laboratories to promote confidence in laboratory activities through accreditation by assuring compliance with international standards and to assure the quality of construction materials.

Evaluation for Initial Laboratory Accreditation

This is the process a laboratory undergoes to first achieve formal recognition of its competence to perform specific tests or types of tests according to defined standards. The process generally follows these steps, typically based on the international standard ISO/IEC 17025: General requirements for the competence of testing and calibration laboratories:

1. *Application*: The laboratory submits a formal application to an accreditation body (AB). Examples relevant to U.S. construction include the AASHTO Accreditation Program (AAP), the U.S. Army Corps of Engineers (USACE) Materials Testing Center (for USACE projects), or other recognized national/international ABs like A2LA, NVLAP. The application details the desired scope of accreditation (the specific tests the lab wants to be accredited for, e.g., ASTM C39 for Concrete Compressive Strength, AASHTO T 180 for Soil Moisture-Density Relations).
2. *Documentation review*: The AB reviews the laboratory's quality manual, procedures, test methods, staff training records, equipment calibration records, and other relevant documentation to ensure they meet the requirements of ISO/IEC 17025 and any specific criteria set by the AB or relevant technical standards like AASHTO or ASTM procedures.
3. *Proficiency testing (PT)*: The laboratory must successfully participate in relevant proficiency testing programs. In these programs, the lab receives samples, tests them, and submits the results to the PT provider. The provider compares the lab's results against known values or the average results from a group of participating labs. Satisfactory performance is crucial evidence of technical competence for the specific tests within the desired scope. For construction materials, bodies like the Cement and Concrete Reference Laboratory (CCRL) and the AASHTO Materials Reference Laboratory (AMRL) run major PT programs.
4. *On-site assessment*: Trained technical assessors visit the laboratory. This is a critical step where the assessors:
   - Verify the implementation of the documented quality management system.
   - Interview staff to assess their competence and understanding of procedures.
   - Observe personnel performing tests included in the scope of accreditation.
   - Examine equipment, calibration records, and maintenance logs.
   - Review test records, reports, and data handling procedures.
   - Check environmental conditions (if relevant to the tests).
   - Evaluate the overall adherence to ISO/IEC 17025 and specific test method requirements.

5. *Reporting and corrective actions*: The assessment team provides a report detailing their findings, including any nonconformities (departures from requirements). The laboratory must provide evidence of effective corrective actions to address these nonconformities within a specified time frame.
6. *Accreditation decision*: An independent committee or decision-maker within the AB reviews the application, documentation review report, PT results, on-site assessment report, and the laboratory's corrective actions. Based on all the evidence, they decide whether to grant accreditation and for which specific scope of tests.
7. *Issuance of certificate and scope*: If successful, the lab receives a certificate of accreditation and a detailed scope listing the specific test methods for which it is deemed competent. This scope is usually publicly available on the AB's website.

Periodic Reviews or Surveillance Audits

Accreditation is not a one-time event. To maintain accreditation status, laboratories undergo periodic surveillance. This ensures they continue to operate in compliance with the standards and maintain their technical competence.

1. *Purpose*:

- Verify ongoing compliance with ISO/IEC 17025 and the AB's requirements.
- Ensure the laboratory's quality management system remains effective and is being maintained.
- Monitor the lab's continued satisfactory performance in proficiency testing.
- Assess any changes in personnel, equipment, facilities, or procedures since the last assessment.
- Follow up on any previously identified areas for improvement or nonconformities.
- Confirm the lab remains competent for the tests listed in its accredited scope.

2. *Frequency*: Typically occurs annually or biennially, as determined by the accreditation body's policies. A full reassessment, similar in depth to the initial assessment, usually happens every few years (e.g., four to five years).

3. *Process*: Surveillance audits are generally less intensive than the initial or reassessment audits but cover key areas. They may involve:

- Review of recent PT results
- Review of internal audit reports and management review records

- Examination of records related to complaints, corrective actions, and preventive actions
- Selective review of recent test reports, calibration records, and staff training records
- Observation of a sample of tests being performed
- Interviews with key personnel
- Review of any significant changes

4. Link to project quality requirements:
   a. *Mandatory requirement*: Many construction projects, especially public infrastructure projects (highways, bridges, government buildings), explicitly require that material testing be performed by laboratories accredited by specific bodies (e.g., AASHTO AAP, USACE validated) for the specific tests mandated in the project specifications or quality assurance plan (QAP).
   b. *Scope verification*: Project owners, engineers, or QA personnel will often verify that the laboratory being used holds current accreditation and that its approved scope includes the exact test methods required for the project (e.g., verifying the lab is accredited for AASHTO T 99 if that specific soil compaction test is required). They can check this via the AB's online directory.
   c. *Audit focus (indirect)*: While project personnel don't conduct the formal AB surveillance audits, the criticality of tests for major projects reinforces the importance of AB surveillance. If a lab consistently performs tests vital to numerous high-stakes projects, the AB recognizes the impact of that lab's continued competence. Problems identified on a project (e.g., questionable test results) could potentially lead to scrutiny or complaints filed with the AB, which might trigger closer examination during surveillance or even a special audit.
   d. *Maintaining relevance*: Surveillance ensures the lab remains competent for the tests projects rely on. If a lab lets its accreditation lapse for a test required by ongoing projects, it can cause significant project delays and contractual issues.

CONCLUSION

The initial accreditation evaluation is a rigorous process to establish competence, while periodic surveillance audits ensure that competence and compliance are maintained over time. Both are crucial in the construction industry, as project quality, safety, and compliance heavily rely on accurate and reliable test results from qualified laboratories whose competence is formally recognized and regularly verified. Project requirements often dictate which accreditations and which specific tests within the scope are necessary.

QUESTIONS

1. Describe the key differences between the process for initial laboratory accreditation and the periodic surveillance audits, emphasizing the purpose and depth of each evaluation.
2. Explain the significance of ISO/IEC 17025 in the context of laboratory accreditation, and give three examples of specific documentation requirements a laboratory must meet to adhere to this standard.
3. Discuss the role of proficiency testing (PT) in maintaining laboratory accreditation, and explain why participation in programs like those offered by CCRL and AMRL is crucial for construction materials testing laboratories.
4. How does the laboratory's scope of accreditation directly relate to project quality requirements in the construction industry, and what steps can project personnel take to verify that a laboratory's accreditation meets those requirements?
5. Beyond ensuring test result accuracy, what are three broader benefits of laboratory accreditation for the construction industry and its stakeholders?

COMPLIANCE MONITORING

> Assess compliance with applicable regulatory requirements, standards, and codes, and conformance to specifications, procedures, and drawings. Document defects and nonconformities, and collaborate with technical authorities and engineering disciplines to review, disposition, and resolve identified nonconformities. (Evaluate)
>
> Body of Knowledge V.D.2

Compliance is the act of validating that a product conforms to specified standards. In the context of construction quality, compliance is validation that the project team is following the QC plan as set forth during QA.

In QA, the project team defines the quality objectives, while in QC, the project team executes those tasks to achieve the objectives (see Chapter 2, "Project Quality Strategy Development"). Once QC is complete, the QA tools validate that the product and the processes that delivered it are also *compliant*. See Table 5.14 for more information.

Table 5.14 Compliance monitoring with QA and QC.

Quality assurance: focus is process compliance	Quality control: focus is product compliance
Compliance monitoring	Monitoring
Evaluates via audits that QC process is functioning	Leverages processes and their associated tools to check that the product is in *compliance*

Assessing Two Types of QC Compliance in Construction

Monitoring is an essential function of QC because it entails actual observations of the QC tasks. Monitoring also involves recognizing trends in the installation of the work. The formal documentation of monitoring is the reporting of the findings from the monitoring efforts (see "Reporting and Analysis" later in this chapter).

Compliance monitoring is the act of continuously assessing whether the organization is adhering to its regulatory requirements, which could include both the specific industry standards it must meet and its internal policies/processes. In the context of quality of construction, compliance monitoring is performed against (1) regulatory requirements and industry standards, and (2) codes, specifications, drawings, and processes and procedures.

Compliance monitoring ensures that the standards and quality objectives are indeed being achieved and identifies corrective actions for processes or materials that do not conform to the specified standard.

Compliance monitoring, through auditing, also identifies trends that need to be incorporated into lessons learned and process improvement efforts (see "Lessons and Improvements").

Assessing Compliance with Regulatory Requirements and Standards

All products must conform to a defined standard or specification. Standards exist to ensure quality, safety, and consistency in the industry from project to project (American Society of Civil Engineers, 2012). Examples of common standards used in the construction industry are shown in Table 5.15 (this table is not all-inclusive). Permits are also examples of standards based on local municipal requirements and code interpretations (see "Required Permits").

Prior to assessing against any standard, the standard must be known, understood, unambiguous, measurable, achievable, and constructible. The project must have a list of the *regulatory* standards that the design must have followed and that construction must achieve. The project's specifications are the baseline for identifying what standards are applicable to the project. In design, the design team must be aware of the project's specific needs and elements to ensure the standards documented in the specification are indeed applicable. During early design and constructability reviews, the specified standards must be reviewed to confirm that they are:

- Reliable
- Applicable
- Achievable
- Realistic

The construction team must also confirm they have the capability (e.g., cost, schedule, labor availability) to achieve the specified standard. An effective preconstruction meeting will cover this by confirming there is a developed work plan, an associated ITP, and that installers are trained and have the tools and resources to complete the work.

Table 5.15 Examples of common standards.

Structural	ACI 318	*Building Code Requirements for Structural Concrete* (American Concrete Institute)
	AISC 360	*Specification for Structural Steel Buildings* (American Institute of Steel Construction)
	ASCE 7	*Minimum Design Loads and Associated Criteria for Buildings and Other Structures* (American Society of Civil Engineers)
Building Codes	International Building Code (IBC)	Comprehensive building safety regulations that are adopted in many jurisdictions
Electrical Standards	NFPA 70	*National Electrical Code (NEC)*, for safe installation of electrical wiring and equipment
	IEC 60364	International standards for electrical installations
Plumbing	Uniform Plumbing Code (UPC)	Governs installation and maintenance of plumbing systems
Mechanical	ASHRAE 90.1	Energy-efficient building design
	ASHRAE 62.1	Ventilation requirements
Fire Life Safety	NFPA 101	*Life Safety Code*, for fire safety design and emergency egress
	Underwriters Laboratories (UL)	Standards for fire-resistant building materials and systems
Accessibility	ADA Standards	Ensures facilities are accessible to individuals with disabilities
	ISO 21542	International accessibility standards for the built environment
Vertical Transportation	ASME A17.1/CSA B44	*Safety Code for Elevators and Escalators*
	EN 81	European safety standards for elevators and escalators
Environmental and Energy	LEED (Leadership in Energy and Environmental Design)	Certification for environmentally responsible building practices
	ISO 14001	Environmental management systems
	IECC (International Energy Conservation Code)	Promotes energy efficiency in buildings
Materials	ASTM C150	Specifications for Portland cement
	ISO 11600	Specifications for sealants

Permits (see "Required Permits") and local AHJ (see "Regulatory Body or Authority Having Jurisdiction") also drive the applicability—as well as the *interpretation*—of standards and codes. Having a full understanding of that standard and how it will be applied to the project is essential in achieving its quality objectives and delivering the client's intent.

Not only do the specified standards need to be relevant; they need to be *current* as well. And for projects of longer duration, project teams may need to confirm that certain codes or standards will not change or have an updated interpretation during the project's timeline. Leveraging local subject matter experts (SMEs) in that geography will help the design and construction teams navigate both the applicability and interpretation of a project's standards. These interpretations need to be made less obscure and more clear, and then input into the project's acceptance criteria to ensure activities are planned to achieve them.

Once the standards are clear and defined, they need to become part of the project's QC procedures in order to assess and ensure compliance. Various tools exist to capture this information, yet their frameworks accomplish the same goal. Examples include the test and inspection matrix; requirements verification matrix; and compliance verification matrix. The CQM must ensure there is a process in place to pull this information into a clear matrix or other tool that all stakeholders and workers have access to.

For example, regarding the Americans with Disabilities Act (ADA), very specific dimensions are noted as part of the standard that must be understood by the contracted work crews. Ensuring these specific dimensions are clear, visible, and understood prior to installing the work ensures the work is installed correctly, while adding these types of dimensions to the associated QC checklist will assess compliance for the specific element of the ADA standard.

The CQM must then assess the matrix to see if all of the requirements are captured, periodically evaluating throughout the life of the project if the project team is complying with the requirements listed in the matrix. However, this can become an administrative endeavor, especially for larger, more complicated projects. In such a case, it is not realistic to assess the entire matrix at every assessment or audit. Different portions of the matrix would be assessed each time. A logical organization of the periodic assessments would meet the intent.

For regulatory requirements and standards, an assessment against regulatory requirements involves looking at agency permit requirements, procurement requirements, disadvantaged business enterprise (DBE) requirements, or similar. Because this naturally involves government regulations, the CQM is not necessarily the individual conducting the audits; rather they must ensure a qualified stakeholder has been assigned the task or that a process is in place (see Chapter 2, "Roles of Quality Personnel and Internal Stakeholders").

Assessing Compliance with Product Codes, Specifications, Drawings, and Procedures

The preconstruction and preinstallation meetings are the critical tools for assessing compliance. Similar to assessing the project's compliance with standards, the CQM must also assess compliance with specifications, drawings, and procedures.

To put it in simpler terms, the CQM asks: "Are we building work in conformance with plans and specifications?" In other words, assessing compliance with the codes, specifications, drawings, and procedures is assessing compliance of the product (see "Construction Plans and Specifications").

The CQM should ensure that ITPs are developed using the drawings, specifications, applicable codes, and procedures as the foundation (see Chapter 3,

"Criticality Levels and Inspection"). Once the ITP is formally issued, the work is to be executed and checked by QC in accordance with that ITP.

- *Codes, specifications, and drawings*: The CQM ensures that periodic assessments are conducted to verify that each work activity or critical element of the work (DFOW) has an ITP, that the ITP has been approved, and is being followed.
- *Procedures*: This involves the normal auditing or assessing of the construction team's processes and procedures for installing the work, specific to the manufacturer's requirements. The submittal is the best source of information. The ITP should include the necessary submittal information to ensure installation conforms to the stated requirements that are not part of the specification or drawings. (See "Construction Submittals" and "Quality Management Processes.")

The CQM must also ensure that associated meetings and coordinating work sessions, such as preconstruction and preinstallation meetings, contain information relative to requirements in the specification, drawings, or submittal for the trade or DFOW. During these work sessions, the CQM must evaluate the specifications and identify what critical elements need to be understood by the installers—those with the greatest ability to impact quality (see "Meetings and Participation"). These requirements then become part of work plans and QC by being part of installation checklists. For example, if the framer is unaware of the dimensions to achieve ADA compliance, they are unable to comply with those stated requirements. The only way they may become aware is by having the information presented to them in a preconstruction or preinstallation meeting where it is clearly communicated. For example, ADA requires that toilets be installed 16 to 18 inches on center from sidewalls. Knowing this dimension will enable the metal stud framing crew to install the framing members and any necessary blocking in the right location.

Assessing Nonconformity versus Nonconformance

Even with the best planning, issues will still arise that require the implementation of NCRs, CARs, and a means by which to document and report nonconformances and their resolutions (see "Quality Control and Monitoring" for the difference between NCR and CAR). The CQM must ensure through quality planning that there is a formal process or procedure in place to document the defects and nonconformances that are identified in the QC inspections, as well as the formal process for corrective and preventive actions. QC would then assess and audit the process to ensure that it is working as intended.

The first and most fundamental question for QC inspections during the construction phase is a distinction that must be formally issued at the beginning of the project as part of the quality plan. This is the distinction between a defect (also called nonconformity or sometimes called deficiency) versus a nonconformance. The following definition must be used as the distinction between the defect and nonconformance:

- *Nonconformity:* A nonconformity or defect is a temporary deficiency in the inspected product. At the time of the QC inspection, the QC inspector may issue a nonconformity if the inspection does not pass, but the construction contractor states that they can correct it without any impact on the schedule or cost. A nonconformity or defect is an element of constructed/inspected work that must be revisited by the QC inspector within the specified time when the contractor states it will be fixed.

  It is extremely important to note that defect or nonconformity is time-based and requires effective resolution within the time the schedule allows. It is in this respect that QC inspections become the centerpiece of quality for the project. Historically, all those projects that allow a nonconformity to stay on a punch list beyond the time that it starts impacting the schedule are the projects that are later forced to accept schedule delays as well as cost impacts. All of this can and should be prevented by a clearly described definition of nonconformity and nonconformance in the project quality plan.

- *Nonconformance:* If, at the time of the QC inspection, the quality inspector determines that the construction nonconformity/deficiency identified by the QC test or inspection is so significant that it cannot be corrected without impacting the schedule or cost, then it can no longer be issued as a nonconformity. The QC inspector must issue it as a formal nonconformance.

  Clarifying this difference between a nonconformity/defect and a nonconformance is perhaps the single most important decision of quality, and it should be formally agreed to by all the stakeholders of the construction project. Knowing when to allow a nonconformity/deficiency on a deficiency log to keep track of the nonconformities/deficiencies makes the difference between an excellent quality program and a substandard one.

The basic questions the CQM would seek to answer in such an assessment or audit include:

- Are defects and nonconformities being identified?
- Are they being identified and resolved within the required time frame?
- Is the project team (and other projects in the organization's portfolio) learning from those mistakes?

The purpose of QC inspection and testing of the products is to ensure that the product is in compliance with the product requirements. For a constructed infrastructure such as a building, the product requirements are defined by the specifications and some by the design plans. It is the primary responsibility of QC inspectors to ensure that the product is in compliance, and if not, to issue an NCR or a CAR to correct the situation.

Leveraging the Expertise of Technical Subject Matter Experts

It is critical that the issues encountered on a project are being reviewed by personnel technically qualified in that specific subject area. These SMEs can be part of the project team, another member of the organization, or even a hired consultant. Leveraging the tools outlined in Chapter 2, "Resource Planning and Analysis," will help confirm that technical knowledge gaps are filled with the necessary SMEs.

This means that NCRs and CARs are being assessed and approved by the appropriate person, helping the team decide if the NCR needs to be removed and replaced, reworked, repaired to acceptable condition, or used as-is. The SME can also incorporate the root causes into the corporate lessons learned framework to be accounted for on other projects (see "Lessons and Improvements"). In some instances, the review and approval of nonconforming work may require approval of the designer, in which case it should be submitted as an RFI. The designer may be the SME who makes the decision to use as-is or else recommend rework.

Best Practices in Compliance Monitoring

Successful compliance monitoring requires active participation by all project team members—not just the CQM.

1. *Foster an empowered, action-driven culture*: Cultivate a workplace culture that is driven by action rather than only by what is stated. Allowing project teams to take a leadership role in monitoring their own compliance and recognizing their efforts will promote the proper execution of QA and QC tasks.

2. *Leverage technology*: Utilize technology solutions, such as modern project management software, to streamline processes and improve efficiency. These tools can assist with tracking compliance metrics, scheduling inspections, and generating reports.

3. *Engage stakeholders*: Involve all stakeholders, including clients, and AHJ, in compliance initiatives. Collaborative efforts can lead to innovative solutions and shared accountability for compliance outcomes (see "Roles of Quality Personnel and Internal Stakeholders").

4. *Regularly review and update compliance monitoring frameworks and policies*: Continuously review and update compliance policies to reflect changes in regulations and industry best practices. Ensure that project teams are informed of any updates and have access to the necessary resources.

5. *Recognize and reward compliance*: Acknowledge project teams and team members who consistently demonstrate compliance, and reward their efforts. Recognition can serve as a powerful motivator to maintain high compliance standards.

Compliance monitoring requires a proactive approach.

1. *Establish clear expectations*: Begin by clearly defining compliance requirements in contracts and agreements. Specify the legal and regulatory obligations that project teams must adhere to.
2. *Conduct thorough prequalification*: Before hiring other project team partners, conduct a comprehensive prequalification process to assess their compliance history and capabilities. Evaluate their quality and safety records, certifications, and previous project performance to ensure they meet the project's specific standards. (See Chapter 4, "Trade Partner Requirement Alignment.")
3. *Implement ongoing monitoring*: Regularly monitor activities to ensure compliance with established guidelines. Conduct site inspections, audits, and assessments to identify potential compliance issues, and address them promptly.
4. *Provide training and education*: Equip project teams with the knowledge and skills needed to maintain compliance. Offer training sessions on QA and QC, understanding building codes and regulations, regulatory updates, and best practices to ensure they are well-prepared to meet compliance requirements.
5. *Establish communication channels*: Maintain open lines of communication with all project team members to facilitate the exchange of tacit and explicit information and feedback. Encourage project teams to report compliance concerns and provide support to address any challenges they may face.

CONCLUSION

Robust compliance monitoring is essential for project success. By establishing clear expectations, conducting thorough prequalification, implementing ongoing monitoring, providing comprehensive training, and fostering open communication channels, organizations can ensure adherence to regulatory requirements, standards, and project specifications. This proactive approach not only mitigates risks and nonconformities but also cultivates a culture of quality and accountability, ultimately leading to the delivery of superior construction projects that meet the client's intent and objectives.

QUESTIONS

1. How is "monitoring" different from "reporting"?
2. What is important about project team culture when it comes to compliance?
3. What is the difference between a defect (nonconformity) and a nonconformance?
4. What is the difference between a standard (includes detailed specifications and references the codes) and a code (enforceable minimum requirement)?
5. Are construction permits considered requirements or AHJ standards?
6. Distinguish between the roles of QA and QC in relation to monitoring and compliance monitoring, as described in Table 5.13. How do these functions differ, and how do they complement each other?

7. Explain the importance of prequalification in compliance monitoring. What specific aspects should be evaluated during this process to ensure a project partner meets compliance standards?
8. Describe the process the CCQM should follow to assess compliance with both regulatory standards (e.g., ACI 318) and project-specific codes, specifications, and drawings. What are the critical meetings and documents involved in this assessment?
9. Discuss the role of subject matter experts (SMEs) in addressing defects and nonconformances. How do SMEs contribute to root cause analysis and continuous improvement within the compliance monitoring process?
10. Beyond audits and checklists, what are three "compliance best practices" that can foster a strong compliance culture within a project team? Explain why each of these practices is important for successful compliance monitoring.

REGULATORY BODY OR AUTHORITY HAVING JURISDICTION (AHJ)

> Conduct regulatory body or AHJ reviews and audits to assess adherence to established criteria and maintain proper documentation of reviews and audits. (Analyze)
>
> **Body of Knowledge V.D.3**

This section details the contractor's and the CQM's ongoing interaction with regulatory bodies during construction. As discussed in Chapter 2, "Project Quality Strategy," the CQM's participation in conversations with AHJs during the planning stages of the project is critical to ensure requirements are accounted for in long-term planning, design, and even procurement considerations. As discussed in this section, during the quality plan deployment phase of the construction process, further collaboration is required to ensure planning is done in accordance with AHJ requirements to obtain permits.

The ability to legally perform construction tasks is typically granted to contractors by these AHJs or regulatory bodies. To procure permits for work, the contractor must demonstrate their plan for adherence to the rules, codes, and standards of those bodies; but to maintain proper control and obtain approval of a project, the contractor must continuously assure them of these adherences.

Main Authority Having Jurisdiction

The most prominent of these AHJs/regulatory bodies for the contractor to have continuous reassuring interactions with is typically the one that has given the contractor the ability to work on a project as a whole, via the construction or building permit. This government agency is referred to as the sole authority having jurisdiction (Trinidad and Strange, 2024). As discussed in previous sections, the AHJ typically dictates and sets out all of the major codes and their interpretations of them and standards by which a contractor must abide. They may also dictate and/or enforce which other agencies' or groups' rules may also apply to the project. These other entities are typically referred to as regulatory bodies, but in some cases may also be referred to as AHJs with respect to specific areas of work or in certain locales. Whether the AHJ for a project is a particular city's department of buildings (DOB) (or a group having an equivalent name, depending on locale) or another agency dictating requirements for its own project, it may require a contractor to abide by numerous regulatory bodies. These bodies may range from local utility companies that answer to that DOB or to local and federal environmental or safety organizations whose regulations the AHJ has no choice but to enforce. In other cases, these adjacent regulatory bodies may receive their mandates to control certain portions of a project from a higher authority than the main AHJ (e.g., a federal environmental agency)—a dynamic a contractor must be aware of throughout the project.

Once permits have been issued and a project has commenced, all of the requirements imparted to the contractor by the AHJs and regulatory bodies become the basis for much of the work that occurs. While in an ideal scenario, all AHJ and regulatory body requirements will have been conveyed to the project designer, there are bound to have been some errors and omissions. Because a contractor is responsible for building a code-compliant project, it is critical to maintain an open line of communication with AHJs and regulatory bodies during the submittal review process and during construction. The CQM should identify and raise potential violations and obtain variances for impossibilities in meeting requirements due to field conditions with the AHJ. Doing so well in advance of the close of the project will prevent a delay in obtaining necessary acceptances and certificates from these bodies.

Various personnel, such as the project manager and field supervision (especially those with pertinent subcontractors) on the construction team will be responsible for interaction with AHJs and regulatory bodies. The CQM's internal audit of a project must contain elements verifying these interactions and the team's preparedness for them. This internal audit should confirm both (1) that the proper discussions and transmission of documentation are occurring regularly and correctly, and (2) that they are correctly ushering in the required interactions for project closeout (as described further in Chapter 6). As discussed in "Required Permits," it is critical for the CQM to review the documents a contractor must submit in order to obtain a permit to commence work, but additional submission of documentation will be required throughout the duration of a project.

An AHJ normally will require components involving structural and/or life-safety aspects to be inspected and tested by individuals certified or trained for those specific segments of work. In some cases, the AHJ may require that these individuals be employed by a company with specific overall certifications for the work, independent of the contractor. Whether these AHJ-required third-party inspections and tests are subcontracted by the contractor or hired directly by the AHJ or owner of a project, it is critical to develop an approximate schedule for them as early as possible. This will ensure that resources can be allocated to these inspections and tests, so they do not prevent future work for which they may be hold points (or predecessor work, in general).

Approaches to planning, performing, and documenting inspections required by the project specifications are discussed later in this chapter, alongside those that might be required by the AHJ. In cases where there are overlaps in these requirements, the more stringent of the two should prevail. Especially if these inspections and tests are conducted by an independent third party, the contractor's direct involvement in conducting them may be minimal and may only require providing access to the work. The CQM's audit of this process should have elements covering ensuring the provision of work schedules to facilitate these in inspections and tests, on-site verification of the qualifications of those performing the tests, and collection of inspection and test reports from those third parties and their submission to the necessary parties. If discrepancies are found and NCRs/CARs issued by these third-party inspectors, the CQM (as the quality function authority on the project) should issue corresponding NCRs/CARs to the necessary parties and follow through on their resolution, as with any other NCR (as discussed later in this chapter). Once NCRs are resolved, the third-party inspectors may return to reinspect the work.

For keeping records of these third-party inspections, some AHJs will require continuous demonstration of successful ongoing inspection and test, while others may only require the contractor to keep records for submission near the closeout phase of the project. Regardless of contractual obligations, maintenance of open communications with the AHJ to ensure they do not see any gaps or lapses in inspections that would hinder closing the project in the future are critical. Sharing results of internal audits performed with AHJs can further build trust and communication between the CQM and the AHJ, and facilitate any final inspections the latter may perform during the closeout phase of the project.

Utility Companies

In addition to the primary AHJ for a project, agencies/companies other than those issuing the construction permit will require documentation and collaboration throughout the course of the project for various other approvals. While they may not be considered regulatory bodies, the most prevalent continuous communication with an outside, controlling entity throughout a project are utility companies. These bodies may be governmental agencies themselves (such as suppliers of clean water) or privately held or publicly traded corporations (such as suppliers of electricity, gas, and telephone/data connections) who operate based on mandates and margins set forth by governmental agencies. These government agencies (sometimes called public service or utility commission) issuing mandates would be considered the true regulatory bodies, but as the contractor rarely interacts with these commissions, we will consider the utility companies themselves the regulatory bodies in question.

It is important that the contractor and CQM understand that utility companies often operate as monopolies within their areas and as such have final say in essentially all matters surrounding their connections to construction projects. Apart from the providers of telephone/data connections in some locales, generally customers have only one option for each utility. For example, virtually nowhere in the world does a homeowner have two adjacent, parallel electrical transformers owned by different companies that could power their house—as such, the homeowner must abide by their sole option's rules. The contractor must be aware of this dynamic and be prepared to operate very strictly within the utility companies' rules and standards.

To further complicate matters, in some cases, the rules and regulations utility companies impose are more strenuous than the codes set forth by the main AHJ. To add even greater burden to the contractor, in many cases, design documents (especially on DBB projects) do not explicitly detail these utility rules; they simply indicate that the contractor must be aware of them and follow them. During the internal audit, the CQM must ensure the contractor is aware of these requirements and has a plan for integrating them into the build. If in doubt, however, the contractor should actively seek to ascertain the interpretations of the company's requirements through open discussions with the utility company. If possible, the contractor should request to hold these conversations with the utility company's personnel who will ultimately be directly responsible for activating the service; this should minimize the possibility of later reinterpretations of earlier approvals. To further minimize potential future issues, records of these conversations, meetings, and reviews should be documented and distributed to the owner, the EOR, and the utility company themselves.

While collaboration with utility companies occurs in the design phase of a project, wherein the EORs prepare utilities with approximate increases or changes to the services, they must supply the project and end product, it is not until the construction phase of the project that the utility company solidifies its commitment to providing these changes. It is typically the responsibility of the contractor to not only reconvey the design intent to the utility companies, but also provide accurate and detailed information for the connections to their services. When construction begins, those involved in these initial discussions with the utility companies during the design phase (even within the designer's organization) may no longer be involved, so reconveying this information may not be simple. For this reason, having individuals knowledgeable about the utility company's requirements is important. During the precontract phase of the project, when selecting specialty subcontractors, one of the criteria the contractor should consider is that specialty subcontractor's experience with their respective utility company (KGDR Consultants, 2024). These seasoned companies will have gone through the utility company approval process on numerous occasions and will be aware of the steps necessary to give the utility the information they require in order to provide service. In most locales, utility companies will require this information to issue their own permits for their respective offered services. These permits differ from finite general construction permits, as in many cases they will be the basis for permanent provision of that utility to the customer. As such, the criticality of the accuracy of this information is high, as in almost all cases, not having all utilities required in a contract will prevent a customer from accepting a project. Specialty subcontractors familiar with the permitting process should provide additional facilitation in their acquisition.

Having understood utility company requirements, it is important for the CQM to audit the project to ascertain the requirements are being translated into shop drawings and other plans the utility companies will eventually require for providing their services. The CQM should specifically be auditing subcontractors who will often be contractually responsible for furthering the relationship with the utility companies and obtaining actionable responses from them. The CQM should ensure whichever combinations of 100% design drawings, approved shop drawings, and equipment specifications the utility company requires are being transmitted to them in a timely manner to not delay construction and/or manufacturing of related service equipment. If the contractor maintains a strong relationship with the utility company throughout the course of the project, this rapport may be leveraged to assure the utility company of the contractor's commitment to a successful project with them, thus maximizing the probability of the timeliest review possible. Maintaining a professional and respectful demeanor regardless of the difficulties in obtaining a timely review is strongly recommended, as especially with utility companies that are the sole provider of a service, there is a general lack of appreciation in being hurried. On completion of the review process, the contractor must be prepared to quickly revise and resubmit documents addressing almost inevitable utility company concerns and comments. The CQM should ensure that all such comments are addressed in the resubmittal in order to prevent multiple revisions. The CQM's audit of the contractor must also ensure all project personnel as well as subcontractors and suppliers are not proceeding with work prior to the receipt of these approvals.

As the project is constructed, the CCQM must ensure that work at the project site as well as at manufacturer facilities pertaining to utility services is progressing per the approvals granted by the utility companies. Auditing work in the field and conducting supplier audits can help to ensure deviations from the "agreements" with the utility companies are being maintained. If during installation or fabrication it appears these agreements are not feasible due to field conditions or otherwise, it is critical for the CQM to ensure the utility company is notified immediately. Specific portions of work affected by these discrepancies should not proceed until the utility company has agreed with the necessary modifications. While this reapproval could delay the project up front, this delay will likely be shorter than a delay on the back end when the utility company performs preservice inspections (KGDR Consultants, 2024).

Accordingly, when work is nearing completion, the contractor must request inspections from these utility companies. Like during the document review process, some utility companies may require a good deal of advance notice in order to schedule necessary inspections. It is important for the CQM to ensure the project team is aware of potential delays and considers a "projection" approach to calling on the utility. While performing these inspections, utility companies will verify whether the approved design and construction documents were applied correctly, and that they believe it is safe for the connection of their particular service to be made. Ideally these inspections will be overseen by the same individual from the utility company who reviewed the work plans. If a different individual or group within the utility company overtakes this responsibility, it is possible for a new interpretation of the company's requirements to be applied to portions (or all) of the work. In such a situation, it is important for the CQM to rely on records kept during the initial documentation review period to demonstrate to the new lead from the utility company that all rules, standards, and interpretations were addressed. If discrepancies relative to the approved documents are found during these inspections, the utility company will enumerate them and provide them to the contractor. The CQM must ensure these issues are distributed to the proper field engineers and tradespeople so that they are corrected in a timely manner. Once these issues are corrected, the contractor must recall the utility company to reinspect the project. Only when all concerns of the utility company are addressed will the utility company make its final connection and establishment of its service.

Safety Regulatory Bodies

Most locales have regulatory bodies dealing with construction safety, and while these are of the utmost importance for ensuring the health and safety of tradespeople and others on-site, interaction with them rarely falls under the purview of the quality function or the CQM. Unfortunately, however, some of these rare interactions the CQM may have with the construction safety bodies can occur after construction accidents. If a construction accident is deemed to have potentially been caused by poor workmanship of permanent work during a project, construction safety bodies could call on the CQM to help examine facts surrounding the occurrence. This regulatory body will likely ask for a full gamut of documentation, from planning, submittals, shop drawings, and inspection and test records. Other sections and chapters in this handbook discuss the importance of retention of documentation for the purposes of superior product delivery to the customer, but in this instance support of worker safety is doubly critical. The CQM should keep records of these interactions but typically need not include them in the project's internal audit.

Prevention of unsafe conditions for end users of a project or the general public, on the other hand (and as previously discussed in this section), is of acute importance to the project's quality function. Aside from the overall AHJ's oversight in overall structural soundness and quality in construction, many locales also put the onus of maintaining public safety on regulatory bodies called to protect the public in case of emergency: fire departments. Fire departments, or more specifically their marshals, are typically given independent authority to inspect and approve components of work related to public safety and to enforce fire codes (QRFS, 2024). (Note that different locales may have other terms for fire marshals; typically, these are individuals separate from the chain of command of firefighters who are responsible for fire prevention, and investigation—a sort of "police" within fire departments). In addition to meeting various building codes, the CQM must ensure the project is progressing in accordance with local fire codes. With respect to construction projects, fire marshals and fire prevention codes are often concerned with some of the same features of the build as building codes (and will frequently reference them), but often focus specifically on their fire prevention aspects (Oden, 2007). While building codes (and subcodes they may reference) provide means for constructing sprinkler and fire alarm systems, doors, walls, and so on, fire prevention codes are specifically concerned with those systems' ultimate performance. In order for a contractor's quality function to satisfy a fire marshal's surveillance of a project, the contractor should be especially concerned with the functionality of these systems, thus necessitating focus on these factors. Because fire departments are often the first responders to building safety incidents, fire marshals and local fire codes may even supersede building codes. The CQM must ensure these requirements are considered by those finalizing submittals and shop drawings. Predominantly due to their responsibilities in numerous

Table 5.16 CCQM contribution for AHJ/regulatory bodies.

	CCQM Contribution for AHJ/Regulatory Bodies			
Phase	**Attend Meetings and Identify Requirements**	**Documentation Review**	**Inspection and Test**	**Internal/ Subcontractor Audit**
Planning	X	x		
Construction	X	x	X	x
Closeout		x	X	X

areas in fire prevention, fire marshals typically are not involved during the construction process until later on; as such, any deviations from fire prevention code requirements could have more dire schedule ramifications than from others. The CQM's internal audit to ensure preparedness for fire marshal inspections should be focused on these differences between codes and ensuring sharp execution of any testing of fire prevention equipment.

In general, the internal audit of the contractor, subcontractors, and suppliers for matters in relation to AHJs and other regulatory bodies (Table 5.16) is of utmost criticality as it deals beyond adherence to a contract—it is a matter, in many cases, of abiding by laws in the locale in which a project is being constructed. Performance of, and report distribution and recordkeeping requirements for, audits should follow guidelines provided throughout this handbook, specifically in "Quality Control and Monitoring" later in this chapter, to foster a means of learning lessons and continual improvement with regard to the interactions with AHJs and regulatory bodies.

CONCLUSION

Maintaining meticulous records, fostering open communication, and ensuring thorough internal audits are paramount to navigating the complex landscape of AHJ and regulatory body interactions. By proactively identifying and addressing potential violations, diligently tracking submissions and approvals, and diligently ensuring all requirements are met, the contractor can minimize delays, foster trust, and secure necessary certifications. This commitment to compliance is not merely a procedural necessity but a cornerstone of project success and a testament to the contractor's dedication to upholding legal standards and delivering quality outcomes.

QUESTIONS

1. How can a general contractor best ensure they are equipped to meet utility company requirements for specific tasks?
2. What are similar and different ways in which building and fire prevention codes are applied?
3. In addition to providing safety for the general public at the completion of a project, how can the quality function and the CCQM help to maximize safety during construction?
4. How can the CQM help to lessen the effects of the sometimes undemocratic approaches AHJs may take during interactions with the contractor?

5. Why is it critical for the contractor to form strong relationships with utility companies, even after the utility companies have approved the project's design?
6. Explain the difference between the main authority having jurisdiction (AHJ) and other regulatory bodies, giving an example of each and how a contractor's interactions with them might differ.
7. Describe the role of the CQM in ensuring compliance with both AHJ and utility company requirements throughout the construction project life cycle.
8. Discuss the potential challenges contractors face when dealing with utility companies, particularly regarding design documents and changes in utility personnel, and suggest strategies to mitigate these challenges.
9. How might discrepancies identified by independent third-party inspectors during AHJ-required inspections be addressed by the CQM, and what steps should be taken to ensure resolution?
10. Contrast the focus of building codes versus fire prevention codes, and explain why the CQM must pay particular attention to fire prevention code compliance during the later stages of a project.

Quality Monitoring and Reporting

ACCEPTANCE CRITERIA DEFINITION AND IMPLEMENTATION

> Establish criteria for accepting work at various stages, such as mock-up inspection and testing, initial inspection and testing, work-in-process inspection, punch list inspection, and final inspection, testing, and commissioning. Integrate material verification at each stage to assess if the materials meet project specifications and quality standards. Apply escalation procedures to address materials that do not meet established acceptance criteria. (Evaluate)
>
> Body of Knowledge V.E.1

Ensuring quality in construction projects hinges on clearly defined and effectively implemented acceptance criteria. This section explores the multifaceted process of establishing these criteria, from reviewing specifications and industry standards to conducting on-site validations. It emphasizes the importance of stakeholder collaboration, clear communication, and the use of technology in creating and distributing validation plans. By outlining various inspection phases and addressing nonconforming materials, this section provides a framework for achieving project quality objectives and delivering successful outcomes.

Quality Control Responsibility

Within the project's quality plan and framework (see Chapter 2, "Project Quality Strategy Development"), there must also be protocols for evaluating the acceptance criteria of a task or activity and how often it will be checked (Hart, 1994).

The project team and its stakeholders need to clearly understand the QC procedures for controlling site materials and also removing nonconforming items from the site. An essential element of the QC procedures is outlining acceptance criteria for the work—what "good" looks like. Just like the project's quality objectives and expectations, this definition of good needs to be translated from the contract documents, client conversations, and stakeholder needs down to those installing the work.

The on-site crews have the greatest impact on QC, so the quality tools they use—checklists, meeting templates, measuring devices, and many others—need to communicate the acceptance criteria in a way that is clear to them. The information contained in that tool needs to be relevant at the time it is being used and through all of the following phase inspections.

QC tools and acceptance criteria need to be organized according to risk per each critical element of the work—by system *and* assembly of materials (see Figure 5.1). Each system and assembly would follow a similar framework for inspection, of which the acceptance criteria must be determined before starting the associated work. As different parts of the project that are contracted within different scopes between disciplines (e.g., mechanical, electrical, architectural, plumbing, etc.) come together as systems, the QC effort needs to be coordinated and communicated among all stakeholders.

Acceptance Criteria Framework

Not only does the acceptance criteria need to be defined, but also the team must evaluate who has the expertise and qualifications to validate the installed work against said criteria. Leveraging RACI in Chapter 2, "Roles of Quality Personnel and Internal Stakeholders," will determine *who* will conduct the validation in QC for each stage of the installation.

- *Review specifications and standards.*
  - *What are the materials and their properties?* For example, does the material need to be installed within a certain temperature range? Is it compatible with adjacent substrates?
  - *What are the installation expectations?* Are those on the project team able to meet those requirements, and do they understand them?
  - *What elements of the project require mock-ups?* While there may be mock-ups outlined in the specifications, there may also be owner-required mock-ups that are not specifically listed but need to be accounted for.
  - *Are testing or acceptance criteria clear and attainable?* Are the criteria acceptable within the climate, and do they account for other project-specific conditions? (See "Construction Plans and Specifications" for more information.)
- *Discuss industry best practices and applicable codes.*
  - Does the material meet code? (ASTM, ISO, others)
  - What guidelines will be followed and referenced? (SMACNA, others)

- *Complete submittal reviews.*
  - Are dimensions correct?
  - What about other physical and/or chemical properties?
  - Will products perform long-term?
  - Are the aesthetic and functional requirements meeting the expectations of the client (client intent)?
- *Perform quality control to validate installation against acceptance criteria.*
  - *Who* will conduct the inspection?
  - *When* will the inspection be conducted? Visual? Invasive?
  - *What* will the inspector (person conducting the inspection) be looking for?
  - *What* is acceptable for the installation?
    - Will air handling unit performance be validated during commissioning, ensuring the discharge air temperature can reach 68 degrees Fahrenheit when the chilled water coils are at the correct design flows and temperatures?
    - Concrete testing, compressive strength tests on cylinders at 7 days, then again at 28 days?
    - Moisture content?
  - *How* will the work be validated for acceptance? (For example, checklist? functional performance test? destructive/nondestructive test? On-site/Off-site?)
  - Is the inspector qualified to conduct the inspection? If not, who will be standing next to them to coach them? Third-party or someone internal to the project team?
  - What are the sampling requirements?
  - What about nonconforming work or materials?
    - What is an acceptable range/variation that is allowable?
    - What will be rejected?
    - How will rejected work/materials be handled (retesting or rework?)

Implementation Strategy

The listed criteria need to be captured within the quality tools on the project. Checklists and other related documentation are often used to document the state of the work according to acceptance criteria. Implementing the above requires a thoughtful approach based on the specifics of the project, yet the rough framework for doing so will be similar from one project to another.

There are two basic strategies:

1. *Using multiple forms of checks or checklists, containing the most critical acceptance criteria for that particular stage.* The efficiencies here are that no one checklist is too long, allowing the checks to be clear and focused, simple enough for any qualified

- This is the first opportunity for inspecting against acceptance criteria for any material or system. The product has not left the manufacturer's facility and often can still be corrected with minimal cost or schedule impact—if caught early enough. For extremely critical elements of the project—those with the highest risk—validation should be conducted at the factory.
- This applies for materials that become part of the individual to use.

2. *Using one checklist that contains all checks for all stages, acting as one complete plan*. The efficiency is that there is one document, but it can be long and cumbersome.

For any construction organization, the checks that validate the installed work need to be documented for insurance purposes and also as a means to control the work. Without documentation, one could argue the check never occurred. To control quality according to the acceptance criteria that results from a review of the project's requirements, the project team must decide what information will be put into each checklist at each stage. Atul Gawandi's *The Checklist Manifesto: How to Get Things Right* is a great reference for creating effective checklists.

Finally, regular review of the acceptance criteria and the associated implementation plan documents must occur to ensure all changes are incorporated and available for installing crews. (This is also where technology can be an essential tool for plan creation and distribution.) Conduct a regular meeting (or have a standing agenda item in an existing meeting with the appropriate stakeholders—see "Meetings and Participation") where the acceptance criteria framework is reviewed to ensure no new information needs to be incorporated. When installing crews or inspectors are unaware that they are using obsolete documentation for QC, inspections are not effective and quality objectives are not achieved. Recall that the team can only create the best plan with the information available to them at that time (see Chapter 2, "Project Quality Plan Development"). Also, portions of the design may be found incompatible, requiring changes to the acceptance criteria (if the acceptance criteria did not drive the change itself). Installing crews, like the owner and other external stakeholders, will learn more as the project progresses, so checklists and other tools in the implementation plan will need to be updated accordingly.

Quality Control Implementation Phases

Submittal review is often the last task in QA and the first step in QC. The following phases are key to a solid implementation plan for validating acceptance criteria, regardless of the project specifics. Each phase will require its own checklist containing checks applicable to that point in time, which is still proactive to resolve issues that are too late or too impactful to the project for validating at the next phase.

- *Factory inspections/witness tests: Does the manufacturer understand the acceptance criteria?*
  - permanent installation or as part of a mock-up.

- *Delivery inspections: Is it the right material/product?*
  - This is the second opportunity to validate material against static acceptance criteria.
  - This also applies for materials that become part of the permanent installation or as part of a mock-up.
- *Mock-up inspection/testing: Does the material/system perform per acceptance criteria before it is permanently installed?*
  - This is the first opportunity to see the products perform as a system.
- *Benchmarking: Does the first instance of the permanent installation meet the acceptance criteria?* (*Note*: The mock-up can also serve as a benchmark inspection, particularly when the mock-up itself becomes part of the permanent installation.)
  - While the term "benchmark" is used here to describe the first instance of permanent installation, other familiar terms include initial inspection, first of a/its kind, and first work.
  - For each high-risk element of the work, it is essential to identify the benchmark inspection during planning.
- *Work in progress/work in place (WIP) inspection: Is the ongoing permanent installation of the work being done per acceptance criteria?*
  - At this point, it is essential also to validate environmental conditions for the installation, ensuring they are per stated requirements. (Adhesives, for example, can fail if environmental conditions are not right during the installation.)
  - It is too late at this point to check if the material is correct or not. That should have been done at the factory or during delivery inspection.
  - One inspection at this phase is not sufficient. Validation here is an ongoing effort, as the crews check the work daily. Installing crews need to understand what will be inspected, not just the supervising or dedicated quality personnel/inspector that will perform the check. This reduces found issues.
  - Installing crews should have the inspection checklist in hand, going over them as the work is progressing, and using that checklist to plan their work in daily huddles.
- *Pre-punch/completion list/punch list inspection: Will the final installation meet acceptance criteria?*
  - Often for static work, the formal/final punch list is too late for correction, at least in mitigating adjacent finished work. A "pre-punch" or equivalent validation can be a proactive means of final validation at a stage when the impact of rework is zero for adjacent finished work.

- *Final inspection: Does the final static assembly/system meet acceptance criteria?*
  - During the early planning stages and when outlining acceptance criteria, the team must determine what "complete" is for that element of work. This is the final stage of validating that the work aligns with the definition of "good" or "acceptable" outlined and communicated prior to starting the work.
- *Pretesting: Will the final system function per acceptance criteria?*
  - For functioning non-static systems—such as mechanical and electrical—the pretest is the equivalent of the pre-punch for static work.
- *Commissioning/other third-party testing: Does the final system function per acceptance criteria?*
  - For functioning non-static systems, commissioning is the equivalent of the punch for static work.
  - Again, what "good" means in terms of functionality must be identified prior to starting the work.
  - This also includes municipal testing against regional code requirements.

Use of Technology and Plan Creation

The implementation plan, including all inspection and acceptance criteria documents, must be distributed to all project stakeholders in an accessible format before work begins. Miscommunication and misunderstanding often hinder effective QC. Issues arise when installation crews are unaware of the acceptance criteria or quality expectations that management has developed, leaving those with the most direct impact on quality uninformed. To prevent this, the plan should be reviewed during preconstruction and preinstallation meetings (see "Meetings and Participation").

Technology can aid significantly in the creation and distribution of the validation plans. When conducting an inspection or test, there's a lot of information readily accessible. Multiple drawings are often required, along with shop drawings and meeting minutes, in addition to the applicable checklist itself. Administrative waste can be eliminated when these plans are digitally created and distributed.

Often, the plans require layering multiple drawings on top of one another to verify coordination between parts of a system. When a team or inspector has to access multiple sources in different locations to conduct an inspection, gaps are not clearly identified and the inspection itself can take longer than required.

QA—the planning meetings and tasks that happen *before* the work starts—is where these plans are made. This flow above is the tactical quality plan for the project, the QC tasks that begin as the work starts and thereafter. These plans are not created in isolation by the CQM alone, but rather are formed during meetings with all available stakeholders. This is how collaboration occurs: everyone understands the expectations, which are communicated to those who need to be aware of them and incorporate them into their work.

Receipt of Nonconforming Materials

As part of outlining a process and acceptance criteria for materials, a procedure must also be established for handling materials that arrive on the jobsite and do not conform to specifications. Within the specifics of the project, the following general protocol should be followed:

- Using the outlined acceptance criteria, note the reason for nonconformance.
- The purchaser of the material (e.g., subcontracting trade partner) must contact the supplier.
- Leveraging the site's logistics plan, immediately remove the material from the site.
- Determine how the issues will be corrected while mitigating cost or schedule implications.

Inspection Frequency

Another critical question is deciding on inspection frequency. Formal inspections should be documented as part of the installation process, using the previously discussed framework. However, this does not mean that formal inspections and validations are the only ones being conducted. As the work is installed, *continuous inspections* should be utilized—the idea that work can constantly be monitored during installation, using the documented checks as guides for installing crews. Inspections that occur only *after* the installed work is completed are often too late. In these situations, even though the incorrect installation is resolved, rework still occurs and schedules are impacted. The core idea is to communicate the acceptance criteria prior to the work starting—during QA—to prevent work from being installed incorrectly.

A great example is a project with many identical "units," such as multifamily. A key strategy to adopt the appropriate frequency is outlining if each unit will require its own separate inspection document or if the same inspection document can be used while noting nonconforming work specific to each unit. In the former instance, the team will manage open inspections, while in the latter instance, teams will manage open deficiencies.

The frequency of formal inspections and associated documenting of the effort should be agreed upon by all stakeholders. Systems of higher complexity and importance, in alignment with the project's quality objectives, will require more formal checks than those of lesser complexity.

Finding the Balance

The project team can't realistically inspect and validate 100% of the project. And each checklist can't contain all the technical information to validate the specifications. If they did, they would be impossibly long and ineffective for validating the work. Implementation of a robust inspection program to validate work against acceptance criteria requires checks for communication as well, and balancing the essential required checks with those that may not have equal benefit for spending energy on. Finding that balance requires open and transparent communication across all project stakeholders (Gawande, 2009).

CONCLUSION

Effective acceptance criteria implementation is crucial for project success. This section outlined a comprehensive framework, from defining criteria through stakeholder collaboration to utilizing technology for plan distribution. By establishing clear validation processes, addressing nonconforming materials promptly, and emphasizing continuous inspection, teams can ensure quality objectives are met. Regular reviews, incorporating learnings, and adapting to project changes are vital for maintaining the plan's relevance. Through diligent execution of these strategies, projects can achieve desired outcomes, mitigate risks, and deliver quality work.

QUESTIONS

1. What are the important elements of acceptance criteria?
2. When must acceptance criteria be identified and discussed?
3. What are the important elements of each QC activity?
4. How can technology help with the implementation of acceptance criteria?
5. Explain the two basic strategies for using checks or checklists within a construction organization to validate the installed work, and discuss the advantages and potential drawbacks of each.
6. Describe the "continuous inspections" concept as presented in the chapter, and discuss why it is considered more effective than relying solely on formal, post-installation inspections. Give an example of a scenario where continuous inspection would be particularly beneficial.
7. According to the chapter, how should nonconforming materials be handled on receipt at the job site? Outline the general protocol and explain why each step is important for maintaining project quality and schedule.
8. Detail the key considerations for determining who will conduct the validation in QC at each stage of installation, referencing the RACI matrix mentioned in the chapter. Explain how leveraging this matrix ensures appropriate expertise and qualifications are utilized.
9. Discuss the role of technology in creating, distributing, and utilizing validation plans, as described in the chapter. Explain how technology can address specific challenges related to inspection processes and information accessibility.

REPORTING AND ANALYSIS

Design periodic reports and conduct statistical analyses to identify quality trends and assess waste elimination. (Evaluate)

Body of Knowledge V.E.2

Reporting and analysis are critical to ensuring the project is proceeding according to the planned quality objectives, that team members are focusing on the right activities, and that waste is not burdening the project and reducing productivity (Rumane, 2019).

Reporting is a function of *monitoring*. Monitoring collects and records the project information, while reporting shares it. This can be done at an organizational level across multiple projects and at the project level. This section will focus on reporting and analysis at the project level.

Framework for Designing a Project-Level Report

1. Identify metrics that will measure performance against the project's quality objectives.
   a. Leverage the project's acceptance criteria (see "Acceptance Criteria and Implementation") to identify what critical metrics to measure regarding materials and installed work.
   b. Utilize the responsibilities matrix (see Chapter 2, "Roles of Quality Personnel and Internal Stakeholders") to determine who will gather this data and compile it for the report.
2. Evaluate the metrics against the project's highest risks (see "Risk-Based Inspection and Testing Strategy") to determine leading indicators required to prevent risks from occurring.
   a. Evaluate how the data will be captured.
   b. Determine if those metrics are *qualitative* or *quantitative*, and select the appropriate statistical tool for analysis.
   c. Identify what statistical tools will be best for analysis and representation.
3. Design a report that enables decision-makers to take action.
   a. Outline specific actions to take, who needs to complete them, and by when.
4. Discuss and distribute the report, and act on its findings.
   a. What waste can be eliminated?
   b. What is the root cause of recurring quality issues?
   c. How is the team being proactive in preventing incidents?
   d. Are the quality objectives being achieved?
5. Monitor the trends to see if the actions were impactful.
   a. Is the report useful and valuable?
   b. Are the actions impactful?

Selecting Data to Report

With modern data capabilities, there is infinite potential for reporting and analysis (Oakland and Marosszeky, 2019). Many project teams move straight to reporting on defects or nonconformances. While this is important information, it is *reactive* (Marsden, 2019).

During the early quality assurance (QA) and planning meetings, the team must agree up front what to report on within the context of the project and its quality objectives (Rumane, 2019). Discussing with all external and internal stakeholders what information is important to them is the first action in designing an effective performance progress report. If a quality objective is to achieve functional temperature control, for example, the team would capture the following information:

- *Process*-related information on the project relating to the QA and QC activities associated with the temperature control systems—that the actions are happening. (What steps in the process are required to achieve the quality objective?)
  - Design reviews
  - Submittal reviews
  - Coordination meetings
  - Startup planning
  - Work-in-place inspections
  - Commissioning
  - Training
  - Turnover
- *Results*-oriented information showcasing issues that need to be resolved to ensure achievement of functional temperature controls. (What issues need to be resolved to achieve the quality objective?)
  - Work-in-place issues that will lead to commissioning failures
  - Procurement issues
  - Commissioning issues

This process outlines behaviors that align with the project's quality objectives, which can be translated into reportable metrics (Table 5.17). Identify those behaviors and actions, and determine what metrics would illustrate their forward progress. Leading and lagging indicators can also be assigned to each, allowing the team to be proactive in mitigating potential problems.

The challenge is utilizing metrics to illustrate the achievement of quality objectives, specific to the construction project, where "widgets" are not available to provide a large sample size, like what can be done in manufacturing (Chung, 1999). It is essential that the team understands how the data is being compiled and that all stakeholders agree with the means and methods of the reporting effort.

Once the metrics are identified, the project team must identify the appropriate sources for the data. The report will only be actionable with accurate and reliable data.

Table 5.17 Example metrics.

Potential Metric
Submittal response times
Number of punch list items
Defect density
Warranty callbacks
Schedule delays
Time to submit as-builts
Hours to process pay applications
Estimate accuracy
Equipment failures
Rework costs*
Material waste
Downtime
Client feedback
Warranty claims/costs
Labor efficiency
Training effectiveness
Accident/incident rates
On-time deliveries
Planned percent complete (PPC)

**Requires that the information is input correctly and accurately. Rework is a lagging indicator and thus purely reactive. The costs are not known until the rework is completed. Many of these costs are rough orders of magnitude (ROM) and therefore are not completely accurate.*

- How will the team get the data?
- Where is the data being stored?
- Is the data reliable?
- Is it entered correctly?

Another concern with reporting is that it is an administrative burden (Kubal, 1994).

Designing the Report

Discussing with all external and internal stakeholders what information is important to them is the first action in designing an effective performance progress report. Who wants to see the report? *How* do they want to see it? How often?

The team needs to design a report that adds value to the project and does not contribute to the process or administrative waste. The report should be short and focused. Reports containing trends that are not relevant to stakeholders or tactical are also not valuable—neither is the effort in creating them. Reports need to drive action and behavior that advance the project's specific quality objectives. A visual that illustrates how the team is performing against the process can help managers identify what areas of the process need attention—or who is responsible for getting that task done.

The report will contain other data related to other facets of the project, such as safety and budget. Quality metrics, which are the focus of this section, should be integrated with the report to prevent the quality activities from becoming isolated. The report should also outline how all metrics tie into one another. Quality should be integrated with the project's other tasks and not seen as something separate.

Aesthetics of the report are also important. How the information is displayed will correlate to the ability for stakeholders to understand the results and thus their ability to make decisions and take action. Based on whether the data and associated metrics are *quantitative* or *qualitative* will determine what type of quality and statistical tool (Table 5.18) will be used to conduct analysis and report. Use of the appropriate chart for the situation can be effective at identifying and communicating the root cause.

Taking Action for Process Improvement and Waste Elimination

Waste includes physical waste from the actual construction work as well as administrative waste from any task on the project. The ultimate goal of the quality process is to not burden the team with administrative overhead and rework due to incorrect installations. For example, the inspection process itself can become wasteful as well if the root causes are not identified and solved, thus eliminating the need for the inspection process. A major concern for the construction industry is that, instead of focusing on processes and causes of problems, it has focused on monitoring and inspection of results.

- What about the process itself can be improved?
- What steps are redundant and what information is not helpful?
- Where are mistakes being made? (Address these immediately.)
- Where are team members/crews spending too much time?
- What obstacles are preventing the achievement of the project's quality objectives?

This requires being proactive and putting forth the due effort in planning. Variability versus the average in the statistical tools is a sign of waste. A combination of tools may be required to identify the root cause of waste and quality issues.

Information Distribution

Designing how the information will be shared is just as important as designing the means for how that information will be analyzed and where the data is coming from. The project team must decide when and how the report will be discussed and how it will be distributed. When outlining the meeting flow and agenda items for the project (see "Meetings and Participation"), evaluate where the report will be reviewed based on the participants. The report must also be available and showcased electronically and in the field on a quality or safety board. Recall that those installing the

Table 5.18 Statistical tools, use cases, and examples.

Statistical Tool	Use Case	Construction Example
Flow chart	Best for analyzing a process and identifying waste	To make the site logistics plan more efficient, the project team can map out the steps for installing a product or material, from delivery to final placement/ installation. Flow charts are effective tools for identifying waste in a process (e.g., how material is delivered to the site and moved to the location for installation). If multiple "stops" are observed, this could be identified as waste and a better path or updated site logistics plan developed to resolve the problem, saving everyone time.
Pareto chart	Best for *qualitative* information but *qualitative* categories, identifying the most critical elements that need focus	To understand what types of nonconformances are impacting the project the most, the team can quantify the types by category and focus their attention accordingly.
Scatter diagram	Best for quantitative information, articulating the relationship between two factors	The project team can identify the relationship between the number of RFIs and change orders to determine if RFIs are a root cause of increasing cost or contingency spend.
Histogram	Best for quantitative information, showcasing trends and frequency of certain incidents	The team can plan for future projects by understanding the range of costs for each scope of work and improve their estimates.
Cause-and-effect diagram (Fishbone/ Ishikawa diagram)	Identifying potential factors contributing to quality issues or process ineffectiveness, and root cause analysis	Teams can identify the root cause in a material defect by outlining all the steps in the manufacturing and delivery process.
Spider chart (radar chart)	Best for quantitative information with multiple variables being assessed on the same scale	The project team can identify what steps in the quality process are being followed for compliance, using each step of the process as a category on the chart.

work have the greatest impact on quality, so they need to have access to the report. If the data is buried in meeting minutes, it won't make its way out to the field (Oakland and Marosszeky, 2017). Finding the balance is critical (Marsden, 2019).

Reporting and Culture

There is an important cultural element to reporting. Often, quality metrics are negative trends, such as the number of defects or nonconformances. Even if this information is used for positive intent, many project teams are hesitant to report their mistakes. While the number of defects and nonconformances is indeed helpful for identifying actions that prevent future mistakes, not just on that project but for other future projects the involved stakeholder organizations will work on, it may not be favorable to showcase mistakes on the project.

In addition, ideas for improvement and elimination of waste should not come from data, reports, and associated analysis alone. Those ideas come again from those doing the work. The culture on the project site needs to be open to sharing team members' ideas so they will be heard, versus being guided by reports alone. Generating dashboard reports can initiate a feeling of negative oversight among work crews. They should have direct input on the report to help support a positive culture of improvement that creates a space for the report itself to be effective—not just among the managing team.

CONCLUSION

Effective reporting and analysis are fundamental to ensuring project success and achieving quality objectives. By establishing clear metrics, utilizing appropriate statistical tools, and fostering a culture of open communication and proactive problem-solving, project teams can identify waste, mitigate risks, and drive continuous improvement. Shifting from reactive defect reporting to proactive process analysis empowers teams to address root causes, enhance efficiency, and ultimately deliver high-quality results. Implementing these strategies ensures that reporting becomes a valuable asset, driving actionable insights rather than an administrative burden.

QUESTIONS

1. What is the difference between reporting and monitoring?
2. Explain the difference between a leading indicator and a lagging indicator in the context of quality reporting, and give a construction-specific example for each.
3. Discuss the potential challenge of reporting becoming an "administrative burden." What strategies can project teams implement to ensure reports add value and drive action rather than just create more paperwork?
4. Beyond merely tracking defects, what kind of data should be collected and reported on to proactively achieve quality objectives in a construction project, and why?
5. How can statistical tools like Pareto charts and scatter diagrams be used in construction quality reporting, and what specific types of information are they best suited to analyze?
6. Explain the "cultural element" of reporting discussed in the chapter. How can project leadership foster a positive reporting culture that encourages open communication and continuous improvement, rather than creating fear or hesitation to report issues?

MEETINGS AND PARTICIPATION

> Organize pre-inspection meetings that include quality meeting sample agenda items that address key topics, such as upcoming inspections, current quality metrics, recent findings, and open issues. (Evaluate)
>
> Body of Knowledge V.E.3

Meetings are an important collaboration tool. When organized and leveraged effectively, meetings become the primary means for information sharing across the entire project team. Project teams often hold required meetings based on specification requirements and owner-requested reporting without evaluating the sequence of meetings and what information should be communicated to whom and when. Meetings then become a frustration for all.

Outlining a Meeting Strategy

The champion facilitating project quality must evaluate what meetings are required and what the flow of information needs to be from stakeholders to those managing and installing the work. Being able to guide the team in determining, based on the collection of project documents, quality objectives, the current status of the schedule and work in the field, what information needs to be translated to the workers in the field is an essential element of a project's success (Elevate Construction, 2024).

There are two means of managing meetings and incorporating quality. First, there can be dedicated meetings for quality-related topics. Second, the quality topics can be integrated within other meetings in the flow of information. A blend is required, as some projects will have different needs than others. Given the significant number of meetings and amount of coordination on any project, establishing an effective balance can save a lot of time and administrative overhead. In addition, selecting who will have clear authority and responsibility for preparing and leading each meeting must be outlined in the project roles and responsibilities matrix (see Chapter 2, "Resource Assessment and Mobilization").

Quality, like schedule and safety, should be an element of every discussion on the project. Some quality topics, like preparation for functional testing for the building automation system or justifying the ITP for the building envelope system, will require deeper discussions outside a general meeting's capacity.

There are two layers to effective meeting management and participation, outlined in Table 5.19.

The transition point between QA and QC in Table 5.19 is the start of the work. The project schedule and upcoming inspections, quality metrics, recent findings

Table 5.19 The two layers of meetings.

Planning (Quality Assurance)	**Execution (Quality Control)**
Example: Preconstruction Meetings	***Example: Preinstallation Meetings***
• Confirming *what* work needs to be inspected, what other trades are involved, and the general quality expectations/requirements for the project. • Understanding the specifications and what the contractual quality requirements are. • Subcontracting trade partner onboarding and orientation.	• Confirming *how* the work will be inspected and by what means the work will be validated/controlled. • Understanding the detailed shop drawing/submittal requirements. • Coordination between subcontracting trade partners.
Per Scope of Work (single trade)	Per Coordinated Work Element (single task)

Table 5.20 Key quality meetings.

Meeting Name	Preconstruction Meetings	Preinstallation Meetings
Interval	Once per subcontracting trade partner prior to mobilization (by contracted scope of work)	As required per high-risk element of the work—DFOWs, for example (by DFOW or element of the work)
Participants	A single trade partner	Multiple trade partners
Agenda	Identification of high-risk elements of the project (e.g., DFOW)	Detailed review of the drawings (including RFIs), specifications, approved shop drawings/submittals, and lessons learned—specific to that element of work
	Review of project-level quality objectives	Review of the quality expectations for that element of the project, tangibly identifying what "success" looks like
	How inspections will be conducted and what tools will be used	What inspections will be conducted, by whom, and at what point in the installation (benchmark, work in place)
	Mock-up requirements for the scope of work	Mock-up requirements for the element of work
		Confirm startup, testing, special/third-party testing or inspections, and commissioning requirements
		Outline how the work will be closed out (punch list, completed testing, other)

and open issues are topics at both phases, but with a differing level of detail. In QA, the schedule and upcoming inspections are discussed but at the trade level over the whole of the project, while in QC, the upcoming inspections are discussed in much deeper detail, such as the next few weeks. The process-related higher-level inspection, quality strategies, and how information will flow to each member of the project team are confirmed prior to the *scope* of work starting. The execution of that flow of information occurs during meetings in the QC phase, prior to that element of work starting.

Key Quality Meetings

There are two groups of meetings essential to quality: the preconstruction and preinstallation meetings. Table 5.20 outlines a high-level agenda for these meetings, focusing on the quality-related topics. (Schedule, safety, budgeting, and other topics could be covered in these meetings as well, but are not shown here.)

In preparation for each of these meetings, the CQM must review the contract documents and specifications—as well as understand the project's quality objectives and the client's intent (cross-reference to planning chapters)—and insert those elements into each of the agendas, both at the project-level and specific to the element of work being constructed by multiple trades.

Preconstruction Meetings

The preconstruction meeting is intended to onboard the subcontracting trade partner, orienting them to the project. In addition to the items in Table 5.19, the quality focus areas should be project-level expectations and outlining how planning occurs and how to prepare for those planning milestones. The preconstruction meeting must occur prior to the subcontracting trade partner mobilizing on-site for each subcontracting trade partner. Example meetings include:

- Electrical
- Mechanical
- Earthwork
- Glazing
- Elevators
- Roofing
- Concrete

Preinstallation Meetings

The preinstallation meeting is for more detailed discussions for each critical element of the work with multiple subcontracting trade partners requiring coordination—all those involved in that particular portion of the work (see Table 5.19). These meetings generally occur after submittals are approved and before the element of the work being discussed begins in the field. Here, the specifications are reviewed in great detail, along with specific details in the drawings of the critical assemblies and other physical elements of the work. Also, shop drawings and submittals are reviewed to confirm the installation expectations. For example, will the installation follow the design documents or the approved shop drawings where the documents are diagrammatic in nature? Also, lessons learned from all stakeholders should be shared to prevent repeating similar mistakes.

Example meetings include:

- Building automation system
- Building envelope
- Underground utilities
- Footings and foundations
- Fire/life safety systems
- Elevators
- Dwelling units

Note that "elevators" fall under both preconstruction and preinstallation meetings. The difference between them is that the elevator preconstruction meeting is held with only the elevator subcontracting trade partner and other associated stakeholders, orienting them to the specifics of the project. No other subcontracting trade partners will attend. For the preinstallation meeting, all subcontracting trade partners are present to coordinate all work elements associated with the elevators.

Depending on the element of work or scope being discussed, some meetings may take longer than others, or require multiple discussions to ensure the work is thoroughly planned. In *How Big Things Get Done*, the authors argue this second phase of planning meetings often gets glossed over or skipped because teams feel short on time, only to discover coordination issues between trades as the work is going on—much too late to be proactive.[2] It is difficult to fully plan the trade coordination, inspection, and testing efforts, for example, for the building automation system or building envelope in 60 minutes or less.

Meeting Participants

The meeting participants for both discussions will be roughly the same, but there will be multiple subcontracting trade partners in the preinstallation meeting versus only one in the preconstruction meeting. The design team stakeholders, owner, municipal stakeholders, special inspectors, and testing agencies are among the participants that should be at any meeting where quality tasks are discussed.

Meeting Coordination and Naming

Preconstruction and preinstallation meetings work together to form a cohesive flow of communication relative to quality and other aspects of the project. Though other meeting names exist in the industry, the overall intent of what needs to be accomplished is relatively the same from project to project, in terms of a framework for quality. In the industry, there is no meeting title standard. The meeting titles "Preconstruction" and "Preinstallation" will vary within the industry from organization to organization. (For example, "Preparatory" is also used, more commonly on public projects, yet the intent of the meeting is the same as what this BoK calls "Preinstallation.") What is noted here as a "Preinstallation" meeting may be known as a "Preconstruction" meeting elsewhere.

Regardless of the title, the information being discussed at that point needs to be clear to all project stakeholders. For example, Atkins and Simpson note the preinstallation meeting (or "conference") is for coordination between subcontractors around critical elements of the work (DFOWs), or high-risk elements of the project (2008). They later state that the preinstallation meeting is for the subcontractors to confirm their plan for a specific system or product. In *Guidelines for a Successful Construction Project*, the authors discuss the "pre-project Inspection Conference," intended for confirming the specifics surrounding inspections. They later note that "on agreement of the specifics of the inspections, another conference is held with the subcontractors, which is the preinstallation meeting previously noted herein. Depending on the complexity of the project, inspections may be discussed as part of the overall preconstruction conference for the project in lieu of the second inspections conference."

The point is that inspections (and other quality expectations) need to be discussed with each subcontracting trade partner prior to the start of their scope of work—where planning and high-level expectations are confirmed—and then in further detail after approved submittals and shop drawings prior to the start of that element of work, being installed by multiple trades.

Running an Effective Meeting

At the beginning of the project, the CQM must establish a meeting culture, being open and transparent regarding what a "good" or "effective" meeting is and communicating that to all stakeholders. Additionally, the CQM should identify what information needs to be transferred and what actions need to be accomplished for each meeting.

Other tips for effective preconstruction and preinstallation meetings include:

1. *Preparation*
   a. Review specifications to determine the title of the meeting and what needs to be accomplished.
   b. Use software to set up project-specific meeting templates for easy distribution.
   c. Send an agenda ahead of time, at least one week prior to the meeting occurring.
   d. Meeting attendees should not be looking at the specifications for which they are responsible for the first time in a preconstruction or preinstallation meeting.
   e. The CQM must work with scope managers to ensure the meeting attendees are prepared ahead of time so the meeting is more efficient. The meeting will be much less effective if team members aren't aware of their scopes of work.
2. *Attendance*
   a. Have the right stakeholders present. If the right team members are not there, do not hold the meeting.
   b. Understand what each participant hopes to get out of the meeting. (For example, the architect may want assurance that the contracting team will install the work correctly and that they understand the nuances of that element of the project.)
3. *Follow-up*
   a. Send minutes within 24 hours.

Meeting Timing

At the beginning of the project, the CQM will need to review the project schedule and determine what quality activities need to be planned earlier than a scheduled preinstallation meeting. For example, concrete-related or earthwork activities may need to have both preconstruction and preinstallation meetings prior to mobilization to ensure information is transferred and expectations are understood before the work starts.

CONCLUSION

Effective meetings are crucial for project success, serving as the primary conduit for information sharing and quality management. By strategically planning and executing preconstruction and preinstallation meetings, teams can ensure clarity on expectations, address potential issues proactively, and foster seamless coordination. Establishing a culture of preparedness, attendance, and timely follow-up ensures that these gatherings translate into tangible results, ultimately contributing to project efficiency, quality assurance, and stakeholder satisfaction. Investing in well-structured meetings is an investment in project excellence.

QUESTIONS

1. What is the difference between a preconstruction meeting and a preinstallation meeting?
2. Describe the key differences in the level of detail and focus between discussions of "schedule and upcoming inspections" during the quality assurance (preconstruction) phase versus the quality control (preinstallation) phase of meetings, according to Table 5.19.
3. Explain why "elevators" is listed as an example for both preconstruction and preinstallation meetings, and what distinguishes the purpose and participants of each of these meetings.
4. Discuss the potential pitfalls of skipping or glossing over the second phase of planning meetings (preinstallation), and how these oversights can lead to coordination issues later in the project, as referenced by Flyvbjerg and Gardner in *How Big Things Get Done.*
5. Beyond the quality-related agenda items, what other critical project aspects are mentioned as potentially being covered in preconstruction and preinstallation meetings, and why is it important to consider these additional elements?
6. Considering the variation in meeting terminology within the industry (e.g., "preparatory" vs. "preinstallation"), how does the chapter emphasize ensuring clear communication and understanding of meeting objectives regardless of the title used?

LESSONS AND IMPROVEMENTS

> Incorporate lessons learned from previous projects into the quality monitoring and reporting process. Document and communicate these lessons learned for future projects. Develop strategies to assess and manage quality risks. Engage in root cause analyses and investigations to understand underlying quality issues. Develop quality improvement programs based on identified needs and insights. (Evaluate)
>
> **Body of Knowledge V.E.4**

Effective knowledge management is essential to the transfer of lessons learned. When discussing "lessons learned," we mean any incident—safety, quality, cost, or other—from a previous project that has an identified root cause that can be applied to a future project to prevent the incident from happening again. As many organizations attempt to be proactive by digitally storing and capturing lessons learned as part of the risk prevention plan for future projects, a data management problem emerges on top of the industry's growing knowledge management problem: There is so much data, the difficulty becomes knowing what to do with it (Howarth and Greenwood, 2018).

The primary issue is that a project team may not know what lesson learned they need to be accounting for at any given time. A lesson learned needs to be delivered to the right project team member, at the right time, and in the right form. A database may not address this directly, because in reviewing a database, true "knowledge transfer" that sustains itself over the long term may not occur. Additionally, it is a huge administrative waste for a team member to search for information they may *think* they need. "The database doesn't *teach* anyone anything. Yes, it *tells* us something, but nothing is *learned. Learning* is evident when people's behavior changes from the last project to the next one. That change in behavior is what actually prevents the problem from happening again" (Dupslaff, 2024).

Note that the BoK for the CCQM examination describes the best practices learned during the design phase in Chapter 3, "Design Lessons Learned." See "Training Needs in Quality" in the same chapter to get a better understanding of knowledge management and lessons learned for the entire project life cycle.

Types of Knowledge

Knowledge is effectively transferred when the content of that knowledge can be categorized as either *tacit* or *explicit*. Tacit knowledge is the knowledge in our own minds that we are unable to express to others (Anheim, 2003). Explicit knowledge is what others can understand that is documented in some fashion. The difficulty of these two knowledge types is that no one can transfer their own tacit knowledge unless the other person is invested in learning the same knowledge.

Transfer of Knowledge

The measure for success when incorporating lessons learned is a change in behavior from the previous project to the future project. A change in behavior is an indication that *learning* has occurred. The question the CQM needs to answer, then, is: "How do we change people's behavior in a way that mitigates future project risks?" Project teams need to learn from the previous project (Anheim, 2003).

For an organization to be proactive, management must create environments where project team stakeholders:

- *Transfer* tacit knowledge into explicit knowledge by incorporating the content into project workflows.

- *Exchange* tacit knowledge between team members by establishing an effective project team organizational chart to ensure that proper mentoring and organic teaching occurs between team members.

At the project level, this means establishing a project team with varied experiences and backgrounds. People and organizations learn from every project (Anheim, 2003).

Documenting and Communicating Lessons Learned for Future Projects

The documentation of lessons learned should be formally done continuously throughout the entire project life cycle. A postmortem or "debrief" is often held at a project's conclusion to document the most impactful lessons learned—which is indeed beneficial. However, certain lessons learned may be forgotten as team members move on to other projects. Incorporating lessons learned as an agenda item on a key project meeting will help facilitate regular conversation and drive a culture of improvement (see "Meetings and Participation" for further discussion).

While the lessons learned effort is often associated with negative project occurrences, this principle applies to positive aspects of a project as well. Understanding "what went well" should also be incorporated into future projects. Tasks such as pull planning or a specific method for sequencing that helped achieve a project's quality goals or improve the project execution must also be documented.

What constitutes a lesson learned is anything a future project could benefit from. This generates a lot of information—as what could benefit a future project is often subjective—thus the organization needs a framework for determining which lessons learned would apply to a certain project and which would not. Figure 5.3 is a formidable start to understanding what data to gather throughout the project—items such as formal changes to the project, quality and safety incidents, and schedule delays. Most project management software has the capability to report or export this data to be analyzed and acted on.

Once incident data is captured, it must be organized and analyzed according to what incidents are the most impactful. (This can be done at the project level or the corporate level.) Using a bar graph (such as Figure 5.3) or similar, the project team and organization can evaluate what lessons learned are most impactful and need to be addressed in similar projects. Every incident category in Figure 5.3 has an associated root cause, but all incidents are not worth investigating if they are not globally impactful. For example, though "Missing electrical" may be a critical category, "Outlets in the wrong location" are much more common, so lessons learned of this type should be accounted for first. Each organization only has so much capacity for change, so thresholds for action based on severity—such as low, medium, and high—help project teams understand which lessons learned to incorporate and which not to incorporate. This helps teams evaluate where to spend their time and energy (full-scale root cause analysis may not be appropriate for all risk categories). The thresholds will vary for each company and project and must account for the risk tolerance and related experience.

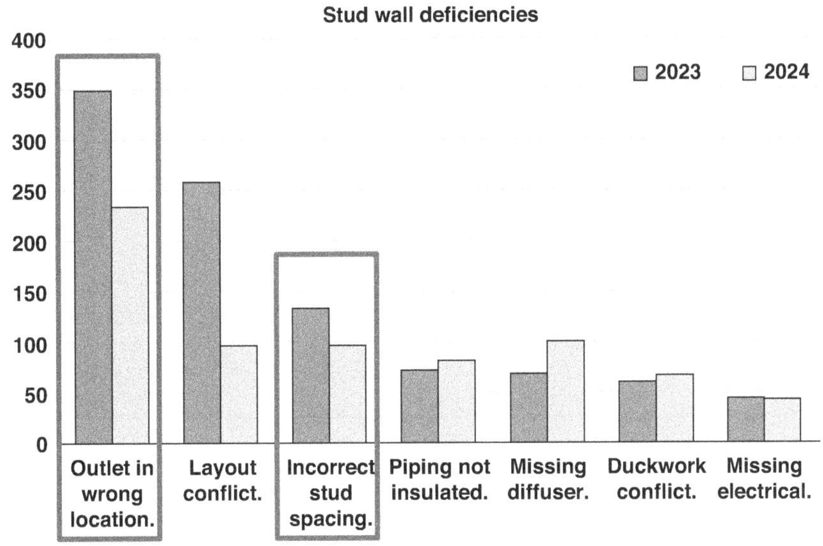

Figure 5.3 Sample project incident data.
Source: Dupslaff, 2024.

Incorporating Lessons Learned

Once the root causes are identified, those actions need to be accounted for in future projects. The first step is organizing the data in an actionable form according to the specifics of the project types that apply to your organization. (See "Reporting and Analysis.")

Let us return to the general project life cycle from Chapter 2, "Project Quality Strategy Development:"

- Precontract
- Planning
- Design
- Procurement
- Construction
- Project turnover and closeout
- Transition to owner operation
- Operations and maintenance

Depending on how a project team organizes a project (around categories such as DFOWs or other system), each of these "systems" proceeds through all the above stages of a project. Within each stage, there is an associated workflow where the critical lessons learned (such as those top items from Figure 5.3) must be captured. Within each project life cycle phase, tools, and processes can act as

Table 5.21 How lessons learned should be transferred, depending on knowledge.

Lessons Learned		
Type of Knowledge	Tacit Knowledge	Explicit Knowledge
Facilitation	Socialization; having the right people in the right place to share their varied experiences	Tools, meeting agendas, checklists, schedule, and budget templates
Timing	Ongoing and continuous throughout the entire project life cycle	At specific points in time as work occurs and tasks are completed
Example	A superintendent teaching a project engineer how to conduct a rebar inspection	An estimator looking at previous profit losses and planning accordingly

vehicles for capturing a specific lesson learned so the stakeholders involved in the project at that time are exposed to it when they need to be made aware.

Both tacit and explicit knowledge (Table 5.21) need to be transferred to the right project team member at the time they need to know. Sharing a lesson learned—knowledge that is either tacit or explicit—too late does not allow the team to plan and prevent the issue from recurring. Knowledge that is transferred too early causes the team to plan, but they may not revisit the planning (or the plan will be obsolete) because the available information is insufficient, or other pressures of the project will take precedence later, increasing the risk the issue would recur regardless.

For explicit knowledge, such as the "Outlet in wrong location" item from Figure 5.3, the associated root cause analysis may look like this during the construction phase:

- Reviewing the location of each outlet in the electrical trade partner's scope of work is a topic in the preconstruction meeting (see "Meetings and Participation").
- During a preinstallation meeting, when the owner, architect, and electrical engineer are also present, the locations of outlets may be reviewed a final time prior to installation.
- On the checklist for electrical rough-in, the inspection is updated to confirm that outlets are installed in the correct location per the approved electrical shop drawings.

Either the CQM, superintendent, or project manager running the associated scope of work (in this case, electrical) would be assigned to ensure the above items are incorporated into the meeting agendas and other tools to be leveraged in the construction phase (see Chapter 2, "Roles of Quality Personnel and Internal Stakeholders").

For a more general approach that applies beyond the construction phase, the following outline could apply:

1. *Predesign and programming:* During an initial kickoff meeting, the client's O&M personnel may share maintenance protocols that worked well for

a specific equipment manufacturer (an example of tacit knowledge being transferred).

2. *Design:* Design review checklists are updated with top lessons learned from a past similar project (an example of explicit knowledge being transferred).
3. *Preconstruction*: Reference class forecasting is used to account for specific costs that were missed on previous projects (Flyvbjerg and Gardner, 2023).
4. *Construction*: Schedule durations may be adjusted based on known weather conditions and delays from a previous project in the geographic region.

For tacit knowledge, as team members move to a new project, they bring their tacit knowledge with them. They need opportunities and a safe environment to express this with someone willing to learn, and the right time so it is relevant. For tacit knowledge, a meeting agenda could serve as a checklist for a conversation, which is an explicit knowledge tool, but triggers tacit knowledge to be shared (Gawande, 2009). See Table 5.20 on how lessons learned should be transferred, depending on knowledge. Every lesson learned cannot be accounted for on paper or documented. The gap to fill between documented lessons learned is with tacit knowledge that can be triggered and shared during conversations among team members.

Note that not all lessons learned can be identified and incorporated into a future project.

Developing Strategies to Assess and Manage Quality Risks

By leveraging the framework shown in the "Risk-Based Inspection and Testing Strategy" section of this chapter, projects can identify and mitigate potential risks for future projects during early planning phases, such as predesign and programming. For a specific project, lessons learned need to be aligned with the project's quality objectives. Not every categorized lesson learned will be applicable or of equal criticality.

When a project's potential risks are identified, the CQM must evaluate what behaviors are required of the project team to mitigate them. "True metrics measure the change in behavior. Identify what success looks like and the corresponding behaviors you're looking for when team members use your new or enhanced quality system and turn those into a metric" (Dupslaff, 2024). While identifying as many possible lessons learned that may apply is beneficial, it may be more effective to think about what changes in behavior are required to achieve the project's quality objectives and also mitigate identified risks.

Developing a Quality Improvement Program from Lessons Learned

According to Dupslaff, "The success of a quality program is specific to each individual company, its culture, and business strategy" (2024). Chapter 8, "Learning and Best Practices Analysis," provides a foundational framework for creating a corporate quality improvement program specific to the incorporation of lessons learned. Chapter 7, "Knowledge Management Analysis," outlines more tactical steps for successful implementation. Efforts such as promoting open communication, setting thresholds for action, creating low-friction reporting methods, and integrating findings into project workflows are all essential tasks for successful utilization of lessons learned.

In addition, for successful implementation, an organization must have a "process enterprise" perspective of its operations (Hammer and Stanton, 1999). This helps drive a culture of improvement. When viewed as a collection of vertical processes over horizontal business units, the incidents experienced by an organization can be prioritized in accordance with their risk potential and severity. Low-impact lessons learned may not require in-depth study or incorporation into project workflows.

Measuring the success of an improvement program requires pilot testing not only the lessons learned across several projects but also the assigned metrics for success. Not only do the incorporated root causes need to resolve the issues, but the assigned metrics need to be tangible as well. ("Reporting and Analysis" provides further information on how to develop metrics for a quality improvement program at both the project and corporate level.)

CONCLUSION

The modern CQM requires a holistic view of the organization and its current challenges to proactively analyze what risks need to be mitigated based on the future projects portfolio. To proactively address such risk, the CQM should be working to answer the following questions:

- What problems are most concerning?
- What types of incidents (lessons learned) are creating the most schedule impacts or profit losses?
- What is the organization's cost of quality? (See Chapter 9, "Cost of Quality Analysis.")
- What are the organization's warranty costs?
- What are clients saying about the organization's performance?

Planning, as illustrated in *How Big Things Get Done* by Bent Flyvbjerg and Dan Gardner, is a critical behavior that will greatly influence a future project's success. "When delivery fails, efforts to figure out why tend to focus extensively on delivery. That's understandable, but it's a mistake, because the root cause of why delivery fails often lies outside delivery, in forecasting, years before delivery was even begun" (2023). A focus on lessons learned, whether at the project or corporate level, must trigger new planning behaviors—in addition to updates to tools used during delivery and execution—to be successful.

Effective knowledge management and the seamless transfer of both tacit and explicit lessons learned are critical for preventing future project risks. By establishing environments that encourage knowledge exchange, documenting valuable experiences, and integrating these insights into project workflows, organizations can drive behavioral changes and continuous improvement. The key to success lies in timely, relevant knowledge delivery, coupled with strategies to assess and manage quality risks. Ultimately, fostering a culture of learning and proactively addressing concerns through well-planned strategies ensure that projects avoid repeating past errors and achieve superior outcomes.

QUESTIONS

1. What is the difference between explicit and tacit knowledge?
2. Explain the challenge of transferring tacit knowledge in a construction project setting, and describe two practical strategies an organization can implement to overcome this challenge.
3. Discuss why simply creating a database of lessons learned is often insufficient for effective knowledge transfer, and describe what key elements are needed for genuine "learning" to occur within project teams.
4. Using an example from the text, illustrate how a single lesson learned could be addressed at multiple stages of the project life cycle, ensuring both tacit and explicit knowledge transfer.
5. Describe how an organization should determine which lessons learned are the most impactful and warrant detailed investigation and incorporation, referencing specific tools or concepts mentioned in the document.
6. Analyze the statement "True metrics measure the change in behavior" in the context of assessing and managing quality risks. Explain how this idea connects to the successful implementation of a lessons learned program.

LEVELS OF COMMISSIONING

> Understand and determine when to use the 5-level (factory acceptance test, component verification, system construction, individual system and major equipment operation verification, integrated system testing) and 8-level (preparation, design, pre-construction, construction, commissioning of services, pre-handover, initial occupation, post-occupancy care) commissioning frameworks. (Evaluate)
>
> **Body of Knowledge V.E.5**

Commissioning is the means by which a building's systems are tested, checked, and confirmed to be in working order according to the design (Patterson, Morton, and Griffin, 2008). It is a quality-focused process that enhances project delivery by validating the facility performs per the owner's objectives (ASHRAE, 2019). While many quality tasks focus on individual components and systems and how they are *installed*, commissioning not only focuses on how those systems operate independently, but more importantly, how they *function* as one integrated system.

Commissioning is commonly understood to be focused on mechanical, electrical, plumbing (MEP), and low-voltage (LV) systems, but other systems are commonly commissioned under a separate process—such as building envelope commissioning—while following the same framework (GSA, 2020).

Commissioning of new facilities yields up to 13% whole-building energy savings, while an existing building could experience up to 10%, while estimated costs of commissioning are less than 1.5% of the total construction cost (GSA, 2020, and Los Alamos National Laboratory, 2002).

Four Types of Commissioning

There are various approaches for commissioning a facility, depending on whether it results from new construction or is an existing building. This section is focused on commissioning as it applies to new construction activities. The general commissioning framework is as follows:

- Predesign
- Design
- Construction
- Occupancy and operations

Commissioning is a best practice as part of new construction, and it is often required as part of certain certifications, such as LEED. Clients may require commissioning contractually, even if the project is not seeking a specific sustainability certification, as the benefits of commissioning are applicable to any project.

Integrating Quality and Commissioning

As referenced in Chapter 2, "Project Quality Strategy Development," the approach to commissioning is another facet of the overall design and construction process that needs to be integrated with the plan to achieve the client's objectives. Seeing quality as the foundation rather than an isolated process ensures that all the project's various plans—schedule, safety, quality, commissioning, and others—integrate with one another to form one cohesive plan that all stakeholders will implement. ASHRAE's Guideline 0–2019 is typically referenced for understanding specific commissioning activities for each phase.

In predesign, where the project's critical elements (e.g., DFOW for public projects) are identified, the commissioning requirements and scope can be outlined for each critical system.

In design, the commissioning activities are further detailed to include specifics for owner training, functional testing, preparation for functional testing, and focused design reviews. Including commissioning system readiness checks (the expectations of the commissioning agent [CxA] for a system or piece of equipment being ready to conduct functional testing) is crucial at this stage. They should be on the work-in-place inspection forms so that only one form is used. This is also where the commissioning requirements are included in the design deliverables.

During construction, much of the commissioning effort is in preparation for functional testing. Key to integration of commissioning with the overall project plan is including the CxA as a key external project stakeholder. The commissioning requirements outlined in the design deliverables are reviewed by all stakeholders during preconstruction buyout and procurement. Submittals are also reviewed by the commissioning agent during the submittal process. The commissioning agent will also conduct site visits to validate readiness of each system to be tested functionally.

The occupancy and operations portion of commissioning begins after major construction activities are completed. This includes validating that the building's systems continue to operate as designed while the building functions per its intended purpose, as functional testing during construction may not be able to simulate typical environmental loads during normal occupancy (e.g., does the air system balancing remain consistent under sustained, normal operating conditions?).

Five Levels of Commissioning in Mission Critical Projects

There are five levels of commissioning internationally recognized for mission-critical projects that occur between the general design and construction portions of a project (Heslin, Orr, and Brown, 2015). While these are often informally part of the overall commissioning plan for a typical project, it is specifically outlined for mission-critical projects, given the criticality and downtime requirements.

- *Level 1: Factory witness testing (FWT)*
  - Testing that occurs at the manufacturer's site to validate intended operation prior to shipping to the project.
- *Level 2: Receipt, installation, and post-installation checks*
  - Includes delivery inspections, work-in-place inspections, and other system readiness checks in preparation for functional testing.
- *Level 3: Functional component testing (FCT)*
  - Testing of the individual system components in accordance with the design and operational intent—for example, the operation of a chiller and its ability to maintain leaving and entering chilled water temperatures.
- *Level 4: Functional system testing (FST)*
  - Testing of the system and its individual components in accordance with the design and operational intent—for example, the operation of the chilled water system, its associated circulating equipment, and modulating valves and other components throughout the facility.
- *Level 5: Integrated systems testing (IST)*
  - Testing of multiple systems, validating their ability to operate in accordance with the design and operational intent. For example, at this level, the chilled water system and downstream air-moving equipment, along with the system's response to emergency power situations, are tested together in a holistic manner.

Commissioning Tactics and Lessons Learned

For any project, the following are helpful tactics to ensure the proper integration of commissioning activities with the project's overall quality plan:

1. *Conduct pretesting as part of preparation for functional testing.* While it may not be noted in the specifications, conducting a pretest of the system prior to formal functional testing with the CxA can reduce the number of issues that may come up and appear on a commissioning issues log (CIL).
2. *Encourage client participation in pretesting and functional testing.* Observing functional performance testing is an invaluable training opportunity for the owner's future operations and maintenance teams.
3. *Review functional performance tests before water-side and air-side test and balance (TAB) activities.* Reviewing functional performance test documents early—even if they are drafts—can streamline the functional performance testing process. This helps stakeholders and installing and testing crews understand expected system performance results and the commissioning agent's interpretations of the sequence of operations.
4. *Incorporate testing activities into the construction schedule according to how each system and component is intended to operate together.* Knowing how all the systems are tied together is essential for creating a proper construction schedule. For example, when testing terminal units, the associated air handling unit must also be functioning, and the water and air flows must be tested, balanced, and commissioned. If the air handling unit has chilled or hot water coils—which the terminal unit may also have—those water flows must be functioning at design flows and temperatures as well. If not, the system cannot be tested fully and must be revisited. This is an added expense for the commissioning agent, and thus a cost to the owner and a schedule impact for the contractor.
5. *Incorporate air and water test and balance (TAB) activities into the construction schedule.* This is often overlooked or considered part of the commissioning process.
6. *Ensure there is a technical subject matter expert (SME) assigned to the project to support the commissioning process.* Commissioning is not commonly understood by all members of the project team. Having an SME—who is separate from the CxA—as part of the responsibilities matrix (see Chapter 2, "Roles of Quality Personnel and Internal Stakeholders") to coach the project team on what to expect for each stage of the project will improve the results of the commissioning process. This SME must also understand the technical details of the systems being commissioned.
7. *Ensure commissioning meetings are part of the overall meeting flow for the project.* See "Meetings and Participation." Adding commissioning-required meetings to the meeting flow will improve awareness of the necessary activities for all stakeholders.

Commissioning is an essential process that brings closure to the static delivery, benchmark, and work-in-place inspections, client training, and turnover procedures.

Note: The CCMQ BoK also describes a different eight-level framework for commissioning (preparation, design, preconstruction, construction, commissioning of services, pre-handover, initial occupation, and post-occupancy care). In the opinion of the authors and based on available research, the use of this eight-level commissioning framework is not industry-wide. Also, the eight-level commissioning framework does not align with the ASHRAE guideline. The recommended commissioning framework and the one that is widely used is the five-level network as described in this section.

CONCLUSION

Understanding and implementing effective commissioning processes is paramount to ensuring project success, particularly for mission-critical facilities. By integrating commissioning into the broader project plan from the predesign phase through occupancy, stakeholders can validate system functionality, enhance quality, and ultimately achieve the owner's objectives. Whether navigating the general framework or the more detailed levels specific to critical projects, meticulous planning, clear communication, and dedicated subject matter expertise are essential. Commissioning is not merely a procedural step but a strategic investment that ensures long-term performance, efficiency, and operational integrity.

QUESTIONS

1. Explain the difference between the general commissioning framework for new construction and the five levels of commissioning specifically recognized for mission-critical projects.
2. Describe how commissioning integrates with the overall project plan, particularly during the design and construction phases. Give an example of how integrating commissioning with the project plan can avoid extra costs for the owner.
3. What is the role of a commissioning agent (CxA), and what are some key expectations the CxA has for a system or piece of equipment being ready to conduct functional testing?
4. Discuss the importance of client participation in the commissioning process, specifically during pretesting and functional testing. Why is this considered an invaluable training opportunity?
5. According to the document, what are some key tactics to ensure the proper integration of commissioning activities with the project's overall quality plan? List at least three and explain their significance.

Chapter 6

Project Turnover and Closeout

Documentation Management

There is an idiom in construction management that states, "Project closeout begins when the project begins." These words, of course, attempt to encourage those involved with a project to not put off project turnover and closeout tasks until just before the project is closed, so as to not delay achieving completion milestones (as discussed in Chapter 5, "Construction"). They also encourage personnel on a project to remain up-to-date on many items that can become more troublesome at the later stages of a project, and to perform all tasks during a project with the intent of leading it toward completion. Many of the preceding sections of this handbook have given CQMs guidance on how to approach various phases of construction to ensure properly verified and validated construction methods. For example, Chapter 5 states the importance of maintaining "redlined" shop drawings while a project is ongoing, as this will aid in creating final as-built drawings. When a project's quality function is poorly managed throughout the project's duration, project turnover and closeout often end up being the most dreaded portion of the job. In such a scenario, project staff members are often "playing catch-up" while other stakeholders are expecting closure. This inevitably leads to unpleasant but avoidable interactions as the project end date nears. To successfully lead a project into these final phases, the CQM needs to learn to apply strategies for proper document maintenance early, as prescribed throughout this handbook.

Despite all of these recommendations, however, project turnover and closeout rarely ever formally commence when a project begins. While there is no exactly defined moment at which they do, it is generally accepted that the turnover and closeout process of a project formally begins when most of the work is done but the project is not yet fulfilling its intended use (Brown and Hamilton, 2024). Project turnover and closeout contain many distinct tasks that can only be accomplished during these later stages of the project, and this section will delve into those requirements. The CQM will learn to evaluate or apply various documentation and deliverables requirements, handling lingering nonconformances, achieving project milestones, and maintaining and executing warranties.

Definitions of "closeout" can vary, but commonly, in construction it refers to a collection of activities performed by the contractor near the end of a project that serve to complete it and provide final contractually required deliverables to the owner. Simply put, it is the final phase of the construction process (Dobberstein, 2022).

IMPACT ASSESSMENT AND COMPLIANCE

> Evaluate the potential impact arising from the demobilization of quality resources. Review and assess inspection and test records related to safety critical equipment to determine if they are compliant with required standards. Confirm how final documentation will be transferred to the client (e.g., digital platforms, electronically, and hard copy). (Evaluate)
>
> Body of Knowledge VI.A.1

For some long infrastructure projects, the project closeout may also be planned in stages. In other words, a portion of the project may be completed, closed out for construction, and turned over to the client. The term "turnover" in the construction industry is the moment in time when the project, after having been under the control of the contractor during construction, is handed over to the owner for their care and custody. Lastly, in some projects, while the project is still ongoing, the transfer of control over to the owner or to the owner's representative can also be phased. Such variations on the project turnover and closeout, if applicable, should be described in the original scope of the contract documents.

IMPACT OF DEMOBILIZATION

When the project turnover and closeout phase begins, a contractor's senior management may propose paring down resources involved with the project. For example, with a majority of the work complete, it is sensible to decrease field personnel to a level far below the level at peak manpower. At the commencement of this phase, it is likely that only testing, commissioning, and some punch list work remain. To keep the same number of tradespeople involved during this phase as there were during significant construction activities would result in a significant loss of capital. Similar considerations may be made for administrative support staff involved with tracking the work's progress, schedule, and commercial status; if there is less work ongoing with fewer tradespeople, a directly proportional decrease in such administrative staff may be logical.

There may be a tendency to think similarly about quality resources. On many projects where a CQM (or position with a similar title) is required, that role is often contractually obligated for the full duration of the project (MTA Construction and Development). On such contracts, a customer may also require or recommend that the contractor provide additional quality staff members reporting to the CQM. These quality staff members may be overseeing various roles: performing daily inspections, performing or managing testing and commissioning, ensuring training is performed, and so on. Depending on the size of a project, the quantity of such quality resources can be sizable. Even on projects where no official CQM or other quality resources are formally defined, some contractors may opt to provide such a role to their customers to provide QA to them, enhance their own image in the marketplace, and reduce their own cost of quality (see Chapter 8, "Cost of Quality").

Still, there are limits to these considerations. How does a contractor determine when to downsize the quality function? What are the risks of cutting quality personnel too soon? Many tasks that are associated with the closure of a project are inherently and uniquely part of the quality function and thus require quality personnel for their completion. This section, however, will shed light on the relationship between quality resources and outcomes.

When project closeout commences, it is quite possible there are systems still needing follow-up or even testing and commissioning for the first time. While testing and commissioning will often be undertaken by specialists in their respective areas, the CQM must ensure their procedures and plans are approved before they commence. The CQM or another member of their staff should also be present during testing to either record or close nonconformances that arose during various rounds of testing. Additionally, the quality function's participation in the verification of calibrated testing and measuring equipment is critical to validating testing and commissioning data. Lastly, the CQM should ensure test reports submitted include all pertinent information and data, and are formatted to highlight this. While more detailed aspects of the CQM's participation in testing and commissioning can be found in other chapters of this handbook, it is important to note that without a well-staffed quality function, the probability of a failed test and/or invalid test result increases. While it is possible during construction that a quality inspector may oversee the work of 20 tradespeople on a daily basis, during the closeout testing and commissioning phase it is likely there is a much lower ratio of craft to quality. As an example, four quality staff were previously overseeing the work of 80 tradespeople, and there are four areas of testing ongoing simultaneously, each requiring only two tradespeople. A contractor's senior management must understand that the quality staff's resources are invaluable to the operation and cannot decrease proportionally.

System Safety-Critical Components

The CQM and quality staff's involvement in the testing and commissioning process can also serve to verify system safety (MTA Construction and Development). There are three modes by which good quality enhances safety in a construction project:

1. via minimization of rework
2. via construction of structures, systems, and components correctly
3. via the construction of features that increase safety for users (this is as an offset of mode 2)

As described briefly in Chapter 5, "Regulatory Body or Authority Having Jurisdiction," mode 1 provides enhancement of safety predominantly to those performing construction: Naturally, the less rework that is done, the less likely an accident becomes. Modes 2 and 3 lean more toward providing a safe work environment for the end user of a project, and in some cases, the public. Thus, their considerations typically involve the components of system safety verifications in the project turnover and closeout phase.

Typically, the procedure for assessing the system safety compliance of a particular structure, system, or component can be broken down into the verification of five steps:

1. Identification of hazards
2. Design
3. Execution
4. Testing and commissioning
5. Training

While steps 1–3 of this procedure typically occur prior to the project turnover and closeout phase, the verification of the safety of a system cannot be considered "closed" until the entire procedure is complete. As such, with verification of steps 4 and 5 for many systems typically occurring during the project closeout phase, the entirety of this system safety adherence is demonstrated.

Step 1 of this procedure is typically assessed by the customer and a designer. For example, a client, along with their EOR, may seek to prove that the proper grounding of a piece of equipment is to be considered system safety-critical, as its failure could pose significant hazards such as injury or death to its personnel. This evidence establishes the verification of step 1 (and was likely performed prior to the start of the project by the client, and in some cases, a design quality manager).

Step 2 of the procedure is then conducted by the design engineer and seeks to prove the designed solution for a safety-critical issue is capable of eliminating or mitigating the identified hazard. Following the example in step 1, here the electrical design engineer may seek to show by utilizing calculational data, code references, and statistical analysis that the design provides proper grounding. This evidence establishes the verification of step 2 (and was likely performed with the support of a design quality manager).

Step 3 of the procedure is conducted by the contractor and intends to prove the designed installation was performed and done correctly and is free of defects that could create safety risks. Following the previous example, the CQM may seek to show that during the construction phase, work plans were adhered to, and the form of the final installation conforms to the design. This evidence establishes the verification of step 3.

Step 4 of the procedure is also conducted by the contractor and intends to prove the functionality of the installation is correct. To follow the example above, here, the CQM may seek to show that all formal inspections, testing, and commissioning of the installation have passed per the approved test criteria. This would establish evidence of the verification of step 4.

Lastly, step 5 of the procedure is also conducted by the contractor and intends to demonstrate the safe operation of various equipment to the customer. In many cases, training itself is performed by a manufacturer's representative, but the CQM may seek to show records of the training, including a list of attendees and results of quizzes that show the efficacy of training. As a final step in the example above, the training component of this safety system may include the manufacturer of the equipment teaching the end user how to understand and operate ground fault detection relaying equipment. This would establish evidence of step 5.

Note that occasionally, during the closeout process, whether due to changes in design or recognition of omission, the CQM may identify new systems or equipment that could be deemed safety-critical. In such a situation, the CQM should recommend that the new "hazard" be analyzed for inclusion in the process (MTA Construction and Development).

After all of the steps have been performed, the CQM should assemble the verifications and validate their completeness before transmitting them. Again, for larger projects or those with many simultaneous safety verifications to perform, the CQM may rely on a dedicated quality staff member to assist in this process, and thus should insist on the maintenance of adequate support. Diminishing quality resources prior to completing steps 4 and 5 may hamper the ability of the contractor to assure the customer that safety-critical systems operate as intended. That could, in turn, slow the closeout process.

Final Closeout Documentation Transference

The CQM and other quality staff are responsible for the identification, completion, collection, and transmission of closeout documentation and deliverables. While the testing, commissioning, and engineering processes and their reports may be completed by experts in those areas, the quality personnel are responsible for ensuring results from those functions are prepared and transmitted in a manner preferable to the client and in conformance with the test procedures and requirements. While final record or "as-built" drawings will ultimately be prepared by a draftsperson or equivalent engineer, quality personnel should ensure the formatting of such drawings will also meet contractor requirements and customer needs. A quality inspector's intimate knowledge of completed work can serve to ensure the accuracy of such record drawings to minimize the client's verification team's work and reduce the probability of rejection. Similarly, operation and maintenance (O&M) manuals typically also require careful formatting and arrangement to ensure they are readable and searchable (see Chapter 7, "Sustaining Quality Requirements"). Generally, verification of the quality and correctness of deliverables to the customer is, again, an inherent way that quality personnel can provide value to a contractor during the closeout process.

At the commencement of the project turnover and closeout process, the CQM and quality staff can provide additional benefit to both the contractor and customer by carefully ascertaining deliverable transmission needs. The quality function should not always rely solely on contractual requirements for this handover process and should begin the process of determining these needs through discussion. On long-term projects, there are often documentation specifications written, citing needs that may exist at the start of a project but may not fit a client's needs by the end. For example, many older construction specifications may still require the printing of record drawings on a polyester film (reminiscent of true "blueprints"). However, many clients indicate that these types of prints have no real utility. The CQM can spearhead the process of having the requirement for these polyester film drawings stricken from the contract in an effort to reduce waste and help divert resources to more important tasks.

Whatever the eventual outcome of these "negotiations," the CQM must be present to document and convey them to all parties involved in their creation and transmission. On larger projects, additional quality resources can ensure the follow-through of all deliverables, regardless of format. (See Chapters 2, 3, and 5 for details on quality deliverables and documentation.)

Lastly, the CQM is responsible for carrying the project through numerous closeout milestones. Achievement of these milestones will hinge on some of the project closeout tasks discussed in this chapter, as well as many to follow. A CQM and an appropriate volume of support staff may be required to ensure these tasks are completed in a timely manner. With project turnover and closeout bringing significant commercial benefits for a contractor, the CQM must ensure there are adequate personnel to expedite this process.

CONCLUSION

Ultimately, whether acting as the sole quality professional on a smaller project or managing a quality staff on larger projects, the CQM must act as a champion for the quality function to elucidate to a contractor's senior management the benefits of maintaining (or in some cases increasing) it and the pitfalls of weakening it. In addition to contractually mandated positions, the CQM must advocate for the importance of as many staff as is needed to enhance safety, customer service and satisfaction, and contractual conformance. The CQM should utilize cost of quality analyses (see Chapter 9, "Cost of Quality") to show that up-front expenditures on the CQM and a staff will result in greater savings for the customer in relation to other project positions. Establishing that a larger quality function can allow the contractor to complete project milestones sooner should provide a very easy means of demonstrating cost reduction. The CQM should finally highlight the relationships the quality function has built and maintained with the customer to senior management. Such relationships provide the client with the reassurance that the project will be closed out and turned over with the same attention to detail that they experienced during initial phases. Construction is a "people business," and strong relationships between the client and the contractor are a great recipe for ensuring repeat business with that customer.

QUESTIONS

1. How might you make a case to your organization to maintain adequate quality function personnel on a project during closeout?
2. What benefits aside from "quality" in itself might maintaining a quality function provide during closeout in relation to future projects?
3. How can the CQM help to streamline the closeout documentation process?
4. How can the CQM provide value during commissioning being performed during the closeout phase?
5. What are the roles of the CQM in ensuring system safety to the customer/end user?

REQUIRED DOCUMENTATION

> Maintain comprehensive documentation throughout the project's duration, including records of all activities and changes. (Evaluate)
>
> Body of Knowledge VI.A.2

To maintain comprehensive documentation throughout the project's duration, three basic steps are essential:

1. Start early with a thorough understanding of the documentation that must be maintained.
2. Implement a robust documentation plan with adequate resources.
3. Maintain regular communication, accountability, and tracking (Gandhi, 2024).

This subsection will focus on these three steps.

Step 1: Starting Early with a Thorough Understanding of Necessary Documentation

To develop a thorough understanding of required documentation, the CQM and team will need to diligently read the specifications governing such documentation. It is necessary to perform this step early in the project. While a good rule of thumb is to do this during the first 10% of the project life cycle, some contractors start their closeout process during the preconstruction phase (Gonzalez, 2024). If this step is performed too late in the project, the CQM and their team risk trying to play "catch-up" every step of the way and consequently may neither be generating nor maintaining the contractually required documentation at the contractually required intervals. This could lead to financial penalties imposed by the owner and likely to prolonged project closeout, which will negatively impact the profitability of the company.

For projects utilizing the Construction Specification Institute (CSI) MasterFormat Specifications (primarily the United States and Canada), such contractually required documentation will be found in the Division 01 specifications—General Requirements (Associated Builders and Contractors, 2024). Depending on the size and scope of the project, the Division 01 specifications can be voluminous. While it is important to perform a thorough read of all such specifications, the requirements for documentation to maintain will likely be found primarily in these four specifications:

1. 01 31 00—Project Management and Coordination
2. 01 32 00—Construction Progress Documentation
3. 01 77 00—Closeout Procedures
4. 01 78 39—Project Record Documents

To perform a diligent read of these specifications, a suggested best practice is to assign individual specifications to team members. Each person's assignment is to read the specifications thoroughly (maybe more than once) and summarize their salient points and contractual documentation deliverables. Such specification reviews should be documented on a matrix (Figure 6.1).

The specification reviewer should pay particular attention to the required documentation identified in the specification and the minimum content of the documentation, as well as the frequency at which it must be provided (if applicable) to the owner, program manager, or other stakeholder(s). It is also recommended that

RAIL LINE PROJECT SPECIFICATION REVIEW NOTES 8/30/2024

Reviewer: Joe Smith

Index Number	Spec Section	Specification Section Name	Subsection	Page	Description of Specification / Requirement	Cost Risk [Y/N]	Schedule Risk [Y/N]	Potential RFI [Y/N]	Submittal Required [Y/N]	Comments/Questions/Explanations
1	01 32 00	Construction Progress Documentation	1.4H 1.10B	3, 7	Material Location Reports (MLR) are to be submitted weekly. MLR is a comprehensive list of all material delivered to and stored at the project site	N	N	Y	Y	Maintain a log of all materials delivered to and stored at the site, and its location, on a weekly basis.
2	01 32 00	Construction Progress Documentation	1.4G 1.10A	3, 7	Daily Construction Reports (DCR) are to be submitted weekly. DCR is a report documenting activities that occurred on the site on a daily basis	N	N	Y	Y	Minimum list of items to be recorded in a DCR is shown in 01 32 00-1.10A.1 through 20.
3	01 77 00	Closeout Procedures	1.2A.3 1.7A	1, 3	Maintain a list of Incomplete Items (Punch List)	N	Y	N	Y	The contractor needs to maintain and submit a Punch List of items for final completion of the Project.

Figure 6.1 Sample specification review matrix.

these reviews be presented by the reviewer internally in a meeting setting. Other internal team members will then have an opportunity to ask questions, make comments, and learn from the reviewer's review. This is a good training opportunity for less experienced members of the team. A specification review as described may also result in identifying illogical or irrelevant specification requirements. When such instances are identified, a request for information (RFI) should be written and issued to the owner for clarification.

Ultimately, this specification review effort should yield a basic listing of the documentation that the contractor is required to maintain. Such documentation may include, but not be limited to, the following:

- Daily construction reports
- Material location reports
- Material receiving reports
- As-built drawings
- As-built specifications
- Inspector daily reports
- Quality inspection reports
- Work plans
- Test plans/procedures
- Field test reports
- Commissioning records
- Audit reports
- Submittals
- Transmittals
- Requests for information

- Certifications
- Change orders
- Safety inspection records
- Safety incident/accident reports
- Subcontractor payment records
- Meeting minutes
- Progress photographs
- Shop drawings
- Product data
- Certificates of compliance (Garcia and Hamilton, 2024)

Following the review of the Division 01 specifications and the identification of required documentation, the next step is to implement a robust documentation plan to generate and maintain the identified documentation.

Step 2: Implementing a Robust Documentation Plan with Adequate Resources

The second step in maintaining comprehensive documentation throughout the project's duration is to implement a robust plan with adequate resources. Such a plan should be a written document that is either stand-alone or a component of another project-specific management plan. As stated in step 1, this plan should be created early in the project; if the plan is created too late, it will be nearly impossible to retroactively create required documentation. Certainly, as the project evolves, the plan can and should be updated to reflect the project's changes and associated impacts to the contractually required documentation. Change management in construction plays a crucial role in the closeout phase of a project by ensuring that all changes made during the project life cycle are properly documented, evaluated, and integrated into the final closeout and turnover documentation. To be truly robust, the project's documentation plan needs to include the various change processes to ensure all required documentation will be captured commensurate with the work.

The essential components of a robust documentation plan are as follows:

- *Scope*: The introduction of the plan should identify its extent and the overall purpose. The scope of this plan should encompass all phases of the project for which the contractor is responsible. This could include design, construction, and closeout/turnover. The plan should address the creation, management, and storage of documents to facilitate transparency, accountability, and compliance with contract requirements.
- *Responsibilities*: This section of the plan needs to identify which team member(s) is responsible for creating, collecting, and as applicable, reporting on the documentation. Additionally, this section should provide specific duties to those responsible for the documentation.

- *Processes*: This section of the plan should document the steps necessary to generate and maintain the identified documentation. Most construction companies utilize a project management software that provides the means to more easily create and manage such documentation. The section should identify the software and explain the process to generate and maintain the required documents. The section should also reference any standardized templates that the project will utilize for certain documents.

Commensurate with creating a robust documentation plan, it is necessary to ensure adequate resources are assigned to execute the plan. Such resources may generally be classified into three categories: human, technological, and administrative.

- *Human resources*: This of course refers to the men and women necessary to execute the plan. It may include but not be limited to a project manager, a document control manager, a quality manager, and applicable document clerks. Each one needs to have roles and responsibilities that are well defined in the plan. Additionally, the plan should contain an organization chart clearly providing the team with the reporting and communication structure
- *Technological resources*: This refers to the software that may be necessary to execute the plan. The plan should identify any necessary software licenses to ensure that the people on the team have the required access to the software to perform their assigned roles and responsibilities. Such technological resources may include project management software, meeting and communication tools, or cloud storage solutions.
- *Administrative resources*: These include a standardized filing system and/or naming convention, regular meetings, key performance indicators (KPIs), regular training of team members, and standard operating procedures (SOPs). Such resources can be viewed as the "skeleton" on which a successful documentation plan is built.

Step 3: Regular Communication, Accountability, and Tracking

The third step in maintaining comprehensive documentation throughout the project's duration is to ensure team members are communicating regularly, are accountable to each other, and are tracking and reporting progress.

To ensure communication among the team, the CQM should hold regular check-in meetings throughout the project to check on the status of required documentation. The frequency of such meetings will vary, depending on the specific project as well as the percentage of completion of the project, but it will most often be between weekly and quarterly. It is important to recognize, however, that the nature of a particular project may necessitate even more frequent meetings. The overall pace of the project and the volume of contractual documentation will influence the frequency. Such meetings should be internal to the contractor's team to address internal issues. But separate meetings can also be held with the owner and/or program manager, serving to inform them of the contractor's efforts toward documentation collection.

Finally, it is essential that the team and its individual members have the means to track and report on their efforts toward generating and maintaining required documentation. Such tracking and reporting often can be performed with various project management software. And there are many different metrics that can be tracked and reported. While project documentation will, of course, vary from project to project, the following is a list of some basic metrics that can be used for some common documents:

- RFIs submitted
- RFIs answered
- RFI response time
- Submittal response time
- Percentage of submittals approved/accepted the first time
- Inspector daily reports generated each day versus the number expected each day
- Percentage of field test reports exhibiting pass; percentage exhibiting fail
- Number of quality checklists created versus the expected number
- Number of audits performed in a given month versus the expected number

CONCLUSION

To maintain comprehensive documentation throughout a project's duration, there are three basic steps:

1. Start early with a thorough understanding of the documentation that must be maintained.
2. Implement a robust documentation plan with adequate resources.
3. Maintain regular communication, accountability, and tracking.

Knowledge and understanding of these three steps will provide the CQM with the criteria on which to evaluate a project team's efforts toward maintaining comprehensive documentation. Such evaluation may take place during an audit. These three steps can provide the basis of an audit scorecard. Such an audit would seek to evaluate the team's performance in three categories:

- Whether the project team's plan was developed early enough in the project to adequately identify all contractual documentation.
- Whether the team has a robust written plan in place with adequate resources.

- Whether the team is regularly communicating in internal and external meetings and if they are tracking and reporting their performance with appropriate KPIs.

QUESTIONS

1. Explain the pitfalls of a contractor not identifying a contract's closeout requirements early enough in a given project.
2. Explain a good practice for performing a diligent review of the closeout procedure requirements that will also serve to train staff members on the requirements.
3. When reviewing a project's closeout procedure specifications, what should the reader be looking for?
4. Explain what a robust documentation plan is.
5. Explain two different ways in which a robust documentation plan may fail to be executed.
6. Given a robust documentation plan with adequate resources, what is required to actually execute the plan?
7. If tasked with auditing a project's closeout documentation plan, what would one look for to assess its effectiveness?

NONCONFORMING WORK

> Ensure all work not conforming to contract specifications (i.e., punch list) is closed. (Evaluate)
>
> Body of Knowledge VI.A.3

The concept of nonconformance is central to the concept of QC. As described earlier, QC inspections are product-based and must result in a pass or fail. Criteria for "acceptable work" are needed to implement a pass/fail QC inspection. These criteria are available in the final design documents and technical specifications for that aspect of work. These two guidance documents are also used to develop the QC plans and ITPs for the project. They must always be consistent with the product requirements described in the design documents and technical specifications.

If the work inspected in a pass/fail QC inspection conforms, then it must be passed; if it does not conform to the requirements, then it must be classified as a failed inspection. The next question is more critical: Is the failed inspection a nonconformity/deficiency, or should it be classified as a nonconformance?

In the construction industry, many contractors prefer to call a failed inspection a deficiency (or nonconformity), mainly because most contracts include a non-payment clause for nonconforming work. Usually, there's a workaround for progress payments; in other words, if the deficient work could be described as still "in progress," then the failed inspection may be classified as a deficiency/nonconformity that will be corrected later. Sometimes a failed inspection is placed on a punch list instead of correcting the issue in a timely manner, reinspecting, and then approving the constructed work *before* further work can be performed. In this case, delaying the correction of the deficiency/nonconformity frequently becomes a larger quality problem later, creating a negative impact on the project in terms of schedule, cost, and reputation. The question then becomes: Who must pay for correcting the failed inspection, and is the rework even possible? Later, if the corrective rework is not possible due to schedule and cost impacts, then the project owner (client) is usually forced to accept a less satisfactory compromise.

In construction QC, corrective rework can become problematic, resulting in claims, conflicts, and client dissatisfaction with the quality of work for many construction projects. As a rule, all deficient items discovered during a QC inspection and placed on a punch list must be corrected in a timely manner so as not to impact the project schedule. For the failed inspections and tests, a re-inspection should be scheduled promptly within the approved project schedule. If the correction is still not made to the failed inspection, then the failure is no longer documented as a deficiency, but as a nonconformance.

The best practice is to clearly define what is a nonconformance versus what is a deficiency/nonconformity. This definition should be included in the technical requirements of quality and throughout the implementation of quality plans at all levels (contractor, subcontractor, owner, and all other stakeholders). All failed QC inspections must be linked with the project schedule. The stakeholders (designer, contractor, owner, and others) must formally agree to the definition of nonconformance and nonconformity/deficiency at the beginning of the project; if the nonconformity/deficiency that's identified by the QC inspection can be fixed without affecting the time (schedule) or the budget (cost) of the project, then it may be called a nonconformity/deficiency, but it should include a time limit. In other words, if the failed inspection cannot be corrected without adversely impacting the schedule or budget, then it must be issued out as a quality nonconformance *not* a nonconformity/deficiency (see Chapter 5, "Quality Control and Monitoring").

Understanding Acceptable Features Versus Nonconformities

- *Acceptable features of work (FOWs)* refer to those aspects of a construction project that meet or exceed the specified requirements. These features align with industry standards, project plans, and client expectations.
- *Defective or nonconforming FOWs* are those aspects of construction inspections that represent a deficiency. The project's QMS must provide detailed guidance on the definition of the defects and deficient parts identified during quality inspections. These are the finished parts or aspects of work that deviate from the established standards. They may fall short of requirements, exhibit flaws, or fail to meet quality expectations. If these can't be corrected according to the existing schedule, then they can't be issued as a deficiency (or defective). They can't be placed on a punch list because they represent a breach of the contract.

- *Tolerances* are predetermined allowances for deviations from the specified requirements. These allowances recognize that minor variations can occur during construction without compromising overall quality. Construction tolerances that are included as the allowable deviations from a specified dimension, location, or other characteristics of a structure must be included in the inspection and test plans. Doing so allows a more accurate QC inspection and ensures there's no ambiguity on the pass/fail criteria. If the inspection shows that the product is outside of tolerance range, then it should be issued as either a defect/nonconformity (if it can be corrected without impacting the schedule), or else it should be formally recorded as a nonconformance.
- *Incomplete work* refers to incomplete portions of the project that are not necessarily nonconforming. For example, an unfinished wall may be acceptable if it aligns with the project plan and design intent. However, most contracts specify that any work identified as incomplete constitutes a breach of the contract and should be noted as nonconformance. A scheduled work may be qualified as incomplete and placed on a punch list by the contractor without being classified as "nonconforming" only under special conditions. In the special cases where this deviation is permitted, it should be approved formally by senior management. Any contracted work allowed as being incomplete and placed on punch lists should *never* be authorized at lower levels of management and supervision, and the approval should be in writing. The absence of formal documentation for these deviations is what leads to construction claims and conflicts—mistakes to avoid in all projects.

The Role of Design Documents

- *Design documents*, such as architectural drawings, engineering plans, and specifications, guide the construction process. Understanding these documents is crucial for identifying nonconformances.
- *Design intent* represents the vision and purpose behind the project. It can be achieved through various means and methods, but consistency with the design documents is essential.

Identifying nonconformances

- *Communication*: Early communication among project stakeholders (including contractors, designers, and quality managers) sets clear expectations. It ensures everyone understands the project's goals and quality standards.
- *Control sampling*: Regular sampling and inspection help establish a *datum* (a reference point) for quality. By comparing actual work to this datum, nonconformances become evident.

- *Adjusting focus*: Once the baseline is established, quality management efforts shift toward identifying and addressing nonconformances. This involves corrective actions, continuous improvement, and adherence to QC processes.

Marking Nonconformances

When a nonconformance is identified (such as the concrete foundation being poured wider than specified), the first step is to clearly mark the affected area. This serves several purposes:

- *Preventing inadvertent use or cover-up*: By marking the area, you prevent accidental use or concealment. For example, if the foundation is wider than intended, it should be marked with barrier tape or other visible signals.
- *Visibility*: Markings make it evident to all project stakeholders that there is an issue. This transparency ensures that corrective actions are taken promptly.
- *Accountability*: Clear markings establish accountability. They communicate that the area needs attention and correction.
- *Documentation*: The marked area becomes a reference point for further actions.

Methods of Marking Nonconformances

Various methods can be used to mark nonconformances:

- *Barrier tape*: For physical barriers like the concrete foundation, barrier tape can be wrapped around the affected area.
- *Paint*: Paint can be applied to surfaces to highlight nonconforming features.
- *Tape*: Adhesive tape (e.g., masking tape) can be used to outline the boundaries of the nonconformance.
- *Tags*: Tags or labels can be attached to equipment, materials, or structures.
- *Other visible signals*: Any visible indication that draws attention to the nonconformance is suitable.

Ensuring Markings Are Not Accidentally Covered Up

It's crucial to ensure that the markings cannot be accidentally covered or removed during subsequent work. Consider the following:

- *Communication*: Inform all relevant personnel about the marked area and its significance.
- *Supervision*: Supervisors and quality managers should monitor the area to prevent accidental cover-ups.

- *Documentation*: Document the location and nature of the nonconformance to prevent confusion.

Follow-Up After Marking Nonconformance

- *Photographic documentation*: Take clear photographs of the marked area. These photos serve as evidence and aid in communication.
- *Project documentation system*: Log the nonconformance details into the project documentation system. Include information such as location, description, and any relevant measurements.
- *Initiating accountability and correction*:
  - Assign responsibility for addressing the nonconformance. This may involve the contractor, subcontractors, or other parties.
  - Define the corrective action needed to rectify the issue.
  - Monitor the status of corrective actions until completion.
- *Timeliness*: Address the nonconformance promptly to prevent delays and additional costs.

Budget Considerations

- Nonconformances can impact project budgets. Addressing them early prevents unexpected expenses.
- *Scope budgeted*: Ensure that corrective actions align with the project's budget scope. If additional resources are needed, communicate this to project stakeholders.

Controlling Work

Assessing the Impact on Quality

When a nonconformance is identified, it's essential to evaluate whether continuing work in the affected area could adversely affect quality. Consider the following scenarios:

- *Quality impact*: If ongoing work could compromise the overall quality of the project or exacerbate the defect, it's crucial to take immediate action.
- *Hidden defects*: Continuing work might inadvertently hide the nonconformance, making it harder to address later.

Stopping Work Until Correction

- If the impact on quality is significant, *stop work* in the affected area until the nonconformance is corrected. This ensures that no further work exacerbates the issue.
- Stopping work prevents additional resources from being invested in an area that requires correction. It also avoids potential rework and delays.

Field Personnel Decision-Making

- Field personnel, including supervisors, forepersons, and quality managers, play a crucial role in decision-making.
- They should assess the situation, considering factors such as safety, project schedule, and quality standards.
- Field personnel decide whether any limitations should be placed on ongoing work. For example:
  - *Partial work*: They may allow work to continue in other areas while addressing the nonconformance.
  - *Temporary measures*: Temporary measures (e.g., covering the affected area) may be implemented to prevent further issues.

Inspecting Previously Completed Work

To ensure consistency and identify potential similar defects, inspect previously completed work of a similar nature.

- *Why inspect?* This step helps verify that the nonconformance is not a recurring issue. It also provides insights into the effectiveness of QC measures.
- *Documentation*: Document the findings from the inspection. If any defects are discovered, address them promptly.

Punch List

A punch list, also sometimes known as a "snag list" or "deficiency list," is a documented list of incomplete (or incorrectly completed) items of work that the contractor needs to address before a construction project is considered fully complete and final payment is made. It essentially represents the remaining tasks or deficiencies that need to be "punched out," or corrected.

Having a contractual definition of deficiency/ nonconformity and its placement on a punch list created during the project's construction phase are essential. Lack of clarity on what constitutes incomplete work and whether incomplete work may be allowed during the construction phase is often the main cause of conflict between contractors and owners. Whenever work that's inspected for quality and notated as deficient is placed on a punch list, but is not recorded as a nonconformance, it results in many subsequent deficiencies, nonconformances, and failures. Subsequent work usually becomes nonconforming as well, causing the amount of rework required at a later stage to increase. For this reason, punch lists must be guarded closely against the approved schedule. CQMs must not allow the provision of punch lists to cause project failures and conflicts. Think of it as a final checklist to ensure everything is built according to the contract documents, plans, and specifications.

The closeout punch list serves several crucial purposes in a construction project:

- *Identifies remaining work*: It provides a clear and organized record of all the outstanding tasks that need attention. This ensures nothing is overlooked before the project is officially closed out.

- *Sets expectations*: It formally communicates the client's or their representative's expectations for completion and quality to the contractor.
- *Assigns responsibility*: It typically indicates which contractor or subcontractor is responsible for rectifying each item.
- *Tracks progress*: It serves as a tool for tracking the completion of these outstanding items. The status of each item can be updated as it is addressed.
- *Facilitates final payment*: Final payment to the contractor is often contingent on the satisfactory completion of all items on the punch list.
- *Ensures contract compliance*: It helps ensure that the project is delivered according to the terms and conditions of the construction contract.
- *Supports project closeout*: Addressing the punch list is a critical step in the formal project closeout process.

A closeout punch list is one that is used in the final stages of a construction project, specifically during the substantial completion phase. Here's a more detailed breakdown of the terminology that is used in the construction industry:

1. *Substantial completion*: This is the point in the project where the work is sufficiently completed in accordance with the contract documents so the owner can occupy or utilize the work for its intended purpose.
2. *Punch list inspection*: Once the contractor believes the project is substantially complete, they often conduct their own internal review. Following this, the owner, architect, engineer, or a designated representative usually conducts a final inspection of the entire project.
3. *Creation of the closeout punch list*: During this final inspection, any remaining incomplete or incorrect items are noted and compiled into the closeout punch list (document).
4. *Contractor rectification*: The punch list is then provided to the contractor, who is responsible for addressing and correcting all the listed items within a specified time frame.
5. *Verification*: Once the contractor believes they have completed all the punch list items, another inspection (often called a "punch list walk-through" or "final completion inspection") is conducted by the owner's representative to verify that all items have been satisfactorily resolved.
6. *Final completion*: After all punch list items are verified as complete, the project reaches final completion, and final payment can be processed.

The punch list (Table 6.1) is a critical document used near the end of a construction project to ensure all remaining work (including nonconformance) is completed to the required quality standards before final payment and project closeout.

Example of a Punch List

Project Name: The Fremont Medical Center—Phase 2 Expansion

Project Address: 123 Main Street, Fremont, CA 94538

Date of Inspection: April 13, 2025
Inspection Conducted by: John Smith, Project Manager, ABC Construction
Weather Conditions: Sunny, 65°F

Recording Nonconformances

Let's break down the process of handling nonconformances in construction projects, focusing on the creation of a nonconformance report (NCR).

Marking and Controlling the Nonconformance Area

- As previously discussed, the first steps involve *clearly marking* the affected area to prevent inadvertent use or cover-up.
- Control measures may include stopping work in the area or implementing temporary solutions.

Creating the Nonconformance Report (NCR)

- The NCR is a formal document that captures details of the nonconformance. Here's how to create one:

  1. *Describe the deviation.*
     - Clearly articulate the nature of the nonconformance. Be specific and concise.
     - Use references from *design documentation* or *product installation guidelines* to remove any ambiguity. These references provide authority and context.
     - For example, if the concrete foundation width deviates from the specified design, mention the relevant section or drawing number from the architectural plans.
  2. *Include relevant information.*
     - *Location*: Specify where the nonconformance occurred (e.g., building section, floor, room).
     - *Description*: Detail the issue—whether it's a dimension discrepancy, material defect, or workmanship problem.
     - *Severity*: Assess the impact on safety, quality, and functionality.
     - *Photographic evidence*: Attach clear photographs of the nonconformance area.
  3. *Give a stakeholder presentation.*
     - Present the NCR to the *proper stakeholders*. This could be the project manager, quality manager, or client representative.
     - *Purpose*: The NCR serves as a formal notification of the issue. It triggers further actions and decisions.
     - *Authority*: By referencing design documents, you give authority to the nonconformity.

Table 6.1 Punch list example of a project.

Item No.	Area/ Location	Item/Description of Deficiency	Responsibility	Due Date	Status	Verification Date	Verified By	Comments
1	Reception Area	Paint touch-up required on the south wall, approximately 3 feet from the floor, small scuff mark.	Painting Subcontractor	April 18, 2025	Open			Mark location clearly with blue tape.
2	Exam Room 102	Door closer is squeaking and needs lubrication or adjustment.	Hardware Subcontractor	April 18, 2025	Open			Ensure smooth and quiet operation.
3	Corridor—Level 2	One ceiling tile above the water fountain is misaligned.	Ceiling Subcontractor	April 16, 2025	Open			Ensure tile is flush with adjacent tiles.
4	Restroom—Ground Floor	Grout missing in a small section (approx. 6 inches) between two floor tiles near the sink.	Tile Subcontractor	April 17, 2025	Open			Match existing grout color and ensure a clean finish.
5	Electrical Room 101	Label missing on the circuit breaker for "Lighting—Exterior West."	Electrical Subcontractor	April 16, 2025	Open			Apply a clear and permanent label.

6	Stairwell B	Handrail on the third floor landing has a slight wobble. Needs tightening.	Metalwork Subcontractor	April 18, 2025	In Progress			Ensure the handrail is securely fastened and stable.
7	Patient Room 315	Window blind is slightly bent at the bottom. Needs replacement or repair.	Window Treatment Vendor	April 21, 2025	Open			Coordinate with vendors for appropriate action.
8	Lobby Entrance	Exterior door sweep is not making full contact with the floor, allowing a draft.	Door Subcontractor	April 17, 2025	Completed	April 19, 2025	Jane Doe	Adjusted door sweep. Verified no draft.
9	IT Room	Data cable hanging loosely behind the server rack. Needs to be properly secured.	Data Cabling Vendor	April 19, 2025	Open			Use appropriate cable management ties or straps.
10	Landscape—West Side	One shrub appears to be dead and needs replacement.	Landscaping Contractor	April 23, 2025	Open			Replace with a healthy shrub of the same species and size.

4. *Make a disposition decision.*
   - The stakeholder reviews the NCR and decides on the appropriate disposition:
     - *Replace*: Replace the nonconforming element.
     - *Repair*: Repair the defect.
     - *Rework*: Rework the work to meet specifications.
     - *Use-as-is*: Use the nonconforming element with customer approval.
5. *Record actions taken.*
   - The NCR serves as a *record* of the entire process:
     - *Corrective actions*: Document the steps taken to identify root cause and corrective action of the issue.
     - *Preventive actions*: Consider preventive measures to avoid similar occurrences of nonconformances in the future.
     - *Timelines*: Note deadlines for corrective actions.
6. *Assign accountability and follow-up.*
   - Assign responsibility for corrective actions.
   - Monitor progress and ensure timely resolution.
   - Update the NCR as needed during the process.
7. *Archive and record lessons learned.*
   - Once resolved, archive the NCR for future reference.
   - Use it as a *learning opportunity* to improve QC processes.

Implementing Corrective and Preventive Actions

Corrective actions address the immediate issue and rectify the specific nonconformance. Preventive actions aim to make systemic changes to prevent similar nonconformances in the future. Both are crucial for maintaining quality standards.

Reinspecting

After corrective actions are taken, reinspect the affected area to ensure the nonconformance has been resolved. This step verifies that the corrective measures were effective and that the work now meets quality standards.

Preventing Recurrence

Finally, focus on preventing recurrence. Analyze root causes, update procedures, and train personnel to avoid similar nonconformances in future projects.

CONCLUSION

A well-defined construction QC plan ensures consistent delivery of quality results and instills confidence in your clients. By addressing nonconformances systematically and timely as the work is performed, the possibility of project failure decreases. Not forcing time-based completion of corrective actions on deficiencies and nonconformances result in conflicts and delays due to excessive rework and quality failures.

QUESTIONS

1. Explain how "design intent" might complicate the identification of nonconformances, and what steps can be taken to mitigate this complication early in the project.
2. Discuss the difference between "tolerances" and "incomplete work" in the context of nonconforming work. Give an example of each and explain why it's important to distinguish between them.
3. Beyond just marking a nonconformance area, what are three critical steps that must be taken to ensure accountability and prompt corrective action? Explain why each of these steps is important.
4. Why is inspecting previously completed work of a similar nature important when a nonconformance is identified? What benefits does this provide beyond addressing the immediate issue?
5. Describe the role of the nonconformance report (NCR) not just as a documentation tool, but also as a trigger for decision-making and a source for future learning and process improvement.
6. While a punch list is typically created at the "substantial completion" phase, what potential issues could arise if contractors do not conduct their own thorough internal review before this inspection with the owner's representative? Explain how these issues might impact the project timeline, costs, and overall perception of the contractor's work.

QUALITY DELIVERABLES VERIFICATION

> Verify the completeness of all quality deliverables. (Evaluate)
>
> **Body of Knowledge VI.A.4**

Effective project closeout hinges on meticulous documentation, particularly the verification of quality deliverables. This section delves into the critical aspects of ensuring all quality-related outputs, from inspection reports to test results, are complete and compliant. Understanding what constitutes a "quality deliverable" within the context of construction is paramount, and this chapter will clarify its definition while providing examples. By verifying the completeness of these deliverables both individually and collectively, CQMs can significantly contribute to a successful and timely project turnover. This proactive approach mitigates risks and ensures accountability throughout the project life cycle.

Earlier in this chapter, during project turnover and closeout, the potential impact of demobilizing quality resources from a project was evaluated. We also identified how to review and assess inspection and test records for compliance with safety-critical processes (see "Required Documentation").

This section also provided ideas, principles, and methods necessary to be successful at maintaining comprehensive documentation. The maintenance of comprehensive documentation can in turn lead to a successful and timely closeout and turnover of a construction project, something that all good contractors strive for. Part of accomplishing this is the verification of all quality deliverables.

Understanding Deliverables and Quality Deliverables

In the context of construction, a *deliverable* is usually a tangible output of the overall construction process, or a subordinate process of the overall construction work (LetsBuild Blog, 2024). Broadly, these outputs could be the physical, completed project itself such as a bridge or a building. But more finitely, the deliverables could be documentation such as design drawings, environmental site assessments, or site evaluation reports, to name a few examples.

Accordingly, it can be logically concluded that a *quality deliverable* is a tangible output of a quality-related process during the construction work. Examples of such quality deliverables include inspector daily reports, installation checklists, nonconformance reports, audit reports, and field test reports. A quality deliverable needs to meet defined quality requirements or expectations. For example, in Figure 6.2, a construction contract may specify that the contractor submit daily construction reports (DCR) by 10 a.m. the following day. The contract may also specify content that should be included in each DCR.

With the minimum content of each DCR established by this specification requirement, one can now evaluate a sample DCR (shown in Figure 6.3) and determine if it contains the minimum content specified.

Based on a review of the sample DCR against the specification's requirements, one can conclude that the minimum content has been provided and it therefore meets the specification requirements. This DCR can be considered a quality deliverable.

Examples of Quality Deliverables

With an understanding of deliverables and quality deliverables in construction, it is important for the CQM to be able to identify common quality deliverables likely encountered on construction projects. Common quality deliverables include:

- Daily inspection reports
- Daily construction reports
- Installation checklists
- Material receiving reports
- Test reports
- Certificates of compliance
- Nonconformance reports
- Corrective action reports (CARs)
- Quality audits and/or assessments
- As-built drawings

1. List of subcontractors at Project site.
2. List of separate contractors at Project site.
3. Approximate count of personnel at Project site.
4. Work performed.
5. Equipment at Project site.
6. Material deliveries.
7. High and low temperatures and general weather conditions, including presence of rain or snow.
8. Accidents.
9. Meetings and significant decisions.
10. Unusual events (see special reports).
11. Stoppages, delays, shortages, and losses.
12. Meter readings and similar recordings.
13. Emergency procedures.
14. Orders and requests of authorities having jurisdiction.
15. Change Orders received and implemented.
16. Services connected and disconnected.
17. Equipment or system tests and startups.
18. Partial completions and occupancies.
19. Substantial Completions authorized.

Figure 6.2 Example of a specification listing out the minimum content of a daily construction report.

Daily Construction Report			
Monday, September 14, 2015	Temperature		Conditions
	H: 83	L: 64	Clear
Subcontractor	Foreman	People	Equipment
R.B. Hinkle	Luis	7	3 pickups

Work Completed Today
R.B. Hinkle - Continued potholes from Station 1105+04 to 1171+00

Meetings/Significant Decisions	Services Connected/Disconnected
No update at this time	No update at this time
Right of Ways	***Stations***
R.B. Hinkle - Yes-N07 (Station 1105+04 to 1171+00)	R.B. Hinkle - Yes-N07 (Station 1105+04 to 1171+00)

Materials Delivered
No update at this time
Change Orders Received & Implemented
No update at this time
Visitors
No update at this time

Other			
Unusual Events	NO	UPDATE:	No update at this time
Stoppages, Delays, Shortages & Losses	NO	UPDATE:	No update at this time
Meter Readings & Similar Recordings	NO	UPDATE:	No update at this time
Accidents/Emergency Procedures	NO	UPDATE:	No update at this time
Orders/Requests of Authorities	NO	UPDATE:	No update at this time
Equipment/System Start-up or Tests	NO	UPDATE:	No update at this time
Partial Completion & Occupancies	NO	UPDATE:	No update at this time
Substantial Completion Authorized	NO	UPDATE:	No update at this time

Figure 6.3 Example daily construction report.

Verifying the Completeness of Quality Deliverables

Verification of something is often defined as "ensuring a product, service, or result complies with the relevant requirements and specifications" (Mosaic, 2017). Verification of the completeness of individual quality deliverables should be performed contemporaneously with the work. As quality deliverables are generated, the CQM or designee should be reviewing them for completeness. If done commensurate with the work, it is much easier to contact the responsible party to get incomplete deliverables corrected, as opposed to waiting until the end of the project when it will likely be much more difficult getting such corrections made. For example, following the erection of structural steel for a given day, a quality inspector may provide the CQM with a steel erection checklist containing multiple blanks. Checklists are quality deliverables, and as such, there should not be any blanks. If fields on the checklist are not applicable, the inspector should mark those as "N/A." Accordingly, the CQM should hand this checklist back to the inspector and ask them to fill in all fields on the checklist.

It is important to evaluate the completeness of quality deliverables on an aggregate level as well. For example, assume that for a given project there will be an anticipated 150 individual concrete placements. The contract requires that an inspection be performed and documented on a preplacement checklist prior to concrete placement. With 150 anticipated placements, the CQM can therefore assume that 150 checklists will be generated. If after the first 30 placements have been made, only 20 checklists have been completed, that likely signifies a problem that needs to be addressed. Perhaps the inspections were indeed performed, but the documentation simply cannot be found—or it may be a sign that concrete has been placed without the proper inspections (a serious problem that warrants immediate attention).

Evaluation of example quality deliverables, both individually and on an aggregate basis, has been summarized in Table 6.2.

Table 6.2 Quality deliverables.

Example Quality Deliverable	Evaluating the Completeness of a Typical Individual Deliverable	Evaluating the Completeness of the Aggregate Quality Deliverables
Installation checklists	Checklists have all spaces filled out, with no blanks; have been signed by the inspector(s); are neat and legible; and checklists are traceable back to the work installed.	An evaluation of all installation checklists should be conducted to verify the anticipated quantity of checklists is reasonably close to the actual number of completed checklists. This will indicate whether the overall level of required inspection was performed per contract requirements or not.
Material Receiving Reports (MRR)	MRRs have all spaces filled out with no blanks; have been signed by the inspector(s); are neat and legible; include evidence that the materials were inspected for any damage or defects; and are traceable back to the work installed.	An evaluation of all completed MRRs should be conducted to validate whether the materials delivered were inspected in accordance with contract requirements or not.

(continued)

Table 6.2 Quality deliverables. (*continued*)

Example Quality Deliverable	Evaluating the Completeness of a Typical Individual Deliverable	Evaluating the Completeness of the Aggregate Quality Deliverables
Test Reports	Test reports have all spaces filled out with no blanks; have been signed by the person conducting the test; are neat and legible; identify pass or fail; are traceable to the work installed; and include evidence that calibrated instruments were utilized.	An evaluation of all completed test reports should be conducted to validate whether the work installed has been tested in accordance with the contract requirements or not. This can be performed on a scope-by-scope basis. The actual number of test reports can be compared against the expected number of test reports.
Certificates of Compliance (CoC)	CoCs clearly state that the supplied product or material complies with the relevant specifications and include evidence of conformance with applicable testing requirements.	An evaluation of all received CoCs should be conducted against the expected number of CoCs typically identified in the specifications.
Nonconformance Reports (NCR)	NCRs should have all spaces filled out with no blanks; include supporting documentation to justify the closure of an NCR; and have been reviewed and approved (as applicable) by the relevant contracting authorities.	It is not possible to know the quantity of NCRs a particular project will generate. However, on an aggregate basis, NCRs that change the as-built condition of a project should be reflected on the as-built drawings.
Corrective Action Reports (CARs)	CARs have all spaces filled out with no blanks and identify the actions necessary as well as documentation substantiating that the actions were indeed taken.	Like NCRs, it is not possible to know the quantity of CARs a particular project will generate. However, on an aggregate basis, it is good practice to ensure CARs are being shared with other stakeholders that may be outside the purview of a given project.
Quality Audits and/or Assessments	The audit/assessment report should include any findings and any necessary actions to take. Audits and assessments should include verification that required actions were taken as well as identification that actions taken yielded improvement.	On an aggregate basis, it is useful to ensure that audits and assessments were performed in accordance with the project's QA plan. For example, if the project's plan stated that a given precast concrete fabricator needs to be audited every 6 months throughout a 24-month fabrication schedule, one would expect to find evidence of four completed audits.
As-built Drawings	As-built drawings should be legible, neat, and accurately reflect the constructed work.	On an aggregate basis, as-built drawings need to accurately capture the as-built condition of the project from a variety of sources. These sources may include but are not limited to: requests for information (RFI), field design changes (FDC), nonconformance reports (NCR), or change orders. It is important to also understand whether the contract requires that shop drawings and technical specifications be as-built as well.

CONCLUSION

The maintenance of comprehensive documentation can in turn lead to a successful and timely closeout and turnover of a construction project—something that all good contractors strive for. Comprehensive documentation includes quality deliverables that will be generated throughout the project. These include but are not limited to checklists, test reports, daily construction reports, nonconformance reports, material receiving reports, and audits.

The CQM and team are expected to continually review these records both on an individual and an aggregate level to ensure complete, accurate, and comprehensive records are being generated. The collection of such records will be a significant step toward the contractor's goal of successfully closing out and turning over a completed construction project to its client in a timely manner.

QUESTIONS

1. Explain what a deliverable is in the context of construction projects.
2. Explain what a quality deliverable is and give an example.
3. Explain why it is important to verify the completeness of quality deliverables commensurate with the work.
4. Give an example of a quality deliverable that is incomplete, and explain the difficulty of correcting the deliverable after the work has been constructed.
5. If during a routine documentation audit the CQM discovers that installation checklists for the prior two weeks of concrete placements are incomplete, what actions should be taken?

FINAL CERTIFICATION

> Facilitate the process for obtaining final certification, ensuring inclusion and maintenance of all documentation related to each closeout milestone and stage (e.g., beneficial occupancy/use, substantial completion, and final completion). Use the current version of Issued for Construction (IFC) documents. (Evaluate)
>
> Body of Knowledge VI.A.5

When is a project "complete"? A project's completeness can mean different things to different individuals involved with it at various points in time. These points in time are called milestones: time limits specified in the contract for the contractor to complete a separate and distinct portion of the scoped work (MTA Construction and Development). Essentially all construction projects have predefined milestones, each with set goals and requirements connected to them and having varying degrees of importance to stakeholders. Those performing the work, those overseeing the work, and those who will ultimately utilize components of the project have a vested interest in seeing project milestones met.

In many cases, project closeout milestones are of utmost importance to some of the most senior project stakeholders, as they tend to have large financial influence over a project's success. If and when project milestones are met can have immense benefit for a project's bottom line or be cause for great loss. In some contracts, financial incentives in addition to a base contract's value are offered if certain milestones are met by a certain date. Conversely, in most contracts, penalties known often as liquidated damages are threatened for failure to meet the same milestones by a mutually agreed-upon date. Early completion incentives and liquidated damages can cause large swings in profit and losses for a contractor, respectively, as they often are sizable relative to a contract's overall value.

For a CQM, it is critical to have these milestones in plans. This ensures the technical and documentational inputs are submitted correctly and expediently to assist the contractor in either reaching milestones early or preventing delays. During construction, it is important for the CQM to ensure that field personnel are supervising building work with the current version of construction documents (IFCs). Similarly, in the closeout phase of a project, it is paramount to verify these technical and documentational inputs meet the most recent requirements and agreements between all stakeholders.

In addition to the financial implications of achieving project milestones, being able to seamlessly provide the necessary documentation to support them is indicative of a robust project quality program (see "Final Turnover and Closeout Documentation" later in this chapter). Being up-to-date on the required inputs necessary for each milestone (as recommended throughout this handbook) should allow for the contractor and customer to seamlessly verify them and keep them as evidential records.

Beneficial Occupancy (or Use)/Temporary Certificate of Occupancy

In many construction contracts, the customer or end user will have a provision enabling them to take possession and begin utilizing a portion of or sometimes seemingly the entirety of the project's scope. The election to take beneficial occupancy will typically denote that the customer believes a given portion of or all of the work is fit for use (MTA Construction and Development). Beneficial occupancy is typically considered the first step on the path to completion of a project; while it may not have many payments or incentives, it is an important target for mitigation of liquidated damages should that topic present itself during a project.

Beneficial occupancy may also not have concrete tasks, documentation, or other criteria associated with it; it may arise out of a need for the customer, rather than a sense of feeling that every portion of the project that was sold to them has been delivered.

Suppose a football team has promised its fan base that they will have the pleasure of seeing their favorite team play in a "brand new, huge stadium" by the start of the upcoming season. Unfortunately, due to an omission found late in the construction process by the owner's designer of record, only 50% of the concession stands and team merchandise stores will be safe to occupy and ready for the home opener. Additionally, the audiovisual equipment is still being tested. What is the more likely scenario in this situation? Will the team decide to continue its games in its old, smaller stadium, or will they opt to move into the otherwise state-of-the-art new stadium that is just about finished? It is clear the team will opt for the latter scenario, as the glamor of the new venue and its increased capacity will generate a great deal more revenue for the team than the loss of its merchandise and some of its concession sales.

In this scenario, the owner will take beneficial occupancy of the stadium. Is the stadium ready for football to be played in? Yes. Will most of the areas be safe for public occupancy? Certainly. Can the team generate a large percentage of the maximum possible revenue at the stadium? Undoubtedly. The owner is gaining benefit from occupying the space.

Typically, on taking beneficial occupancy of a project, a customer issues such a certification to the contractor and begins to assume risk of loss or risk of damage for the portion of work it has taken over, provided the open items are not the root cause of the loss or damage. Let us look back at our stadium example: If, once the season starts, fans begin to use and cause minor damage to the stadium seating, should the contractor be responsible for repairing or replacing the seating just because they have not yet completed the entire project? As written in most contracts, certainly not. Now, suppose during testing of one of the yet-to-open concession stands while the stadium is empty, the contractor causes a fire that carries into a restroom and destroys the fixtures there. Is the contractor liable even though the team had taken beneficial occupancy of the restroom? Most likely, yes.

Beneficial occupancy can signal almost a "sigh of relief" for a contractor in terms of lessening their liabilities on a project. Some involved with other facets of the project believe this milestone will prevent a customer from returning to the contractor with additional punch list items and other concerns, but the CQM must not be so easily assuaged (and should discourage other members of their team from it). While the above scenarios are clear-cut in terms of risks and liabilities associated with beneficial occupancy, other, more opaque issues could arise wherein a customer may associate a failure in an area they are utilizing with a failure on the contractor's part. Even when a punch list is provided by a customer at beneficial occupancy listing outstanding items to be completed, most contracts will not allow for that list to be considered as relieving the contractor from completing all contractually required scope. Awareness of the most current IFCs to use as justification or a rebuttal against punch list items is a means for the CQM to coalesce the project team toward the goal of closeout while also providing strong customer service to the owner.

Even in the example of promises made to fans for a stadium's grand opening, in order for beneficial occupancy to take place, the team must generally feel content with the product it is planning to occupy. The CQM must be the primary guide in ensuring satisfaction, by proving to the customer that all necessary inspections, tests, commissioning, and related documentation are in good order and that the customer is not stepping into a potentially disastrous situation by taking on beneficial occupancy and embarrassing themselves with a subpar experience for its fan base—or worse, potentially risking their well-being. The CQM should be engaged in project closeout meetings tied to this milestone and participate in the needed inspections. The CQM must record any observations made by the customer during this phase to continue the process of completing the project.

Typically, only government agencies can declare a project "fit for use," as regulations for public safety, safety of users, as well as proper documentation for taxation purposes fall under their responsibilities. When the government agency is the owner or end user of a project, they will typically utilize the beneficial occupancy mechanism. When a tenant of a building is not a governmental agency—for example, on residential, commercial, or industrial projects—the owner of a project likely will not have the sole authority to take beneficial occupancy. In such cases, a client (or a subagency or one of its arms) is not its own AHJ, and that duty falls on a governmental agency such as a department of buildings (DOB) or some equivalent moniker, which varies by locale. Such an AHJ must often give the customer additional permissions to occupy the space.

Before the project has been fully closed, frequently a *temporary certificate of occupancy* (TCO) from the AHJ must be granted to the customer for beneficial occupancy. Like the client's own requirements to ensure that beneficial occupancy is the right decision, the "external" or stand-alone AHJ will often have similar requirements specifically regarding the suitability of a space for occupancy. Unlike a client's commercial considerations, this AHJ is typically only concerned with whether that portion has been sufficiently inspected and tested to be compliant with necessary codes and standards (see Chapter 2, "Project Quality Strategy Development," and Chapter 5, "Quality Management Processes" and "Regulatory Body or Authority Having Jurisdiction." Verification of successful inspections and tests for the portion of work completed (sometimes called a temporary sign-off) will be required by the AHJ in order for the granting of the certificate (New York City Department of Buildings, 2024).

In a TCO scenario, as CQM, it is important to not only have the customer's internal requirements in mind but also be the primary facilitator between the customer and AHJ to ensure that both contractual and legal requirements are met. While some of these legal requirements may be better known by a code compliance expert, it is important for the CQM to recruit such experts in order to collectively ensure the customer's efforts are not impeded. In our football stadium scenario, a TCO would almost certainly be required by the AHJ for thousands of spectators to be allowed to enter; as such, the CQM's inputs toward obtaining one are equal in importance to those that bring on beneficial occupancy.

Note that the TCO is not a mechanism utilized in every jurisdiction. Some jurisdictions may have a comparable certificate with different nomenclature, and some may not provide temporary allowances for occupation. The CQM needs to recognize where such opportunities exist and where they do not.

Substantial Completion

In many cases on a project, substantial completion is a milestone that is utilized synonymously with the previous milestone: beneficial occupancy As the term hints, *substantial completion* is the level of completion of all the work at which, in the customer's judgment, the work is complete and in accordance with the contract (with the exception of punch list items); is fit for its intended purpose; and has been sufficiently inspected, tested, and documented to verify this conformance with the contract (MTA Construction and Development). In many cases, the granting of substantial completion will occur simultaneously with the taking of beneficial occupancy, and as such the terms are often used synonymously.

Typically, the key procedural difference between the two is who typically is the initiator of their consideration. Beneficial occupancy typically occurs when the customer has a need to utilize a portion of the project, so it follows that they would make that announcement to the contractor. For substantial completion to be considered, the contractor typically must believe in good faith that a majority of the work has been accomplished and must request that the necessary inspections for this agreement take place.

The reason is made clear when the benefits of each milestone for each of the parties involved are considered. With beneficial occupancy, it is typically the client or end user who has substantial gains, usually with the ability to generate some sort of revenue from the handover. While the contractor does experience a decrease in risk of loss liabilities, it is clear the client or end user is the primary beneficiary of beneficial occupancy.

When the substantial completion milestone occurs (and the certificate of substantial completion is issued), however, the contractor experiences essentially all of the visible benefits in the process. For one, on most projects, substantial completion is tied to a predetermined percentage payment of the contract's overall value. Additionally, fiscal incentives for the achievement of substantial completion earlier than the agreed-on date may exist for the contractor. Lastly, in most contracts, substantial completion also signifies the cessation of accumulation of all contractually defined liquidated damages (MTA Construction and Development). While the client receives the benefit of knowing there will likely be less construction occurring on their property and knowing the project is closer to completion, it is clear the contractor is the primary beneficiary of substantial completion.

In most contracts, the granting of substantial completion by the client is also an acknowledgment that a significant enough portion of the project is complete, such that if they were to take beneficial occupancy, they would not experience any real detriment while utilizing it. Because of this, control over whether substantial completion is granted belongs to the client. In most contracts, the client can unilaterally determine whether substantial completion has been achieved, so it is of the utmost importance for the CQM to provide all necessary reassurances, via documentation and verbal means, of the project's relative completion in relation to the most recent IFCs.

The CQM must be current with all documentation required for verifying the contract agreement. In addition to completed inspection and test documentation, the CQM should also review final record ("as-built") drawings to ensure they meet contract requirements in terms of both accuracy to the completed work relative to the current IFC drawings, and acceptable formatting and required content. While the CQM should ensure that record drawings have been properly redlined and kept current digitally for the duration of the project, a final review of the product to be submitted must also be performed. On a large project, the review of final record drawings can be the most time-consuming endeavor in the closeout process and can therefore be the last impediment to receiving substantial completion on a project.

When substantial completion is granted, the final punch list of outstanding work must be given to the contractor. For the contractor to achieve the ultimate milestone on the project, it is important that the CQM fully reviews this punch list and works with other project staff to ensure the plan to close the punch list items addresses them correctly.

In this interim period, it is also important for the CQM to continue to maintain a dialog with the end user and, to the greatest extent possible, be aware of their operations. If failures occur on completed work, it is critical for the CQM to be kept up to date.

Beneficial occupancy and substantial completion only occur simultaneously in some cases. Returning to our football stadium scenario: Suppose the contractor finishes all but a few punch list items, submits, and is given a substantial completion certificate. However, suppose there is a collective bargaining agreement issue resulting in a strike in the football league. In such a situation, it is possible the team will not take beneficial occupancy of the stadium, and in most contracts, risk of loss remains with the contractor. On any project, the contractor must take all necessary precautions to prevent damage or loss to work already completed. The CQM must ensure the contractor has sufficient measures and plans in place to protect installed work, and if necessary, perform maintenance on installed work to ensure it does not lose verifications or validations performed on it until the point that it is fully ready when the customer takes beneficial occupancy. Typically, only when a customer takes beneficial occupancy of a project does the risk of loss get transferred to them. The only other scenario that alleviates risk of loss is described next: final completion.

Final Completion Certification of Occupancy

Final completion denotes the absolute conclusion of a project (AIA Contract Documents, 2024). It signals the achievement of closure or completion of all contractually required work and deliverables by the contractor to the client, including all punch list items established during the taking of beneficial occupancy, or the granting of a substantial completion certificate. Like substantial completion, final completion must typically be initiated by the contractor; when the contractor believes they have achieved this milestone, they must request that it be considered (MTA Construction and Development). The final completion milestone typically has a sizable payment (sometimes referred to as the "full release of retainage") associated with it; thus its achievement is typically of great importance. As such, the CQM must exercise great expediency after substantial completion is achieved to reach final completion.

Once the request for consideration of final completion has been issued, final inspections and review of documentation will take place; the CQM should be party to all such events. The CQM should take part in pre-inspections to ensure all required punch list items have been addressed by the field to prevent a customer's later inspections from uncovering further deficiencies. Additionally, in some projects where beneficial occupancy has been taken by a customer, additional needs not previously realized during design may arise, resulting in additional work.

Even at this stage, additional work will likely result in a requirement for modifications to be made to the IFCs. It is critical that the CQM remains abreast of these ad hoc changes to ensure final completion inspections and documentation conform to them. Once all required final inspections and documentation verifications occur, the contractor will typically be guaranteed a certificate of final completion.

While final completion typically signifies the true, final closure of a project, there may still be scenarios where the contractor is called on by the customer to perform further work. The most common such scenario is for any rights to warranties or guarantees the customer has as stated in the contract (see "Warranties" later in this chapter).

Two other areas wherein the project's closure may not be fully guaranteed per contract are: (1) latent defects (those that may not have been readily observable by the customer), and (2) other errors the contractor may have known about but did not correct (MTA Construction and Development). For the former, such defects may include items that were arguably difficult for the customer to observe during its inspections. This category of potential further issues can be a contentious topic of discussion due to its subjective nature, but the CQM's best response to such a claim will be the strong verification and validation information recorded and provided to the client. It is critical for the CQM to be continuously involved with inspection and testing to prevent those not directly involved with the quality function from ignoring issues, no matter how minor.

And again, on projects where an owner is not an AHJ, final completion may not take place until a third-party AHJ has deemed it as such (again, after full verification of applicable codes and standards). Whether or not a TCO has been issued, in such scenarios for a tenant to be allowed to occupy the space, a certificate of occupancy (CO) or equivalent certification must be issued. As its name suggests, a TCO almost always has a set expiration date, with constant renewals required if the project does not reach its final state before that date, and each renewal having an associated cost. To that end, it is critical that a CO is pursued as expediently as possible. Obtaining each TCO comes not only with a cost, but more importantly, is not guaranteed, as the granting of a TCO by the AHJ once does not guarantee a renewal should further significant deficiencies be found in the interim. To pursue the CO, it is critical for the CQM to ensure all AHJ requirements are met, including those related to applicable codes. These requirements will typically be identical to those required for the TCO but will be denoted as being "final" in nature.

QUESTIONS

1. How can the CQM ensure work quality is maintained during various contract milestones that other project personnel may not?
2. What procedures should the CQM implement in case of beneficial occupancy not being taken?
3. Why is it critical for the CQM to be involved in punch list closure?
4. Why is it for beneficial occupancy that the CQM helps to reassure the customer of relative completeness?
5. Why should even minor issues that the customer or end user does not notice during milestone inspections not be ignored by the CQM?

FINAL TURNOVER AND CLOSEOUT DOCUMENTATION

> Audit the final turnover plan and related closeout documentation, including permit closeout, operations and maintenance manuals, as-built drawings, attic stock/spare parts records, commissioning reports, and owner training materials. Confirm all quality inspection and test records are reviewed and validated for fitness for use. (Evaluate)
>
> Body of Knowledge VI.A.6

What Is Closeout?

As indicated in "Impact Assessment and Compliance," definitions of closeout in the construction industry can vary. Indeed, the term can be a noun or a verb. Put simply, closeout is the final phase of the construction process and involves the compilation and turnover of a multitude of records generated during the performance of the work (Dobberstein, 2024).

Importantly, the closeout phase happens on a construction project, regardless of the industry sector. Indeed, each of the common construction industry sectors—commercial, residential, heavy civil/infrastructure, industrial, and environmental—all have closeout activities (Adbuilcon). However, as the matrix in Table 6.3 illustrates, the extent of closeout activities can vary widely among construction industry sectors.

The owner or client operating in any one of those construction industry sectors hires a contractor to perform work. Naturally, there would be a contractual agreement between the owner/client and the contractor. As with any contractual arrangement, a series of contractual deliverables are expected to be turned over to the owner/client on completion of the work. The completion and turning over of these deliverables happens during the closeout phase.

Now, it is important to walk through what the closeout phase would look like in each common construction industry sector.

- *Commercial construction*: Examples of commercial construction may include healthcare buildings, schools, retail construction, movie theaters, and theme parks. As an owner or client of one of these facilities, one can certainly expect a punch list phase performed by the contractor as the work nears completion. One also would expect the contractor to turn over any spare parts, attic stock, or extra materials during the closeout phase, as such materials will be useful during the ownership and operation period in the years ahead. Additionally, the owner/client would expect to receive as-built drawings of the constructed facility from the contractor, as they will be useful should any enhancements or additions be made to the facility. The owner also would expect

Table 6.3 Closeout matrix by industry verticals.

Industry Sector	**Commercial Construction**	**Residential**	**Heavy Civil or Infrastructure**	**Industrial**	**Environmental**
Examples	***Healthcare, education, entertainment***	***Single family or multifamily***	***Transit, roads, highways, bridges, marine***	***Oil, gas, chemical, manufacturing***	***Solar, wind, remediation***
Closeout Phase Rigor	Moderate	Low	High	High	Moderate
Parties Involved	General contractor, Subcontractor, architect, engineer, permitting agencies, authorities having jurisdiction, owner	Contractor, Architect, owner	General contractor, Subcontractor, architect, engineer, consultants, permitting agencies, authorities having jurisdiction, owner	General contractor, Subcontractor, architect, engineer, owner	General contractor, Subcontractor, architect, engineer, owner
Level of requirements	Moderate	Low	High	High	Moderate
Closeout duration	Months	Weeks	Years	Years	Months

to receive the operation and maintenance (O&M) manuals for any equipment installed during the work from the contractor. Examples may include air handling equipment, pumps, boilers, electrical switchgear, security monitoring, and so on. During the closeout phase of the project, the contractor will be expected to submit draft copies of such documentation to the owner/client for review and approval.

- *Residential*: Examples of residential construction commonly include single-family homes and multifamily dwellings. In both cases, an owner/client can also certainly expect any spare parts, attic stock, or extra materials during the closeout phase. Just as with commercial construction, such materials will be useful to the owner of the residence to perform any future maintenance.
- *Heavy civil or infrastructure*: Examples of such construction include roads, highways, bridges, railroads, dams, marine, and airports. Such construction naturally is more complex and of a longer duration than some other sectors, and certainly the closeout phase of an infrastructure project follows suit. However, fundamentally, the closeout phase of an infrastructure project is the same. The owner/client of an infrastructure facility will need as-built drawings, O&M manuals, training, completed punch lists, and final permit agency inspections performed.
- *Industrial*: Examples of such construction include manufacturing facilities, gas-fired power plants, chemical plants, and food and beverage processing facilities. An owner would expect that O&M manuals and as-built drawings will be very important as construction nears completion and the facility becomes ready for occupancy and operation. Punch list completion also is an important item for the contractor during the closeout phase. Prior to acceptance of the facility, the owner/client will expect that the work, and any punch list work, has been completed. While O&M manuals certainly are important, the contractor may have installed some specialized equipment for which the owner's employees need to be trained to operate and maintain. Therefore, the owner would expect the contractor to conduct such training prior to commencing operation of the facility.
- *Environmental*: Examples of such construction include solar panel fields, wind turbines, and hazardous cleanup/remediation. Closeout of an environmental-related project is similar to the others. An owner of a solar panel field or a series of wind turbines would expect the contractor to provide spare parts, attic stock, or extra materials during the closeout phase (for example, additional solar panels). Should any get damaged during ownership and operation of the facility, extra panels would be on hand to swap out the damaged ones. Similarly, the contractor would be expected to provide O&M manuals and training on the operation and maintenance of any equipment installed during the construction phase.

What Is Turnover?

As indicated in "Impact Assessment and Compliance," *turnover* in construction is commonly known as the point in which the control of the project is turned over to the owner for care and custody. It is noteworthy to understand the characteristics of successful and poor turnover so that one can be more proactive.

The matrix in Table 6.4 provides typical closeout and turnover elements of any construction project, as well as characteristics of both poor and successful closeout and turnover.

Projects with successful closeout and turnover phases are often the result of an effective closeout and turnover plan developed by the contractor with input from the owner early in the project. A successfully executed closeout and turnover process can improve a contractor's profit margins and, importantly, lead to a very satisfied client.

And although the term "closeout and turnover plan" sounds like something developed at the end of the project, it is recommended that the CQM not fall into that trap. It is best to develop the plan early in the project. Indeed, a 2021 study by the National Cooperative Highway Research Program (NCHRP) found that even U.S. Departments of Transportation reap benefits from beginning the closeout process early. The NCHRP Synthesis 570 concluded that "the sooner that a project team considers project closeout during the design and construction phases, the easier it is to collect and organize required documents and information as the work progresses" (Harper et al, 2021). Generally speaking, a closeout and turnover plan should be developed within the first 10% of the project. As the project progresses and it becomes necessary to revise the plan based on changes made to the project, then changes may be made. A good and effective closeout and turnover plan is one that: (1) is developed early in the project, (2) has been developed with input from the owner, and (3) does not stay stagnant throughout the project.

Auditing the Final Closeout and Turnover Plan

It is important that the CQM understands their role will likely include auditing the closeout and turnover plan to ensure processes are in place and being followed that will result in a successful and efficient turnover. Such an audit may be conducted virtually, as a desk/paperwork audit, or in person at the project site. Such audits may be conducted multiple times over the life of the project, depending on its duration.

To conduct an audit of a final turnover plan, there are a variety of documents to read and understand. These documents will provide the CQM with the content needed to develop the audit checklist. They may include, but are not necessarily limited to, the following:

- Project's specification for closeout and turnover
- Closeout responsibility matrix
- Records closeout matrix

Table 6.4 Typical closeout and turnover elements comparison.

Closeout and Turnover Element	Characteristics of Poor Closeout and Turnover	Characteristics of Successful Closeout and Turnover
Contractor's Closeout Plan	Contractor's general lack of understanding of closeout requirements until too late	Contractor developing a closeout plan early in the project, with input from the owner, which then is executed by the contractor
End-of-Project Submittal Deliverables	Very late end-of-project submittals from the contractor	Quality submittals that are correct, complete, and submitted on time
Punch list	A very lengthy punch list the contractor struggles to complete in any reasonable length of time	Minimal punch list that the contractor completes timely
Operation and Maintenance Manuals	Seemingly endless and preferential comments from the owner on O&M manual submittals	O&M manual submittals being developed right the first time and subsequently approved by the owner
Commissioning Records	Commissioning of vital project equipment was not performed as specified, so equipment does not perform as the owner expected. Commissioning records are incomplete, and the equipment does not perform as specified.	Commissioning procedures submitted for approval to the owner early and subsequent commissioning of vital equipment performed as specified. Commissioning reports document the required and expected performance of the commissioned equipment.
Warranties	Financially troubled subcontractor(s) that are unable to provide a warranty for their work	The contractor being able to quickly and efficiently turn over all project warranty documentation in the manner that both contractor and owner previously agreed to
Contractor's Staffing	Unusually high staff turnover from the owner or the contractor	A contractor who can draw down his staff from the project and reallocate them to other projects
Quality Records (checklists, test reports)	Incomplete quality documentation for the work	The contractor quickly and efficiently being able to compile all quality documentation in an organized manner and provide to the owner in the manner that both contractor and owner previously agreed to.
Certificates of Occupancy	Untimely inspections from the AHJ, which lead to delays in obtaining certificates of occupancy (CO)	Contractor timely notifies the AHJs of required inspections only after all required work is performed per code requirements. AHJ then timely inspects, approves the project, and issues CO.

(*continued*)

Table 6.4 Typical closeout and turnover elements comparison. *(continued)*

Closeout and Turnover Element	Characteristics of Poor Closeout and Turnover	Characteristics of Successful Closeout and Turnover
As-Built Record Drawings	Contractor's failure to record the as-built condition of the project during construction	Contractor contemporaneously records the as-built project on the contract drawings. On completion of the work, as-built drawings are provided to the owner in the agreed-on format and timely approved.
Owner Satisfaction	A severely dissatisfied owner who will remember the poor closeout process when selecting another contractor	An owner who is so satisfied with how efficiently the closeout process was performed they have become a "raving fan" of the contractor. This can lead to more work from the same owner.

- Closeout kickoff meeting minutes conducted with the owner
- Subcontractor and supplier closeout checklist
- Punch list procedure
- Punch list tracking log
- Attic stock or spare parts inventory control plan
- Project demobilization plan
- Periodic closeout status meetings

With a basic understanding of how closeout and turnover can positively or negatively affect a project, it is important to understand the components of an effective closeout and turnover plan. Such components are described next.

Project's Specification for Closeout and Turnover

One of the first steps toward developing a successful closeout and turnover plan is reading the project's specifications to develop a master list of all required closeout and turnover activities and deliverables. This process can be somewhat automated with the use of optical character recognition (OCR) in software. Use terms such as "closeout," "turnover," "handover," "deliverable," "final inspection," "warranty," and so on to collate a master list of closeout and turnover requirements. These requirements should be pulled into a matrix-style document. Each row in this matrix would be used to identify the base requirement. Columns in the matrix would be used to identify things such as: (1) due date of requirement, (2) who is tracking the requirement, (3) which contractor is responsible for fulfilling the requirement, (4) how often the requirement needs to be fulfilled, and so on.

Such a master list of requirements should be developed early in the project, then reviewed and agreed to by the owner. As the project progresses, it is the contractor's responsibility to track the closeout process and update the matrix on a regular basis. The contractor should hold regular meetings, both internally and with the owner, to keep the closeout requirements fresh in the minds of all project stakeholders.

Closeout Documentation

Records Closeout Matrix

The records closeout matrix is a listing of all items required to be turned over to the owner with key milestones from the contract, including but not limited to: quality records, as-built drawings, environmental records, project engineering records, O&M manuals, final payment and retention release, and others. The matrix is developed during the specification review and discussed with the owner during the closeout kickoff meeting. The contractor will assign a responsible staff person to monitor overall progress, and designated people will be assigned to each item on the matrix.

Closeout Responsibility Matrix

Following development of the master list of closeout requirements, it is necessary for the contractor to then assign responsibilities for fulfilling those requirements. It is recommended that each requirement in the matrix be assigned to a single primary responsible person on the contractor's project staff (who may also have secondary support staff).

Closeout Kickoff Meeting with Owner

A documented meeting should be held with the owner no later than at 25% completion of the project. During the meeting, the contractor should review the overall closeout plan as well as the matrix of closeout requirements. Both contractor and owner should also discuss risks to closeout, past experiences/lessons learned, closeout preferences, third-party requirements/risks, punch list process, and so on.

Subcontractor and Supplier Closeout Process

The process to close out a subcontractor or a supplier on a project can vary widely for different general contractors. It is important for the contractor to understand the internal requirements to close out a subcontractor or supplier's work. Moreover, it is equally important for the project personnel to understand the subcontractor or supplier scope of work and contractual requirements in order to know when their work is complete. A checklist should be developed to ensure that subcontractors and suppliers have completed all contract and change order work, satisfied all requirements of the prime contract and their contract, and all documentation has been received prior to final payment and/or releasing retainage.

Punch List Procedure

A fundamental feature of nearly all construction projects at completion is a punch list, the list of generally minor items of work that the contractor must complete.

The contractor must understand the contractual procedure and timelines for everything related to the punch list on the project. To do this, the contractor must review and understand the contractual specifications governing the punch list, and ideally develop flowcharts further detailing the punch list procedure. Such flowcharts can be used to communicate the procedure to both internal staff and externally to owner participants in the punch list process. Flowcharts would reflect the specified punch list procedure for a given project and often entail timelines, durations, and notifications to specific stakeholders to perform punch list inspections.

Punch List Tracking Log

After the punch list procedure has been understood by all parties, the contractor should develop a tracking log to log and monitor all identified punch list items. Such a log, often maintained in a spreadsheet format, is used to assign responsibility to each item as well as track eventual completion of the item. Punch list data maintained in such a format can easily be visualized in various tracking charts.

Spare Parts Inventory or Attic Stock Control Plan

Many projects, on completion, require that the contractor turn over specified extra materials to the owner. Such materials, commonly called spare parts or attic stock, are used by the owner to perform future maintenance on the project. The contractor, having first reviewed the specification requirements for such, should maintain a log of all required spare parts or attic stock. Such a list enables the contractor to ensure the required materials are on hand when the project needs to be turned over to the owner, so it the contractor should know if there are any long-lead time materials that may impact the planned turnover date to the owner.

Project Demobilization Plan

It is important for the contractor to develop a plan to demobilize from the project, and as applicable, restore the site to the same or better condition as it was at the beginning. Again, the contractor must review the contractual specifications to identify all requirements pertaining to site demobilization. The contract likely specifies that a condition precedent to site demobilization is that all applicable permits be closed out.

Periodic Closeout Status Meetings

Periodic closeout status meetings, both internal and with the owner, are to be held to review the status of closeout. Each component of the closeout and turnover plan should be reviewed and planned dates used to determine if tasks are on track.

Other Turnover Audit Activities

In addition to auditing the project's conformance to the aforementioned documents, other activities should be conducted during the audit to assess the project team's efforts toward an efficient and effective turnover. These activities are discussed below.

Review and Quality Inspection and Test Records

The CQM will conduct a review of a sample of quality inspection and test records for completeness and fitness for use. Such records may include installation checklists, field test reports, commissioning reports, or daily inspection reports, to name a few. It is important that these records are contemporaneously completed during the progress of the work. Such records, if found to be incomplete or inaccurate, would not qualify as fit for use and need to be corrected before being provided to the owner.

It is best practice to ensure such quality records are filled out completely, with no blanks. Any fields on these records that are not applicable should be marked "NA" rather than be left blank.

An audit of these records should also identify their accuracy and correctness. An auditor of the records ought to be able to identify clearly incorrect or inaccurate information. Such issues should be corrected prior to being turned over to the owner/client.

Interviews with Closeout and Turnover Staff

An effective audit of the contractor's turnover plan will also include interviews of personnel engaged in the turnover process. The overall purpose of these interviews is to identify: (1) whether the staff members understand the closeout and turnover requirements, (2) if the project is staffed appropriately to handle the turnover demands, (3) if the owner/client is satisfied with the progress of closeout and turnover, and (4) opportunities for improvement.

QUESTIONS

1. Define closeout in the context of a construction project.
2. Define turnover in the context of a construction project.
3. Identify two construction industry sectors whose closeout processes are often the most rigorous, and explain why this might be the case.
4. As a CQM on a gas-fired power plant, what two types of closeout documents would likely be very important to the owner?
5. Give examples of three characteristics that likely indicate a poorly executed closeout and turnover process.
6. Give examples of three characteristics that likely indicate a successfully executed closeout and turnover process.
7. What are two characteristics of inspection and test records that a CQM should look for in order to assess their fitness for use?

WARRANTIES

> Implement effective communication regarding the warranty program and documentation or warranty work to all relevant stakeholders. (Evaluation)
>
> Body of Knowledge VI.A.7

So far in this project closeout section, we have discussed all the verifications and validations via inspection and testing, and the necessary documentation the CQM must understand. The CQM is responsible for ensuring these items during the project closeout process and understanding how they are tied to requirements during other phases of the construction process. These "checks" are all steps in assuring the customer that project requirements are met. All this alone, however, does not suffice for a customer as it does not provide a true guarantee that what was installed does and will continue to operate as intended. This guarantee of sorts is referred to as a *warranty*.

Most people are familiar with warranties as they relate to consumer goods they purchase. Consumer products are often stamped with some duration of warranty—months or years—after the customer "opens the box." This provides the customer with some reassurance that if the product they purchase fails within a given time frame, a refund or replacement may be due to them. On a construction project, the warranty system is similar in concept.

Throughout the course of the project, during its closeout phase and after substantial and final completion (see quality deliverables and verification of this section) milestones are met, warranties provide greater assurance to the customer that they will receive that which they have purchased as well as peace of mind that, should their purchased project fail to serve its intended purpose during a specified period in any way, the cost to repair or replace will not fall on their shoulders.

Warranty Requirements

Construction contracts often contain predefined warranties that the contractor must affirm adherence to at time of award. These warranties will typically require that the contractor's work:

1. Conforms to the contract (including both in its appearance and utility)
2. Be of equivalent grade to existing or replacement items (in cases of rehabilitation/refurbishment contracts)
3. Is in accordance with good industry practices
4. Be of good general quality
5. Be of the most suitable grade for its intended use (MTA Construction and Development)

In almost all cases, the first two items above will be written directly into contracts; any warranty requirements stated in a contract are referred to as explicit warranties. Items 3, 4, and 5 may or may not be written into contracts, but are generally recognized as legally mandated assurances. In cases where they are not in a construction contract, these items are known as implied warranties (Killough, 2024). The CQM must be aware of all of these requirements not only to be cognizant that they may be justification for later corrective actions during this closeout stage, but also to promote good practices and contract conformance throughout the duration of the project.

As with consumer goods, warranties are not forever. A construction contract will mandate that the contractor maintain the above conditions of its work for a specified period, typically from when the work commences until sometime after the contract is closed. In some cases, these warranty expiration dates will be sometime after substantial completion, and in some cases after final completion (see "Quality Deliverables and Verification").

The CQM must also be familiar with all warranty requirements and timelines to ensure necessary corrective actions are performed on work, should they arise at various points throughout the project and following completion. We will look at examples of warranty cases during both time frames, but first we should establish how the CQM classifies and keeps records of the two main types of warranties: general and specific (or specified).

Types of Warranties

Typically, construction contracts will contain a general warranty timeframe for all work as well as different warranty durations for specific work or equipment relative to when that equipment is installed. It is critical (if not required by contract) for a CQM to create a spreadsheet or database of all DFOWs, or eventual "assets" of the client that can independently have warranty requirements (specific warranties). Table 6.5 provides an example of a possible excerpt, and this type of table should be given to the clients, subcontractors, and suppliers for their buy-in and records.

In some contracts, the specific and general warranties may commence on the day a piece of equipment is tested successfully in the factory, while on others they may commence on the day the product is installed. For work performed on the project job site, warranty almost always commences on completion of the work. The CQM must be cognizant of all general contractor, subcontractor, and supplier requirements for warranty and relay that information via their warranty log and record books.

The CQM must be aware of the general warranty period in the contract and must research individual technical specification sections to determine if any equipment, materials, services, or other DFOWs have a greater warranty period than the general warranty (MTA Construction and Development). As the CQM will often be the representative of a project's general contractor, it is important to be cognizant of which subcontractor and suppliers' work is subject to war-

Table 6.5 Warranty excerpt.

GENERAL CONTRACTOR: GLOBAL BUILDERS　　INDOOR GENERATOR INSTALLATION PROJECT　　Final Completion Date 6/2/28

Asset Name	Asset Abbreviation	Asset Description	Main Asset Characteristic	Manufacturer Name	Asset Serial Number	Manufacturer Warranty Contact	Manufacturer Warranty Email	Manufacturer Warranty Duration (Months)	Manufacturer Warranty Start Date	Manufacturer Warranty End Date	(Sub) Contractor Name	(Sub) Contractor Warranty Contact	(Sub) Contractor Warranty Email	Contractor Warranty Duration (Months after Final Completion)	Contractor Warranty Start Date	Contractor Warranty End Date
Exhaust Fan 1	EF-1	Exhaust Fan	2000 CFM	Mechanical Widgets	MO74832	John Smith	[illegible]	48	5/5/25	5/5/29	Global Mechanical	Steven Wing	[illegible]	12	5/5/25	6/2/29
Emergency Generator 1	EG-1	Emergency Generator	13,200 Volts	Electrical Widgets	EG753621	Mary Perez	[illegible]	72	4/3/24	4/3/30	Global Electric	Jana Patel	[illegible]	12	4/3/24	6/2/29
Generator Building	GB	Metal Building Roof	Standing Seam	N/A	N/A	N/A	N/A	N/A	N/A	N/A	Global Roofers	Suri Nazari	[illegible]	120	7/8/23	7/8/33

ranty requirements. The CQM must be able to determine when specific warranties may commence and conclude, to know which parties are responsible. In cases where a specific warranty has expired, the CQM must be aware of the project's general warranty and be prepared to enforce that on responsible parties, as well.

In some cases, a contract may not contain an explicit specific warranty for supplier-provided equipment; however, it is likely the contract will state that, in lieu of a customer-defined specific warranty, the specific warranty of the equipment will be equivalent to the manufacturer's typically provided warrant (MTA Construction and Development). The CQM must identify and evaluate all such "typical" warranties offered by suppliers or manufacturers on a general basis to ensure they are upheld and include them in any tracking mechanisms.

Warranty Time Frames

Some complex pieces of equipment with multiple subcomponents, each from a different manufacturer, may have equally complex warranty considerations. For example, a manufacturer of EG-1 from Table 6.5 may be required to provide a 72-month warranty as shown, but suppose the starter of that generator is a standard, commercially available piece of equipment manufactured by a second-tier supplier, who only provides a 36-month warranty on it. Most contracts stipulate that the general contractor must pass down all warranty requirements to subcontractors and suppliers. If the contract requirements were written correctly into subcontractor/supplier agreements, this should allow the general contractor's CQM to pursue these items from just the first-tier subcontractor or supplier, regardless of that first-tier party's own subcontracts with lower-tier suppliers. Ideally, the CQM will have been party to the creation and review of subcontracts with such suppliers and subcontractors to ensure these requirements were passed on to them (MTA Construction and Development). In rare cases where that was not done, the CQM must be the agent within their own company to take on all associated costs with meeting the warranty.

When seemingly well-built construction or an apparently well-functioning piece of equipment fails prematurely, it is often a reflex for a CQM to write an NCR or CAR (see Chapter 5), and they are sometimes right to do so. But what drives these issues? These types of nonconformances may not be as clear-cut as NCRs written for an item that clearly arrives damaged at receiving, or a failure to construct or assemble work per contract requirements or manufacturer's recommendations—or one that simply fails to operate as intended on first start-up. In these former cases, it is clear the contractor has failed to deliver on their contractual obligations. What about a piece of equipment that has seemingly been delivered in good condition, was assembled correctly, and has operated well during initial testing and commissioning, but suddenly fails while training the end user on its functions? In such scenarios, a contractor may in frustration assert that this is not their fault, because if it were, the failure would likely have been noticed during the commissioning stage. The CQM must now evaluate the contractor's liabilities per warranty expiration dates and evaluate circumstances to determine if any extenuating circumstances caused the failure or if the machine simply failed to meet its stated life expectancy or mean time to repair. If the NCR/CAR is ultimately issued, the CQM must communicate that with all applicable stakeholders, and carry out evaluating the root cause analysis/corrective and preventive action plans submitted by the responsible party as with other deficiencies requiring NCRs/CARs.

Because warranty claims can (and often do) initiate after substantial or final completion of a project, the CQM may no longer be assigned to the project—or even be employed by the contractor. In Table 6.5, you will notice fields for contact information for those individuals a contractor or supplier may designate as warranty points of contact. It is critical that the CQM keep this information current via communication with all subcontractors/suppliers throughout the project (and especially during the closeout phase), because if a warranty claim is issued after completion of the project, those individuals will be responsible for fulfilling such claims. After a project is completed, there may be no way to track warranty claims via NCRs/CARs—and if the CQM is no longer involved with a project, it is important that the warranty tracking documentation be as accurate as possible so the CQM's company personnel can ensure the right equipment or builds are done by the correct parties.

When it is determined that a warranty claim or corrective action is applicable, it is typically up to the party responsible for absorbing all costs related to the necessary repairs/replacements. On completion of claim work, the CQM (or a post-completion designee, if the CQM is no longer involved) must continue to track warranty information, as many contracts may stipulate a reinstatement/prolongment of warranty after such work is completed (MTA Construction and Development). The CQM must log all warranty work performed and ensure manufacturer or contractor warranties are extended by whatever the required durations are and are communicated and logged accordingly. If, for example, this extension of warranty period equals 12 months and the exhaust fan in Table 6.5 undergoes a warranty repair on June 1, 2029, its warranty expiration date will no longer be June 2, 2029, but June 1, 2030.

Difficulties in Substantiation of Warranty Claims

Because warranties are a guarantee of sorts, and are most often called on after the completion of a project and final payment, they can often be contentious. In some cases, when a contractor believes they have completed their project obligations and would like to assert that failures occurred at no fault to them, they may be less likely to comply with warranty claims readily. In such cases, warranty claims can result in complex legal proceedings. In some jurisdictions, statutes of limitations may hinder a customer's ability to make a claim, and breach of warranty statutes may also amplify the level of controversy in these matters (Killough).

The CQM need not be familiar with all such laws regarding warranties. While the CQM will not typically be a legal professional, it is important that meticulous records are kept of all failures and deficiencies so they can help any legal professionals in determining the validity of all claims. The CQM may be called on as a witness to provide testimony to nonconformances throughout the duration of the project and to provide expertise on how quality planning, and inspection and testing may have impacted the current state of disrepair. The CQM must fulfill ethical duties to the customer and their employer in such cases.

Ultimately, in best-case scenarios, warranties claims will not be a factor on a project, but it is important for the CQM to be prepared to take responsibility for executing, communicating, and tracking them both during the closeout of a project and in some cases after its completion.

QUESTIONS

1. What are the two main types of warranties, and what are the CQM's responsibilities in case each is called on in terms of determining party responsibility?
2. Why is it important for the CQM to be familiar with warranty requirements even after they have ceased work on a project?
3. Why should the CQM keep a log of dates for all tested and subsequently installed equipment?
4. What are some complications in the issuance of NCRs for work that fails after some time?
5. Why should the CQM keep a log of warranty repairs, even when initial warranty periods elapse?

Chapter 7

Operations and Maintenance

Operations and maintenance (O&M) are critical phases in the life cycle of any construction project, particularly in ensuring sustained quality after the project's completion. This section discusses the application of quality management principles in the O&M phase, ensuring that the operations do not deviate from the client's quality expectations set during the project's inception. This enduring commitment to quality is vital for maximizing the asset's value, ensuring safety, reliability, availability, and fulfilling regulatory requirements.

The application of quality management principles in O&M goes beyond routine checks and repairs; it involves a dynamic approach to capturing and integrating feedback from all stakeholders, leveraging technologies for monitoring and diagnostics, and adapting to changing environmental and usage conditions. In sectors such as renewable energy (solar and wind) and data centers, O&M is often outsourced to third-party service providers through service level agreement (SLA) contracts. By maintaining strict adherence to the customer's quality expectations, the O&M phase helps in realizing the true value of the infrastructure investment, promoting sustainability, and safeguarding the asset's value over time.

SUSTAINING QUALITY REQUIREMENTS

> Prepare quality objectives required for operations and maintenance (O&M) based on customer expectations. Understand how in-contract and warranty maintenance programs are used throughout the project life cycle. Report future O&M inspections and testing strategies needed (e.g., building component health, risk-based) to the design team. (Apply)
>
> Body of Knowledge VII.A

When O&M activities are outsourced, sustainability becomes a key focus area. Third-party service providers are typically required to adhere to strict environmental and sustainability guidelines as part of their SLA contracts. For example, in renewable energy projects, providers might use advanced monitoring systems to ensure optimal performance and longevity of solar panels and wind turbines, while also implementing practices that minimize environmental impact. These could include using eco-friendly cleaning solutions for solar panels and employing predictive maintenance techniques to reduce waste (infrastructure downtime).

This proactive approach in the O&M phase ensures that the project remains resilient, functional, and relevant, meeting both current and future demands. By leveraging technology, adhering to stringent quality and sustainability standards, and integrating continuous feedback, outsourced O&M activities can effectively support long-term sustainability goals and enhance the overall value of the infrastructure.

Quality objectives for O&M should be explicitly aligned with customer expectations. These objectives are critical because they set standards for maintaining the performance and functionality of the infrastructure over time. Quality objectives must be specific, measurable, achievable, relevant, and time-bound (SMART) (see Chapter 8). Aligning these objectives with customer expectations ensures that the operational phase maintains the quality standards promised to the customer.

Aligning Quality Objectives with Client Expectations

The CQM should follow three steps to align the project team's quality objectives with client expectations.

Step 1: Identify Customer Expectations

Engage directly with customers through surveys, interviews, and feedback forms to understand their expectations for the project's operational phase. This can include aspects like safety, functional quality, sustainability, reliability, efficiency, and aesthetics (visual quality). An example would be conducting a focus group with residents of a new housing development to understand their expectations for maintenance services, safety features, and amenities.

Step 2: Define Quality Objectives

Based on customer feedback, define clear and measurable quality objectives. These objectives should address various aspects of infrastructure O&M to ensure comprehensive quality management. Examples of quality objectives for infrastructure O&M include:

- *Equipment availability*: Ensure that critical equipment maintains a high level of availability, such as targeting 99% uptime (e.g., mission-critical infrastructure like a data center may require 5–9 availability that is 99.999%) for essential systems through preventive maintenance and rapid response to issues.
- *Cost of poor quality (COPQ) reduction*: Aim to reduce costs associated with unplanned breakdowns of equipment by 15% annually through improved predictive maintenance techniques and regular performance assessments.

- *Regulatory compliance*: Achieve full compliance with relevant regulatory standards and environmental guidelines, ensuring all maintenance activities meet or exceed these requirements.
- *Safety and reliability*: Establish objectives to enhance safety and reliability, such as reducing incident rates (both recordables and near miss) by implementing rigorous safety protocols and regular training for maintenance personnel.
- *Customer satisfaction*: Increase customer satisfaction scores by 20% through proactive communication, timely maintenance interventions, and continuous improvement initiatives based on customer feedback.

By defining these diverse and targeted quality objectives, the O&M phase can more effectively address customer needs, enhance operational efficiency, and ensure the long-term sustainability and reliability of infrastructure systems.

Step 3: Implement Monitoring Systems

Use monitoring systems and sensors (e.g., advancements using the Internet of Things [IoT]) to continuously track performance against the defined quality objectives. This can include automated real-time alerts for deviations from set parameters. For example, in managing a commercial building, the O&M team should ensure the HVAC systems operate within parameters defined for energy efficiency and air quality. Adherence not only reflects the client's sustainability goals but also complies with local environmental regulations, ensuring efficient energy use and minimizing environmental impact.

Application of In-Contract and Warranty Maintenance Programs

In-contract and warranty maintenance programs are essential for maintaining the operational integrity and safety of equipment and systems throughout their life cycle. These programs are specifically designed to ensure that all components function as expected, thereby preventing major failures and extending the lifespan of the infrastructure. The effectiveness of these programs hinges on several key elements, each designed to uphold the infrastructure's quality and reliability from installation through to end-of-life: developing strong schedules, compliance with warranty terms, and careful documentation.

Maintenance Schedules

Developing robust maintenance schedules is critical for the systematic upkeep of infrastructure systems. These schedules should be crafted meticulously based on manufacturer recommendations and tailored to align with industry best practices and specific operational conditions. The objective is to establish routine processes that ensure all systems are inspected, serviced, and maintained regularly to prevent the onset of larger issues. As an example, for HVAC systems, a comprehensive maintenance schedule might include quarterly inspections to assess and clean components such as filters, coils, and ducts, along with annual performance testing to evaluate system efficiency and adjust operational settings accordingly.

Similar maintenance may be required for aspects of infrastructure like electrical systems, generators, filtration plants, effluent treatment, and so on. Regular upkeep helps to maintain optimal air quality and energy efficiency, reducing operational costs and environmental impact.

Compliance with Warranty Terms

Warranties protect against manufacturing and construction defects and failures, but they also impose specific maintenance requirements to remain valid. Monitoring compliance with these terms is crucial to ensure that warranties are not inadvertently voided. This involves detailed recordkeeping of all maintenance activities, verifying that repairs and inspections are performed by authorized, qualified personnel, and using only approved parts and procedures and closely monitoring warranty expiry.

Effective spare parts management is also essential for maintaining compliance with warranty terms. Keeping an inventory of critical spares, particularly those with long lead times, ensures that necessary parts are available when needed, thereby preventing prolonged downtime and potential warranty violations. For instance, in the context of elevator maintenance, it is crucial to ensure that all service activities, such as safety checks and component replacements, are performed by certified technicians according to the manufacturer's guidelines. Maintaining an inventory of key spare parts, like control panels and safety devices, helps facilitate timely repairs and replacements.

Regular audits of maintenance logs and technician credentials help verify compliance, ensuring that warranty protections remain in force and that elevators operate safely and efficiently. Additionally, tracking the usage and availability of spare parts through an inventory management system can prevent delays in maintenance activities and support adherence to warranty requirements. By integrating these practices, organizations can uphold the integrity of their warranties and maintain the reliability and safety of their equipment.

Document Maintenance Activities

Comprehensive documentation of maintenance activities is not only a requirement for warranty compliance but also a best practice for effective facility management. Detailed records should include dates, descriptions of performed actions, parts replaced, and the identity of personnel involved. Documentation serves multiple purposes: It provides a clear audit trail for compliance, aids in the troubleshooting of recurring issues, and supports strategic planning for future maintenance or upgrades. Utilizing a computerized maintenance management system (CMMS) to log and manage all maintenance activities offers a centralized platform where performance data can be analyzed to identify trends, forecast potential issues, and optimize maintenance schedules. This system can generate detailed reports that provide insights into system performance, operational efficiencies, and areas needing attention.

Example: Implementation in a New Residential Area

In a newly developed residential complex, the CQM might establish a biannual comprehensive electrical inspection schedule. This includes testing of safety systems like fire alarms and emergency lighting, along with routine checks on residential wiring and communal area lighting systems to ensure they adhere to national safety standards. This proactive approach not only prevents potential electrical hazards but also ensures that all residents live in a safe environment. Regular audits and updates to the maintenance schedule are conducted based on the latest safety guidelines and technological advancements, maintaining the developer's reputation for delivering high-quality, reliable housing.

Identifying and Reporting Future O&M Needs

Proactive management of future O&M requirements is crucial for the longevity and reliability of infrastructure systems. This approach involves detailed forecasting and strategic planning, which focuses on the health of building components and employs a risk-based methodology to prioritize maintenance activities. By anticipating and preparing for future maintenance challenges, organizations can ensure infrastructure safety, optimize functionality, and manage maintenance costs more effectively.

By adopting systematic steps in forecasting, planning, and reporting on future O&M needs, facilities can maintain a high standard of operational integrity and safety. A structured approach supports strategic decision-making regarding maintenance investments and operational adjustments, thereby enhancing the overall asset value and operational readiness of the infrastructure. To report future O&M needs, follow the steps below.

Step 1: Conduct Risk-Based Assessments

Risk-based assessments are essential for identifying potential future issues that could adversely affect the structural integrity and operational efficiency of building components. These assessments involve a detailed analysis of each component's failure probability and the potential consequences of such failures, factoring in variables like environmental conditions, usage intensity, historical performance, and expected life cycle.

Begin with a comprehensive inventory of all components. Analyze each component's exposure to risk factors such as mechanical wear, environmental exposure, operational overload, and so on. Employ statistical tools and historical data to predict the likelihood of failure and its potential impact on overall operations.

Examples of Risk Management and Statistical Tools

- *Failure mode and effects analysis (FMEA)*: This tool helps identify potential failure modes, their causes, and their effects on system operations. For instance, in an HVAC system, FMEA can be used to assess the risk of compressor failure due to mechanical wear and determine the impact on system performance and maintenance needs.
- *Reliability-centered maintenance (RCM)*: This tool involves analyzing failure data to develop a maintenance strategy that ensures system reliability. In a data center, RCM might be used to assess the reliability of backup power systems, ensuring they meet operational demands during power outages.

- *Weibull analysis*: This statistical tool evaluates the probability of failure over time, which is useful for components with varying failure rates. In a chemical processing plant, Weibull analysis can predict the failure rates of valves and pumps handling corrosive materials, helping in scheduling timely maintenance and replacements.

For example, materials might be assessed for corrosion risks and mechanical degradation. Predictive analytics, such as using Weibull analysis to evaluate failure rates, could be employed to determine the expected lifespan of these components under current operating conditions. This proactive approach helps in planning maintenance activities and mitigating risks effectively.

By integrating these statistical tools into risk-based assessments, organizations can enhance their ability to predict potential failures and prioritize maintenance activities, ensuring the continued operational efficiency and structural integrity of their infrastructure. Learning from industry best practices may also be another source of understanding risks and opportunities. It is wise to attend industry conferences and workshops, subscribe to industry journals, and so on.

Step 2: Develop Targeted Inspection and Testing Plans

Following the risk assessments, develop specific inspection and testing plans that are tailored to address the identified risks of each component. These plans should detail the techniques to be used, the frequency of inspections and tests, and the criteria for action should issues be detected.

Outline specific protocols for each type of inspection (e.g., direct physical inspection, the use of specialized diagnostic tools like ultrasonography, or advanced techniques like predictive modeling). Define clear thresholds for each test that determine the conditions under which maintenance or replacement actions are triggered.

For example, for electrical systems in high-rise buildings, implement infrared thermography to detect overheating issues and electrical faults during biannual checks. Establish guidelines that if a particular junction shows a temperature anomaly exceeding 20% of the norm, it triggers an immediate maintenance response.

Step 3: Report and Communicate Inspection Needs and Strategies

Effective communication of inspection strategies and schedules is vital to ensure preparedness across the maintenance teams and other stakeholders. Regular reporting of these plans facilitates transparency, allows for better resource allocation, and sets clear expectations for all parties involved.

Use dashboards and automated reporting tools to provide real-time updates on inspection schedules and results. Conduct regular review meetings with all stakeholders to discuss the findings and refine the strategies as needed.

Integrating Feedback into the Project Life Cycle

Integrating feedback from O&M activities into the design and planning phases of future construction projects is a cornerstone of continuous quality improvement in infrastructure management. This systematic feedback loop is instrumental in refining design standards and operational procedures to better meet established quality benchmarks and respond to user needs more effectively.

Collecting Feedback

Effective feedback collection is foundational to understanding how current projects perform in real-world scenarios. Establishing robust mechanisms to gather this feedback ensures that valuable data on the functionality and user satisfaction of facilities is captured systematically.

Set up diverse channels to collect feedback, including digital platforms such as mobile apps where users can directly report maintenance issues or suggest improvements. Additionally, implement routine surveys and maintain detailed logs that track day-to-day maintenance activities and user interactions.

For example, a residential complex could implement a resident portal app that allows tenants to report issues in real time, schedule maintenance visits, and provide feedback on the service received. This app could also gather user suggestions for amenities and features they feel would enhance their living experience.

For sophisticated infrastructure, such as semiconductor facilities and data centers, automated systems can provide continuous feedback. These systems monitor various parameters in real time, such as temperature, humidity, and equipment performance, and automatically alert maintenance teams to any anomalies. This continuous feedback loop ensures that potential issues are identified and addressed promptly, maintaining optimal operational efficiency and reducing downtime.

By leveraging both manual feedback mechanisms and automated systems, organizations can ensure comprehensive feedback collection that supports effective maintenance and continuous improvement.

Analyzing Feedback

Analyzing feedback involves more than just collecting data; it requires a structured approach to identify trends, pinpoint recurring problems, and understand user satisfaction levels. This analysis helps in recognizing the strengths and weaknesses of current design and maintenance approaches.

Utilize data analytics tools to sift through feedback for patterns that indicate common issues or areas for improvement. Schedule regular review meetings with project teams to discuss these findings and consider their implications for current and future projects.

For example, feedback from the monitoring of a newly constructed dam could reveal unexpected stress points or erosion issues. Analyzing this feedback helps engineers refine their models and predictive analyses for future hydrological projects.

Implementing Improvements

The final step in the feedback loop is to apply the insights gained from the analysis to make informed improvements in the design and maintenance protocols of future projects. This ensures that each project iteration is better suited to the needs and expectations of its users and that any defects are addressed during commissioning to prevent them from escaping into O&M phases.

Develop a formal process for incorporating feedback insights into the project design and commissioning phases. This might include revising material specifications, updating installation procedures, and enhancing user interfaces based on feedback. Additionally, implementing thorough commissioning processes can help identify and rectify defects before they impact the O&M phase.

For example, if feedback indicates frequent failures of wiring insulation in specific environments, future projects might use higher-grade materials or enhanced protective measures during installation. During commissioning, rigorous testing and quality checks can be performed to ensure these enhancements are effective, thereby preventing such issues from recurring in the O&M phase.

By integrating feedback into both the design and commissioning stages, organizations can improve the overall quality and reliability of their infrastructure projects. This proactive approach not only meets user needs but also reduces the likelihood of defects and maintenance issues, leading to more sustainable and efficient operations.

CONCLUSION

In summary, O&M are critical for ensuring sustained quality in construction projects. By aligning quality objectives with customer expectations, implementing effective maintenance programs, and integrating feedback into the project life cycle, organizations can enhance the value and reliability of their assets. A proactive and systematic approach to O&M ensures that infrastructure remains resilient, functional, and relevant, meeting both current and future demands.

QUESTIONS

1. How can in-contract and warranty maintenance programs contribute to sustaining quality requirements throughout the project life cycle?
2. Describe the importance of risk-based assessments in O&M planning, and give examples of how statistical tools can assist in predicting potential failures.
3. How can integrating feedback from O&M activities into the design phase of future projects improve the overall quality and reliability of infrastructure?
4. Beyond simply adhering to manufacturer recommendations, how can an O&M team tailor maintenance schedules to specific operational conditions, and what data should they collect to inform these adjustments?
5. Review the use of IoT advancements in monitoring systems. Explain how the integration of such technology not only helps with real-time alerts but also contributes to achieving other quality objectives mentioned, like reducing the COPQ and enhancing long-term sustainability goals.

BUSINESS CONTINUITY

> Provide inputs for facility shutdown plans, considering both planned (e.g., O&M-required) and unplanned events (e.g., weather or disaster-related incidents). Align the quality plan with the business continuity plan for operational resilience. Apply drills and simulations to test these plans. Monitor the quality-related impacts resulting from shutdowns. (Apply)
>
> **Body of Knowledge VII.B**

Business continuity refers to the ability of an organization to maintain essential functions during and after a disaster or unexpected disruption. It involves planning and preparation to ensure that an organization can continue to operate despite facing adverse conditions. The importance of business continuity cannot be overstated, as it ensures that critical services and operations remain functional, protecting the organization's reputation, financial stability, and stakeholder confidence. In the context of construction, business continuity is vital for ensuring that projects are completed on time, within budget, and to the required quality standards, even in the face of unforeseen challenges.

Disruptions can come in many forms, including natural disasters, equipment failures, supply chain interruptions, and other unforeseen events. Without a solid business continuity plan, these disruptions can lead to significant delays, cost overruns, and compromised quality. Effective business continuity planning helps organizations prepare for these challenges, enabling them to respond quickly and effectively to minimize their impact. It also fosters a culture of preparedness, ensuring that all team members are trained to handle emergencies, understand the chain of command, and know their roles in recovery efforts.

Business continuity planning specifically refers to the detailed procedures and instructions that an organization follows to maintain continuity in the face of disruptions. It covers critical business processes, assets, human resources, and business partners, detailing the steps to manage and mitigate risks. For CQMs, business continuity planning extends beyond the active phases of building and includes post-completion operations and maintenance. This comprehensive approach ensures that the constructed facility remains operational and maintains its integrity long after the project has been handed over.

Importance of Business Continuity

Disruptions, whether scheduled or unforeseen, can significantly derail organizations. By breaking down risks into core elements and analyzing their relationships, effective business continuity planning ensures the maintenance of essential functions and a swift recovery to normal operations. A business continuity plan (BCP) helps build a culture of preparedness by ensuring that all team members are trained to handle emergencies, understand the chain of command, and know their roles in recovery efforts.

Construction Quality Managers (CQMs) play a pivotal role in this process by integrating continuity planning with quality management. By identifying critical quality processes that need protection during disruptions, CQMs can establish robust mitigation strategies to prevent quality degradation.

Developing Business Continuity Plans

A BCP encompasses detailed procedures and instructions that an organization follows to maintain continuity in the face of disruptions. It covers critical business processes, assets, human resources, and business partners, detailing the steps to manage and mitigate risks. In the context of construction, this plan extends beyond the active phases of the building and includes post-completion operations and maintenance. This comprehensive approach ensures that the constructed facility remains operational and maintains its integrity long after the project has been handed over.

Key Elements of BCP in Operations and Maintenance

1. *Critical business processes*: In the post-completion phase, critical processes include regular maintenance, emergency repairs, and the management of essential services (such as power and water supply). The BCP should detail protocols for continuing these services in the face of disruptions, ensuring that there are no interruptions that could affect the building's functionality or the safety of its occupants.
2. *Assets*: The plan must catalog all critical assets required for safe construction and operations—such as heating, ventilation, and air conditioning (HVAC) systems; elevators; structural components; and utility controls—that require regular checks and rapid restoration in the event of failure. Documentation should include maintenance schedules, the location of spare parts, and contact information for repair services.
3. *Human resources*: Effective BCPs incorporate strategies for managing personnel involved in operations and maintenance. This includes training in emergency procedures, establishing communication chains, and arranging backup staffing solutions. The continuity plan should ensure that there is always sufficient skilled staff available to manage and maintain the facility safely.
4. *Business partners*: Identify and collaborate with essential business partners, including suppliers, service providers, and emergency response teams. Agreements should be in place to guarantee quick response times and priority service, which are crucial during unplanned disruptions. These relationships are vital for expedited recovery and the continued operation of the facility.
5. *Risk management*: Conduct thorough risk assessments using the criteria provided in Table 7.1 as guidance to identify potential threats to operations and maintenance activities. Assess the frequency and impact of risks ranging from routine maintenance to catastrophic events like natural disasters. Focus on critical areas such as safety compliance, legal ramifications, financial implications, and reputational damage. Quantify the potential impacts to prioritize planning efforts. The plan should outline specific mitigation strategies for each identified risk, focusing on minimizing downtime and protecting critical infrastructure.
6. *Continuity strategies*: Develop detailed action plans for maintaining or quickly restoring services and operations. These strategies might include installing backup generators, maintaining redundant systems, and employing smart technology for remote monitoring and control of building systems.

Table 7.1 Sample risk management summary.

Risk Type	Potential Impact	Frequency	Mitigation Strategies
Routine Maintenance Issues	Minor delays, increased operational costs	High	Regular maintenance schedules, quick-response teams
Equipment Failures	Downtime, repair costs, safety hazards	Medium	Preventive maintenance, equipment inspections, redundancy
Natural Disasters	Significant downtime, damage to infrastructure	Low	Emergency response plans, structural reinforcements, insurance
Safety Compliance	Legal issues, fines, shutdowns	Medium	Regular safety audits, compliance training, safety protocols
Financial Implications	Increased costs, budget overruns	Medium	Contingency funds, budget monitoring, cost control measures
Reputational Damage	Loss of trust, reduced stakeholder confidence	Low	Transparent communication, PR management, quality assurance
Legal Ramifications	Lawsuits, fines, operational restrictions	Low	Legal reviews, compliance checks, proactive legal consultations
Supply Chain Interruptions	Project delays, increased material costs	Medium	Alternative suppliers, inventory management, flexible contracts
Health and Safety Incidents	Injuries, fatalities, work stoppages	Low	Safety training, protective equipment, emergency procedures

7. *Communication plans*: Clear communication is crucial, particularly in large facilities where many stakeholders, including tenants, maintenance staff, and management, must be informed about disruptions and recovery activities. The BCP should specify protocols for communicating with all relevant parties, utilizing various channels to ensure messages are received and understood.

Planning for Planned Shutdowns

Planned shutdowns are necessary, particularly when conducting essential maintenance to prolong the life of critical infrastructure. However, these shutdowns must be carefully strategized to minimize operational disruptions while maintaining quality standards. CQMs play a vital role in ensuring that planned shutdowns are handled effectively. They oversee quality protocols, as well as support communication and resource planning, to ensure successful maintenance activities. To ensure planned shutdowns are handled effectively, consider the following strategies.

Developing a Maintenance Schedule That Minimizes Operational Disruptions

- *Avoid critical phases*: Plan shutdowns during project phases that are less critical to minimize their impact on overall project timelines.

For example, schedule maintenance during periods of low activity or when construction is progressing through nonessential tasks.

- *Cross-disciplinary coordination*: Work closely with all departments involved, such as engineering, procurement, and operations, to ensure that planned shutdowns align with other key milestones and do not interfere with other critical activities.
- *Buffer time*: Allocate extra time in the maintenance schedule to account for unforeseen challenges or delays, providing a buffer without compromising the project deadline. This buffer should be flexible and consider the type and complexity of maintenance tasks. For instance:
  - *Complex maintenance tasks*: More intricate tasks, such as high-tech equipment, integrated systems, or structural repairs, may require longer buffer times due to the need for specialized equipment and personnel. Consider potential complications, such as unexpected issues discovered during the maintenance process.
  - *Routine maintenance*: Simpler tasks, like regular inspections or minor repairs, may require shorter buffer times but still need consideration for potential delays.
  - *Seasonal factors*: Account for seasonal factors that might affect maintenance, such as weather conditions that could delay outdoor work.
  - *Resource availability*: Ensure the availability of necessary resources, including materials and skilled labor.
  - *Risk assessment*: Conduct a risk assessment to identify potential issues that could arise during maintenance and plan buffer time accordingly. For example, if there's a high likelihood of discovering additional repairs once work begins, increase the buffer time to accommodate these potential delays.

Creating a Comprehensive Checklist of Maintenance Tasks

- *Task prioritization*: Prioritize tasks based on their criticality to project quality and functionality, ensuring that the most important maintenance activities receive priority attention.
- *Compliance checks*: Verify that each maintenance task complies with relevant safety, legal, and environmental standards.
- *Safety protocols*: Include safety protocols that should be followed during the shutdown, detailing protective gear, emergency procedures, and hazard identification measures.

Formulating Strategies to Ensure Continuity of Operations During Maintenance Periods

- *Backup systems*: Establish backup systems that can temporarily take over essential functions during maintenance. For instance, utilize secondary power generators or alternate communication systems.

- *Workforce allocation*: Arrange for personnel to monitor and manage operations during the shutdown period, ensuring that workflows continue smoothly in unaffected project areas.
- *Client communication*: Maintain open lines of communication with clients to inform them about the scope, duration, and expected impacts of the shutdown.

Setting Explicit Quality Objectives to Be Maintained During Maintenance Shutdowns

- *Performance metrics*: Define clear performance metrics and quality thresholds that must be met even during operational downtime, such as minimum standards for structural integrity or acceptable limits for equipment wear.
- *Realistic and measurable*: Ensure that quality objectives are realistic, measurable, and aligned with stakeholder expectations, allowing for consistent monitoring throughout the shutdown.
- *Continuous monitoring*: Establish a continuous monitoring plan that uses technology like sensors or automated reporting systems to track progress against set objectives.

Outlining Resource Management Strategies

- *Availability of resources*: Identify all necessary resources required for maintenance, including personnel, equipment, spare parts, and safety gear, ensuring they are available in sufficient quantities ahead of time.
- *Contingency planning*: Develop contingency plans to address resource shortages or equipment failures, such as sourcing alternate suppliers or having backup machinery on standby.
- *Efficient utilization*: Monitor the use of resources closely to minimize wastage, ensuring that each task is completed efficiently and that surplus supplies are stored securely for future use.

Planning for Unplanned Shutdowns

Unplanned shutdowns can result from various emergencies, such as natural disasters, equipment failures, supply chain disruptions, and other unforeseen incidents. These unexpected events demand proactive planning and a rapid response strategy to minimize their impact on operations, timelines, and quality. CQMs are pivotal in ensuring that business continuity and quality management plans are fully integrated to handle unplanned shutdowns effectively. To effectively navigate unplanned shutdowns, consider the following strategies.

- *Emergency response*: Establish an emergency response team with clearly defined roles and responsibilities. Include a communication strategy that encompasses all key personnel and emergency services, ensuring rapid activation and coordination.

- *Critical function analysis*: Identify and secure critical operations that must remain functional during unplanned shutdowns. Develop strategies to isolate or protect these functions, such as maintaining emergency power supplies or implementing robust data backup solutions.
- *Communication plans*: Craft detailed communication plans for notifying employees, stakeholders, and the public during various incident types. Develop templates for rapid dissemination of information to streamline communications during emergencies.

Aligning Quality Plans with Business Continuity

Integration of Plans

Align quality plans with business continuity strategies to ensure cohesive action during disruptions. This alignment should focus on upholding quality standards and project specifications under all conditions.

Quality Control During Shutdowns

During disruptions, targeted QC measures are crucial to safeguard ongoing work and project specifications. Implement targeted QC measures designed for shutdown scenarios, including:

- *Increasing the frequency of inspections*: Schedule additional inspections of critical project areas, such as unfinished structural components or newly installed systems. Implement checklists for rapid but thorough inspections of safety equipment, material storage, and temporary structures.
- *Utilizing alternative assessment measures*: Consider remote sensing technologies, such as drones and IoT sensors, to gather data where physical access is limited. Develop a risk-based assessment matrix that prioritizes inspections according to each component's criticality.
- *Establishing temporary QC stations:* Simulate normal operations and equip these stations with essential testing tools, protective gear, and communication equipment.

Adjustments and Modifications

Develop a protocol for making rapid adjustments to quality plans, detailing:

- *Decision-making authorities*: Define a clear chain of command outlining who has the authority to approve plan adjustments. Assign decision-making roles based on the criticality of the adjustments required and the expertise of each stakeholder.
- *Criteria for adjustments based on varying shutdown scenarios*: Set predefined criteria that trigger adjustments based on the type of disruption, such as power outages, floods, or material shortages. Differentiate criteria for temporary and permanent modifications, ensuring that temporary adjustments can easily revert to regular quality standards.

- *Methods for implementing and communicating these adjustments swiftly*: Develop standardized templates and forms for implementing adjustments, allowing team members to follow a consistent process. Ensure communication channels are open to inform all relevant stakeholders of quality plan modifications swiftly.

Testing and Simulating Plans

Testing and simulating BCPs is crucial for ensuring they are effective in upholding quality standards during disruptions. By linking these exercises directly to quality plans, CQMs can better understand how various challenges impact critical quality objectives, identify gaps in current strategies, and refine their response measures accordingly.

Conducting Drills and Simulations

Regular drills and simulations provide essential opportunities to validate the effectiveness of BCPs and ensure that quality standards can be maintained. To achieve this:

- *Schedule and frequency*: Plan these exercises at regular intervals, such as semiannually or quarterly, while ensuring the timing aligns with critical project milestones. This helps identify emerging vulnerabilities affecting both continuity and quality.
- *Realistic scenarios*: Simulate real-world challenges that could affect quality, like power outages, supply chain interruptions, or adverse weather conditions. These scenarios should represent typical risks faced by the project to train teams on maintaining quality during disruptions.
- *Inclusive involvement*: Ensure all relevant personnel are involved, including quality managers, operations teams, and external partners. This inclusive approach ensures every team understands their role in protecting quality.
- *Assessment and debrief*: After each drill or simulation, assess QC measures and response times to identify areas needing improvement. Conduct debriefing sessions to discuss findings and plan corrective actions.

Scenario Planning

Scenario planning tests quality standards across various situations by designing comprehensive scenarios that reveal potential gaps in the BCP.

- *Planned shutdown scenarios*: Develop scenarios for planned shutdowns due to scheduled maintenance, equipment replacements, or system upgrades. Assess how quality protocols are maintained and improved during these periods.
- *Unplanned shutdown scenarios*: Create scenarios for unplanned shutdowns due to unexpected incidents like natural disasters, equipment malfunctions, or labor strikes. Determine if quality objectives can be met or require adjustment.

- *Combination scenarios*: Combine planned and unplanned scenarios to assess how multiple disruptions affect quality and reveal weaknesses in current plans. For instance, simulate a planned HVAC system upgrade occurring during a sudden material shortage.

Learning and Adaptation

Incorporate insights from drills and real incidents into the continuous improvement of business continuity plans and quality management systems.

- *Document lessons learned:* Record findings from exercises and incidents, including delays in quality inspections or resource constraints affecting standards.
- *Refine quality plans*: Integrate lessons into ongoing training and use them to update quality protocols and BCPs. For instance, revise quality thresholds or monitoring processes to reflect practical feedback.
- *Gather feedback*: Collect feedback from all participants and stakeholders to ensure that quality adjustments align with project expectations.

Monitoring and Managing Quality Impacts

Effective quality monitoring during disruptions is crucial to maintaining control over the construction process and ensuring the final product meets the required standards. Continuous monitoring during disruptions is essential to manage potential impacts on project quality and make real-time adjustments. Quality monitoring may include the following elements:

- *Implement monitoring systems*: Implement real-time monitoring systems that continuously track critical project parameters, such as structural stability, environmental conditions, and resource consumption. Utilize technologies such as IoT sensors, drones, and real-time data logging devices to monitor critical quality metrics continuously. For example, sensors can be used to measure concrete-curing conditions, environmental conditions, and structural loads to ensure they remain within acceptable limits during disruptions. Develop systems that allow remote observation and control. This includes the use of surveillance cameras and remote access to system controls and sensor data, enabling managers to oversee site conditions without being physically present.
- *Impact analysis*: Regularly analyze the collected data to evaluate the extent of disruptions to project quality. Adjust schedules, resources, and quality controls accordingly.
- *Documentation and continuous improvement*: Document all disruptions and their impacts in detail. Analyze these events to identify trends and refine response strategies for future incidents.

- *Integration with quality management systems (QMS)*: Ensure that data collected during the monitoring process is integrated seamlessly with existing QMS platforms. This integration should provide real-time alerts and automated reports that help in making swift decisions based on the latest data.
- *Data analysis tools:* Utilize data analysis and visualization tools to interpret the vast amounts of data collected. These tools can help identify trends that may indicate potential quality issues before they become critical. The seven quality tools are particularly useful in this context:

  1. *Cause-and-effect diagram (Ishikawa or fishbone diagram)*: For example, during a disruption, use this diagram to identify potential causes of a drop in concrete-curing quality, such as temperature fluctuations, material inconsistencies, or improper mixing procedures.
  2. *Check sheet*: For example, develop a check sheet to record occurrences of equipment malfunctions during disruptions. This data can help identify patterns and prioritize maintenance activities.
  3. *Control charts*: For example, implement control charts to monitor structural stability parameters like load-bearing capacity over time. These charts can highlight any variations outside acceptable limits, prompting immediate corrective actions.
  4. *Histogram*: For example, utilize histograms to visualize the distribution of data collected during monitoring. Or create histograms to display the frequency of different types of defects or deviations encountered during disruptions.
  5. *Pareto chart*: For example, apply Pareto charts to identify the most frequent causes of delays during disruptions. For instance, it may reveal that most delays are due to supply chain interruptions, guiding efforts to mitigate this specific issue.
  6. *Scatter diagram*: For example, create scatter diagrams to analyze the relationship between environmental conditions (like humidity) and structural load stability. This can help determine if specific conditions correlate with reduced stability.
  7. *Flowchart*: For example, develop flow charts to map out the processes involved in quality monitoring and control. Create a flowchart to detail the steps taken when a quality issue is detected, from initial detection to resolution.

Impact Assessment

Thoroughly assessing the impact of any disruption on project quality is key to understanding its effects and preparing for future incidents. Impact assessment can be done in many ways, but the following are key components of an impact assessment:

- *Developing a standard impact assessment protocol*: Create a detailed protocol that defines how to measure the impact of different types of disruptions on project quality. This should include specific metrics such as work delays, cost overruns, material wastage, and any deviations from the project specifications.
- *Scenario analysis*: Use data from past disruptions and predictive analytics to simulate various scenarios and their potential impacts on project quality. This analysis helps in understanding the most vulnerable areas and preparing mitigation strategies accordingly.
- *Continual improvement*: Based on the impact assessment, identify areas for improvement in both the construction process and the BCP. Implement changes designed to mitigate the identified impacts in future projects.
- *Documentation and reporting*: Ensure all findings from impact assessments are thoroughly documented and reported to all relevant stakeholders. This documentation should include detailed analysis, conclusions, and recommended actions.

Feedback Loops

BCPs must remain dynamic, evolving documents that can adapt to emerging challenges and shifting project requirements. A crucial aspect of maintaining their relevance and efficacy is establishing feedback loops. These loops are continuous improvement mechanisms that gather insights from various stakeholders, identify weaknesses in the existing BCP, and suggest data-driven enhancements to the planning process.

Effective feedback loops are particularly vital where the scale, complexity, and unique interdependencies mean that disruptions can have a substantial ripple effect across timelines, quality benchmarks, and financial outcomes. By systematically collecting and analyzing feedback from multiple sources, CQMs can understand the impacts of disruptions on different facets of the project, allowing them to refine their strategies. The steps involved in setting up and maintaining effective feedback loops for BCPs are as follows:

1. *Stakeholder engagement:* Regularly engage with all stakeholders, including project teams, clients, suppliers, and subcontractors, to gather feedback on the effectiveness of the BCP and its impact on quality.
2. *Feedback collection tools:* Implement tools such as digital surveys, mobile apps, and interactive meetings to collect feedback efficiently. These tools should be designed to gather both quantitative and qualitative feedback.
3. *Review meetings*: Schedule regular review meetings to discuss feedback and any incidents or disruptions that occurred. These meetings should aim to analyze the effectiveness of the response and identify lessons learned.

4. *Action plans*: Develop action plans based on the feedback received. These plans should focus on addressing shortcomings in the current business continuity plan and implementing suggestions to improve QC measures.
5. *Feedback integration*: Ensure that feedback is systematically integrated into the business continuity planning process. This should include revising the plan to incorporate new insights and strategies derived from stakeholder feedback.
6. *Training and development*: Based on feedback and the outcomes of impact assessments, identify training needs for the team. Implement training programs to enhance skills related to quality management and crisis response.

CONCLUSION

This chapter equips CQMs with strategies to analyze and implement comprehensive business continuity plans that ensure operational resilience and uphold quality standards during both expected and unexpected disruptions. By integrating continuity planning with quality management, CQMs can identify critical quality processes that need protection during disruptions and establish robust mitigation strategies to prevent quality degradation. Through effective business continuity planning, CQMs can ensure that construction projects are resilient, adaptable, and capable of maintaining high standards of quality regardless of the challenges they face.

QUESTIONS

1. What are the primary benefits of integrating quality plans with business continuity strategies during both planned and unplanned disruptions?
2. How can CQMs effectively assess and address quality impacts that arise during a shutdown?
3. Describe the role of drills and simulations in maintaining quality standards during disruptions. How can these exercises help identify potential gaps in business continuity plans?
4. What feedback mechanisms are essential for continually improving a BCP, and how can these insights be effectively integrated into future planning?
5. How can a CQM strategically balance the need for comprehensive buffer time during planned maintenance shutdowns with the potential for increased project costs and schedule delays, while still ensuring quality objectives are met? Give examples of factors that should be considered in this balancing act.

Chapter 8

Quality Manager Skills

The purpose of the American Society for Quality (ASQ) Code of Ethics is to establish global standards of conduct and behavior for its members, certification holders, and anyone else who may represent or be perceived to represent ASQ. In addition to the code, all applicable ASQ policies and procedures should be followed. Violations of the Code of Ethics should be reported. Differences in work style or personalities should be first addressed directly with others before escalating to an ethics issue.

The ASQ Professional Ethics and Qualifications Committee, appointed annually by the ASQ Board of Directors, is responsible for interpreting this code and applying it to specific situations, which may or may not be specifically called out in the text. Disciplinary actions will be commensurate with the seriousness of the offense and may include permanent revocation of certifications and/or expulsion from the society.

ASQ CODE OF ETHICS

> Demonstrate appropriate behaviors and actions that align with the ASQ Code of Ethics. (Analyze)
>
> Body of Knowledge VIII.A

ASQ requires its representatives to be honest and transparent. Avoid conflicts of interest and plagiarism. Do not harm others. Treat them with respect, dignity, and fairness. Be professional and socially responsible. Advance the role and perception of the quality professional.

Act with Integrity and Honesty

- Strive to uphold and advance the integrity, honor, and dignity of the Quality profession.
- Be truthful and transparent in all professional interactions and activities.

- Execute professional responsibilities and make decisions in an objective, factual, and fully informed manner.
- Accurately represent and do not mislead others regarding professional qualifications, including education, titles, affiliations, and certifications.
- Offer services, provide advice, and undertake assignments only in your areas of competence, expertise, and training.

Demonstrate Responsibility, Respect, and Fairness

- Hold paramount the safety, health, and welfare of individuals, the public, and the environment.
- Avoid conduct that unjustly harms or threatens the reputation of the Society, its members, or the Quality profession.
- Do not intentionally cause harm to others through words or deeds. Treat others fairly, courteously, with dignity, and without prejudice or discrimination.
- Act and conduct business in a professional and socially responsible manner.
- Allow diversity in the opinions and personal lives of others.
- Safeguard Proprietary Information and Avoid Conflicts of Interest
- Ensure the protection and integrity of confidential information.
- Do not use confidential information for personal gain.
- Fully disclose and avoid any real or perceived conflicts of interest that could reasonably impair objectivity or independence in the service of clients, customers, employers, or the Society.
- Give credit where it is due.
- Do not plagiarize. Do not use the intellectual property of others without permission. Document the permission as it is obtained.

QUESTIONS

1. According to the ASQ Code of Ethics, what is the initial step someone should take when facing a difference in work style or personality with another individual before escalating to an ethics issue?
2. The ASQ Code of Ethics emphasizes avoiding conflicts of interest. Explain what constitutes a "perceived conflict of interest" in the construction industry and why it is also addressed in the code.
3. What are the potential disciplinary actions that the ASQ Professional Ethics and Qualifications Committee can take for violations of the Code of Ethics, and what determines the severity of these actions?

4. The code states quality professionals should "offer services, provide advice, and undertake assignments only in your areas of competence, expertise, and training." Describe a scenario where a quality professional might unintentionally violate this principle and what steps they could take to rectify the situation.
5. Beyond simply not plagiarizing, what specific action does the ASQ Code of Ethics require regarding the intellectual property of others?

STAKEHOLDER ANALYSIS AND ALIGNMENT

> Identify the perspectives, needs, and objectives of various internal and external stakeholders. Integrate collaboration and negotiation techniques to align the organization's strategic objectives and stakeholder expectations. (Apply)
>
> Body of Knowledge VIII.B

Effective stakeholder analysis and alignment are crucial for successful project delivery in the global construction industry. Identifying the perspectives, needs, and objectives of various internal and external stakeholders is paramount. This includes understanding the distinct viewpoints of owners/clients, architects/engineers, general contractors, subcontractors, suppliers, regulatory agencies, local communities, and financial institutions. By employing collaboration and negotiation techniques, organizations can align their strategic objectives with stakeholder expectations, mitigating risks and fostering positive relationships. This ensures project goals are met efficiently and sustainably throughout the entire project life cycle.

Who Are the Stakeholders?

ISO 9000 international standard for quality management systems fundamentals and vocabulary defines a *stakeholder* as "a person or organization that can affect, be affected by, or perceive itself to be affected by a decision or activity made by you or your organization" (ISO 9000). As an example, customers (clients), owners, general contractors, subcontractors, tradespeople, employees in an organization, other service providers, bankers, regulators, unions, partners, or society can include competitors or opposing pressure groups. Key stakeholders in the global construction industry are listed below:

Owner/Client

- *Perspective*: Stakeholder commissioning the construction project and responsible for its outcome. They are invested in the project's outcome and want it to meet their vision, and business needs, and stay within budget.

- *Needs*: A functional, safe, and aesthetically pleasing construction completed on time and within budget. They also need clear communication and transparency throughout the project.
- *Objectives*: Maximize return on investment, minimize risks, and ensure quality and timely completion.

Architect/Engineer

- *Perspective*: Responsible for the design and technical aspects of the project. Their focus is on creating a functional and aesthetically pleasing design that meets the client's requirements and adheres to building codes and regulations.
- *Needs*: Clear project scope, client collaboration, and timely feedback to ensure design aligns with expectations.
- *Objectives*: Deliver a creative, innovative, and sustainable design that is also practical and cost-effective.

General Contractor

- *Perspective*: Responsible for overseeing the construction process and managing subcontractors. They are focused on delivering the project on time and within budget while maintaining quality and safety standards.
- *Needs*: Clear project plans, adequate resources, and effective communication with all stakeholders.
- *Objectives*: Efficient project execution, cost control, risk mitigation, and timely completion.

Subcontractors

- *Perspective*: Specialized tradespeople responsible for specific aspects of the construction, such as electrical, plumbing, or heating, ventilation and air conditioning (HVAC). They are focused on completing their work to the required standard and on schedule.
- *Needs*: Clear instructions, timely payments, and a safe working environment.
- *Objectives*: Deliver quality workmanship, meet deadlines, and maintain profitability.

Suppliers

- *Perspective*: Provide materials and equipment necessary for construction. Their focus is on delivering the right products on time and at the agreed-on price.
- *Needs*: Clear orders, timely payments, and accurate delivery schedules.
- *Objectives*: Maintain good relationships with contractors, ensure timely delivery, and achieve profitability.

Government/Regulatory Agencies

- *Perspective*: Ensure compliance with building codes, zoning laws, and environmental regulations. They are focused on public safety and welfare.
- *Needs*: Accurate project information, adherence to regulations, and timely permits and inspections.
- *Objectives*: Protect public safety, ensure compliance, and promote sustainable development.

Local Community

- *Perspective*: This may be impacted by the construction project in terms of traffic, noise, or environmental concerns. They are interested in the project's potential benefits and drawbacks for their community.
- *Needs*: Information about the project, opportunities for input, and mitigation of any negative impacts.
- *Objectives*: Minimize disruption, ensure the project benefits their community, and have their concerns addressed.

Financial Institutions

- *Perspective*: Provide funding for the project. Their focus is on ensuring the project is financially viable and that their investment is secure.
- *Needs*: Detailed project plans, financial projections, and risk assessments.
- *Objectives*: Protect their investment, ensure timely repayment, and achieve a return on investment.

Additional Stakeholders

- *Labor unions*: Protect workers' rights and interests.
- *Insurance companies*: Provide coverage against risks and liabilities.
- *Consultants*: Offer specialized expertise in areas like public relations, environmental, sustainability, acoustics, or security.

Effective stakeholder management is crucial for successful project delivery in the global construction industry. Understanding the diverse perspectives, needs, and objectives of each stakeholder helps ensure collaboration, mitigate risks, and achieve project goals.

Stakeholder Analysis

Stakeholder analysis is a technique used to identify and understand the needs and expectations of parties involved in a project, system, or organization (Furterer and Woods, 2021; Ramu, 2016). It evaluates the relative importance of different stakeholders and their interests and prioritizes allocating resources to best address their needs. It provides a way to visually evaluate how well an organization's efforts align with stakeholder needs and interests.

The first step is to identify all relevant stakeholders (see "Key Stakeholders in the Global Construction Industry"). The next step is to identify the needs and interests of each stakeholder. This information can be gathered through surveys, interviews, focus groups, and other methods.

Once the needs and interests of each stakeholder have been identified, they can be prioritized. This involves assessing the relative importance of each stakeholder and the potential impact they can have on the organization.

The final step is to develop strategies for addressing the needs of each stakeholder. This may include making changes to products, services, processes, or policies. It may also involve communicating with stakeholders to keep them informed about the organization's plans and progress.

Stakeholder Collaboration and Negotiation

Collaboration opportunities for construction stakeholders span the entire project life cycle, from conception to completion and beyond. Here's a breakdown of these opportunities at each stage.

During the PreConstruction Phase

Collaboration begins with the preconstruction phase, as described next.

Project Definition and Planning

- *Collaborative workshops*: Involve the owner, architect, engineer, and potential contractors to define project goals, scope, budget, and schedule.
- *Feasibility studies and value engineering*: Jointly assess project viability, explore alternative designs, and optimize cost-effectiveness.
- *Risk assessment and mitigation planning*: Collectively identify and develop strategies to manage project risks.

During the Design Phase

- *Integrated design process*: Foster close collaboration between the architect, engineer, and relevant consultants for a cohesive and efficient design.
- *Building information modeling (BIM)*: Use a shared 3D model to facilitate collaboration, clash detection, and visualization.
- *Design reviews and constructability analysis*: Involve contractors early in the design process to identify and resolve potential construction challenges.

During the Procurement Phase

- *Procurement planning*: Collaborate with project planners, design engineers, and CQMs to identify, select, and qualify suppliers for material, equipment, and maintenance. Negotiate with suppliers on pricing, on-time delivery, and quality, reliability, safety, and sustainability requirements.
- Conduct audits to qualify and periodically monitor the quality and service level of suppliers throughout the project. CQMs may have to

Table 8.1 Power and influence grid.

Quadrant	Description
High Power, High Influence	These stakeholders are key players and decision-makers. They have a significant impact on the project's success. *Strategy*: Manage closely and keep them actively engaged.
High Power, Low Influence	These stakeholders have the power to affect the project but may not be highly involved. *Strategy*: Keep them satisfied and informed.
Low Power, High Influence	These stakeholders are highly involved but have limited power to affect the project. *Strategy*: Keep them informed and consult regularly.
Low Power, Low Influence	These stakeholders have minimal impact on the project and are not highly involved. *Strategy*: Monitor them with minimal effort.

also collaborate and negotiate requirements with subcontractors of suppliers where critical materials are provided.

Stakeholder Management

There are a variety of tools that can be used for stakeholder management. The power and influence grid (Table 8.1), also known as the power/influence matrix, is a tool used in stakeholder management to categorize stakeholders based on their level of power and influence over a project. This helps project managers prioritize their efforts and develop appropriate strategies for engaging with different stakeholders. Here's a breakdown of the grid.

Power and Influence Grid

The salience model is also popular. The salience model is a tool used in stakeholder management to identify and prioritize stakeholders based on their importance to a project. This model was developed by Ronald K. Mitchell, Bradley R. Agle, and Donna J. Wood in 1997. It uses three key attributes to classify stakeholders:

Key Attributes of the Salience Model

- *Power*: The ability of the stakeholder to influence the project or its outcomes.
- *Legitimacy*: The appropriateness of the stakeholder's involvement in the project, based on legal, moral, or contractual grounds.
- *Urgency*: The degree to which the stakeholder's needs require immediate attention.

Stakeholder Types in the Salience Model

- *Dormant stakeholders*: Have power but lack legitimacy and urgency. They may become more active if their legitimacy or urgency increases.
- *Discretionary stakeholders*: Have legitimacy but lack power and urgency. They are often supportive but have limited influence.
- *Demanding stakeholders*: Have urgency but lack power and legitimacy. They may be vocal but have little impact on the project.

- *Dominant stakeholders*: Have both power and legitimacy but lack urgency. They are influential and should be kept satisfied.
- *Dangerous stakeholders*: Have power and urgency but lack legitimacy. They can be disruptive and need careful management.
- *Dependent stakeholders*: Have legitimacy and urgency but lack power. They rely on others to advocate for their needs.
- *Definitive stakeholders*: Have power, legitimacy, and urgency. They are the most important stakeholders and require close attention.

Using the Salience Model

1. *Identify stakeholders*: List all individuals and groups affected by the project.
2. *Assess attributes*: Evaluate each stakeholder's power, legitimacy, and urgency.
3. *Classify stakeholders*: Place each stakeholder into one of the seven types based on their attributes.
4. *Develop engagement strategies*: Tailor your communication and engagement strategies based on the stakeholder's classification.

The salience model (Figure 8.1) helps project managers prioritize their efforts and ensure the most critical stakeholders receive the attention they need.

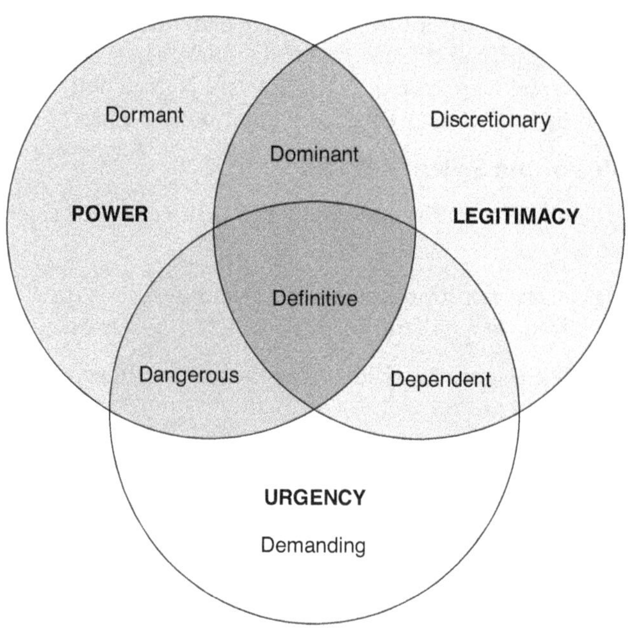

Figure 8.1 Salience model for stakeholders' assessment.

During the Construction Phase

Collaboration continues in the construction phase, as described next.

Project Management

- *Lean construction principles*: Employ collaborative techniques like the last planner system (see the Lean Construction Institute, https://leanconstruction.org, for more information) to optimize workflow, minimize waste, and improve productivity.
- *Collaborative project management tools*: Utilize digital platforms to centralize communication, document sharing, and progress tracking.
- *Regular progress meetings*: Hold meetings with all stakeholders to review progress, discuss challenges, and ensure alignment.
- *Change management*: Establish a transparent process for managing design changes and their impact on project scope, budget, and schedule.
- *Quality and safety management*: Work together to create a safe working environment, share best practices, and conduct regular safety inspections. Collaborate on QA and QC measures to ensure the project meets specifications and standards.

During the Closeout and Turnover Phase

- *Testing and inspection*: Jointly conduct comprehensive testing and inspections to verify system functionality and compliance.
- *Training and documentation*: Collaboratively develop user manuals and provide training for building operation and maintenance involving client, third-party service providers, original equipment manufacturers (OEMs), and other relevant stakeholders.

During the Operations and Maintenance Phase

- *Building performance monitoring*: Share data and insights on building performance to optimize energy efficiency and identify potential issues.
- *Asset management*: Collaborate on maintenance plans and life cycle management strategies to ensure long-term asset value with client's operation staff, third-party maintenance service providers, OEMs, and spares suppliers.

During the Project Evaluation Phase

Conduct post-project reviews with all stakeholders to identify lessons learned and implement improvements for future projects.

By embracing a collaborative approach throughout the entire construction project life cycle, stakeholders can achieve better project outcomes, optimize resource utilization, and foster long-term relationships built on trust and mutual success.

All relevant stakeholders are brought together through various meeting opportunities to collaborate on the accomplishment of organizational objectives.

Negotiation Techniques

Here are some effective negotiation techniques to align an organization's strategic objectives with stakeholder expectations in the context of construction projects.

Collaborative Approach

- *Principle*: Focus on creating win-win solutions rather than a zero-sum game. Encourage open communication and active listening.
- *Application*: Hold workshops or brainstorming sessions with key stakeholders to explore their expectations and understand their underlying interests. This fosters mutual respect and trust.

Clear Communication

- *Principle*: Ensure all parties have a clear understanding of project goals, constraints, and individual expectations.
- *Application*: Use visuals like project timelines, budget breakdowns, and design presentations to clearly illustrate project scope. Maintain regular communication channels to update on progress and address any emerging concerns.

Identify and Prioritize Interests

- *Principle*: Distinguish between stakeholders' positions (what they want) and their interests (why they want it).
- *Application*: Engage in open dialogue to uncover stakeholders' underlying interests, which may offer more flexibility for finding mutually agreeable solutions.

Flexibility and Creative Problem-Solving

- *Principle*: Be open to exploring alternative options and solutions that can satisfy multiple interests.
- *Application*: Brainstorm creative options, offer trade-offs, or consider phased implementations to bridge gaps between strategic objectives and stakeholder expectations.

Transparency and Trust Building

- *Principle*: Foster an environment of open communication and honesty. Share relevant information and acknowledge potential challenges.
- *Application*: Be transparent about project constraints, risks, and any potential trade-offs. This demonstrates respect for stakeholders and builds trust.

Objective Criteria

- *Principle*: Use objective data and industry benchmarks to evaluate proposed solutions and facilitate decision-making.
- *Application*: Utilize cost estimates, market analysis, or sustainability metrics to support proposed solutions and demonstrate their alignment with strategic objectives.

Compromise and Concessions

- *Principle*: Be prepared to make reasonable compromises and concessions to reach an agreement that satisfies key interests.
- *Application*: Identify areas where flexibility is possible and offer concessions that are aligned with strategic priorities while addressing stakeholder concerns.

Documentation and Agreement

- *Principle*: Document agreements clearly to avoid misunderstandings and ensure all parties are aligned.
- *Application*: Prepare a comprehensive project charter or agreement outlining project goals, stakeholder expectations, and agreed-on solutions.

Ongoing Stakeholder Engagement

- *Principle*: Maintain regular communication and engagement with stakeholders throughout the project life cycle.
- *Application*: Hold periodic review meetings to track progress, address concerns, and ensure alignment with evolving expectations.

Dispute Resolution Mechanisms

- *Principle*: Have a clear process for resolving disputes if they arise.
- *Application*: Consider using mediation or arbitration to address conflicts in a constructive and efficient manner.

By applying these negotiation techniques, organizations can navigate complex stakeholder relationships in the construction industry, fostering collaboration and achieving project success while aligning with their strategic objectives.

CONCLUSION

Remember, successful negotiation requires a combination of preparation, communication, flexibility, and a focus on mutual benefit.

Robust stakeholder analysis and alignment are essential for navigating the complexities of the global construction industry. By meticulously identifying, understanding, and prioritizing the needs and objectives of diverse stakeholders, organizations can foster collaboration, mitigate risks, and enhance project outcomes. Effective negotiation, clear communication, and the use of tools like the power and influence grid and salience model ensure that strategic objectives are aligned with stakeholder expectations. Ultimately, these practices pave the way for successful, sustainable, and mutually beneficial project completion and long-term relationships.

QUESTIONS

1. How would you identify all relevant stakeholders for your project or organization?
2. What are the methods to identify the needs and expectations of relevant stakeholders?

3. What are the corroboration techniques with stakeholders to develop strategic objectives?
4. What are the negotiation techniques used while collaborating with stakeholders, and when are they applied?
5. Why is a CQM involved in the identification of the needs and expectations of stakeholders? What is their scope in this process?
6. Explain how the salience model differs from the power and influence grid in categorizing and prioritizing stakeholders, and provide a scenario where using the salience model would be more advantageous.
7. Describe the potential conflicts that might arise between the objectives of an owner/client and those of a local community in a large-scale construction project, and outline strategies a CQM could employ to mitigate these conflicts.
8. Analyze the statement "Distinguish between stakeholders' positions and their interests" in the context of negotiation techniques within a construction project. Give an example of how this distinction can lead to a more favorable outcome for all parties.
9. Discuss the role of building information modeling (BIM) as a tool for stakeholder collaboration throughout the design and construction phases. How does BIM address specific needs of various stakeholders, such as architects, engineers, and subcontractors?
10. Considering the concept of "definitive stakeholders" in the salience model, explain why it is crucial for a CQM to identify and closely manage these stakeholders. What are the potential risks of neglecting the needs and expectations of definitive stakeholders?

Team Building

VISION DEVELOPMENT

> Collaborate with stakeholders to adopt quality goals and a clear vision for a team or project. (Apply)
>
> **Body of Knowledge VIII.C.1**

Effective team building begins with a shared vision and clearly defined quality goals. In the realm of construction, these foundational elements ensure projects align with stakeholder expectations and industry standards. This chapter explores how collaborative approaches, encompassing early engagement and transparent communication, are pivotal in establishing and adopting these goals. Quality goals, ranging from conformance to specifications to sustainability, guide project execution and success. The chapter will delve into making goals SMART: specific, measurable, achievable, relevant, and time-bound (Figure 8.2), transforming broad aspirations into actionable strategies.

We discussed the identification of stakeholders and their perspectives, needs, and objectives in the previous section. In this section, we will discuss collaboration with stakeholders.

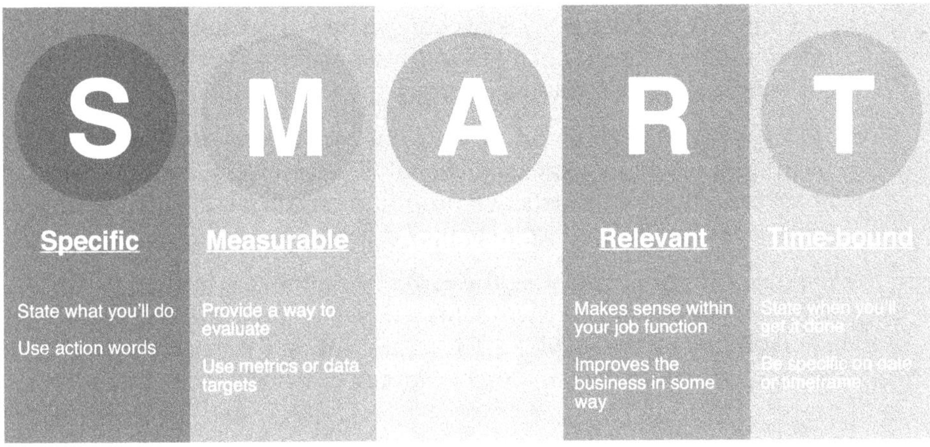

Figure 8.2 SMART goal expectations.

Quality goals are established during the planning phase of a construction project. Inputs to quality goals come from contractual requirements, market needs, and project-specific expectations. Quality goals are initially addressed in the project charter, which is part of the initiation phase and are progressively elaborated during the planning phase of a project. Quality goals are checked routinely during execution and then evaluated during closure and reviewed during lessons learned.

Goals are both short-term and long-term. Short-term may be limited to the length of the project. Long-term goals relate to the performance of the project during operations and maintenance. Examples of quality goals include reducing waste across the construction project through all sources of waste (see "Lean Wastes" later in this chapter)—for example, transportation, delay, waiting, defects, rework, and others. (*Note*: The eighth waste, "unused creativity," was later added to the previously well-known seven wastes.) Other examples of waste include cost spent on storing high inventory of materials due to supplier's quality variability; movement of materials from a distant warehouse; repair, rework, or regrade materials; and so on. Quality goals can include conformance to client requests, compliance to regulatory requirements, reduction of waste by XX%, or improvement in productivity and efficiency by YY%.

The CQM should ensure that quality goals are drafted with the SMART approach in mind. So instead of saying "Reduce construction defects," present the goal as "Reduce busbar torquing errors from current 15% to 7% on the project A by Q2 of YYYY." Reduction is the action; busbar torquing error is specific; 15% to 7% is measurable; the achievable or actionable component comes from resource availability; torque error reduction is relevant to project A; and Q2 of YYYY is time-bound.

Typical Quality Goals in the Construction Industry

- *Conformance to specifications and standards*: This is the most fundamental quality goal, ensuring that the final product meets all the requirements outlined in the design documents, building codes, and industry standards. Lack of conformance can result in direct cost impacts like rework, scrap, retesting, repair, and restoration. Other costs include warranty and legal costs.

- *Zero defects or defect-free work*: Striving for zero defects or minimizing them significantly is a key quality goal. This involves ensuring that every aspect of the construction process is executed correctly the first time, reducing the need for rework and delays.
- *On-time delivery*: Completing the project within the agreed-on schedule is essential for client satisfaction and cost control. Quality management processes play a role in preventing delays caused by rework, poor planning, and other unforeseen issues.
- *Cost control*: Staying within the budget is another critical goal. Effective quality management helps avoid cost overruns by preventing errors and minimizing waste.
- *Zero safety incidents*: Ensuring the safety of workers and the public is paramount. Safety incidents happen due to a lack of controls in QMSs. Safety protocols and procedures minimize accidents and injuries.
- *Client satisfaction*: Meeting or exceeding client expectations is the ultimate measure of success.
- *Sustainability*: Incorporating sustainable practices and materials into construction projects is increasingly important. Quality goals may include reducing environmental impact and promoting energy efficiency.
- *Continuous improvement*: Striving for ongoing improvement in processes, techniques, and materials is essential to remain competitive and adapt to changing industry demands.

These goals collectively contribute to delivering successful construction projects that meet the highest standards of quality, safety, and sustainability while also satisfying clients and controlling costs.

Lean Wastes

The concept of lean manufacturing seven wastes, although originating in manufacturing, has applicability to the construction industry as well. They highlight areas of inefficiency that can lead to delays, cost overruns, and reduced quality.

1. *Transportation*: Unnecessary movement of materials, equipment, or workers around the construction site. *Example*: Materials stored far from the point of use, requiring frequent movement.
2. *Inventory*: Excessive storage of materials or components on-site, leading to potential damage, loss, or obsolescence. *Example*: Over-ordering materials due to inaccurate forecasting or poor communication.

3. *Motion*: Unnecessary movement of workers or equipment during tasks. *Example*: Poor site layout that requires workers to walk long distances between work areas.
4. *Waiting*: Idle time for workers or equipment due to delays or lack of coordination. *Example*: Crews waiting for materials to arrive, inspections to be completed, or previous tasks to be finished.
5. *Overproduction*: Producing more than is needed or doing work before it's required. *Example*: Building components or sections before the design is finalized or subsequent tasks are ready.
6. *Over-processing*: Performing unnecessary tasks or adding more complexity than required. *Example*: Using overly complex construction methods or materials when simpler alternatives would suffice.
7. *Defects*: Errors or rework due to poor workmanship or materials. *Example*: Incorrect installations, faulty components, or damage during construction.

By identifying and addressing these wastes, construction companies can improve efficiency, reduce costs, enhance quality, and deliver projects on time and within budget. Lean principles and practices offer valuable tools for streamlining construction processes and achieving better outcomes.

Project stakeholders in construction engage in one or more approaches described below to ensure the adoption of vision and quality goals.

- *Early engagement and periodic communication*: Establish open channels of communication from the project's outset. Regular meetings, workshops, and other forums enable stakeholders to share their perspectives, expectations, and concerns.
- *Clear definition of stakeholder roles and responsibilities*: Ensure everyone understands their role in the project and how their contributions affect overall quality. This clarity minimizes confusion and fosters accountability.
- *Shared understanding of project goals*: Develop a comprehensive project brief that outlines the vision, objectives, and key performance indicators (KPIs). This shared understanding aligns everyone's efforts toward a common purpose.
- *Collaborative decision-making*: Encourage active participation from all stakeholders in decision-making processes. This fosters a sense of ownership and commitment to the project's success.
- *Transparent information sharing*: Maintain transparency by providing access to relevant project information, progress updates, and performance data. This builds trust and facilitates informed decision-making.

- *Conflict resolution mechanisms*: Establish clear procedures for resolving conflicts and disagreements that may arise. This ensures that issues are addressed promptly and constructively, minimizing disruptions.
- *Continuous feedback and improvement*: Regularly solicit feedback from stakeholders and use it to refine processes and improve outcomes. This demonstrates a commitment to learning and adaptation.
- *Celebration of successes*: Acknowledge and celebrate milestones and achievements to maintain motivation and reinforce the shared vision.

CONCLUSION

The development and adoption of a clear vision and quality goals are fundamental to the success of any construction project. By actively collaborating with stakeholders, establishing SMART objectives, and addressing potential wastes, teams can ensure projects meet specifications, timelines, and budgetary constraints. The emphasis on conformance, defect-free work, safety, and sustainability underscores a holistic approach to quality management. Moreover, continuous improvement and transparent communication are essential for adapting to evolving industry demands and fostering client satisfaction. Ultimately, a robust QMS, driven by shared vision and well-defined goals, is the cornerstone of delivering successful, high-quality construction projects.

QUESTIONS

1. How are quality goals set for construction projects and an organization?
2. What quality goals would you set for a construction project across its life cycle?
3. How would you formalize quality goals to be adopted by the project and an organization?
4. Explain how the SMART framework (specific, measurable, achievable, relevant, time-bound) can be applied to transform a broad quality goal, such as "Reduce construction defects," into an actionable strategy. Give an example.
5. Discuss the interconnectedness between quality goals and the identification of "lean wastes" in construction projects, illustrating with at least three examples of lean wastes and their potential impact on quality goals.
6. Describe how stakeholder engagement and communication throughout the project life cycle contribute to the establishment and adoption of quality goals, and what measures can be implemented to ensure effective collaboration.
7. Analyze the difference between short-term and long-term quality goals in a construction project, and how each type impacts the project's life cycle from initiation to operations and maintenance.
8. Elaborate on how the various "typical quality goals" listed (e.g., conformance to specifications, zero defects, on-time delivery, sustainability) can either reinforce or potentially conflict with each other, and how project management can balance these goals to ensure overall project success.

STAGES OF TEAM DEVELOPMENT

> Identify and distinguish between the stages of team development: forming, storming, norming, performing, and adjourning. (Analyze). VIII.C.2. Body of Knowledge—Stage of team development

As team members work together on a program, project, or even a new team, they progress through different stages of growth and maturity. Understanding these developmental stages is crucial for effective team management. The intensity and duration of these stages can vary based on the program, project and team type, work environment (face-to-face, virtual, hybrid), and individual team members. A generic model for team development phases, proposed by Tuckman, is illustrated in Figure 8.3.

Stage 1: Forming

In this stage, as team members assemble, they carry their unique identities, values, social norms, and priorities from their respective backgrounds. Each team presents a novel experience, even for seasoned team members. Initially, individuals approach this situation cautiously, unsure of their role and performance, power

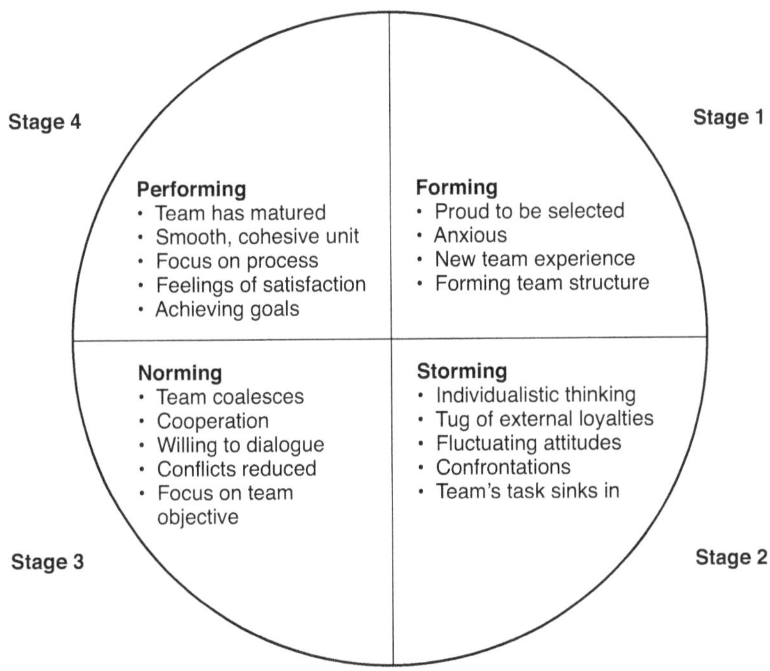

Figure 8.3 Team development phases.

dynamics within this new environment. During the forming stage, the team typically clarifies its purpose, assigns roles to each member, and establishes acceptable behavioral guidelines. At this stage, it is important to engage each member so that they feel included and welcome.

Stage 2: Storming

In this phase, team members gradually come to grips with the reality of their task. Initially, they might think predominantly as individuals, trying to make decisions based on personal experiences rather than seeking input from others. Collaboration is yet to become the norm, and attitudes regarding the team's chances of success can fluctuate. Behaviors during this period may include arguments, personality clashes, power struggles, challenging the leader's authority, attempting to redefine goals, and engaging in competitive or defensive tactics. This stage is unavoidable and essential for team members to understand each other's strengths and persistence. However, it's crucial not to dwell in this stage for too long, as it may jeopardize overall objectives.

Stage 3: Norming

During this phase, individuals transform into a cohesive team by shifting their focus from personal interests to addressing team-related challenges. Interpersonal conflicts and external loyalties diminish as team members recognize the futility of continued conflicts and misalignment. They engage in constructive discussions to resolve differences, fostering greater cooperation and dialogue. The sooner a team reaches this stage, the better, as it increases the likelihood of achieving project goals.

Stage 4: Performing

At this stage, the team has matured to the point where it is working as a smooth, cohesive unit. Team members have a good understanding of each other's strengths and weaknesses and how they support the mission, and are now able to work through group problems. There is an overall appreciation of the team's processes and a general satisfaction with team membership.

During the performing phase, the team is making significant progress toward achieving its goals. Although the four stages of development indicate a logical sequence that occurs over time, the actual progress made by each team will vary. For example, a team that has progressed to stage 3 or 4 may fall back to stage 1 or 2. This may occur if they learn that previous assumptions about one another are not true, information they have been using for decision-making is found to be inaccurate, or team membership changes due to job transfers. Some teams might never progress beyond an earlier stage because of limited project duration or poor resolution of group dynamics issues. This is one of the reasons why not all projects succeed.

In 1977, Tuckman and Jensen added the "adjourning" stage to address the final phase of a team's life cycle, which involves completing the task and the team's disbandment. It is the process of closure that occurs when the team has accomplished its mission and is to disband. The team should take time to review the lessons learned and what was accomplished, celebrate, and formally close the project.

Team development can be enhanced by making sure that team members have a basic understanding of how to: (1) interact in positive ways, (2) deal with difficult people or situations, (3) contribute to accomplishing the team's goals, and (4) give or receive constructive feedback.

Note: Consider examples of participation in a recent program, project, or team from your own organization and evaluate if you went through similar stages (and what the durations were).

With the emergence of virtual and hybrid team models since the COVID-19 pandemic, it is important to consider the effect of various types of engagement on team stages (Table 8.2). On-site construction staff engage and interact face-to-face. However, other project team members and people from other construction sites may engage in virtual or periodic visits to a designated location (hybrid) to make key decisions.

Table 8.2 Effect of various engagement models on team stages.

Stage	Interaction Type	Impact	Examples
Forming	Face-to-Face	Faster rapport building through nonverbal cues, shared experiences, and informal conversations	Team-building lunch, coffee breaks, or social events
	Virtual	Slower trust building due to lack of physical presence; requires intentional virtual icebreakers and frequent video calls	Virtual coffee breaks with personal interest sharing, online team games, or virtual tours
	Hybrid	Combines benefits of both. Initial in-person meetings accelerate relationship building, followed by virtual tools for ongoing communication.	In-person kickoff meeting followed by regular virtual check-ins, virtual water cooler channels for informal interaction, online team-building activities
Storming	Face-to-Face	Conflicts addressed more effectively through direct communication, immediate feedback, and nonverbal cues	Face-to-face conflict resolution meetings, open discussions, or brainstorming solutions together
	Virtual	Misinterpretations can arise due to lack of nonverbal cues; requires open communication and use of video conferencing for emotional cues	Shared documents for listing and discussing concerns, virtual brainstorming sessions, video calls for sensitive discussions
	Hybrid	Combines approaches; sensitive conflicts addressed in person, less critical ones virtually	Virtual brainstorming followed by in-person decision-making meetings, virtual feedback channels for minor concerns, in-person meetings for major disagreements

(*continued*)

Table 8.2 Effect of various engagement models on team stages. (*continued*)

Stage	Interaction Type	Impact	Examples
Norming	Face-to-Face	Shared experiences and regular interactions facilitate the development of group norms and shared values.	Team traditions like celebrating milestones with lunch, informal conversations, team activities
	Virtual	Establishing shared norms takes longer due to less frequent informal interaction; requires clear communication guidelines and virtual tools	Regular virtual meetings, shared documents, virtual "water cooler" channels, online team-building activities
	Hybrid	Offers both formal and informal interactions. In-person meetings strengthen team cohesion, virtual tools maintain communication and norms.	In-person retrospectives followed by virtual updates on action items, regular virtual stand-ups, occasional in-person social gatherings
Performing	Face-to-Face	Close collaboration, immediate feedback, and shared understanding lead to high performance.	Frequent in-person meetings, brainstorming sessions, design reviews, team workshops.
	Virtual	High performance relies on clear communication, defined roles, and effective collaboration tools. Regular check-ins and celebrating successes.	Project management software for tracking progress and celebrating milestones, virtual stand-ups, virtual team-building activities, online recognition programs.
	Hybrid	Combines strengths of both modes; adapts interaction style to the task	In-person workshops for complex problem-solving, virtual stand-ups for daily progress, virtual brainstorming followed by in-person decision-making, in-person celebrations for major milestones, virtual feedback channels for ongoing communication

Stage 5: Adjourning

Adjourning Stage

- *Description*: This stage occurs when the project or task is completed and the team disbands. The focus shifts to wrapping up activities, celebrating achievements, and reflecting on the team's journey.

- *Behavior*: Team members may experience mixed emotions, such as pride in their accomplishments and sadness about the team disbanding. It's important to acknowledge the contributions of all members and provide a sense of closure.

Activities

- *Reflection*: Reviewing the project's success and the team's performance. Discussing what went well and areas for improvement.
- *Recognition*: Celebrating achievements and acknowledging individual and team contributions.
- *Documentation*: Finalizing reports, documenting lessons learned, and ensuring all project deliverables are completed and handed over.
- *Transition*: Supporting team members as they move on to new projects or roles. Providing any necessary transition plans or knowledge transfer.

Importance of the Adjourning Stage

- *Emotional closure*: It provides an opportunity for team members to process the end of the project and their experience working together.
- *Learning opportunity*: Reflecting on the team's journey helps identify lessons learned that can be applied to future projects.
- *Celebration and recognition*: Celebrating achievements boosts morale and recognizes the hard work and contributions of all team members.

CONCLUSION

Understanding the adjourning stage helps leaders ensure a smooth transition for team members and conclude the project on a positive note.

Team development is a dynamic process, vital for achieving project success. Understanding and navigating the forming, storming, norming, performing, and adjourning stages allows for effective management and fosters a cohesive, high-performing unit. Adapting to various interaction types, whether face-to-face, virtual, or hybrid, further enhances team dynamics and ensures goals are met. By prioritizing positive interactions, conflict resolution, and constructive feedback, teams can progress through these stages, leading to successful project outcomes and valuable lessons learned.

QUESTIONS

1. Identify top reasons why it is beneficial for any team to go through these stages.
2. Have you been in a team that went through these stages? If so, which stage took the shortest and which the longest, and why?
3. How would you address the challenges emerging from the storming stage from affecting the team's ability to meet project milestones?
4. What are the challenges to a team that is global and meets virtually to accomplish project objectives?

5. What are the skills required for a CQM to be an effective team member and also to lead a team?
6. Explain how the shift from "personal interests to addressing team-related challenges" distinguishes the Storming stage from the Norming stage of team development. Give an example of potential behaviors that would indicate a team is transitioning from one stage to the other.
7. Discuss the potential challenges of maintaining progress through Tuckman's stages of team development in a hybrid work environment, specifically highlighting how different interaction types can impact the Storming and Performing stages.
8. Analyze why a team might regress from the Performing stage back to an earlier stage like Forming or Storming. Describe two scenarios that could cause this regression, and suggest how a quality manager could address these situations.
9. Considering the Adjourning stage, describe why it is not just about completing the task and disbanding, but also an important phase for team development and future projects. What key activities should be undertaken during this stage and why?
10. Evaluate the statement: "Some teams might never progress beyond an earlier stage because of limited project duration or poor resolution of group dynamics issues." Explain the implications of this statement for project success and suggest strategies to mitigate the risk of a team remaining stuck in an early developmental stage.

EFFECTIVE TEAM DYNAMICS AND COMMUNICATION

> Employ various communication techniques (e.g., verbal, nonverbal, written, and visual) and active listening. Collaborate with multiple groups (e.g., designers, project management, and craft teams). Implement the Responsible, Accountable, Consulted, Informed (RACI) Matrix to clarify roles and responsibilities and streamline decision-making processes. Understand decision-making pitfalls (e.g., groupthink and power dynamics) to promote healthy team dynamics and achieve project objectives. Resolve conflicts through established procedures. (Apply)
>
> Body of Knowledge VIII.C.3

Effective communication across diverse teams (e.g., designers, project managers, and construction crews) requires adapting techniques to various factors: the situation's urgency, the environment (on-site, virtual, or offline), and the stakeholder's knowledge and engagement level. Formal, contract-related communication necessitates written documentation for legal purposes. In contrast, swift, direct verbal instruction is more effective for providing immediate guidance during construction. Visual communication, such as safety signage, is crucial in construction settings. Active listening is employed to understand and address feedback from construction workers, demonstrating empathy and fostering a collaborative environment. Nonverbal cues are consistently observed to gauge how stakeholders interpret and respond to communication efforts.

Verbal communication is advantageous when fast and efficient communication is required, facilitates rapid information exchange, and prompts responses. It makes communication personal and engaging. The tone and inflection of one's voice are most important when conveying emotions and fostering connections. It is also flexible and adaptable: Messages can be modified effortlessly in light of a listener's response. However, it has disadvantages as well. There is no permanent record. It may also result in misunderstandings and misinterpretations due to lack of documentation. It is difficult to control. Emotional responses can sometimes hijack conversations. It can only have limited audience reach as it typically only reaches those present at the time.

When to Use Verbal Communication

- When immediate feedback is necessary
- When building relationships is crucial
- When complex or sensitive information requires discussion

Nonverbal communication (aka body language) has enhanced emotional expression using body language (see Table 8.3), facial expressions, and tone of voice

Table 8.3 Understanding body language.

Body Language	Description
Crossed Arms	Suggests defensiveness, discomfort, or resistance, but can also indicate feeling cold or relaxed.
Leaning Forward	Shows interest and engagement, especially in conversations.
Avoiding Eye Contact	Indicates nervousness, shyness, or discomfort, but may also mean the person is being thoughtful.
Eye Contact	Demonstrates confidence, attentiveness, and sincerity when used appropriately.
Tapping Fingers or Feet	Reflects impatience, boredom, or nervous energy.
Smiling	Usually conveys friendliness, warmth, or approval, but can be forced or insincere at times.
Shrugging Shoulders	Denotes uncertainty or lack of concern.
Hands on Hips	Expresses confidence or readiness but may also come across as aggressive or assertive.
Mirroring Someone's Actions	Indicates rapport, agreement, or a subconscious connection with the other person.

that can convey emotions and attitudes more effectively than words alone. It acts as a supportive communication to verbal. Nonverbal cues add depth and meaning to spoken words, enhancing overall communication. Nonverbal is universally understood. Certain nonverbal cues, such as facial expressions, are commonly understood across cultures, facilitating communication beyond language barriers. There are also some disadvantages, like the potential for ambiguity as interpretations of nonverbal cues can vary, leading to misunderstandings or miscommunication. Nonverbal cues may unintentionally convey messages you did not intend to send. Nonverbal cues can have different meanings in different cultures, requiring awareness and sensitivity to avoid misinterpretation. For example, in India, people nod their heads sideways for yes. Internationally, people get confused by interpreting it as no.

When to Use Nonverbal Communication

- *Emphasizing a point or emotion*: Nonverbal cues can reinforce and emphasize verbal messages, making them more impactful.
- *Building trust and rapport*: Nonverbal communication helps establish trust and rapport by conveying warmth, sincerity, and attentiveness.
- *Cross-language communication*: Nonverbal cues can bridge language barriers, facilitating communication with individuals who speak different languages.

When to Use Written Communication

Written communication provides a record. It serves as documentation for future reference, ensuring important information is preserved. It allows for careful thought while giving the opportunity to craft and refine messages, ensuring clarity and accuracy. It can reach a wider audience by being easily distributed to multiple recipients, enabling efficient information dissemination. However, it can be time-consuming, requiring effort to write clearly and concisely, and potentially leading to delays in communication. Written communication may feel like it's lacking in personal touch. It may feel impersonal and distant, lacking the warmth and spontaneity of face-to-face interactions. With no immediate feedback, communication may be delayed as recipients take time to read and respond to written messages.

Use written communication when:

- Formal communication is required
- A permanent record is needed
- Communicating complex or detailed information

When to Use Visual Communication

Visual communication is engaging and memorable. Visuals like images and graphics can capture people's imagination, attention, and enhance understanding. It helps to convey complex information by simplifying data and concepts. It offers a universal understanding as visuals transcend language barriers, making them accessible to diverse audiences. Disadvantages include interpretation challenges. Visuals can be ambiguous and open to interpretation. To develop visual communication, design skills may be required: creating effective visuals requires specialized design skills. There may be limitations in visually depicting certain information. Not all information is best conveyed visually; for example, people-sensitive information. Some topics may require other communication methods.

Use visual communication when:

- Presenting data or statistics
- Explaining complex concepts
- Creating a lasting impression
- Creating signage for safety and other important alerts that are not language dependent

Active Listening

Active listening provides an enhanced comprehension: It ensures accurate reception and understanding of the speaker's message, helps build stronger relationships, and demonstrates respect and fosters trust and rapport. Further, it helps to minimize misunderstandings and disagreements. The challenge is that it demands focus and requires sustained attention and mental effort. It also requires an increased time investment as it takes more time to thoroughly process information. It has situational limitations and is not always the most suitable approach, especially when a more direct style is needed.

When to Use Active Listening

It is important to remember that these communication techniques are not mutually exclusive. Often, the most effective communication involves a combination of methods. Choosing the right technique depends on the context, your audience, and the message you want to convey.

Use active listening when:

- Building relationships
- Resolving conflict
- Gathering information

The RACI Matrix

As specified by the Project Management Institute (PMI) in the PMBOK Guide, the RACI matrix is a widely used communication tool in project management. It maps responsibility, accountability, stakeholders who are consulted prior to approval, and those who are informed onto a list of key project activities and milestones that are typically arranged sequentially. Most projects managed by trained and

Table 8.4 RACI matrix.

Task/Activity	Project Manager	Site Supervisor	Architect	Contractor	Client
Project Planning	A	R	C	I	I
Design Approval	C	I	R	I	A
Site Preparation	I	A	C	R	I
Foundation Construction	I	A	C	R	I
Structural Framing	I	A	C	R	I
Electrical & Plumbing	I	A	C	R	I
Interior Finishing	I	A	C	R	I
Final Inspection	A	R	C	I	I
Project Handover	A	R	C	I	I

qualified project managers diligently develop a RACI matrix (Table 8.4). After you investigate the issues throughout the project, however, you understand that key stakeholders are not fully aware of their role within the scope of the project. You may notice that the deliverables are either delayed or not executed, that stakeholders performing the same responsibilities across the project are making their own final decisions without any central accountability, a lack of coordination on responsibilities, missing stakeholders from consultation, or stakeholders who are uninformed and surprised by project activities that are already completed. Let us explore reasons why these failure modes emerge in a project.

Please note that another variation of this chart is referred to as RASIC. It incorporates one additional component, "Supportive." The main difference between RACI and RASIC is the addition of the *supportive* role in RASIC (Table 8.5). This role is intended to identify team members who assist or provide support to those responsible for the task but are not directly responsible themselves.

Lack of Collaboration

The project manager may be working in a silo to draft the RACI matrix and submitting it just to satisfy documentation requirements. No one person in a project knows everything about the project's complexities and interdependencies, so project managers must collaborate with key stakeholders to develop a meaningful RACI matrix.

Table 8.5 RASIC matrix.

Task/Activity	Project Manager	Site Supervisor	Architect	Contractor	Client
Project Planning	A	R	C	S	I
Design Approval	C	I	R	S	A
Site Preparation	I	A	C	R/S	I
Foundation Construction	I	A	C	R/S	I
Structural Framing	I	A	C	R/S	I
Electrical & Plumbing	I	A	C	R/S	I
Interior Finishing	I	A	C	R/S	I
Final Inspection	A	R	C	S	I
Project Handover	A	R	C	S	I

Not Updating the RACI Matrix

Like any organizational document, nothing is written in stone. Projects change. There may be scope changes, impact to budgets, schedule changes, new and evolving regulatory requirements, and other external factors such as weather conditions, labor availability or disputes, or negotiations.

Organizations should apply effective change management strategies during these situations and update relevant documents, including the RACI matrix. Changes may have triggered new activities to account for, new responsibilities, or additional stakeholders to consult and inform.

Not Aligning with Business Process

The RACI matrix cannot be a mutually exclusive document; it must align with existing business processes. That said, you do not want to assume following the existing business process is the right approach. You must verify whether the existing business process related to the activities in the RACI matrix is current and accurate. Next, you must ensure the responsibility and accountability are consistent with that of the business process.

Remember that a RACI matrix should not override a business process or be used with that intent as a shortcut. If the RACI matrix contradicts an existing business process, it will cause more confusion and disagreement among stakeholders.

Too Many Cooks in the Kitchen

A RACI matrix should be developed so that one person is accountable for the outcome and one person is responsible for the execution of each task/activity or deliverable. Identifying too many people as accountable for a task/activity is a red flag. The project manager should discuss this with stakeholders to identify one person who is accountable. Typically, this is someone with authority. Identifying too many people as responsible for a task/activity is plausible in situations in which the task/activity must be performed at different locations by different roles or functions. But in most situations, it should be one role.

If the task/activity is too large in scope, break it down so one role is responsible. In smaller organizations and projects, it is possible that the same role is responsible and accountable. In this situation, consult stakeholders to ensure objectivity in decision-making.

Identifying too many people as "consulted" also is inefficient. Identify who, at minimum, must be consulted. Consider moving others from "consulted" to "informed." This simple due diligence while developing a RACI matrix will save a lot of heartburn during project execution. A RACI matrix is not for archives. It should not be developed just for documentation compliance and archiving to fulfill contractual obligations. It must be used during daily project operations when a task/activity is planned and performed.

If the reality on the ground is different than what was planned earlier, changing the RACI matrix roles, feeding them back to the project manager, and making those changes formally will help you learn about the situation and be proactive in future projects. A RACI matrix is a handy reference for project managers and management to oversee resources and their availability. For example, if a key SME responsible for performing a task/activity is asking for a two-week vacation, the project manager can ensure there is continuity during the SME's absence. Not all stakeholders have an equal stake in the project. This is where it may be highly beneficial to perform a stakeholder analysis before developing a RACI matrix. This will help identify those stakeholders with power and interest, activity and attitude, and ensure these stakeholders play key roles in the project task/activity.

Finally, an ideal RACI matrix is a 7×7 matrix and should not exceed 10×10. Large matrices are difficult to follow and implement on the ground. They also are challenging to maintain and sustain. By addressing the failure modes and putting the RACI matrix to use, you make the best of this valuable project management tool

Understanding and Reducing Decision-Making Pitfalls

Construction project teams constantly face decisions, from routine operational choices to impactful ones affecting project outcomes like scope changes, budget adjustments, supplier shifts, or altered monitoring controls. To ensure objective and effective decision-making, CQMs must be vigilant about potential pitfalls that can lead to undesirable results.

Construction projects are complex and involve many stakeholders. This makes them prone to decision-making pitfalls that can derail progress, impact budget, and even compromise safety. Table 8.6 shows how to identify and mitigate some common ones.

Table 8.6 Decision-making pitfalls and mitigation strategy.

Decision-Making Pitfalls	Mitigation strategy
Groupthink: When the desire for harmony in a decision-making group overrides a realistic appraisal of alternatives. During the team selection process and daily operations, it's crucial to prevent groupthink. This occurs when team members align behind an idea or decision without thorough exploration, potentially harboring unspoken disagreements. The priority becomes preserving harmony over engaging in challenging discussions.	*Mitigation*: Encourage dissenting opinions, appoint a "devil's advocate," and use anonymous feedback mechanisms. • Brainstorm various alternatives before choosing a path. • Encourage members to voice concerns openly. • Dedicate sufficient time to explore, analyze, and understand all ideas and suggestions. • Establish guidelines for evaluating each alternative.
Power dynamics: Decisions can be skewed to favor those with more authority, regardless of their expertise.	*Mitigation*: Foster a culture of open communication where all voices are heard, and use structured decision-making processes that evaluate ideas based on merit, not the source.
Anchoring bias: Over-relying on the first piece of information received, even if it's irrelevant.	*Mitigation*: Seek diverse perspectives and data sources, challenge initial assumptions, and consider alternatives before making a decision.
Confirmation bias: Favoring information that confirms existing beliefs.	*Mitigation*: Actively seek out information that challenges your viewpoint, consider the opposite, and be open to changing your mind.
Overconfidence bias: Overestimating one's ability to make accurate predictions or estimates.	*Mitigation*: Use data and historical information to inform decisions, consult with experts, and build in contingency plans.
Status quo bias: A preference for the current state of affairs, leading to resistance to change.	*Mitigation*: Clearly articulate the benefits of change, involve the team in the decision-making process, and address concerns openly.

Promoting Healthy Team Dynamics

Effective teamwork is essential for successful construction projects. Here are some strategies to promote healthy team dynamics:

- *Clear communication*: Establish open and transparent communication channels. Encourage regular updates, active listening, and constructive feedback.
- *Defined roles and responsibilities*: Ensure everyone understands their role and responsibilities within the project. This minimizes confusion and promotes accountability.
- *Shared goals and objectives*: Align the team around common project goals. This fosters a sense of shared purpose and encourages collaboration.

- *Mutual respect and trust*: Cultivate a culture of respect and trust among team members. Encourage diverse perspectives, value contributions, and recognize achievements.
- *Conflict resolution mechanisms*: Establish clear processes for addressing and resolving conflicts constructively. This helps to prevent disagreements from escalating and damaging team dynamics.
- *Regular team building*: Invest time in team-building activities to strengthen relationships, improve communication, and foster a sense of camaraderie.

CONCLUSION

Effective team dynamics and communication are crucial for project success. By employing diverse communication techniques, implementing the RACI matrix, and understanding decision-making pitfalls, quality managers can foster healthy team environments. Open communication, defined roles, and conflict resolution mechanisms further enhance collaboration. Ultimately, choosing the right communication method based on context and audience ensures clarity and understanding, driving projects toward successful completion. Remember, continuous improvement in team dynamics and communication is essential for achieving project objectives and fostering a positive work environment. By understanding and addressing decision-making pitfalls and promoting healthy team dynamics, construction projects can achieve their objectives more effectively, improve efficiency, and enhance overall project success.

QUESTIONS

1. What are various communication methods and techniques appropriate for different stakeholders and under what circumstances?
2. Why is a RACI matrix beneficial to a construction project team? When is the best opportunity to develop a RACI matrix?
3. How might a CQM ensure that a RACI matrix is current and relevant to their organization?
4. What are the effective methods to resolve conflicts?
5. Why are disagreements not a bad thing for a team? How do they help or hinder decision-making by a team?
6. Explain how the choice of communication technique (verbal, nonverbal, written, visual) should be adapted to different factors in a construction project setting, providing specific examples of when each technique would be most effective.
7. Describe the potential pitfalls of using the RACI matrix in project management, and outline strategies to ensure its effectiveness as a communication and accountability tool.
8. Discuss the concept of "groupthink" in decision-making, and describe mitigation strategies that a CQM can implement to prevent it from negatively impacting team dynamics and project outcomes.
9. Compare and contrast the RACI and RASIC matrices, explaining the additional element included in RASIC and its intended purpose within a project team.

10. Analyze why disagreements within a team are not necessarily negative, and explain how they can contribute to or hinder decision-making processes, providing examples from a construction project context.

Change Management

ORGANIZATIONAL CULTURE

> Define and explain the factors that shape an organization's culture. (Apply)
>
> Body of Knowledge VIII.D.1

Organizational culture is the bedrock on which a company's success is built. It encompasses the shared values, beliefs, and behaviors that define "how things are done around here." More than just a mission statement, it's the lived experience of employees, shaping interactions, decisions, and overall performance. A strong culture fosters collaboration, safety, and engagement, driving quality outcomes and competitive advantage, particularly vital in industries like construction. Understanding and cultivating a positive organizational culture is crucial for sustained success in today's global landscape.

ISO 9004 defines culture as "beliefs, history, ethics, observed behavior and attitudes that are interrelated with the identity of the organization." Organizational culture refers to the shared values, beliefs, assumptions, and behaviors that shape the way people interact and work together within an organization. It's the "personality" of a company, influencing everything from communication styles to decision-making processes to how employees feel about their work.

The famous quote "Culture eats strategy for breakfast" is often attributed to Peter Drucker. What does it mean? There are several attributes that are synonymous with culture.

Factors That Shape Organizational Culture

Organizational culture can be observed in a variety of ways. Employees and visitors alike can get a sense of the culture of a workplace by observing the physical space, dress code, communication style, decision-making, employee interactions, and company celebrations or rituals. These visual and behavioral cues can be very telling.

- *Values*: Core principles and beliefs that guide the organization's actions and decisions can be explicitly stated (e.g., in a mission statement) or implicitly understood through everyday actions and interactions.
- *Norms*: Unwritten rules and expectations about behavior within the organization dictate how people dress, communicate, interact with each other, and approach their work.

- *Assumptions*: Deeply held beliefs that are often taken for granted and rarely questioned, assumptions shape how people perceive the world and their place within the organization.
- *Symbols*: Objects, logos, or rituals may represent the organization's values and identity. These symbols can serve as powerful reminders of what the organization stands for.
- *Language*: The specific jargon, phrases, and communication styles used within the organization can reflect the organization's culture and reinforce shared values.
- *Stories*: Narratives about the organization's history, successes, and failures transmit cultural values and create a sense of shared identity.
- *Physical environment*: The layout of the office, the décor, and the overall ambiance can provide clues about the culture. A bright, open office with collaborative spaces might suggest a culture of teamwork and innovation.
- *Dress code*: How people dress can reflect the formality or informality of the culture. A casual dress code might suggest a relaxed and creative environment.
- *Communication style*: Whether people communicate formally or informally, directly or indirectly, can reveal a lot about the culture.
- *Decision-making*: Who makes decisions, and how those decisions are made, can be a key indicator of the culture. A centralized decision-making process might suggest a hierarchical culture.
- *Employee interactions*: How employees interact with each other, with customers, and with management can provide insights into the culture. A friendly and supportive atmosphere might indicate a positive and collaborative culture.
- *Celebrations and rituals*: The way an organization celebrates successes, marks milestones, or handles difficult situations can reflect its values and culture.

Culture in the Construction Industry

Organizational culture plays a significant role in the global construction industry in various ways, impacting project success, safety, employee engagement, and overall company performance.

Project Success

A positive and collaborative organizational culture fosters effective communication and collaboration among team members, leading to improved project outcomes such as on-time completion, zero safety incidents, cost control, quality assurance, and improved customer satisfaction. Conversely, a negative or toxic culture can lead to miscommunication, conflicts, delays, accidents, cost overruns, and negative customer perception.

Safety

Safety in the construction industry is a key operational metric. Quality of construction is key to public safety. A culture that prioritizes safety and emphasizes risk management leads to a safer work environment. This includes clear communication about safety protocols, training programs, and a focus on preventing accidents. The absence of a culture that neglects safety or prioritizes productivity over safety can result in higher accident rates and injuries.

Employee Engagement and Retention

A positive organizational culture that values employees, provides opportunities for growth and development, and promotes a healthy work-life balance leads to higher employee engagement, job satisfaction, and retention rates.

Company Performance

A strong and positive organizational culture enhances a company's reputation, attracts top talent, and improves its competitive advantage in the global construction market.

The design of the organization has an impact on culture. If a hierarchical structure restricts cross-functional communications, then close relationships between interrelated functions will be more difficult. Similarly, in a unionized organization, the relationship between organization management and union management will set the stage for how cooperatively people can work together toward common goals.

At an even deeper level are the shared tacit assumptions of which the organization's members may not be consciously aware. Failure to identify and consider these assumptions can result in serious consequences—for example, a failure to understand the mutual level of demonstrated trust needed for management to introduce major change in a unionized organization where an adversarial relationship persists.

Juran's Five Steps for Changing to a Quality Culture

1. Create and maintain an awareness of quality.
2. Provide evidence of management leadership on quality.
3. Provide self-development and empowerment.
4. Provide participation as a means of inspiring action.
5. Provide recognition and rewards. (Juran, 1999)

Changing an organization's culture is difficult and requires time. Fear of change must be removed, poor labor-management relations must be resolved, and the company's focus must change from the status quo. Employees should be convinced of the benefits a quality management approach will provide and should buy into the changes. This may imply that employees at all levels will need to adapt and make changes to cultural behaviors to perform tasks differently. If strong leadership, motivation, and enthusiasm are lacking, frustration and stress will result.

In the Global Context

In the increasingly interconnected and diverse landscape of the global construction industry, the impact of organizational culture extends far beyond national borders. The ability to navigate cultural diversity within project teams and adapt to the ever-changing global environment are becoming essential components of a successful organizational culture.

- *Cultural diversity*: Construction projects often involve teams from different cultures. A company that fosters a culture of inclusion and respects cultural differences can facilitate effective collaboration and avoid misunderstandings.
- *Adaptability*: The construction industry is constantly evolving. Companies with a culture that embraces innovation and adaptability are better positioned to navigate the challenges and opportunities presented by globalization and technological advancements.

Organizational culture plays a crucial role in shaping the success and sustainability of companies operating in the global construction industry. By cultivating a positive culture that prioritizes safety, values employees, encourages collaboration, and embraces innovation, companies can create a competitive advantage and achieve long-term success.

Organizational culture is a potent force that profoundly influences every aspect of a company, particularly in the complex construction industry. By understanding and actively shaping its culture, an organization can drive project success, ensure safety, boost employee engagement, and enhance its competitive edge. Recognizing the multifaceted elements that constitute culture, from values and norms to communication styles and physical environment, is crucial. Ultimately, cultivating a positive, inclusive, and adaptable culture is not just beneficial but essential for sustained success in the dynamic global landscape.

QUESTIONS

1. Explain in plain English what organizational culture means to you.
2. What are the key attributes of an organizational culture?
3. How would you assess your organization's current culture? How do you think it could improve?
4. Why is culture important to an organization?
5. What is the relationship between an organization's culture and its quality outcome?
6. Explain how Peter Drucker's quote "Culture eats strategy for breakfast" relates to the construction industry, giving examples of how a strong or weak culture can impact project outcomes and safety.
7. Describe three ways that an organization's physical environment can reflect and reinforce its organizational culture. For each way, give an example of how it might manifest differently in a highly hierarchical versus a highly collaborative construction company.
8. Analyze the relationship between organizational culture and employee engagement and retention in the global construction industry. What cultural elements are most likely to foster engagement and retention, and how can a company measure the effectiveness of these elements?

9. Discuss the challenges of changing an organization's culture, particularly in a unionized environment within the construction industry. What steps can a quality manager take, based on Juran's five steps, to facilitate a shift toward a quality culture, and what potential obstacles might they encounter?
10. Considering the global context of the construction industry, how does cultural diversity within project teams impact organizational culture? What strategies can companies implement to leverage this diversity positively and avoid potential misunderstandings or conflicts?

CHANGE MANAGEMENT STRATEGIES

> Analyze and provide input to various change management strategies to overcome organizational roadblocks, adapt to global changes, assess the achieved change levels, and review outcomes for effectiveness. (Analyze)
>
> Body of Knowledge VIII.D.2

The McKinsey Report states that 70% of change initiated by organizations fail. And why is that? Not all changes can be implemented without planning, preparation, and careful execution.

Change management is a structured approach to making a change to a product, process, system, individual's role, team priorities, and organization from a current state to a desired future state. A change could be as simple as shifting to a new format for reporting monthly metrics. It can also be as large as an organization's product or service offering. Change management involves a set of processes, tools, and techniques to manage the people's side of change, ensuring that changes are implemented smoothly and successfully.

Managing Change Versus Change Management

There is a common misconception that managing change and change management are the same thing. While they are related, there is a key difference.

Managing change is the overall process of dealing with change, from inception to completion. It includes all the activities necessary to successfully implement a change, such as planning, communication, training, and support.

Change management is a more specific discipline that focuses on the people's side of change. It is about helping people to understand, accept, and adopt changes in their work environment. It is a proactive and structured approach to implementing planned changes. It involves a set of processes, tools, and techniques to manage the people's side of change, ensuring that transitions are smooth and successful.

Another common misconception is that change management is only about communication (Creasey, 2020). While communication is a critical component, change management is about supporting people through planning, engagement, training, support, and evaluation.

Strategic Versus Tactical Change Management

Strategic change management sets the direction and long-term goals, while *tactical change management* focuses on the practical implementation of specific changes to achieve those goals.

Strategic Change Management

- *Focus*: Long-term and holistic view
- *Objective*: Aligning organizational change with overall business strategy and vision
- *Scope*: Organization-wide impact, often involving significant shifts in culture, processes, or structures
- *Planning*: High-level, comprehensive plans with broad goals and objectives
- *Leadership*: Driven by top executives and senior management
- *Examples*: Mergers and acquisitions, major technological overhauls, restructuring, or entering new markets

Tactical Change Management

- *Focus*: Short-term and specific initiatives
- *Objective*: Implementing specific projects or changes within the framework of the overall strategy
- *Scope*: Limited to specific departments, teams, or processes
- *Planning*: Detailed, actionable plans with clear steps and milestones
- *Leadership*: Managed by middle management and project leaders
- *Examples*: Process improvements, software implementations, or team reorganization

Start with planning and preparation by defining the change, its scope, and impact, and developing a strategy to implement it.

Change Control Versus Configuration Management

Change control and configuration management are two essential aspects of managing changes within an organization, especially in the context of IT and software development. While they are related, they serve different purposes. Here's a breakdown of the key differences:

Change Control

- *Focus*: Managing changes to a system or project in a controlled and systematic manner
- *Objective*: Ensure that changes are introduced in a way that minimizes disruption and ensures traceability

- *Scope*: Covers changes to any aspect of the project, including software, hardware, processes, and documentation
- *Activities*:
  - Identifying and documenting proposed changes
  - Evaluating the impact and risk of changes
  - Approving or rejecting changes
  - Implementing approved changes
  - Monitoring and reviewing changes to ensure they are effective
- *Examples*: Adding new features to software, updating hardware components, or modifying project schedules

Change Control Board (CCB)

- *Focus*: Overseeing and managing changes to a project or system
- *Objective*: Ensure that changes are evaluated, approved, and implemented in a controlled manner to minimize risk and disruption
- *Scope*: Changes related to project scope, schedule, budget, and technical aspects
- *Activities*:
  - Reviewing and approving change requests
  - Assessing the impact and risk of proposed changes
  - Prioritizing changes based on their importance and urgency
  - Monitoring the implementation of approved changes
- *Leadership*: Usually consists of project managers, technical leads, and key stakeholders
- *Examples*: Software development projects, infrastructure changes, or process improvements

Steering Committee

- *Focus*: Providing strategic direction and oversight for a project, program, or organization
- *Objective*: Ensure that the project or program aligns with the organization's overall strategy and objectives
- *Scope*: Broader oversight, including strategic decisions, resource allocation, and high-level project monitoring
- *Activities*:
  - Setting the project's vision, goals, and objectives
  - Approving major milestones and deliverables
  - Allocating resources and budgets
  - Reviewing project progress and addressing high-level issues
- Leadership: Typically composed of senior executives, department heads, and key stakeholders

- Examples: Large-scale projects, organizational change initiatives, or strategic programs

In summary, the change control board (CBB) focuses on managing specific changes within a project or system, ensuring they are handled systematically and with minimal disruption. The steering committee provides strategic direction and oversight, ensuring that the project or program aligns with the organization's overall strategy and objectives.

Ask the following questions.

1. Have you defined your change clearly and objectively?
2. Are you reviewing and applying any applicable past learnings proactively?
3. Have you evaluated the current state and planned future state?
4. Have you scoped the affected stakeholders and details for people/teams in each group?
5. Have you assessed the potential impact of your change?
6. Have you gathered stakeholder input for each impacted group through interviews and/or surveys?
7. Do you have a feedback mechanism planned?
8. Do you have approval: are the organization's managers on board to drive this change for their teams?

Once stakeholders are identified, an analysis is conducted to assess the level of engagement with stakeholders. See "Stakeholder Analysis and Alignment" for more details.

Implementing

Implement planned actions to execute the change and obtain the desired results.

Ask the following questions:

1. Have you communicated the upcoming change to impact groups?
2. Is your communication and execution plan complete?
3. Have you surveyed your stakeholders and impact groups for feedback (timelines, resources, priority conflicts, etc.)?
4. Have you actively moved your stakeholders through the adoption curve?

Communication: Keeping stakeholders informed about the change, its purpose, and its benefits

Engagement: Involving employees and other stakeholders in the change process, to build support and reduce resistance

Training and support: Providing the necessary training and support to help people adapt to the change

Monitoring and evaluation: Tracking progress, measuring results, and making adjustments as needed

Ask the following questions:

1. Do you have the success measurements (outcome for effectiveness) assessed from implementation?
2. Have you actively solicited feedback and lessons learned from stakeholders and impact groups?
3. Have you communicated the lessons learned and action plans with leadership and broader teams?
4. Have you applied and communicated the lessons learned for any improvements to ongoing change implementation?

Applying Change Management to Various Types of Changes Within an Organization

- *Strategic changes*: Major shifts in direction, such as mergers, acquisitions, or new product launches. For example, enter a new construction vertical like renewables.
- *Technological changes*: Implementation of new technologies or systems. For example, shift to digital, artificial intelligence.
- *Process changes*: Improvements or modifications to existing processes or workflows. For example, implement RFI, design changes, and process improvements.
- *Cultural changes*: Shifts in organizational values, beliefs, or behaviors. For example, bias for action, blameless problem-solving, and so on.

The goal of change management is to minimize disruption, ensure a smooth transition, and maximize the benefits of the change for the organization and its people.

It is important to note that not all change management efforts by organizations are successful. Failure can be attributed to several causes:

- *Lack of planning*: Change initiatives are more likely to fail if they are not properly planned. This includes having a clear vision for the change, identifying the stakeholders who will be affected by the change, and developing a plan for how the change will be implemented.
- *Lack of communication*: Communication is essential for any change initiative. People need to understand why the change is necessary, what it will mean for them, and how they can be involved in the process. If communication is not effective, people will be more likely to resist the change.
- *Lack of support*: Change can be difficult for people, and they may need support during the change process. This support can come from

managers, peers, or change management professionals. If people do not receive the support they need, they are more likely to resist the change.

- *Lack of flexibility*: Things don't always go according to plan, so it is important to be flexible during the change process. If the change is not flexible, it is more likely to fail.
- *Lack of leadership*: Change requires strong leadership. Leaders need to be able to articulate the vision for the change, build support for it, and overcome resistance to it. If leaders are not effective, the change is more likely to fail.

Table 8.7 shows some common change management roadblocks and strategies to overcome them.

Table 8.7 Change management roadblocks and strategies.

Roadblocks	Proactively address resistance by:
Resistance to Change	1. Communicating the need for change and its benefits early and often 2. Involving employees in the change process and soliciting their feedback 3. Addressing concerns and providing support 4. Celebrating successes and recognizing individuals' contributions
Lack of Leadership Buy-in and Support	1. Demonstrating the business case for change and its alignment with strategic goals 2. Ensuring leaders are actively involved in communicating and championing the change 3. Providing leaders with the necessary tools and training to support their teams
Poor Communication	1. Developing a clear and consistent communication plan 2. Using multiple channels to reach all stakeholders 3. Providing opportunities for two-way communication and feedback 4. Tailoring messages to different audiences
Inadequate Training and Support	1. Identifying the skills and knowledge gaps 2. Designing and delivering targeted training programs 3. Offering ongoing support and resources 4. Creating a safe environment for learning and experimentation
Unrealistic Expectations and Timelines	1. Conducting a thorough impact assessment 2. Developing a detailed project plan with milestones 3. Building in buffer time for unexpected delays 4. Regularly reviewing progress and adjusting plans as needed

(continued)

Table 8.7 Change management roadblocks and strategies. *(continued)*

Roadblocks	Proactively address resistance by:
Lack of Resources	1. Identifying the resources required for successful implementation 2. Making a strong business case for the investment 3. Allocating resources effectively 4. Exploring alternative resourcing options, such as outsourcing or partnering
Organizational Culture and Structure	1. Assessing the current culture and its impact on change readiness 2. Identifying and addressing any structural barriers to change 3. Fostering a culture of continuous improvement and learning 4. Aligning incentives and rewards with desired behaviors
Competing Priorities	1. Clearly defining the change initiative's priority relative to other projects 2. Aligning the change with strategic goals 3. Ensuring dedicated resources and focus 4. Communicating the importance of the change to all stakeholders

CONCLUSION

Effective change management is paramount for organizational success. By prioritizing planning, communication, engagement, training, and evaluation, organizations can navigate transitions smoothly and achieve desired outcomes. Recognizing the differences between managing change and change management, and understanding the nuances of strategic versus tactical approaches, are crucial. Addressing common roadblocks proactively, as outlined in Table 8.7, ensures resilience. Ultimately, robust change management minimizes disruption, maximizes benefits, and fosters a culture of continuous improvement, driving organizations toward their strategic goals.

QUESTIONS

1. Why is change management important to a construction organization?
2. How do you go about planning to make a change?
3. How do you ensure a change is adopted?
4. What are the aspects to consider during the planning for a change?
5. How would you measure and monitor the effectiveness of a change?
6. Explain the distinction between "managing change" and "change management," and discuss why understanding this difference is important for a quality manager.
7. Compare and contrast strategic change management with tactical change management, providing examples of when each approach would be most effective in a construction organization.
8. Describe the roles and responsibilities of a change control board (CCB) versus a steering committee in the context of change management within a project. How do their scopes and objectives differ?

9. Identify and discuss at least three common roadblocks to successful change management, and outline proactive strategies a quality manager can implement to overcome each of these roadblocks, referencing Table 8.7.
10. Discuss how the key aspects of change management (planning, implementation, communication, engagement, training, and evaluation) can be applied to a specific type of change within a construction organization, such as a technological change or a process change.

Cost of Quality

COST OF QUALITY (CoQ) ANALYSIS

> Examine the methods and sources for collecting CoQ data. Interpret the CoQ data to derive meaningful insights. Report and communicate findings from the CoQ data analysis. (Analyze)
>
> Body of Knowledge VIII.E.1

Effective management of quality directly impacts an organization's bottom line. This chapter delves into cost of quality (CoQ) analysis, a crucial framework for understanding and optimizing quality-related expenditures. By examining the methods for collecting CoQ data, interpreting it for meaningful insights, and communicating findings effectively, organizations can drive significant improvements. Standardized terms and definitions, as outlined in ASQ TR2–2024, are essential for consistent data collection and analysis, ensuring alignment and actionable strategies across all projects and departments.

First, it is important to understand the terms and definitions to be consistent in data collection, analysis, interpretation, and implementation of improvement actions across organization and projects.

Even though the idea of CoQ has been around since the early 1950s, there are varying definitions in practice. Some practitioners interchangeably use CoQ and cost of poor quality (COPQ) as one and the same. Over the decades, ideas about CoQ have developed and evolved to our current understanding today. For the first time in 2018, CoQ was published as a technical report (TR), ASQ TR2–2024 *Guidance on CoQ for development, implementation, and monitoring to improve quality and overall organizational performance.* This publication's objective is to standardize the terms, definitions, CoQ categories, calculations, and analysis methods. This report is broadly applicable to all industry sectors, including construction.

Cost of Quality

CoQ is the total cost of COPQ and the cost of good quality (COGQ) (see Figure 8.4). CoQ is often expressed as a percentage of a cost base. The base could be the cost of goods and services (COGS), cost to deliver a project to a client, or cost of revenue

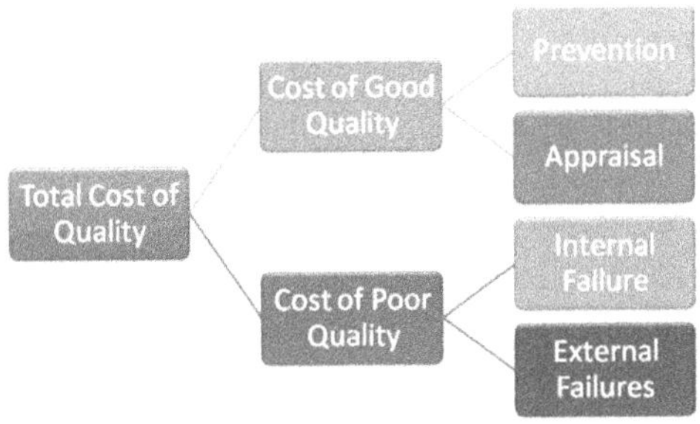

Figure 8.4 CoQ breakdown by categories.
Source: ASQ TR2-2024 *Guidance on CoQ for development, implementation, and monitoring to improve quality and overall organizational performance.*

or other bases as relevant to an organization. More specifically, COGQ spent at the source has higher returns by preventing COPQ downstream. The CoQ relative cost improvement measure is more meaningful than the absolute measurement value. The relative cost improvement measure in CoQ refers to the reduction in these costs over time as quality improvement initiatives are implemented. CoQ provides a framework for various strategic improvement programs.

Cost of Good Quality Versus Cost of Poor Quality

Cost of poor quality (COPQ) is internal and external failure costs incurred when stakeholder needs and expectations aren't met (Ramu, 2021). Cost of good quality (COGQ) is appraisal and prevention costs incurred in failure control, assurance, and prevention (in meeting stakeholder needs and expectations).

- *Prevention Costs*: These costs (before operation) are incurred during error or waste prevention and control; creating new systems, processes, tools, or training; or building competencies.
- *Appraisal Costs*: These costs (during operation) are incurred during measurement and inspection related to assuring conformance to quality expectations.
- *Internal Failure Costs*: These costs are incurred (before customer delivery) from discovery of errors/wastes before shipping products and offering services to customers or external stakeholders.
- *External Failure Costs*: These costs are incurred (after customer delivery) from errors/wastes after shipping products and offering services to customers or external stakeholders.

Cost of Good Quality (COGQ)

Prevention costs (before operation) include those incurred during error or waste prevention and control; creating new systems, processes, tools, or training; or building competencies.

Examples of prevention cost are a competency assessment of the CQM during the hiring process, process qualification and capability assessment prior to construction, material quality measurement equipment qualification for capability and repeatability, qualification of key supplier and material prior to shipping to project site, and creation of controlled environment to ensure required outcome. Construction-related prevention costs are preparation of work plans, just-in-time (JIT) training, training in QMSs/QMPs, training and credential verification for special processes, holding planning phase, pre-activity (sometimes called preparedness or readiness review) meetings, creating and reviewing work plans during meetings with all stakeholders.

Appraisal costs (during operation) are those incurred during measurement and inspection related to assuring conformance to quality expectations. Examples include continual monitoring of skills and proficiency of all key personnel, process monitoring through periodic sampling, equipment regular maintenance, incoming material inspection, testing and approval for use, internal audit, supplier and subcontractor surveillance audits, and monitoring of controlled environments to ensure required outcome.

Cost of Poor Quality (COPQ)

Internal failure costs (before project turnover and closeout, and discovered by contractor's quality function and not customer's oversight team) are costs of poor quality. COPQ also refers to costs incurred from discovery of errors/waste before delivering projects and offering services to customers or external stakeholders. Examples include rework, retest, repair, regrade, retrofit, remediation, scrap, quarantine, obsolescence, breakdown maintenance, accelerated degradation of equipment spares, cost of spares inventory to address breakdown, damaged, expired, obsolete, unusable, and the cost of excess material storage to shield against supplier's quality variability. Other examples include out-of-control environments requiring shutdown and restart of the facility, scrap of stored, work-in-progress materials affected by temperature/humidity excursion.

External Failure Costs (After Project Turnover and Closeout)

Examples include loss of market share, revenue from public opinion, unfavorable public relations, customer dissatisfaction, customer expectations not aligned with internal capability of process, construction not meeting customer expectations, shifts in market conditions, regulatory requirements, and litigations and negative impact to brand due to construction defects.

Methods and Sources for Collecting CoQ Data

Costs mentioned above in each category must be first standardized throughout an organization. Provide the following for each cost (e.g., rework, scrap, reinspection, etc.):

- Operational definition
- Formulae
- Assumptions
- Applicable counting rules to include or exclude certain situations

For example, for the cost component filed under internal failure cost, rework means:

- *Operational definition*: Activities performed to nonconforming material to meet and conform to specifications and requirements.
- *Formulae*: Time and material spent on rework. Standard loaded labor cost × time taken to bring material to conformance including reinspection.
- *Assumptions*: Productivity loss from labor time spent on rework not accounted for.
- *Applicable counting rules to include or exclude certain situations*: Where cost is reimbursed from supplier, subcontractor, or client for nonconformance responsibility, cost is recorded but not included in the CoQ analysis.

These standardized CoQ components should be reviewed and validated by the site finance staff to ensure there are no disagreements from the leadership stakeholders (recipient of the reports) as to the data collection method and analysis.

After each cost is standardized through formal internal documentation, staff who are responsible for collection of data should be trained in using the spreadsheet, and software applications used for processing data should be coded with formulae. Methods for collecting data may differ by organization. Organizations that have invested in infrastructure may be able to acquire data in real time through an internal system. However, other organizations may choose to collect the data manually and enter it into a spreadsheet or software. Unless the data collection is straightforward, like scrap material or nonconforming service, it may be easy to acquire relevant data readily. But when an evaluation is required to determine whether a material is reworked or repaired to conform to requirements, additional follow-up on the data is needed to confirm if the rework was authorized, and whether reworked material was reinspected for fitness after the correction.

Data collection typically starts with the inspection of incoming materials and continues through construction at various phases where QA and QC activities are performed all the way to final inspection and handover to client. CoQ data collection opportunities also exist past handoff at the operations and maintenance phase by monitoring maintenance costs, spares management, downtime, and so on. CQMs should identify the QA and QC activities for capturing costs and keep a record of quality costs incurred through the construction. A document management team assigned to PM typically supports this effort by capturing these costs and preparing periodic reports.

While capturing CoQ data, the CQM should take into consideration the cost of data acquisition against its estimated potential impact. Consider brainstorming with your construction team and creating a 2 × 2 matrix grid to populate the cost components as your team identifies them.

Table 8.8 shows examples of CoQ cost categories comparing measurement difficulty and its impact on an organization.

The construction team should focus their early reporting of CoQ on items that are low-effort and high-impact first.

Table 8.8 CoQ cost components presented by level of effort (LOE) versus impact.

	Easy to measure	Hard to measure
High Impact	Prevention cost • Customer perception survey, product/service qualification, training, equipment calibration, etc. Internal failure cost • Scrap, rework, retest, repair, SW bug fix, etc. External failure cost • Customer returned material, repair, rework, retest, etc.	Prevention cost • Design review, regulatory review, risk management, FMEA development, supplier development, quality planning, proactive engineering changes, etc. Internal failure cost • Resources spent on failure investigation, lost productivity, regrading of products, additional processing to restore product and services External failure cost • Market share, lost revenue from customer dissatisfaction (loss of goodwill), recall, liability, reactive engineering changes from field failures, etc.
Low Impact	Appraisal cost • Inspection, testing, auditing, measuring test equipment, headcount expenses in quality control and assurance	Internal failure costs • People management—Dealing with absenteeism, people performance issues, disciplinary proceedings

Source: Ramu, 2021.

Challenges in Implementation in the Construction Industry

- *Increased costs*: Implementing preventive and appraisal measures can be perceived as overly burdensome for contractors without a robust quality program, leading to increased costs.
- *Craftspeople perception*: Increased preventive measures can be perceived as hindering productivity by craftspeople.
- *Oversurveillance*: Increased appraisals can be perceived as excessive monitoring aimed at finding faults, leading to distrust and resentment.
- *Lack of buy-in*: Without proper training on the importance of CoQ across all categories, achieving buy-in for the program can be challenging.

Overcoming Challenges

- Providing comprehensive training and building awareness on the importance and benefits of CoQ across all categories can foster understanding and acceptance among all stakeholders.

Interpret the CoQ Data to Derive Meaningful Insights

CoQ data as an absolute value is difficult to interpret and draw meaningful insights. It must be normalized with the cost of the project as a proportion. There are no international guidelines on the acceptable proportion. Every organization determines what percentage of COPQ or CoQ is acceptable and sets as a goal not to exceed that number. Even though an increase in COGQ is a good thing for an organization, increasing it disproportionate to benefits realized is not good. Having a consistent approach to measuring and monitoring the cost components helps a construction organization compare the trend through the reporting period.

Report and Communicate Findings from the CoQ Data Analysis

Reporting CoQ can get tricky in some organizations. Anytime a dollar value is attached to any report, there will be more scrutiny, specifically by the leadership. This report helps to put quality efforts in terms of financial and economic impact to the organization. CQMs should obtain review and verification of the analysis and interpretation by finance staff and key stakeholders before presenting to the leadership. The CoQ report itself is not an output by the CoQ program. The effectiveness of CoQ should be linked to reduction in organization operational expenses (OPEX), gross margin, cash flow, customer satisfaction, and so on. Further, the CoQ report should make recommendations for improvement in areas where they can make a high impact.

CONCLUSION

CoQ analysis is not merely an accounting exercise; it is a strategic tool for organizational improvement. By diligently collecting, interpreting, and communicating CoQ data, construction organizations can identify areas for optimization, reduce waste, and enhance efficiency. Overcoming implementation challenges through training and fostering a culture of quality is crucial. Ultimately, the success of a CoQ program translates into tangible benefits: reduced operational expenses, increased customer satisfaction, and a stronger bottom line. Embracing CoQ is an investment in sustainable quality and long-term organizational success.

QUESTIONS

1. How would you identify CoQ cost components for your organization's program? Which ones would you consider first and why?
2. What are the aspects of standardizing CoQ cost components? Why is it important to standardize these costs before measuring, monitoring, and reporting?
3. Why is it not meaningful to compare CoQ metrics among different organizations?
4. Why should a CQM be interested in the CoQ program?
5. What are the benefits of the CoQ program to an organization?
6. Describe the relationship between cost of good quality (COGQ) and cost of poor quality (COPQ) within the overall CoQ framework, and give construction-related examples of each category.
7. Outline the steps involved in standardizing CoQ cost components within an organization, and discuss why this standardization is crucial for accurate data collection, analysis, and reporting.

8. Discuss the potential challenges in implementing a CoQ program in the construction industry, and propose strategies for overcoming these obstacles and fostering buy-in from stakeholders.
9. Explain why simply having absolute CoQ data is not sufficient for drawing meaningful insights, and describe how this data should be normalized and interpreted to drive strategic improvements within a construction organization.
10. Given that COGQ is often seen as an investment, how would an organization determine the optimal level of COGQ spending to avoid overinvestment while still effectively minimizing COPQ, and what metrics could be used to track this balance?

IMPROVEMENT PLANS

> Incorporate CoQ as a foundation for improvement methodology, alongside Lean, Six Sigma (e.g., DMAIC), and Deming's System of Profound Knowledge (SoPK). Identify areas for improvement from CoQ data analysis, and participate in the improvement plan to address these, integrating the plan with existing business processes. Evaluate the effectiveness and applicability of CoQ tools and methodologies. (Evaluate)
>
> Body of Knowledge VIII.E.2

Driving meaningful change within an organization requires more than just identifying problems; it demands a structured approach to improvement. This section delves into the critical role of CoQ as a foundational element for strategic improvement initiatives. We explore how CoQ analysis guides the selection and application of methodologies like lean, Six Sigma, and Deming's System of Profound Knowledge (SoPK). By understanding and leveraging these tools, organizations can effectively reduce waste, enhance efficiency, and ultimately achieve a sustainable culture of continuous improvement integrated into existing business processes.

Cost of Quality Analysis and Improvement Planning

In the previous section's discussion of the CoQ program and its analysis, we emphasized that the true value of CoQ lies not just in reporting but in taking concrete action. CoQ analysis determines the proportion of our chosen base (e.g., cost of goods sold, cost of revenue) that is attributable to CoQ. What constitutes a "healthy" or "excessive" percentage is ultimately a decision for the organization itself, as there is no formal international guidance on absolute CoQ targets. This lack of standardization stems from the absence of internationally agreed-on approaches to CoQ component selection, definition, formulas, and assumptions. Historical literature suggested a range of 3–5% of total revenue, but this guidance is not particularly meaningful given the variability in how organizations calculate CoQ.

The key is to strategically increase the "cost of conformance" (prevention and appraisal costs) while decreasing the "cost of nonconformance" (internal and external failure costs). However, it's crucial to avoid excessive investment in conformance without a clear return on investment (ROI), as this would waste valuable organizational resources. The challenge lies in identifying the right improvements that will effectively reduce COPQ and bring the overall CoQ percentage to an acceptable level.

There are many improvement methodologies, like lean and Six Sigma, as well as philosophies, like Deming's System of Profound Knowledge (SoPK), to address the areas requiring improvement. CoQ acts as an overarching program to evaluate the costs of various categories and components identified for measurement and monitoring. CoQ by itself is not an improvement methodology. However, CoQ as a foundational program provides direction as to which areas to improve and how much.

Lean

At its core, lean aims to minimize waste within the value stream and deliver optimal value to customers. In essence, it promotes the philosophy of "doing more with less." Over time, organizations have adopted various industry practices that prioritize maximizing output from our investments, such as machinery and workforce. However, we have often neglected to question whether these outputs truly align with the quality and quantity desired by our customers. This has resulted in the production of excess goods that accumulate in inventories. Traditionally, managers were rewarded for overproduction, leading to pressure on sales managers to offload these inventories. Gullible customers, influenced by marketing strategies, would purchase these products only to discover they did not meet their needs. This created a society where consumers were oversaturated with undervalued products, leaving no clear winners. Lean concepts address these weaknesses prevalent in conventional business models and industry practices. They offer methodologies that prioritize delivering exceptional value to customers while ensuring the organization's economic sustainability.

Within an organization, a value stream represents the sequential activities undertaken, encompassing processes like contract review, planning, design, construction, commissioning, delivery, operation, and maintenance. Typically, a value stream begins with the supplier's supplier and concludes with the customer's customer. Along the value stream, waste can manifest in both obvious and hidden forms. Value stream mapping and analysis serve as valuable tools in identifying value-added activities and distinguishing them from non-value-added ones.

Lean enterprise methodology uses a number of lean tools to accomplish value to customers and the organization. Value stream mapping (VSM), 5S, 8 Wastes identification, visual management, kanban, standard work, and error proofing are widely used. These tools, although originated in manufacturing, can be effectively applied in construction projects.

Six Sigma

Six Sigma, a well-structured methodology, aims to enhance processes and resolve issues systematically. At its core, Six Sigma employs a rigorous approach known

Table 8.9 PDCA and Six Sigma mapping.

Plan	Define
Plan	Measure
Plan	Analyze
Do / Check	Improve
Act	Control

Source: The *ASQ Six Sigma Yellow Belt Handbook*, Second Edition.

as DMAIC (Define, Measure, Analyze, Improve, Control). DMAIC outlines the sequential steps that a Six Sigma practitioner must follow, beginning with the identification of the problem and culminating in the implementation of sustainable solutions. While there are alternative Six Sigma methodologies, DMAIC stands as the most extensively adopted and recognized structured approach within the Six Sigma framework.

The DMAIC model closely resembles the plan-do-check-act (PDCA) or plan-do-study-act (PDSA) model, which you may already know. Table 8.9 shows the alignment between these models.

A crucial element for successful implementation of DMAIC and continuous improvement is management's commitment to providing the necessary time and resources for each phase. This factor sets Six Sigma apart from other quality improvement programs.

Other driving forces that differentiate Six Sigma include involving everyone in the organization, ensuring timely access to data for all, and incorporating financial data analysis, such as the CoQ, into the improvement process.

Within the organization, each individual will be encouraged to actively participate in the Six Sigma model, assuming the responsibility of identifying and pursuing ongoing improvement opportunities within their respective work areas. Basically, practitioners will be following in each step:

1. *Define*: Identify an issue causing decreased customer satisfaction, a reduced bottom line, safety incidents, supplier quality failure, and so forth.
2. *Measure*: Collect data from the process and verify the validity of the data.
3. *Analyze*: Study the process data to identify root cause(s).
4. *Improve*: Act on the root causes to change the process for the better.
5. *Control*: Monitor the system to sustain the gains.

Deming's System of Profound Knowledge (SoPK)

Deming emphasized that transformation of organizations begins with the individual. Organizational transformation is accomplished, in part, by gaining more exposure to other people, groups, and cultures. This will require understanding the system of profound knowledge, which consists of four major components:

1. *Appreciation for a system*: An organization is a system of interrelated components with a common purpose. Changing one part of the system affects other parts.
2. *Knowledge about variation*: Everything varies, but understanding whether the variation is due to chance cause or random cause will change what action should be taken, if any.
3. *Theory of knowledge*: No learning has occurred if there is no theory that allows prediction. Operational definitions are necessary to allow theories to be useful to many.
4. *Psychology*: We all want to be appreciated. People have different needs, however, and what will make one happy might negatively impact another.

Integration of the Improvement Plan with Existing Business Processes

CoQ is the fundamental program on which an organization's overall improvement strategy is built. It provides the foundation for a QMS to standardize business processes, lean enterprise methodologies to minimize waste and enhance effectiveness, and Six Sigma programs to reduce variability and boost efficiency.

To effectively enhance a business, lean, Six Sigma, and other methodologies must be integrated into existing business processes. This involves mapping current processes, identifying gaps and waste, and developing improvement plans that directly address resource limitations. These resources encompass people's competencies, equipment and process capabilities, material quality and availability, and measurement capabilities. The ultimate goal is to establish a process that consistently performs well, even when faced with uncontrollable changes in the operating environment.

Effectiveness Metrics and Monitoring

Process audits are a valuable tool for establishing a process baseline, identifying areas for improvement, and verifying the efficacy of any changes made. By addressing gaps and inefficiencies, organizations can reduce operating expenses associated with recurring errors, waste, and excessive appraisal costs (e.g., excessive inspections). Key performance indicators (KPIs) for tracking improvement include:

- Reduction in the COPQ
- Achievement of CoQ targets
- Improved efficiency and productivity
- Fewer safety incidents
- Higher employee and customer satisfaction

An example of tools and their application for effective use are provided in Table 8.10.

Table 8.10 Six Sigma tools overview by stage.

Tool	Description
	Define
SIPOC (supplier-input-process-output-customer)	Tool to describe and understand a process more clearly. Supplier can also be interpreted as "source of inputs" or "who supplies the inputs," and customer can be interpreted as the "recipient of the outputs."
Is-is not analysis	Tool helpful in defining a problem. Oftentimes, understanding what is not a problem will help the team zero in on areas that are a problem. This saves a significant amount of time during problem-solving.
Timeline of events	Tool used to understand the problem through a sequence of events (one event leads to another).
Trend chart	Tool used to understand process trends (favorable and unfavorable) from historical events of data.
Run chart	Tool used to understand process patterns from historical data.
Process flowchart	Graphical representation of the sequence of interrelated activities.
Current state map	A value stream map that represents the current state of a process.
GANTT chart	Management of schedule, resources, cost, and risk used in project management.
Stakeholder analysis	Tool used to understand the people who are influential to a project and those who offer resistance coupled with their level of involvement in a project.
	Measure
Data collection plan	Plan describing what, why, where, when, who, and how of data collection (5W1H). Prior to collecting data, having a detailed collection plan can help make collection more effective.
Measurement systems analysis (MSA)	Analysis of the capability of a measurement system (tools, methods, people, environment, etc.).
Benchmark	Start by setting the current baseline for the process.
Process capability	Ability of the process to meet expected output. Often represented by indices Cp and Cpk (the higher the indices, the better).
	Analyze
Why-why	Method of continuing to ask "why" on a problem symptom until the root cause is reached.
Hypothesis tests	Statistical methods to test the hypothesis on process improvement.
Statistical sampling	Statistically valid sampling (e.g., power and sample size, acceptance sampling).
	Improve
Design of experiments	Tool for understanding cause and effect in a process and for optimizing process variables to achieve improved performance.

Continued

Table 8.10 Six Sigma tools overview by stage. *(continued)*

Tool	Description
Future value stream map	Creates a "should be" future state value stream with non-value-added process steps removed.
Theory of constraints	Tool for understanding bottlenecks.
FMEA	Tool to evaluate potential risk and prioritize actions.
Poka-yoke (mistake proofing)	Concept that prevents errors from happening in processes and products by "designing in" controls.
	Control
Process behavior chart	Tool to help understand the behavior of a process by analyzing historical patterns and trends.
Control plan	Document that provides basic information on what is required to manage a process during the manufacturing or service delivery.
Process audit	Tool to verify compliance of a process as per the specification.
Cost of quality	A measure that captures the cost of conformance and the cost of nonconformance; often represented as a ratio comparing to the revenue or cost of goods sold of an organization.
	Tools applicable to all stages
Brainstorming	Tool (and a process) for collecting inputs and ideas from team members without making any instant judgments.
Seven quality tools	Check sheet, histogram, flowchart, cause and effect diagram, Pareto chart, run chart, scatter diagram (see Chapter 5 for a description and application of these tools).

The success of tools like lean and Six Sigma hinges on factors such as their suitability for the task, user adoption, training, and real-time feedback loops for measuring improvements. For instance, a well-executed 5 Whys analysis is often sufficient for root cause analysis, but its effectiveness diminishes when practitioners overlook the complexity of real-world problems with multiple, interconnected causes. Improvement professionals need broad training in various tools and the discernment to choose the right tool for each situation.

Successful implementation of improvement plans hinges on a holistic approach that integrates CoQ analysis with methodologies like lean, Six Sigma, and SoPK. By understanding the nuances of each tool and its applicability, organizations can effectively target and reduce the COPQ. Continuous monitoring through process audits and KPIs ensures sustained improvement and validates the efficacy of implemented changes. Ultimately, a culture of improvement, driven by informed decision-making and strategic resource allocation, is paramount for achieving organizational excellence.

QUESTIONS

1. What are the factors to consider while deciding which methodology is suitable for driving continuous improvements?

2. What are the applications where lean tools can be effectively utilized in a construction industry?
3. Explain how a construction quality manager could apply the Six Sigma (DMAIC) approach to a construction project?
4. How does Deming's SoPK play a vital role in driving improvement?
5. Identify key methods to assess effectiveness of application of improvement tools?
6. Explain how CoQ analysis acts as a foundational program for improvement strategies, and differentiate its role from methodologies like lean and Six Sigma.
7. Describe the challenge organizations face in balancing the "cost of conformance" and the "cost of nonconformance" within a CoQ framework. What are the potential risks of overinvesting in conformance?
8. Compare and contrast the DMAIC methodology of Six Sigma with Deming's System of Profound Knowledge (SoPK). How does each approach contribute to organizational improvement, and in what scenarios might one be preferred over the other?
9. Discuss the significance of integrating improvement plans with existing business processes. What are some potential barriers to this integration, and how can organizations overcome them?
10. Identify and explain three key performance indicators (KPIs) that can be used to track the effectiveness of improvement initiatives derived from CoQ analysis. How do process audits complement the use of these KPIs?

ORGANIZATIONAL GOALS

> Assess how CoQ improvement efforts align with organizational goals. Monitor and measure the progress made towards achieving organizational goals. Evaluate the steps taken to achieve organizational goals based on CoQ improvements. (Evaluate)
>
> **Body of Knowledge VIII.E.3**

To achieve their vision and mission, every organization has a set of goals. While vision is long-term, short-term goals are also necessary. During the annual strategic planning process, the leadership team identifies and agrees on these goals, which are then deployed annually. Organizational goals can span various areas, such as financial performance, operational efficiency, supply chain management, customer engagement, and employee development.

Organizations have numerous goals (Table 8.11) in different areas, and they must determine how much they can realistically accomplish each year, both in the short term and long term. For instance, an organization might set a goal of reducing operational expenses (OPEX) by 10% year over year and by 50% over a five-year plan without affecting revenue. This can be achieved by streamlining organizational overheads and eliminating waste throughout the business value stream.

By conducting a COQ analysis, organizations can gain valuable insights into their improvement efforts and align them with their overall goals. The CoQ analy-

Table 8.11 Organizational goals by function.

Functional area	Organizational goal
Finance	Improve • EBITA • Cash flow • Earnings per share (EPS) Reduce operational expenses
Operational	Improve • Efficiency • Productivity • Inventory turns • On-time delivery (customer) • Agility • Business continuity • Regulatory compliance Reduce • Safety incidents • COPQ
Supply Chain	Improve • Lead time • On-time delivery (supplier) • Incoming quality • Sustainability
Customer facing	Improve • Customer experience • Loyalty • Market share
Employee	Improve • Morale • Satisfaction • Working condition • Compensation par with living condition • Competency • Retention

sis offers a detailed breakdown, highlighting areas where costs are concentrated. A thorough review of the costs associated with nonconformances reveals the extent of the organization's losses due to internal and external failures.

Internal failures may show up as costs involved in design changes, repair, rework, scrap of materials, regrading an existing product or service to a lower cost, productivity losses from quality failures, equipment breakdown, work stop from safety incidents, cost of addressing and resolving safety incidents, and so on. *External failure* may show up as costs involved in customer complaint investigation, product returns, service failure, penalty for not meeting contract conditions, cost of resolving noncompliance with regulators, and so on.

Many organizations prioritize addressing design modifications, repairs, reworks, scrap, and customer complaint resolution as low-hanging fruit. Improvement strategies in these areas could align with organizational goals such as reducing operational expenses and the COPQ, and enhancing customer experience and loyalty. Implementing lean, Six Sigma initiatives or utilizing a structured plan-do-check-act approach can align with these objectives.

To achieve organizational goals, the following steps can be taken based on CoQ improvements and their effectiveness:

1. Identify the cost components within each cost category that have the greatest impact on organizational goals.
2. Investigate the cost components and their respective reporting sources. Utilize Pareto analysis to identify the critical few to focus on.
3. Identify opportunities for improvement with clearly defined objective statements.
4. Develop a financial calculation, such as net present value (NPV) or internal rate of return (IRR), to present leadership and secure sponsorship for improvement initiatives by aligning them with organizational goals.
5. Appoint a program manager and establish a cross-functional team to oversee the implementation of improvement initiatives and establish a reporting structure.

To ensure that improvement plans reach milestones and achieve their objectives, project management is essential. Program managers create dashboards, infrastructure for measuring and monitoring improvements, and periodically reporting to sponsors and leadership. To confirm the efficacy of the actions taken and the capacity to maintain long-term benefits, the sponsor may designate a quality auditor.

Usually, these improvement initiatives are finished within 90 days. Even if a complex improvement initiative is proposed, it is divided into 90-day periods to demonstrate progress steadily. This aims to respond more quickly to waste, stay aligned with organizational priorities, and prevent long-term resource allocation for a few initiatives.

In large, complex organizations, leadership and organizational changes are likely. If improvement initiatives are carried out over longer periods, such as one to two years, there is a risk of disruption or even cancellation before the goals are achieved.

CONCLUSION

Aligning CoQ improvement efforts with organizational goals is paramount for achieving both short-term and long-term objectives. By meticulously identifying, investigating, and prioritizing cost components, organizations can effectively implement targeted improvement initiatives. Utilizing financial calculations like NPV or IRR to secure leadership sponsorship ensures these initiatives are strategically aligned. Short, 90-day improvement cycles further enhance agility and responsiveness, mitigating risks associated with prolonged projects. Ultimately, effective project management, monitoring, and auditing are crucial for realizing sustained benefits and driving organizational success.

QUESTIONS:

1. How would you prioritize the costs from the CoQ report?
2. What is the relationship between CoQ and organizational goals?
3. What are the challenges in pursuing improvement to address CoQ costs?
4. How are improvement programs assessed for their effectiveness?
5. Why is it important to plan improvements to be achieved within a short duration, for example, 90 days?
6. Explain how utilizing net present value (NPV) or internal rate of return (IRR) calculations can aid in aligning CoQ improvement initiatives with organizational goals, and why this is crucial for securing leadership sponsorship.
7. Describe the potential risks associated with implementing CoQ improvement initiatives over extended periods (e.g., one to two years) in large, complex organizations, and explain why the document suggests 90-day improvement cycles.
8. Based on Table 8.11, how might a reduction in the COPQ in the operational functional area positively influence goals in the finance- and customer-facing functional areas? Give examples.
9. Discuss the steps outlined in the document for achieving organizational goals based on CoQ improvements. Focus on the importance of identifying and prioritizing cost components, and how this relates to the use of Pareto analysis.
10. Explain the role of a quality auditor in ensuring the sustained benefits of CoQ improvement initiatives, and how their involvement contributes to effective project management and monitoring.

Customer Satisfaction

CUSTOMER FEEDBACK SYSTEM

> Apply feedback collection tools, such as listening posts, focus groups, complaint logs, warranty data, surveys, and interviews to capture positive and negative feedback. (Apply)
>
> Body of Knowledge VIII.F.1

Customer satisfaction is defined as a measurement that determines how happy customers are with a company's products, services, and capabilities. Customer satisfaction information, including surveys and ratings, can help a company determine how to best improve or change its products and services (ASQ Glossary, 2025). Organizations, both for-profit and nonprofit, must prioritize their customers, who are their most crucial stakeholders. Customers include anyone the organization supplies with products or services. The existence of a business hinges on its ability to meet customer needs, whether they are internal or external. Businesses that fail to satisfy their customers risk losing them to competitors, leading to decreased revenue, profitability, and ultimately, business closure.

In recent decades, organizations have recognized that mere customer satisfaction is insufficient. To ensure long-term success, businesses must strive for customer delight and loyalty. While satisfied customers may still explore other options, loyal customers consistently return, making repeat purchases. This emphasis on customer loyalty has led many organizations to adopt a core value of "customer obsession."

To truly grasp customer needs and wants, a formal system for gathering feedback is essential. This should begin early in the process by soliciting customer expectations, rather than solely relying on postdelivery feedback. Feedback should be gathered continuously at key stages: preplanning, planning, design, preconstruction, handover, and operation and maintenance.

Early feedback integration allows for proactive planning and design. For instance, a client may desire a reliable construction design and for that design to utilize sustainable materials. If existing technology cannot meet both requirements, this must be addressed early on. Solutions could involve developing new materials that are both durable and sustainable, ensuring compliance (regulatory, legal, etc.). This proactive approach prevents such issues from becoming afterthoughts, thus dissatisfying the customer.

For decades, leading companies have been measuring and tracking customer satisfaction levels. These results are then used to pinpoint problems and opportunities, evaluate the performance of managers and executives, and reveal how they stack up against competitors. These companies have also successfully integrated these findings—often obtained through customer surveys—into their core business strategies. This includes areas like new product development, manufacturing quality, product and service delivery, and competitive positioning. By using statistical analysis to understand the key factors influencing their customers' buying decisions, these companies make informed marketing choices.

Customer satisfaction measurement has become a cornerstone of quality management processes. Companies leverage customer satisfaction survey results to drive improvement initiatives. However, assessing customer satisfaction levels is impossible without first understanding their expectations, priorities, and needs. This crucial process is often referred to as "listening to the voice of the customer."

Cutting-edge customer satisfaction systems incorporate several key elements:

- *Formal processes for data collection and analysis*: These systems have structured methods for gathering and analyzing customer data. The results are then shared with the relevant business functions for action. This data can be used for company or market segment analysis to develop targeted marketing strategies. It can also be used at the account level to gain a deeper understanding of individual customer opinions.
- *Multifaceted feedback collection*: Feedback on organizational performance is gathered using a combination of methods, which are described in more detail later. Most companies employ several formal methods

for collecting customer satisfaction data. By using multiple sources, companies can better validate results and enhance their credibility.

- *Strategic combination of methods*: Each method of gathering customer feedback has its own set of advantages, disadvantages, and specific applications. To get a complete picture, most companies opt for a combination of methods. Once the information is gathered, sophisticated techniques can be used to analyze it.
- *Accessible data storage*: To be useful, data must be stored appropriately and made readily available to those who need it.

Customer focus teams enhance an organization's ability to implement customer relationship management strategies. These teams often serve as a central point of contact for customers, coordinating across various functions. Typically, a customer focus team includes an account manager, customer service representative, field engineer or technical support specialist, and a quality manager. The team's goal is to provide a seamless customer experience and ensure that all key stakeholders within the supplying organization are aligned in their strategy and actions on behalf of the customer.

Feedback Collection

Organizations can use both reactive and proactive feedback methods to understand the voice of their customers. Reactive methods include complaint monitoring, 800-type numbers, and warranty systems, while proactive methods include surveys, phone calls, and customer visits. Proactive methods like customer surveys have become more popular. The pressure to be competitive and financially successful has also driven many organizations to focus on improving performance and customer satisfaction.

Businesses can collect customer data using various tools and from different parts of the organization. To effectively use data, it's important to have systems in place for collecting, analyzing, and sharing it. Customer satisfaction data can come from sources like product warranty registrations, complaint logs, customer surveys, transaction data, established listening posts, and lost-customer analysis. When this information is combined, it becomes very valuable for the organization's stakeholders.

However, individual departments might not always share this information with others who could benefit from it. This can lead to missed opportunities, as data analyzed separately might suggest a different course of action than when considered collectively. Organizations need to integrate, process, and track trends in the data, and then share the resulting information in a way that maximizes its positive impact.

Data Collection Techniques

Survey Research

Survey research is frequently used to gauge customer satisfaction, track changes in these perceptions over time, and understand the reasons behind customer attrition. It encompasses both quantitative and qualitative approaches. Quantitative research, often employing rating scales and multiple-choice questions, yields numerical data suitable for statistical analysis and generalization to larger populations. Qualitative research, on the other hand, delves deeper into the motivations and opinions behind customer responses, providing richer insights but without the same statistical rigor.

Survey Methods

- *Written/mail surveys*: This traditional method involves sending questionnaires directly to customers. While offering the advantage of low cost and respondent convenience, mail surveys often suffer from low response rates and potential bias. Techniques like prenotification, compelling cover letters, and follow-up mailings can improve participation. Although mail surveys can incorporate both quantitative and qualitative questions, the latter may yield less detailed responses compared to other methods. To enhance effectiveness, these surveys should be concise and focused.
- *Focus groups*: These may be one of the most effective methods of gathering feedback because you are bringing together a group of people.

Steps to conduct a focus group:

1. *Define objectives.*
   - Clearly outline the purpose of the focus group and the specific information you aim to gather.
2. *Recruit participants.*
   - Identify and invite individuals who match the target demographic criteria.
   - Offer incentives, if appropriate, to encourage participation.
3. *Prepare materials.*
   - Develop the discussion guide with key questions and prompts.
   - Prepare any materials or products that will be reviewed or discussed during the session.
4. *Conduct the session.*
   - Begin with a brief introduction, explaining the purpose and ground rules.
   - Use the discussion guide to facilitate the conversation, encouraging all participants to contribute.
   - Take detailed notes or record the session (with participants' consent) for later analysis.
5. *Analyze data.*
   - Review the notes or recordings to identify common themes, patterns, and insights.
   - Summarize the findings and present them in a clear, actionable format.

- *Suggestion/comment boxes or response forms*: These may also prove valuable when there are circumstances that make it challenging to obtain feedback through a survey or focus group. A comment box can be set up on a website to collect feedback electronically or as a physical receptacle using a standard short form or card.
- *Other survey methods*: Other common approaches include telephone surveys, in-person interviews, and online surveys. Each method carries its own set of advantages and disadvantages in terms of cost, response rates, depth of information, and logistical considerations. The choice of method should align with the specific research objectives and target audience.

Key Considerations for Survey Design

Regardless of the method chosen, effective survey design is critical for gathering meaningful data. Key aspects to consider include:

- *Clear objectives*: Defining precise research questions ensures the survey instrument is tailored to capture the necessary information.
- *Careful question wording*: Unclear or biased questions can lead to inaccurate or misleading results.
- *Appropriate sampling*: Selecting a representative sample is essential for generalizing findings to the broader population.
- *Thorough analysis*: Employing appropriate statistical and qualitative analysis techniques ensures that insights are extracted effectively from the collected data.

By considering these elements, businesses can leverage survey research to gain valuable insights into customer perceptions and preferences, ultimately informing strategic decision-making and driving improvements in customer satisfaction and loyalty.

Telephone Surveys

Telephone surveys are a common tool for gathering customer feedback, often used alongside or instead of written surveys. They are particularly effective for collecting information about a company's overall performance or specific areas like sales and service. These surveys involve interviewers directly contacting customer decision-makers and influencers to understand their needs and assess how well the company is meeting them.

Advantages

- *Speed*: Surveys can be conducted and results analyzed relatively quickly.
- *Lower nonresponse bias*: Compared to mail surveys, telephone surveys typically have higher response rates, leading to more accurate representation of the target customer population.
- *Smaller sample sizes*: Higher response rates allow for smaller, more manageable sample sizes.

- *Clear respondent identification*: Companies know exactly who answered the survey questions.
- *Flexibility*: Experienced interviewers can adapt their questions based on customer responses, allowing for deeper exploration of specific topics.
- *Quality open-ended feedback*: With proper interviewer training, telephone surveys can yield valuable insights through open-ended questions.

Disadvantages

- *Interviewer bias*: Inconsistent interviewing techniques can introduce bias into the results.
- *Limited complexity*: Complex concepts may be difficult to evaluate over the phone without visual aids.
- *Time constraints*: While generally longer than mail surveys, phone calls have practical time limits.
- *Higher cost*: Compared to well-executed written surveys, telephone surveys can be more expensive per response.
- *Interruptions and resistance*: Phone calls can be interrupted, and there's growing resistance to unsolicited calls, especially among consumers.
- *Accessibility Issues*: Some customers may have unlisted numbers or no phone service.

Despite these challenges, telephone surveys remain a popular and valuable method for gathering customer feedback and driving quality improvement in sales and customer service. However, it's crucial to consider the potential downsides, such as increasing customer resistance to unsolicited calls and the impact of caller identification technology.

Many customers resent the intrusion of organizations seeking survey responses, and the increasing number of requests can lead to frustration. The use of caller ID allows recipients to screen calls, and the lack of face-to-face interaction may make it easier for potential respondents to end the call prematurely. This is particularly relevant for cell phone users, who may incur charges for incoming calls.

In Japan, a technique called "observation murmurs" is used to gather information about customer behavior. In the late 1980s, Honda sent a team of engineers to Disney World's parking lot to study how customers used their car trunks. The team observed people entering and exiting the theme park, noting how they packed and unloaded their belongings. This research led to several new design ideas for future Honda trunks, making them more convenient and beneficial for customers. To gain inspiration for the design of their family van and pickup truck products, Japanese manufacturers closely observed American consumers' preferences and behaviors at Home Depot and other hardware stores.

Listening post: Numerous individuals within organizations interact with customers periodically or occasionally, including engineers, salespeople, CEOs, and delivery personnel. During these interactions, customers often express opinions, suggestions, complaints, or compliments regarding the supplier's organization, product or service quality, delivery, price, and even the personal attention they receive. While severe negative expressions may be addressed, these casual and informal comments are typically not captured. Without a formal process to collect and analyze these interactions, an organization may miss the early stages of emerging customer problems and fail to acknowledge positive feedback that could motivate the responsible individuals.

Construction businesses can effectively implement various feedback collection tools to gather both positive and negative feedback from current and potential customers.

Listening Posts

To facilitate customer feedback, establish designated areas at construction sites or offices. These areas can include physical suggestion boxes, digital kiosks, or online forms that can be accessed using QR codes.

Capturing Feedback

- *Positive*: Encourage customers to share their appreciation for aspects of the project, such as the quality of work, adherence to schedule, or the professionalism of the crew.
- *Negative*: Provide a clear and easily accessible method for customers to voice concerns, complaints, or suggestions for improvement.
- *Example*: At the entrance of a construction site, place a suggestion box with prompts such as "What are we doing well?" and "How can we improve?"

Focus Groups

Organize small group discussions (6–10 participants) with existing and potential customers, facilitated by a moderator, to explore their experiences and expectations.

Capturing Feedback

- *Positive*: Ask open-ended questions to identify the aspects of the construction process that customers value the most and the reasons behind their decision to choose (or consider choosing) your company.
- *Negative*: Facilitate discussions to uncover pain points, challenges, and areas where the company can enhance its services.
- *Example*: Conduct focus groups specifically with homeowners who have recently completed renovation projects to gather feedback on communication, project management, and their overall satisfaction.

Complaint Logs

Create a centralized system for recording and tracking customer complaints, using a spreadsheet, dedicated software, or an online platform.

Capturing Feedback

- *Positive*: Complaint logs can indirectly reveal positive aspects by highlighting areas where the company consistently meets or exceeds expectations (e.g., lack of complaints in certain areas).
- *Negative*: Establish a clear process for customers to submit complaints, including details like the issue, date, time, and location. Categorize complaints to identify recurring problems and patterns.
- *Example*: Implement a customer service portal on the company website where customers can conveniently submit complaints, track their resolution status, and provide additional information as needed.

Warranty Data

- Analyze warranty claims and repair requests.
- Identify trends and potential areas for improvement in construction quality or materials.

Capturing Feedback

- *Positive*: Low warranty claims for specific projects or materials indicate customer satisfaction and high work quality.
- *Negative*: A high volume of warranty claims for a particular issue signals a recurring problem that needs proactive attention.
- *Example*: Track warranty claims related to roofing installations to identify potential issues with materials or installation practices.

Surveys

Feedback Collection

- Utilize online survey platforms like SurveyMonkey or Google Forms, or paper-based surveys, to gather construction process feedback.

Capturing Customer Perspectives

- *Positive*: Incorporate questions that enable customers to rate their satisfaction with project aspects, including communication, timelines, and quality.
- *Negative*: Offer open-ended response options for customers to elaborate on their concerns and provide suggestions.

Post-Project Survey

- Send out a post-project completion survey to gather feedback on overall satisfaction, potential improvements, and the likelihood of recommending the company.

Interviews

Conduct one-on-one interviews with current and potential customers to gather in-depth feedback. These can be done in person, over the phone, or via video conferencing.

Capturing Feedback

- *Positive*: Ask questions that encourage customers to share their positive experiences and what they value most about the company.
- *Negative*: Use of probing questions to uncover any underlying concerns or challenges that customers may have faced.
- *Example*: Conduct phone interviews with potential clients who requested a quote but did not proceed with the project to understand their decision-making process and identify areas for improvement in the sales process.

Key Considerations for Designing Customer Feedback Systems

- *Target audience*: Tailor feedback collection methods to specific customer segments (e.g., homeowners, commercial clients).
- *Timing*: Collect feedback at various stages of the construction process (e.g., preconstruction, during construction, post-completion).
- *Confidentiality*: Ensure customer anonymity and data privacy when collecting and analyzing feedback.
- *Actionable insights*: Analyze collected feedback to identify trends, prioritize areas for improvement, and implement changes to enhance customer satisfaction.
- *Follow-up*: Close the feedback loop by acknowledging customer input and communicating any actions taken in response to their feedback.

By implementing a comprehensive feedback collection system, construction businesses can gain valuable insights into customer perceptions, identify areas for improvement, and build stronger customer relationships. It is important to use the insights gathered from customer feedback as the input to lessons learned.

Cultivating robust customer feedback systems is paramount for organizational success, particularly in the construction industry. By implementing a multifaceted approach encompassing listening posts, focus groups, surveys, and complaint logs, businesses can gain invaluable insights into customer perceptions. This data, when analyzed and acted on, not only enhances customer satisfaction but also drives continuous improvement, informs strategic decision-making, and ultimately strengthens the organization's bottom line. Embracing a culture of customer obsession is essential for long-term viability and competitive advantage.

QUESTIONS

1. What are different types of customer feedback systems used in your industry? How might they be applicable to the construction business?

2. Define what type of feedback tools and techniques are applicable to which type of customer.
3. In the current competitive world, why Only customer satisfaction measurement may not be just adequate?
4. Design a customer feedback system for small to midsize construction organizations in residential building business and justify your rationale.
5. How can an organization's bottom line be impacted (positively or negatively) from customer feedback? *Hint*: Take into consideration data sources from both organization designed and social media.
6. Describe the strategic importance of gathering customer feedback at multiple stages of a construction project (preplanning, planning, design, preconstruction, handover, and operation and maintenance) rather than solely relying on postdelivery feedback. Explain how early feedback integration can proactively address potential issues and enhance customer satisfaction.
7. Compare and contrast reactive and proactive feedback methods in the context of construction businesses. Give specific examples of each method and discuss the advantages and disadvantages of relying primarily on one type over the other.
8. Explain how different data collection techniques, such as survey research, focus groups, and listening posts, can provide unique insights into customer satisfaction. Discuss the factors a construction business should consider when choosing the most appropriate combination of these techniques for their specific needs.
9. Discuss the role of customer focus teams in enhancing an organization's customer relationship management strategies. What key roles should typically be included in such a team, and how do they contribute to providing a seamless customer experience?
10. Analyze the statement "mere customer satisfaction is insufficient for long-term success." Explain the concept of "customer obsession" and how it relates to customer loyalty and repeat purchases. Give examples of how a construction business can cultivate a culture of customer obsession within its organization.

CUSTOMER VALUE ANALYSIS

> Analyze a customer value analysis to determine the financial impact of customer retention and the potential consequences of customer attrition. (Analyze)
>
> **Body of Knowledge VIII.F.2**

Customer value analysis (CVA) is a strategic process crucial for understanding how customers perceive the value of products or services relative to competitors. It illuminates the factors driving customer decisions and reveals their core priorities. By identifying key value attributes such as functionality, quality, cost, and timeliness, businesses can tailor their offerings to enhance customer satisfaction. This chapter delves into how CVA can be performed in the construction industry, its impact on customer retention and attrition, and its connection to financial metrics, ultimately driving long-term success.

Before embarking on a CVA, let's delve into its essence. CVA is a strategic process designed to show how customers perceive the value of your products or services relative to your competitors. It aims to shed light on the key factors that influence customer decisions and uncover what truly matters to them. Armed with this knowledge, you can refine your offerings and enhance customer satisfaction, fostering stronger connections with your target audience.

The next sections explain how CVA can be performed in the construction industry.

Identifying Key Value Attributes

- *Functionality*: This includes the basic requirements of a construction project, such as meeting building codes, providing structural integrity, and delivering the intended purpose (e.g., a livable home, a functional office building).
- *Quality*: Customers value quality workmanship, durable materials, and attention to detail. This can encompass aspects like the finish of the construction, the quality of fixtures and fittings, and the longevity of the structure.
- *Cost*: Price is always a significant factor in construction. Customers seek value for their money, balancing cost with the desired quality and functionality.
- *Timeliness*: Project schedules and completion dates are crucial. Delays can cause disruptions and increase costs for customers.
- *Reliability and trust*: Customers value contractors who are reliable, trustworthy, and communicative. This includes transparency, responsiveness, and adherence to agreements.
- *Safety*: A safe construction site is paramount. Customers value companies that prioritize worker safety and minimize risks.
- *Sustainability*: Increasingly, customers are interested in sustainable construction practices, including energy efficiency, environmentally friendly materials, and waste reduction.
- *Aesthetics*: For many projects, aesthetics plays a significant role. This includes the design, appearance, and overall visual appeal of the finished structure.

A Certified Construction Quality Manager (CCQM) can significantly influence key value attributes in construction. Their expertise ensures that functionality is achieved by adhering to building codes and standards, while quality is upheld through rigorous QC processes and material inspections. By implementing efficient project management practices, CCQMs can help ensure timeliness, preventing costly delays. Their focus on clear communication and ethical practices builds reliability and trust with clients and stakeholders. Additionally, a CCQM's emphasis on safety protocols and risk management directly addresses the safety attribute. Furthermore, their knowledge of sustainable practices and materials can contribute to meeting sustainability goals, while their understanding of design and aesthetics can enhance the overall aesthetics of the project. By effectively managing these attributes, a CCQM adds substantial value to construction projects and boosts customer satisfaction.

Assess Attribute Importance

- *Surveys*: Use surveys to quantify the relative importance of each attribute to different customer segments (e.g., homeowners, commercial developers).
- *Focus groups*: Conduct focus group discussions to explore customer priorities and understand the reasoning behind their preferences.
- *Interviews*: Conduct one-on-one interviews with customers to delve deeper into their individual needs and expectations.

Compare Your Performance to Competitors'

- *Market research*: Gather data on your competitors' offerings, pricing, and customer satisfaction levels.
- *Customer feedback*: Analyze customer feedback (from surveys, reviews, and testimonials) to understand how your company is perceived compared to competitors.
- *Benchmarking*: Compare your company's performance on key attributes against industry benchmarks and best practices.

Analyze Segment-Specific Value

- *Segmentation*: Divide your customer base into distinct segments (e.g., residential vs. commercial, budget-conscious vs. luxury).
- *Targeted analysis*: Assess how each customer segment perceives the value of specific attributes and how your company performs in meeting those needs.

Revise Customer Values Over Time

- *Monitoring*: Continuously track changes in customer preferences, market trends, and emerging technologies.
- *Adaptation*: Adjust your offerings and strategies to align with evolving customer values and maintain a competitive advantage.

Applying CVA in Construction

- *Bidding and proposals*: Use CVA insights, including value engineering suggestions and options, to develop competitive bids and proposals that highlight the value you offer to clients.

- *Project management*: Prioritize project aspects that customers value most, such as communication, transparency, and on-time delivery.
- *Marketing and sales*: Tailor your marketing messages to emphasize the specific value propositions that resonate with different customer segments.
- *Customer relationship management*: Use CVA data to improve customer relationships by addressing their needs and exceeding their expectations.
- *Continuous improvement*: Identify areas where your company can enhance its offerings and deliver even greater value to customers.

By conducting a thorough CVA, construction businesses can gain a competitive edge by understanding customer priorities and delivering projects that truly meet their needs and expectations.

CVA provides crucial insights into what customers truly value, and this knowledge can be directly applied to assess the financial impact of both customer retention and attrition, as described next.

Quantifying the Value of Retention

Customer lifetime value (CLTV): CVA helps determine the CLTV by identifying the key value drivers for different customer segments. By understanding what keeps customers coming back, you can estimate their long-term revenue potential. This includes factors like:

- *Repeat purchases*: In construction, this could be repeat business from developers, homeowners who need further renovations, or property management companies with ongoing projects.
- *Referrals*: Satisfied customers are more likely to recommend your services to others, leading to new business acquisition. CVA can identify which attributes drive referrals (e.g., quality, trust).
- *Price premium*: Customers who perceive high value in your services may be willing to pay a premium price, increasing profitability. CVA can reveal which attributes justify higher prices (e.g., specialized expertise, sustainable practices).

Reduced costs: Retaining existing customers is generally more cost-effective than acquiring new ones. CVA can highlight the cost savings associated with retention, such as:

- *Lower marketing and sales costs*: Repeat customers require less marketing effort and sales resources.
- *Increased efficiency*: Working with repeat customers often leads to smoother project execution and fewer misunderstandings, reducing overhead costs.

Assessing the Cost of Attrition

- *Lost revenue*: CVA helps estimate the direct revenue loss from customer churn. By understanding the CLTV of different customer segments, you can calculate the financial impact of losing those customers.
- *Increased acquisition costs*: Replacing lost customers requires investing in marketing and sales efforts to acquire new ones, which can be expensive.
- *Reputational damage*: Customer attrition, especially due to negative experiences, can harm your company's reputation and lead to lost future business opportunities. CVA can identify the attributes that contribute most to customer satisfaction and prevent attrition-related reputational damage.
- *Increased project risks*: Working with new customers often involves a learning curve and higher potential for misunderstandings, which can lead to project delays, cost overruns, and disputes.

Linking CVA to Financial Metrics

- *Profitability analysis*: By combining CVA data with financial metrics, you can analyze the profitability of different customer segments and identify the most valuable customer relationships.
- *Return on investment (ROI)*: CVA can help calculate the ROI of customer retention initiatives, such as loyalty programs or customer satisfaction surveys.
- *Risk assessment*: CVA can be used to assess the financial risks associated with customer attrition and develop strategies to mitigate those risks.

Construction Example

Let's say a construction company specializes in building custom homes. Through CVA, they discover that "quality of craftsmanship" and "effective communication" are the top two value drivers for their target customers.

- *Retention*: By focusing on delivering exceptional craftsmanship and maintaining clear communication throughout projects, they can increase customer satisfaction, leading to repeat business and referrals. This translates to higher CLTV and increased profitability.
- *Attrition*: If they neglect these key attributes, customers may become dissatisfied and choose competitors for future projects. This results in lost revenue, increased acquisition costs, and potential damage to their reputation.

By linking CVA insights to financial data, the company can quantify the impact of customer retention and attrition on their bottom line. This allows them to make informed decisions about resource allocation, prioritize customer-centric initiatives, and optimize their business strategy for long-term success.

Determining return on investment (ROI) for construction projects.

1. *Identify costs.*
   - *Initial costs*: Include land acquisition, design, permits, and initial construction costs.
   - *Ongoing costs*: Consider labor, materials, equipment, and any other recurring expenses.
   - *Indirect costs*: Account for overhead, administrative expenses, and financing costs.
2. *Estimate revenue.*
   - *Project revenue*: Calculate the expected revenue from the sale or lease of the completed project.
   - *Additional revenue*: Include any other sources of income, such as tax incentives or grants.
3. *Calculate net profit.*
   - $\textit{Net profit} = \textit{total revenue} - \textit{total costs}$
4. *Determine ROI.*
   - $ROI = (\textit{net profit} / \textit{total costs}) \times 100$

Example calculation: Let's say a construction project has the following financials:

- *Initial costs*: $1,000,000
- *Ongoing costs*: $500,000
- *Indirect costs*: $200,000
- *Total costs*: $1,700,000
- *Project revenue*: $2,500,000

$$\textit{Net profit} = \$2{,}500{,}000 - \$1{,}700{,}000 = \$800{,}000$$

$$ROI = (\$800{,}000 / \$1{,}700{,}000) \times 100 = 47.06\%$$

Factors influencing ROI.

- *Project duration*: Longer projects may incur higher costs and affect ROI.
- *Market trends*: Fluctuations in the real estate market can impact revenue.
- *Risk management*: Effective risk management can minimize unexpected costs.
- *Technological innovation*: Adoption of new technologies can improve efficiency and reduce costs.

CONCLUSION

CVA is an indispensable tool for construction businesses aiming to thrive in a competitive market. By systematically identifying, assessing, and comparing key value attributes, companies can align their offerings with customer expectations. Understanding the direct correlation between CVA insights and financial metrics, such as customer lifetime value and the cost of attrition, allows for data-driven decision-making. Ultimately, leveraging CVA to launch quality improvement projects and refine business strategies ensures long-term success and sustainable growth.

QUESTIONS

1. Explain how you interpret CVA and its importance in enhancing potential customer value.
2. How do key value attributes differ in different verticals in the construction section?
3. How is customer lifetime value calculated for business that provides operations and maintenance services?
4. What are the financial metrics measured and analyzed for estimating customer value?
5. How might you leverage CVA to launch quality improvement projects?
6. Describe the process of conducting a CVA in the construction industry, detailing each step and providing examples of how a CCQM contributes to enhancing key value attributes.
7. Explain how the insights gained from a CVA can be directly linked to financial metrics, specifically detailing how it impacts CLTV and the cost of customer attrition.
8. Discuss the significance of segmenting customer bases when performing a CVA. Give examples of different segments within the construction industry and how their perceived value of attributes might vary.
9. Beyond identifying key value attributes, how can a construction company effectively assess and compare its performance against competitors in a CVA? Give specific methods and their benefits.
10. Given the dynamic nature of customer preferences and market trends, how should a construction company ensure that its CVA remains relevant and effective over time? Describe the process of revising and adapting customer values in the CVA framework.

CUSTOMER EXPERIENCE ENHANCEMENT

> Evaluate corrective strategies and proactive methods to enhance customer satisfaction, loyalty, and retention. (Evaluate)
>
> **Body of Knowledge VIII.F.3**

Customer experience (CX) encompasses both the immediate and ongoing interactions between a company and its customers. It's a journey that includes how customers engage with a company's employees, physical spaces, and marketing efforts, across both physical and digital channels. While traditional quality approaches often prioritize meeting product and service specifications, organizations can achieve greater success by focusing on understanding and improving the overall customer experience. This involves going beyond simply delivering products and services to create a positive and memorable experience at every touchpoint.

As discussed earlier, customer satisfaction, loyalty, and retention are closely related concepts that drive business success, both in B2B (business-to-business) and B2C (business-to-customer) contexts. Businesses often engage with other businesses to accomplish their goals. For instance, a client might hire a general contractor for an infrastructure project, who then subcontracts portions of the work to other businesses. This illustrates a B2B relationship. In contrast, a B2C relationship occurs when a homeowner hires a construction company to build a single-family residence.

The next section discusses how they relate to each other.

Customer Satisfaction

- *Foundation*: Customer satisfaction is the baseline measurement of how well a product or service aligns with customer needs and expectations.
- *Drivers*: Satisfaction is influenced by several key factors in both B2B and B2C contexts. These include:
  - Product quality
  - Service reliability
  - Communication effectiveness
  - Problem resolution
- *Impact*: High customer satisfaction levels directly correlate with increased customer loyalty, repeat business, and positive word-of-mouth recommendations.

Customer Loyalty

- *Beyond satisfaction*: Customer loyalty, which surpasses mere customer satisfaction, indicates an emotional bond with the brand or company, resulting in a strong preference for their offerings. While a satisfied customer may switch to a new service provider if they find something better, a delighted customer remains loyal even during challenging times and becomes a brand ambassador, promoting the product and services to others. (See ASQ's Quality Resources: the Kano Model to understand delighters.)

  In the construction industry, a supplier can earn customer loyalty by taking measures to recover significant schedule delays, such as expediting materials and services. Additionally, they can demonstrate a commitment to ESG by providing sustainably sourced materials and employing local workers for projects. These actions will improve the client's perception of the supplier as a partner dedicated to ESG principles and excellent service.
- *Key factors*: Building customer loyalty depends on consistent satisfaction, positive experiences, trust, and perceived value.

- *Outcomes*: Loyal customers exhibit the following behaviors:
  - Repeat purchases
  - Reduced price sensitivity
  - Forgiveness for occasional errors
  - Advocacy for the brand

Customer Retention

Customer retention is the ability of a company to keep its customers over time. It is a direct result of customer satisfaction and loyalty.

Benefits

- *Increased revenue and profitability*: Retaining existing customers is often more cost-effective than acquiring new ones—for example, in the construction of nonpublic projects (or public projects that are design-build) becoming listed as a preferred contractor or subcontractor so that owners can advise general contractors to hire from their preferred subcontractors.
- *Reduced customer acquisition costs*: The expenses associated with attracting new customers can be significant. High retention rates lessen the need for constant new customer acquisition.
- *Improved brand reputation*: Satisfied, loyal customers are more likely to recommend a company to others, enhancing its reputation.
- *Valuable customer insights*: Long-term customer relationships provide opportunities to gather valuable feedback and insights for improving products and services.

Relationship Dynamics

- *Customer satisfaction and loyalty*: When customers consistently have positive experiences, they become satisfied. This satisfaction nurtures loyalty toward the company and its products or services.
- *Loyalty and customer retention*: Customers who are loyal to a company are more likely to continue doing business with them over an extended period.
- *The cycle of retention and loyalty*: The positive interactions that come with customer retention further strengthen their loyalty, creating a continuous cycle.

B2B Versus B2C Customer Engagement: Key Differences

While B2B and B2C customer engagement strategies share some common principles, there are important distinctions to consider.

B2B

- *Relationship-driven*: B2B interactions often involve complex, ongoing relationships between businesses. Building trust and fostering personal connections are essential for success.
- *Rational decision-making*: B2B purchasing decisions are typically based on objective factors such as cost-effectiveness, efficiency, and ROI.
- *Longer sales cycles*: B2B sales cycles tend to be longer than B2C, requiring sustained engagement and relationship-building efforts.

B2C

- *Emotional connection*: B2C relationships often rely on building an emotional connection with the consumer, influenced by brand image, personal preferences, and social trends.
- *Impulse purchases*: B2C purchases are more likely to be driven by immediate needs or desires, leading to more impulsive buying behavior.
- *Shorter sales cycles*: B2C sales cycles are generally shorter, with quicker purchasing decisions.

Understanding these nuances is critical for tailoring effective customer engagement strategies in both B2B and B2C contexts.

Customer satisfaction, loyalty, and retention form a virtuous cycle. Satisfied customers are more likely to become loyal, and loyal customers are more likely to be retained. By focusing on all three elements, businesses can build strong customer relationships, increase profitability, and achieve sustainable growth in both B2B and B2C markets.

Evaluating corrective strategies and proactive methods to enhance customer satisfaction, loyalty, and retention in the construction industry requires a structured approach that combines quantitative and qualitative data. See below a structured approach for implementation.

Define Clear Objectives and Metrics

- *Specific goals*: Clearly outline what you aim to accomplish. For example, you might want to increase customer satisfaction scores by 10%, reduce customer churn by 5%, or improve online reviews by one star.
- *Measurable metrics*: Determine how you will track your progress toward these goals. Examples of metrics include customer satisfaction (CSAT) surveys, net promoter score (NPS), customer retention rate, online review ratings, and number of referrals.

Refer to "SMART Goals" for additional guidance.

Identify Corrective and Proactive Strategies

- *Corrective strategies*: These strategies address existing problems or weaknesses. Examples in the construction industry include:

- *Complaint resolution*: Implementing a robust system to handle complaints quickly and effectively.
- *Warranty claim analysis*: Analyzing warranty data to identify recurring issues and implementing preventative solutions.
- *Project audits*: Conducting regular project audits to identify deviations from standards and taking corrective action.

- *Proactive strategies*: These strategies aim to prevent problems and enhance the overall customer experience. Examples include:
  - *Proactive communication*: Keeping customers informed throughout the project with regular updates, progress reports, and prompt responses to inquiries.
  - *Personalized service*: Tailoring communication and service delivery to meet individual customer needs and preferences.
  - *Exceeding expectations*: Going above and beyond to deliver a positive and memorable customer experience.
  - *Loyalty programs*: Offering incentives for repeat business and referrals.
  - *Continuous improvement*: Regularly seeking customer feedback and using it to improve processes and service delivery.

Implementation and Monitoring

- *Rollout*: Implement the selected strategies in a systematic way. Ensure that employees receive proper training and communication.
- *Data tracking*: Continuously track the chosen metrics to monitor the effectiveness of the strategies.
- *Regular reviews*: Conduct periodic reviews to analyze the data, identify trends, and make necessary adjustments.

Evaluation Methods

- *Quantitative analysis*
  - *Key metrics tracking*: Monitor and measure fluctuations in crucial indicators such as CSAT, NPS, retention rates, and other pertinent metrics. CSAT measures how much customers agree on the quality of products and services provided by suppliers. It considers all aspects that customers value, such as quality, delivery, cost, service, and other implicit factors. A higher score indicates higher satisfaction. NPS, on the other hand, measures customer loyalty, indicating that a customer is not only satisfied but would also recommend the product or service to others.
  - *Statistical analysis*: Employ statistical techniques to scrutinize data and ascertain the significance of observed changes.

  - *Cost-benefit analysis*: Assess the financial implications of your strategies by comparing implementation costs against realized benefits (e.g., increased revenue, decreased expenses).
- *Qualitative analysis*
  - *Customer feedback*: Collect qualitative insights through various channels like surveys, interviews, focus groups, and online reviews.
  - *Sentiment analysis*: Examine customer feedback to gauge the overall sentiment toward your company and its offerings.
  - *Employee feedback*: Gather input from employees regarding the efficacy of implemented strategies, and pinpoint areas for enhancement.

Iterate and Improve

- *Continuous improvement*: Regularly evaluate your strategies and use the results to pinpoint areas for enhancement.
- *Adapt to change*: Stay informed about evolving customer expectations, market trends, and industry best practices to ensure your strategies remain effective and relevant.

Customer Experience in the Construction Industry

The construction industry presents unique challenges for customer experience due to the inherent nature of its projects. These projects are often complex, involving numerous stakeholders, and can span extended durations. Such complexities necessitate effective communication and coordination to ensure customer satisfaction. Additionally, the substantial financial investments involved in construction projects underscore the importance of fostering trust and confidence with clients.

To navigate these challenges successfully, construction businesses must prioritize several key aspects of the customer experience:

- *Proactive communication and relationship management*: Essential for maintaining customer engagement throughout lengthy project timelines.
- *Safety and compliance*: Adhering to safety regulations and industry standards is vital for building customer trust and mitigating risks.
- *Reputation management*: Given the significance of online reviews and word-of-mouth referrals in the industry, actively managing your online reputation and promptly addressing customer feedback are crucial.

By implementing a robust evaluation framework that encompasses these factors, construction businesses can effectively measure the success of their customer experience initiatives. This enables them to implement corrective actions as needed and proactively enhance customer satisfaction, loyalty, and retention, ultimately leading to long-term success in the industry.

QUESTIONS

1. What are the differences between customer satisfaction and loyalty? How do these affect customer retention and attrition?
2. What are the strategies to improve customer experience and loyalty in a B2B and B2C?
3. What are the benefits of proactive strategies in comparison to corrective strategies? Give specific examples for each strategy.
4. Discuss some of the key challenges for improving customer experience in the construction industry. Suggest approaches to overcome.
5. How do digital presence and social media play a role in enhancing customer experience and loyalty?
6. Explain the "virtuous cycle" of customer satisfaction, loyalty, and retention. How do changes in one element of this cycle impact the others, specifically in both B2B and B2C contexts?
7. Compare and contrast corrective and proactive strategies for enhancing customer experience. Describe a detailed construction industry example where a business might utilize both types of strategies to address a single customer issue.
8. Discuss the role of ESG (environmental, social, and governance) factors in building customer loyalty within the construction industry. How can a supplier's commitment to ESG principles influence a client's perception and lead to increased retention?
9. Analyze the differences in customer engagement strategies between B2B and B2C models, focusing on the decision-making processes and the length of sales cycles. How should a company tailor its approach to each model to maximize customer satisfaction and loyalty?
10. Describe a structured approach for evaluating and improving customer experience in the construction industry, including the use of both quantitative and qualitative analysis. How can metrics like CSAT and NPS be used in conjunction with customer feedback to drive meaningful change?

Quality Management Tools

QUALITY TOOLS

> Select the appropriate tool and evaluate the output from the seven classic quality tools (Pareto charts, cause and effect diagrams, flowcharts, control charts, check sheets, scatter diagrams, and histograms) as preventive measures, analyze root causes, and problem solve. (Evaluate)
>
> **Body of Knowledge VIII.G.1**

Mastering quality management is crucial in construction, where intricate processes and diverse variables can lead to defects. This chapter delves into the seven classic quality tools vital for preventing issues, analyzing root causes, and driving continuous improvement. These tools, ranging from simple check sheets to complex statistical analyses, empower quality managers to proactively address challenges. By effectively utilizing Pareto charts, cause-and-effect diagrams, flowcharts, run charts, check sheets, scatter diagrams, and histograms, professionals can streamline operations, mitigate risks, and ensure project success.

These are some of the most common problem-solving tools across all construction phases. These tools may be a simple check sheet to collect data or a complex analysis to identify the effect of critical variables in a process. Most improvement projects can be conducted with these easy-to-use basic quality and statistical tools.

Let us start with the seven quality tools that are considered "basic;" they are applicable in many problem-solving and continual improvement projects:

1. Pareto charts
2. Cause-and-effect diagrams
3. Flowcharts
4. Run charts
5. Check sheets
6. Scatter diagrams
7. Histograms

Recurring quality defects in the construction industry (Table 8.12) can be analyzed effectively using these quality tools.

Table 8.12 Construction defect category and types.

Defect Category	Type	Defects
Structural Defects	Foundation issues	Uneven settling, shifting soil, and inadequate support can cause cracks, instability, and equipment misalignment.
	Steel erection errors	Improperly welded joints, incorrect bolt torques, and inadequate bracing can compromise the structural integrity of the building.
	Concrete defects	Poorly mixed concrete, inadequate curing, and improper formwork can lead to cracking, spalling, and reduced strength.
Mechanical, Electrical, and Plumbing (MEP) Defects	Piping and instrumentation issues	Leaks, improper insulation, and incorrect installation of valves and fittings can lead to safety hazards, inefficiencies, and costly downtime.
	Electrical system failures	Short circuits, ground faults, and inadequate wiring can cause fires, equipment damage, and safety hazards.
	HVAC system malfunctions	Inadequate insulation, improper ductwork, and malfunctioning equipment can lead to energy inefficiency, temperature control issues, and reduced productivity.
Process Equipment Defects	Misaligned equipment	Improper alignment of rotating equipment (e.g., pumps, compressors) can lead to vibration, premature wear and tear, and reduced efficiency.

Continued

Table 8.12 Construction defect category and types. *(continued)*

Defect Category	Type	Defects
	Control system errors	Faulty sensors, malfunctioning controllers, and inadequate instrumentation can disrupt production processes and lead to safety hazards.
	Corrosion and erosion	Inadequate corrosion protection and improper material selection can lead to equipment failure, downtime, and costly repairs.
Fire and Safety Defects	Inadequate fire protection	Insufficient fire suppression systems, inadequate fire exits, and lack of proper fire-resistant materials can increase the risk of fire and endanger personnel.
	Hazardous material handling issues	Improper storage and handling of hazardous materials can lead to leaks, spills, and explosions.
Environmental Compliance Issues	Soil and groundwater contamination	Improper handling of construction waste and hazardous materials can contaminate soil and groundwater.
	Air quality violations	Inadequate dust control measures and emissions from construction equipment can lead to air quality violations.

Pareto Charts

The Pareto chart has been widely used in recent years. It is not uncommon to hear from managers about "Pareto" data for presentations. Some people who are not familiar with Pareto charts interchangeably use a bar graph for "Pareto" data. The true Pareto chart, however, is unique. It shows the data arranged in descending order of frequency of occurrence, the "trivial many" data are often pooled together as "miscellaneous" or "other" (represented by the last bar of the Pareto chart), and the chart contains a cumulative percent line with a secondary *y*-axis. These characteristics make the Pareto chart more informative and useful than an ordinary bar graph (Figure 8.5).

The Pareto chart helps us visualize the items charted as "vital few" and "trivial many" using the famous 80:20 principle of twentieth-century Italian economist Vilfredo Pareto. Joseph Juran is credited with being the first to apply this principle in quality improvement.

In construction, Pareto charts can be effectively applied in many situations. Examples include:

1. Construction defects
2. Safety incidents
3. Audit findings (finding location, process)

By applying another variation of Pareto, we can also analyze the data by criticality of construction defects, safety incidents, and audit finding by multiplying with appropriate "weights" (e.g., X1, X1.1.5, X2). This useful variation is called weighted Pareto analysis.

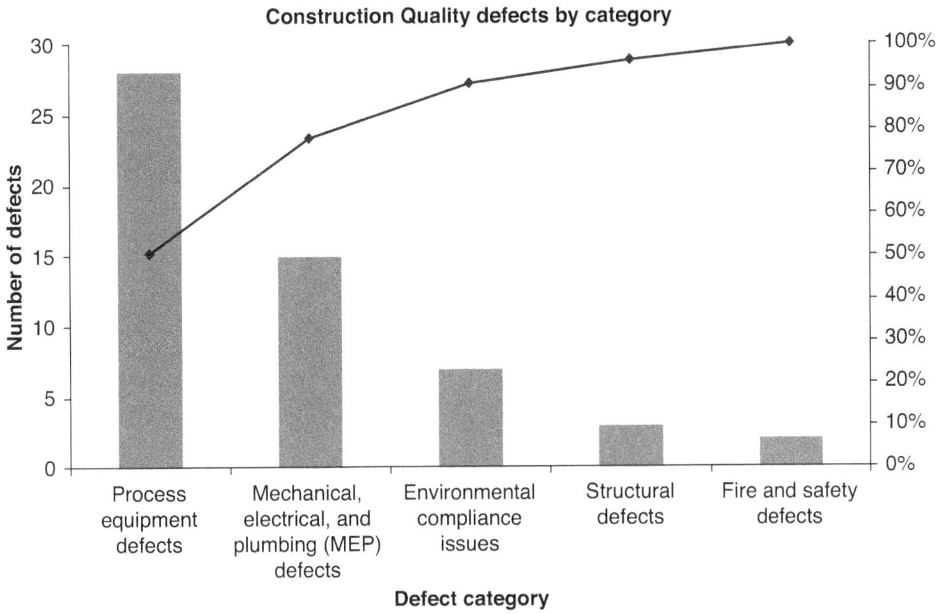

Figure 8.5 Pareto chart of construction quality defects.

Cause-and-Effect Diagrams

A thorough understanding of process inputs and outputs and their relationships is a key step in process improvement. The cause-and-effect diagram (also called an Ishikawa diagram or fishbone diagram) divides causes into several general cause categories. A large empty diagram with a fishbone skeleton is often drawn on a whiteboard or flip chart.

This diagram is used as a next step to document the final list of causes from the brainstorming session. The participants in the session should include people with a working knowledge of the process as well as those with a theoretical background. For example, suppose a machining operation is producing surface defects. After a few steps of typical brainstorming, the cause-and-effect diagram would be fully populated with effect, causes categories, causes, and sub-causes (Figure 8.6).

CEDAC (cause-and-effect diagram with addition of cards) is an alternative approach employed by some organizations where the fishbone diagram is displayed on a huge wall or board and employees are encouraged to identify causes by writing on it or using sticky notes. The success of this approach depends on organizational culture and communication.

Flowcharts

A flowchart is a quality tool (and a business process tool) that helps individuals visualize a process in discrete steps arranged in the sequence of events. There are variations of flowcharts that can accommodate complexities in the process—for example,

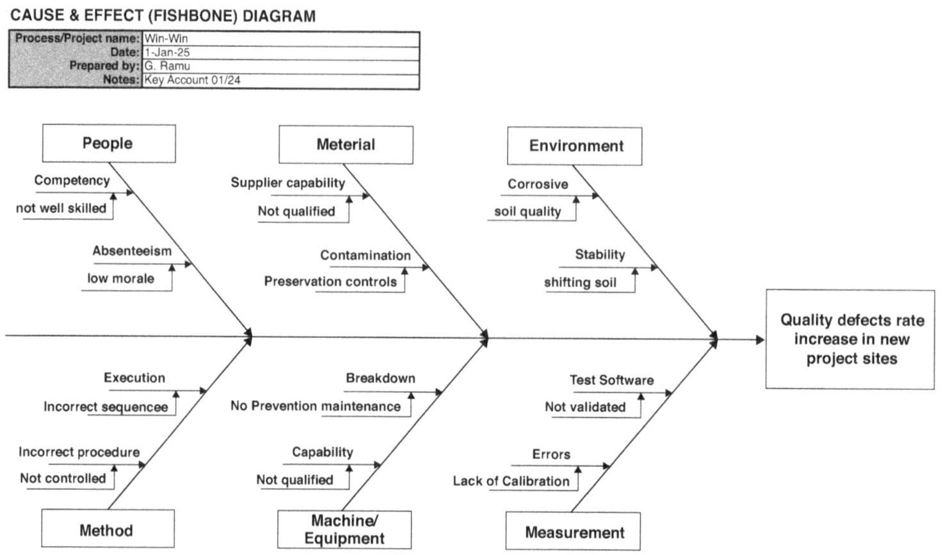

Figure 8.6 Cause-and-effect diagram in construction.

interactions between process steps, ownership, and process measures. Flowcharts can also be used as an improvement tool to understand current complexities, identify opportunities for removing redundancies and waste, and ensure the process is simplified and standardized. Once the process has been simplified and standardized, the flowchart may be used as a training tool and for auditing processes.

Processes

A process is a series of steps for producing products and/or services. A process is often diagrammed with a flowchart depicting inputs, a path that material or information follows, and outputs. Understanding and improving processes is a key part of every problem-solving or improvement project.

Steps in the development of a flowchart are as follows:

1. Determine the scope of the flowchart with boundaries.
2. Write down the steps in the process.
3. Rearrange where required to reflect the current sequence of steps.
4. Draw the flowchart (use appropriate flowcharting symbols for ease of understanding).
5. Walk through the process to test the flowchart.
6. Identify opportunities for removal of redundancies, waste reduction, simplification, and standardization.
7. Communicate the flowchart to all stakeholders.

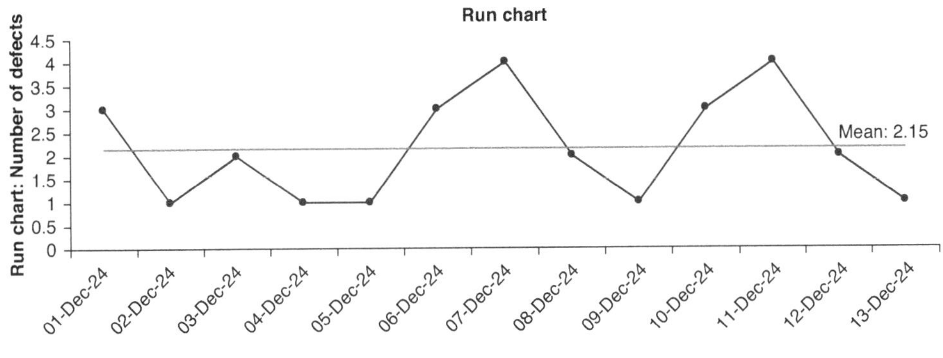

Figure 8.7 Run chart of daily defects.

Run Charts

The run chart is used to identify patterns in process data. All of the individual observations are plotted in a time sequence, and a horizontal reference line is drawn at the median. A run chart (Figure 8.7) is used when the subgroup size is one. When the subgroup size is greater than one, the subgroup means or medians are calculated and connected with a line, similar to a control chart. However, run charts are different from control charts (e.g., X bar and *R* charts); run charts do not have statistical control limits to monitor variation. There are also related statistical tests that can be performed to detect any nonrandom behavior.

Run chart tests can detect trends, oscillations, mixtures, and clustering of sequential data. These are nonrandom patterns and suggest that the variation observed can be attributed to special causes. Lack of steadiness in a process can cause oscillation. A trend can be either upward or downward due to, for example, tool wear or loosening of a holding fixture in a piece of equipment.

Check Sheets

Check sheets are used to observe or review a process, usually during its execution. Check sheets pre-categorize potential outcomes for data collection using sets of words, tally lists, or graphics. In tabular format, they can be used to collect data related to a paint-mixing process. This simple tool provides a method of easy data collection. By collecting data on a check sheet, common patterns or trends can be identified.

The basic steps in making a check sheet are as follows:

1. Identify and agree on the causes or conditions that are to be collected.
2. Decide who will collect the data, over what time period(s), and how the data will be collected. (Determine whether any training is required for staff regarding the types of data and the data collection process.)
3. Create a check sheet that will work within the operation where it will be used.

4. Conduct any training or briefing for effective data collection.
5. Collect the data as designed to ensure consistency and accuracy of the information.

Check sheets can be the basis for other analytical tools and are incorporated into attribute process behavior charts. Creating and using a check sheet helps keep a focus on continual improvement, and the simple act of using it may foster changes. Data collection is the basis for analysis and subsequent improvement, and its importance should not be underestimated.

Check sheets can be very handy in capturing data from the construction site. For example, a construction worker/inspector can use it to tally the defects by defect type on the rows and defect locations across construction sites on the columns. The user can visually look at the tally marks and decide which type of defect happens the most in which location of the construction.

Scatter Diagrams

A scatter diagram is used to explore the relationship between two sets of variables. This diagram provides a visual display of correlation between the variables. It should be used in conjunction with the cause-and-effect diagram. The scatter diagram (Figure 8.8) is created with the independent variable as the x axis and the dependent variable as the y axis. If one is cause and the other is effect, then effect is the dependent variable and cause is the independent variable. In this diagram, the data are obtained in pairs (X, Y). Once the data pairs are plotted for the range of the axis, the plotted data can be interpreted.

Is there any correlation between construction worker work hours' duration and number of defects? [Likely cause being human fatigue]. This scatter diagram analysis helps you explore just that. Other examples include correlation between number of change requests and project cost overrun. The example shown in the figure is analyzing if there is any correlation between temperature during the day and measurement of the spacing. The analysis shows a positive correlation between the two variables.

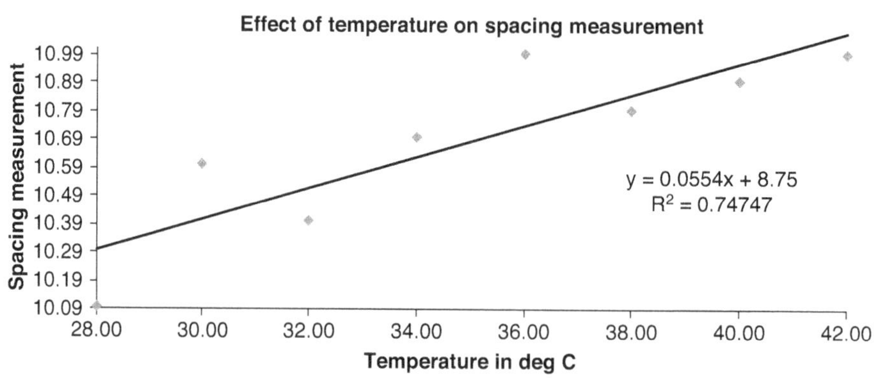

Figure 8.8 Scatter diagram of temperature vs. measurement.

Histograms

The histogram (Figure 8.8) is an easy-to-use tool for graphically summarizing continuous data collected from a process. The frequency of data occurrence is reflected as bars of various heights. Since data can be unique, they are grouped by intervals. The histogram shows the shape of the data, centering of the process, variation, and a rough distribution of the data. The histogram is helpful for comparing "before and after" improvement readily while waiting for a more detailed analysis. An initial indication of process capability may be seen by comparing the center and width of the distribution with the specification limits.

The following are steps for developing a histogram:

1. Identify the process parameter to be analyzed.
2. Collect the data to the required precision.
3. Divide data into number of intervals (as a guideline, use the square root of number of samples n).
4. Arrange the collected data by interval.
5. Draw a histogram based on the frequency of data falling at every interval.

Histograms are used for variable data (measurement like weight, time, distance, etc.). In the construction, this tool can be effectively applied to monitor the torque strength of fasteners and verify if there is a drift in the average torque value. Histograms, as seen in Figure 8.9, show that there may be two different groups of spacings in the construction measured. More data is required to establish this inference. However, visually it indicates bimodal, or two different groups. Histograms can also be used to compare concrete cube compacting strength of materials from different sources, and so on.

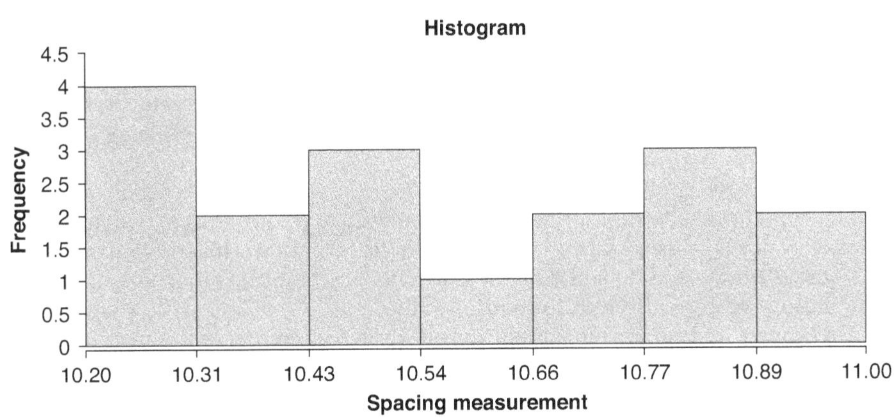

Figure 8.9 Histogram of space measurement.

CONCLUSION

These seven classic quality tools provide a robust framework for managing and improving quality in construction. By strategically employing Pareto charts, cause-and-effect diagrams, flowcharts, run charts, check sheets, scatter diagrams, and histograms, construction professionals can effectively identify, analyze, and resolve quality issues. These tools facilitate data-driven decision-making, enhance process understanding, and promote continuous improvement initiatives. Mastery of these tools is essential for ensuring project success, mitigating risks, and delivering high-quality outcomes in the complex construction environment.

QUESTIONS

1. You have a number of quality defects types from inspection of the construction site. Your management has asked to prioritize your improvement and provided a limited budget. Which tool would you use to narrow down quality defects to drive improvements?
2. From question 1, you have narrowed the quality defects to drive improvement. You want to understand the root causes of the prioritized quality defects. Which tool would you use?
3. You notice at the construction site, people have been doing things ad hoc. You are made responsible as the CQM to formalize how things would be done. Which would be your go-to tool to get this accomplished?
4. A quality inspector is asked to collect data of various types of quality defects for a week each day. What tool would you advise the inspector to use for such data collection?
5. The CQM is collecting the temperature and humidity of their test lab daily for a month and wants to monitor and analyze the data. Which tool would be most effective?
6. You have been given the daily concrete mixture test results for a month and asked to provide average, minimum, maximum, and range of the test results for making a decision. Which of the seven quality tools is most appropriate to represent and visualize the data useful for decision-making?
7. Your leadership is asking you to investigate if the quality defects are increasing due to the absenteeism rate of craftspeople. You collect the data of daily quality defects for 30 days and corresponding absenteeism of craftspeople. How do you go about making an investigation?
8. Explain the distinction between a standard bar graph and a Pareto chart, emphasizing the unique characteristics that make a Pareto chart more informative for quality improvement in construction.
9. Describe the application of a weighted Pareto analysis in construction, and describe a hypothetical example of how it could be used to prioritize quality defects.
10. How does the use of CEDAC (cause-and-effect diagram with addition of cards) differ from a traditional Ishikawa (fishbone) diagram, and what organizational factors influence the effectiveness of CEDAC?
11. Compare and contrast the use of run charts versus control charts in monitoring process data in construction, highlighting the specific situations where each tool is most appropriate.
12. Discuss how scatter diagrams and cause-and-effect diagrams complement each other in analyzing quality issues, and give an example of a construction scenario where using both tools together would provide deeper insights.

LEARNING AND BEST PRACTICE ANALYSIS

> Incorporate after-action reviews and root cause analysis (RCA) meetings to critically assess project outcomes and identify lessons learned from all phases of the project, identifying both successes and areas for improvement. Analyze the critical success factors that led to best practices being implemented within the project. Use key performance indicators (KPIs) to systematically track and assess the impact of implemented improvements and evaluate strategies to reinforce and replicate these improvements across future projects. (Evaluate)
>
> Body of Knowledge VIII.G.2

Effective learning and best practice analysis are pivotal for organizational growth, particularly in complex construction projects. While capturing lessons learned is valuable, implementing closed loop learning ensures these insights translate into actionable improvements. This section explores the critical distinction between these concepts, emphasizing the need for a formal program to systematically capture, analyze, and share knowledge. By addressing underlying issues at the source, organizations can prevent recurring errors, enhance safety, and drive continuous improvement, ultimately leading to significant cost savings and increased project success.

There's a difference between lessons learned (LL) and closed loop learning (CLL). CLL happens once a lesson is captured, validated, and curated. Most organizations do not call out CLL explicitly, although it is always implied. Even our ASQ CCQM BoK does not explicitly call out CLL. However, when we perform LL, CLL is always required to ensure value is derived from learning. Many organizations conduct LL activities, such as post-project reviews, post-mortems, agile retrospectives (in software development) or reflective discussions about what went well and what could be improved. While these sessions are valuable for gathering insights, they often fall short of driving meaningful change.

CLL, on the other hand, takes the key LL and promptly addresses the underlying issues upstream, following the principle of "quality at the source." This proactive approach aims to prevent the recurrence of problems and drive continuous improvement. By implementing CLL, organizations can ensure that the lessons learned are not merely academic exercises but are translated into practical actions that add value to the organization.

The limitations of traditional learning resources, such as textbooks and publications, necessitate a focus on experiential learning. Product, project launches, service implementations, and process improvements offer valuable feedback from key stakeholders, including customers, throughout the life cycle. This feedback holds immense potential for organizational growth and improvement. However, the typical approach of capturing, archiving, and then neglecting this information leads to missed opportunities. Organizations often repeat the same mistakes and fail to leverage best practices identified through feedback. This oversight results in productivity and efficiency losses, unsafe working conditions, and overall poor quality, leading to significant costs for the organization. To prevent this loss of learning, organizations should establish a formal LL and CLL program. This program would ensure that feedback is systematically captured, analyzed, and shared across the organization, promoting continuous improvement and preventing the recurrence of mistakes.

Here is a recommended approach to how a lessons learned program is typically implemented on a construction project.

Making a Proposal

To: Leadership and Key Stakeholders

Proposal: Implementing a Formal Lessons Learned and Closed Loop Learning Program

I propose the establishment of a formal LL and CLL program. This program will systematically capture insights from past projects, enabling us to identify recurring errors, safety incidents, and sources of customer dissatisfaction. By addressing these issues proactively, we can significantly reduce costs and improve project outcomes.

Impact of Past Issues

Our organization has experienced recurring challenges in previous projects, including [list specific examples of errors, safety incidents, and customer dissatisfaction]. These issues have resulted in [quantify the financial impact of these issues, e.g., project delays, rework, lost business].

Projected Savings

A formal LL program will enable us to avoid these recurring problems in future projects. By implementing corrective actions and sharing best practices, we can conservatively estimate a potential savings of [estimated savings amount] across a portfolio of large, complex construction projects.

Implementation Requirements

To successfully implement this program, we will need:

- *Dedicated resources*: Program leadership, a team or individual responsible for collecting, analyzing, and disseminating lessons learned
- *Standardized processes*: Clear procedures for capturing and sharing insights throughout the project life cycle
- *Technology infrastructure*: A centralized system for storing and accessing lessons learned data

> - *Cultural shift*: Encouraging a learning mindset and open communication across teams
>
> Investing in a formal lessons learned program is a strategic decision that will improve our project performance, reduce costs, and enhance customer satisfaction. I am confident that this program will deliver significant benefits to our organization in the long run.
>
> Thank you for your consideration. I look forward to discussing this proposal further.
>
> Sincerely,
> [Construction Quality Manager, ASQ CCQM]

The program manager for the LL program should be a qualified individual with a deep understanding of construction project phases and strong relationships with stakeholders. Excellent facilitation, communication, and presentation skills are essential. Collaborating closely with site project managers, subject matter experts, site construction personnel, and other functions like environmental, health, and safety, the program manager disseminates relevant lessons learned to the appropriate stakeholders at the optimal time. Additionally, they work with leadership to secure resources, ensuring that lessons are implemented to prevent recurring issues.

Establish a collaborative process with key project stakeholders to create a lessons learned and CLL system that can be efficiently implemented across all projects. Develop a comprehensive training program that encompasses both the process itself and the necessary tools for effective execution, such as root cause analysis techniques.

Before the construction project begins, identify and train all relevant stakeholders on the processes and tools required for a successful implementation. This training should include communication methods for reporting incidents in near real time. By sharing lessons learned across all project sites, we can raise awareness and take prompt action to prevent similar incidents.

Project sites report all incidents, as well as events that happen during construction, by providing information with the following metadata:

- *Date of submission*
- *Construction site ID*
- *Submitting employee ID*
- *Construction phase* (precontract, planning, design, procurement, construction, commissioning and closeout, operations and maintenance)
- *Lesson type*: Unfavorable incident: Failure/error, safety, OR best practice
- *Impacted trade* (mechanical, electrical, plumbing, civil, architectural, chemical, EHS, sustainability)
- *Description of the event*: Answer the following and combine the answers to create an objective statement.

- What happened?
- When did it happen?
- Where did it happen?
- Who were impacted? (key stakeholders)
- How much is the impact? (dollar, quantity, etc.)
- Why did it happen? (root cause analysis, e.g., 5 why, or why-why analysis).

Peel the incident event until you arrive at the system process–level breakdown that leads to the incident. Often it is 5 Whys deeper from the apparent cause.

How Might We Prevent This From Happening Again?

Outline improvement actions that address the root causes of the incident and prevent future occurrences. For each action, specify:

- *What*: The specific action to be taken.
- *Who*: The individual or team responsible for completing the action.
- *When*: The target completion date for the action.

In addition, provide a measurable indicator of the action's effectiveness. For example, you might expect to see a reduced percentage of errors, fewer safety incidents, or improved efficiency and productivity after the action is fully implemented.

Now, for best practices, the above sections are different. Here is a summary of best practice description:

- What is the best practice?
- How did the best practice help meet project goals?
- How much did the success benefit the organization (dollars—cost saved or avoided)?

Critical Success Factors and Constraints

Critical success factors (CSF) are those that are critical to the success of this best practice and ability to replicate across multiple construction sites—for example, custom-designed infrastructure, competency, timing of the implementation within a project, and so on.

Constraints are those that reduce the probability of success of this best practice and ability to replicate across multiple construction sites—for example, lack of budget, labor turnover, supply chain issues, and so on.

How might people be able to replicate across all (applicable) construction sites?

- What is the plan to replicate success?
- Who is responsible for implementing this best practice in each project?
- When or by when should this be implemented to realize maximum benefit?

- How do we measure the effectiveness of best practice replication across all (applicable) sites?

Site personnel are to receive consistent training on delivering the aforementioned information. This information is to be gathered by a designated staff member, who reports to the LL program manager, and then they enter this information into a spreadsheet for database management. The program manager, on receiving incident reports, designates an owner and team members to conduct an RCA. Once a systemic or process-level root cause is identified, the team collaborates to determine improvement actions and timelines. An independent staff member, assigned by the program manager, verifies the implementation and effectiveness of these actions. The primary goal of the lessons learned program is to enable prompt, upstream closure of the learning loop (quality at the source), ideally during phases like planning, design, or construction kickoff. Additionally, lessons learned may lead to the establishment of controls for early error detection.

To facilitate in-depth analysis and drive systemic improvements across the organization, metadata collected for each incident, encompassing root causes and improvement actions, is encoded with concise data strings. This enables analytics, such as Pareto analysis, to be performed based on various parameters, including incident type, site, trade, failure symptom, root cause, and improvement action.

For instance, if the Pareto analysis uncovers a correlation between labor turnover and ineffective retention of process knowledge, leading to recurring errors, it cannot be resolved solely at the construction site level. Instead, it necessitates addressing the issue at the organizational level through thorough investigation and policy-level remedies, such as enhancing working conditions, improving compensation, providing training and development opportunities, and implementing reward and recognition programs.

In construction organizations, KPIs are often employed to assess the efficacy of initiatives such as LL, best practice sharing, and CLL. Tracking KPIs is crucial for construction companies to measure success and improve efficiency. Here are some examples of KPIs specific to the construction industry.

Project Progress

- *Schedule variance*: Measures the difference between the planned project timeline and the actual progress.
- *Cost variance*: Compares the budgeted cost to the actual cost incurred.

Safety Metrics

- *Total recordable incident rate (TRIR)*: Number of recordable incidents per defined work hours.
- *Lost time injury frequency rate (LTIFR)*: Number of lost time injuries per defined hours worked.

Financial Performance

- *Profit margin*: Measures profitability of a project or the company as a whole.
- *Cash flow*: Tracks the inflow and outflow of cash within the business.

Quality Control

- *Rework costs*: Measures the costs associated with correcting defective work.
- *Defect rates*: Tracks the frequency of defects in construction projects.

Resource Management

- *Equipment utilization*: Measures the usage rate of construction equipment.
- *Labor productivity*: Assesses the efficiency of the workforce in completing tasks.

Customer Satisfaction

- *Customer feedback*: Collects and analyzes feedback from clients regarding project outcomes.
- *Net promoter score (NPS)*: Measures the likelihood of customers recommending the company.

Environmental Impact

- *Waste management*: Tracks the amount of waste produced and how it is managed.
- *Energy consumption*: Monitors the amount of energy used during construction projects.

These KPIs help construction companies monitor their performance, identify areas for improvement, and make data-driven decisions. They encompass various organizational aspects, including customer satisfaction, supply chain on-time delivery, operational efficiency, productivity, safety, regulatory compliance, employee satisfaction, and the cost of poor quality. By effectively implementing LL and CLL, organizations can minimize waste, save costs, and prevent unnecessary expenditures. This directly or indirectly influences numerous KPIs within the organization. Reducing repetitive construction errors can lead to direct cost savings, improved regulatory compliance, enhanced efficiency, and a reduction in the cost of poor quality. Indirectly, the same error reduction can result in fewer safety incidents and improved employee satisfaction. Therefore, this program holds significant operational importance for construction organizations.

CONCLUSION

Establishing a robust lessons learned and closed loop learning program is essential for driving continuous improvement in construction projects. By systematically capturing, analyzing, and implementing insights, organizations can prevent recurring errors, enhance safety, and boost efficiency. This proactive approach, coupled with the use of KPIs to track progress and effectiveness, ensures that knowledge gained from experience is translated into tangible benefits. Ultimately, this investment in learning fosters a culture of improvement, leading to significant cost savings, increased project success, and enhanced organizational resilience.

QUESTIONS

1. Explain how learning lessons (LL) from our construction project is beneficial to an organization. How might an organization make use of it?
2. How would you identify a best practice? How are best practices shared in an organization?
3. What are the questions you would ask while developing a problem statement from the lessons identified from your organization?
4. Explain your approach to verifying effectiveness of LL and best practice deployment.
5. Identify five to seven key process indicators (KPIs) that are aligned and integrated with your organization's strategic goals.
6. Explain the distinction between lessons learned (LL) and closed loop learning (CLL). Give an example of how an organization might implement CLL to address a specific issue identified through LL in a construction project.
7. Describe the key metadata fields that should be captured when reporting an incident or best practice on a construction project, and explain the purpose of each field in facilitating effective analysis and implementation of lessons learned.
8. Discuss the role of KPIs in evaluating the effectiveness of an LL or CLL program. Give three construction-related KPIs, and explain how they can be influenced by a well-implemented LL/CLL program.
9. Outline the steps involved in conducting a root cause analysis (RCA) for an incident reported during a construction project, and explain why it is crucial to "peel the incident event until you arrive at the system process–level breakdown."
10. In the context of implementing a formal LL/CLL program, why is a "cultural shift" toward a learning mindset and open communication emphasized? What potential challenges might an organization face in fostering this cultural shift, and how could they be addressed?

KNOWLEDGE MANAGEMENT ANALYSIS

> Document the strategies and outcomes of knowledge management throughout the project. (Apply)
>
> **Body of Knowledge VIII.G.3**

Knowledge management (KM) in the construction industry is about enabling informed decision-making and effective job performance by ensuring the right people have access to the right information at the right time. This process involves creating, sharing, using, and managing organizational knowledge, and information, which is especially critical in a complex and fast-paced sector like construction. By implementing effective KM practices, construction firms can enhance compliance (regulatory, legal, etc.), better meet client needs, and improve their ability to deliver projects on time and within budget.

There are five ways in which KM can be implemented in a construction organization, as follows.

Identifying Knowledge Needs

Collaborate with key stakeholders to identify critical knowledge areas. This encompasses project plans, building codes, safety protocols, task-specific best practices, insights from previous projects, and individual employee expertise. Additionally, identify the knowledge beneficiaries, which may include project managers, site supervisors, subcontractors, estimators, and clients.

Capturing and Storing Knowledge

To effectively manage and leverage organizational knowledge, a centralized repository should be established. This could take the form of a database, document management system, or a dedicated knowledge management platform. Implementing standardized documentation practices, such as templates and consistent formats for project documents, reports, and lessons learned, enhances knowledge organization and accessibility.

As discussed previously, capturing lessons learned is crucial for preserving project and process knowledge. Teams should be encouraged to document successes, failures, and recommendations after each project. This can be achieved through various methods, including project management software, mobile apps for on-site documentation, and video recordings of expert interviews. Additionally, leveraging technologies like machine learning and artificial intelligence can help extract actionable insights from captured knowledge, further amplifying its value.

Sharing Knowledge

To foster a knowledge-sharing culture, encourage teamwork and transparent communication. For example, this is increasingly important for virtual teams where there may be challenges with relationship building due to factors such as culture, time zones, or geography. Recognizing and rewarding contributions can further incentivize this behavior. Importantly, ensuring job security helps alleviate concerns about sharing valuable insights. This enables the transition of implicit, or "tribal," knowledge into explicit, documented forms. Addressing fears around job security encourages open collaboration and knowledge exchange.

Promote seamless information exchange by implementing user-friendly collaboration tools, such as shared workspaces, intranets, and online forums. Encourage knowledge sharing and the dissemination of best practices by establishing communities of practice and connecting individuals with similar roles or expertise. As an example, ASQ promotes communities of practice using the MyASQ web page platform and encourages design and construction professionals to share an ASQ CCQM BoK pertinent topics webinar or a publication. Facilitate knowledge transfer through mentorship programs and training initiatives, pairing experienced employees with those who are less experienced.

Using Knowledge

Enhance workflows by integrating KM, enabling employees to easily access and utilize knowledge in their daily tasks. Leverage data analytics and modeling to develop decision support tools that inform project planning, risk management, and resource allocation. Promote a culture of innovation by encouraging the application of lessons learned and best practices to optimize processes and generate creative solutions.

Evaluating and Improving

Enhance your KM system by:

- Gathering employee feedback.
- Tracking key metrics to pinpoint areas for improvement.
- Staying current on KM best practices and technologies.
- Adapting your KM approach to align with the organization's changing needs.

By implementing these strategies, construction organizations can encourage a knowledge-driven culture that leads to better decision-making, improved project outcomes, and increased competitiveness.

Knowledge management in construction is about capturing, sharing, and effectively using knowledge to improve project outcomes. Table 8.13 shows a breakdown of strategies and their associated benefits.

Table 8.13 Strategy at various phases.

KM Phase	Relevant Strategy
Knowledge capture	• Project documentation • Lessons learned databases • *Expert interviews*: Extracting knowledge from experienced individuals through structured interviews • *Site photos and videos*: Visual documentation of progress, issues, and solutions
Knowledge sharing	• Collaboration platforms • Communities of practice • Knowledge repositories • Training and mentoring
Knowledge use	• *Decision support systems*: Using data analytics and modeling tools to inform decision-making • *Risk management*: Applying past project experiences to identify and mitigate potential risks • *Standardization*: Developing and implementing best practices and standard operating procedures • *Innovation*: Encouraging the generation and application of new ideas and technologies

Outcomes

Improved Project Performance

- *Reduced costs*: Avoiding rework, preventing errors, and optimizing resource allocation.
- *Enhanced schedule adherence*: Improved planning and execution based on past project data.
- *Increased quality*: Applying best practices and lessons learned to enhance construction quality.

Increased Efficiency

- *Streamlined processes*: Optimizing workflows and communication channels.
- *Reduced rework*: Learning from past mistakes and implementing preventive measures.
- *Improved decision-making*: Accessing and analyzing relevant information in a timely manner.

Enhanced Innovation

- *Promoting creativity*: Encouraging knowledge sharing and collaboration to generate new ideas.
- *Promoting best practices*: Identifying and implementing innovative solutions from other projects.

Organizational Learning

Building a knowledge base: Creating a valued asset of project experiences and lessons learned.

Developing expertise: Facilitating continuous learning and professional development.

By effectively managing knowledge throughout the project life cycle, construction companies can achieve significant improvements in project delivery, efficiency, and overall success.

Measuring the effectiveness of a knowledge management system (KMS) is crucial to ensure it is meeting its objectives and providing value to the organization. The next section shows a breakdown of key metrics and methods.

Usage and Engagement Metrics

- *Active users*: Monitor the number of employees who are actively using the KMS. A high adoption rate indicates that employees find the system valuable and relevant.
- *Usage frequency*: Track how often employees are accessing the KMS. Frequent usage suggests that the knowledge resources are useful and easy to access.

- *Content consumption*: Measure metrics such as page views, downloads, and time spent on specific content to identify popular resources and areas where content may be lacking.
- *Search effectiveness*: Analyze search queries and success rates. A high percentage of unsuccessful searches may indicate that the content is poorly organized or insufficient.
- *Collaboration*: Track user interaction, including comments, questions asked, and contributions to knowledge-sharing platforms. High levels of interaction indicate active engagement with the KMS.

Metrics for Evaluating Performance and Outcomes

- *Project performance*: Evaluate if the KMS contributes to positive project outcomes, such as cost reductions, faster completion times, and improved quality of deliverables.
- *Decision-making*: Assess whether the KMS facilitates better and faster decision-making by providing employees with timely access to relevant information.
- *Problem-solving*: Measure how effectively the KMS supports employees in finding solutions to problems and challenges.
- *Innovation*: Monitor the number of new ideas or improvements generated because of knowledge sharing and collaboration enabled by the KMS.
- *Employee satisfaction*: Gather employee feedback through surveys to gauge their satisfaction with the KMS and identify areas for enhancement.

Learning and Development Metrics

- *Training effectiveness*: Evaluate the impact of knowledge management on employee onboarding and training.
- *Skill development*: Assess if the KMS improves employee skill development and expertise.
- *Knowledge retention*: Measure the KMS's effectiveness in retaining organizational knowledge, especially during employee turnover.

Qualitative Feedback

- *Surveys and feedback forms*: Collect user opinions on the KMS's usability, content quality, and overall effectiveness.
- *Focus groups and interviews*: Facilitate detailed discussions with employees to understand their KMS experiences.

- *Observations and case studies*: Analyze how employees utilize the KMS in their daily tasks and record successful instances of knowledge sharing and application.

Methods for Analyzing and Improving Knowledge Management

- *Analytics dashboards*: Utilize dashboards to monitor KPIs and visualize data trends.
- *Surveys and polls*: Gather user feedback and assess satisfaction levels through regular surveys.
- *A/B testing*: Conduct experiments to compare different knowledge management strategies and content formats.
- *Return on Investment (ROI) analysis*: Evaluate the financial benefits of knowledge management initiatives, including cost savings, efficiency gains, and improved project outcomes.

Organizations can gain a complete picture of how well their KM initiatives are working by analyzing both quantitative metrics and qualitative feedback. This data-driven approach enables them to make targeted improvements, optimizing how knowledge is shared and used within the organization.

CONCLUSION

Effective KM is paramount for construction organizations seeking sustained success. By systematically capturing, sharing, and utilizing knowledge, firms can enhance project performance, streamline processes, and foster innovation. Implementing robust KM strategies, coupled with continuous evaluation and improvement, transforms organizational learning into a tangible asset. This knowledge-driven approach empowers teams to make informed decisions, mitigate risks, and deliver projects with greater efficiency and quality, ultimately solidifying a competitive edge in the dynamic construction industry.

QUESTIONS

1. What is the difference between explicit and tribal knowledge? Why is knowledge management (KM) important to an organization?
2. Why would people be unwilling to share their knowledge? How might an organization address this issue?
3. What are the different approaches to capturing knowledge across a construction project?
4. What are the benefits of having structured KM? What types of infrastructure are required to sustain it?
5. How would you assess the effectiveness of KM implementation in a construction organization?
6. Explain the difference between explicit and implicit (tribal) knowledge in the context of a construction organization. Give examples of each and discuss the challenges associated with transitioning implicit knowledge to explicit forms.

7. Describe the steps involved in implementing a knowledge management system (KMS) within a construction firm. Detail the strategies for knowledge capture, sharing, use, and evaluation, including specific technologies or tools that could be employed at each stage.
8. Analyze the potential reasons why employees in a construction organization might be unwilling to share their knowledge. Propose strategies that leadership could implement to foster a culture of knowledge sharing, and address concerns related to job security or competition.
9. Evaluate the effectiveness of different methods for measuring a KMS's success in a construction setting. Discuss the importance of balancing quantitative metrics (e.g., usage frequency, content consumption) with qualitative feedback (e.g., surveys, interviews) to gain a comprehensive understanding of the KMS's impact.
10. Considering the unique challenges of virtual teams in construction, how can KM strategies be adapted to ensure effective communication and knowledge transfer? Discuss specific tools and practices that can overcome barriers related to culture, time zones, and geography.

Certified Construction Quality Manager (CCQM) Body of Knowledge

The topics in this Body of Knowledge include additional detail in the form of subtext explanations and the cognitive level at which test questions will be written. This information will provide guidance for the candidate preparing to take the exam. The subtext is not intended to limit the subject matter or be all-inclusive of what might be covered in an exam. It is meant to clarify the type of content to be included in the exam. The descriptor in parentheses at the end of each entry refers to the maximum cognitive level at which the topic will be tested. A complete description of cognitive levels is provided at the end of this document.

I. Pre-contract Phase (13 Questions)

A. Project delivery, contract, and payment models

1. Project delivery methods
Identify various types of clients (e.g., private, commercial, federal, state, municipal, and international) and their use of project delivery methods, such as Engineering, Procurement, Construction Management (EPCM), Design-Bid-Build (DBB), Design-Build (DB), Design, Build, Finance, and Maintain (DBFM), Construction Manager at Risk (CMAR), and Construction Manager for Fee (CMFF). Understand project lifecycle phases (e.g., project initiation, project planning, project execution, monitoring & control, and project closeout) align with the project delivery method. **(Understand)**

2. Contract types
Understand various types and sources of contracts (e.g., Public-Private Partnership (P3), Design-Build Institute of America (DBIA), American Institute of Architects (AIA), International Federation of Consulting Engineers (FIDIC), lump sum, subcontract, consultant agreement, consultant staffing agreement, joint venture contracts, and purchase orders) along with their quality requirements. Recognize contract terms and conditions (e.g., liquidated damages [LDs], liquidated and ascertained damages [LADs], and general damages [GDs]). **(Understand)**

3. Payment models and methods
Understand the risks associated with and the differences between payment models (e.g., unit rates and reimbursable costs) and methods (e.g., Guaranteed Maximum Price (GMP). **(Understand)**

4. Payment documentation requirements
   Understand payment documentation requirements, such as schedule of values, percent completion determination (cost complete vs. work complete), retainage, billing methods, and billing terms. **(Understand)**

B. Contractual requirements

1. Scope, schedule, and resource planning
   Define a project's scope components (e.g., boundaries), develop the schedule and resource plan, track milestones, and identify roles and responsibilities in terms of quality of different entities (e.g., trade partner, contractor, owner, and quality manager). (Analyze)
2. Quality management inputs
   Determine and analyze quality management inputs (e.g., Certificate of Authorization (CoA), input for proposals, and Request for Proposal (RFP)) to adhere to a project's quality standards. (Analyze)
3. Stakeholders
   Identify and acknowledge internal and external stakeholders and engage them. (Apply)
4. Local contract opportunities and contract values
   Understand the impact of local contract (e.g., DBE, SWaM, local small businesses and suppliers) opportunities and contract values (e.g., allocation percentages to local resources) in relation to the quality of a project. (Understand)
5. Procurement and contracting methods
   Understand procurement methods, such as sole source, single source, competitive bid, lowest bid, technical bid, and contracting methods, such as general contracting, and subcontracting. Identify and understand the best value proposal. (Understand)
6. Delegated design
   Apply delegated design or specialized engineering to transfer specific design responsibilities for a project to a contractor or subcontractor and understand the associated risks. (Apply)

C. Bid evaluation and finalization

1. Prequalification process
   Analyze the role of audits in the prequalification process, align the prequalification process to meet organizational needs, and explain the prequalification process for short-listed bidders. (Analyze)
2. Criteria development
   Review the criteria for assessing and comparing bids (e.g., dollars, schedule, lead times, workforce, products, assemblies, safety, sustainability) to integrate the bid criteria with quality requirements, critical to quality (CTQ) concepts, and examine bid criteria for the specific contract type in use. (Analyze)

3. Bidding process
   Distinguish the role of quality requirements in the context of different contract types during the bidding process. Identify strategies and techniques for building consensus when finalizing quality requirements. (Analyze)

II. Planning Phase (30 Questions)

A. Strategic quality planning

1. Project quality strategy development
   Formulate a comprehensive project quality strategy, integrating stakeholder requirements (stated and implied), procurement requirements, contracting approaches, and regulatory compliance based on Authority Having Jurisdiction (AHJ), and securing management commitment. Establish key performance indicators (KPIs) to align with each element of the quality strategy. (Create)
2. Environmental, social, and governance (ESG) practices
   Apply sustainable practices and programs (e.g., LEED®, ENERGY STAR®, and Cradle to Cradle Certification), social responsibility, and governance standards to enhance the overall quality and long-term viability of projects. (Apply)
3. Alignment of quality plan with other project plans
   Create the quality plan, ensuring alignment with the quality strategy and technology strategy. Align the quality plan with other project plans, such as the Health, Safety, and Environment (HSE) Plan, Execution Plan, Communication Plan, project schedule, and Risk Management Plan. (Create)

B. Quality management systems (QMS)

1. Industry standards
   Analyze the quality management system (QMS) to ensure it follows industry standards and understand how it complies with international standards (e.g., ISO 9001, ISO 10006, ISO 14001, and ISO 45001) to deliver quality outcomes. (Analyze)
2. Quality Management Plan
   Create a project-specific quality management plan (QMP) that complies with the appropriate QMS, incorporating policies and procedures (e.g., nonconformance control procedures, inspection and testing procedures, and method statements). Integrate PMBOK® Guide principles and system assurance practices into the development of the QMS and QMP. (Create)
3. Quality assurance (QA)
   Develop and implement quality assurance (QA) practices to verify the project meets quality specifications and is compliant with applicable standards by establishing standardized quality policies and procedures. (Create)

4. Quality control (QC)
   Develop and implement quality control (QC) measures such as an inspection and test plan to ensure compliance with requirements. (Create)

C. **Resource planning and analysis**

1. Resource assessment and mobilization
   Identify the specific quality resources and timely availability needed for a project and develop a mobilization plan including allocation and a gap analysis of current and necessary resources. (Create)
2. Roles of quality personnel and internal stakeholders
   Ensure all relevant quality personnel and internal stakeholders understand the quality plan and have the necessary knowledge and skills to carry out their responsibilities. Formulate the roles and expectations of quality personnel, contribute to the personnel qualification process, and ensure all personnel understand their roles and responsibilities. (Create)
3. Training needs in quality management
   Identify a project team's quality-specific training needs to achieve project deliverables, develop a training plan, deploy the training, and evaluate the effectiveness of the training. (Create)
4. Third-party inspection and outsourcing
   Identify the need for third-party inspection services as applicable and develop criteria for selecting and integrating outsourced quality services into the project. (Create)
5. Measurement systems analysis
   Assess the measurement system and its requirements to verify if equipment accurately and reliably measures specific parameters. Ensure calibration and certification requirements are met and that all equipment, including equipment used by subcontractors, is accurate, precise, and compliant with relevant standards and regulations. (Evaluate)
6. Quality management tools and infrastructure
   Identify and assess quality tools (e.g., data analysis tools, statistical process control, dashboards, construction pre-control checklists, and budgets) and quality infrastructure essential for effective quality management. (Evaluate)

D. **Quality audits**

1. Audit types
   Identify the appropriate type of audit (e.g., first-party, second-party, and third-party) for different situations, such as personnel independence from the area being audited. (Apply)

2. Quality audit plan
Create and evaluate a quality audit plan that aligns with strategic objectives, key project controls, prime contractor audits, and compliance checks for areas at risk using specific, measurable, attainable, relevant, and time-bound tactics. (Create)

3. Audit performance
Conduct a quality audit to evaluate the management system's effectiveness, review adherence to standards, and identify nonconformances through detailed observations. (Evaluate)

III. Design Phase (20 Questions)

A. Design inputs

1. Quality objectives
Examine the quality objectives set by the client, owner, or user from the initial scope agreement and determine how these objectives align with design inputs. Consider the delivery method and contract when reviewing design phase / development planning. (Analyze)

2. Coordination of functional and nonfunctional requirements
Explain the coordination between functional (e.g., foundation) and nonfunctional (e.g., finish) requirements and illustrate their application at various levels, such as system and component. Relate these requirements to industry standards and demonstrate their relationship to overarching quality objectives. (Apply)

3. Design lessons learned and best practices
Apply lessons learned from previous projects and industry best practices to design inputs. Understand the impact of value engineering on design. (Apply)

B. Design process

1. Stakeholder involvement
Determine key stakeholders involved in the design process for each discipline. Collaborate with the subject matter experts from each discipline responsible for making design decisions. Analyze the impact of these design decisions on quality. (Analyze)

2. Application of QMS in design feedback
Apply QMS procedures and interdisciplinary reviews to address feedback related to quality requirements in the design process. (Analyze)

3. Functionality and constructability reviews
Participate in reviews that assess the design's functionality and constructability. Provide quality inputs that focus on various aspects, such as complexity, inspectability, testability, reliability, durability, survivability, interoperability, compatibility, maintainability, and comparative cost of ownership. (Analyze)

C. Design outputs

1. Quality objectives review
   Apply verification and validation to design outputs to evaluate whether quality objectives are met. (Apply)

2. Cross-disciplinary design review
   Support the design review process across multiple disciplines to identify and address gaps in design outputs using a design review checklist. Determine and address conflicts within the overarching system design, among individual component teams, and in interactions between component teams. (Analyze)

3. Criticality levels and inspection
   Analyze the proposed criticality levels for equipment and material. Identify appropriate inspection levels and methods (e.g., in-process inspection and third-party design review) based on criticality and budget considerations. (Analyze)

D. Quality controls in design

1. Project controls
   Evaluate the effectiveness of project controls, including the timeline, in meeting the quality expectations of stakeholders, such as the client, owner, or user. (Evaluate)

2. Design review output
   Examine the design review output received at each phase (e.g., 30%, 60%, 90%) to determine its alignment with the expectations of stakeholders, such as the client, owner, or user. (Analyze)

3. Design changes
   Understand sources of design change communication, including requests for proposals (RFPs), change order directives, addenda, and requests for information (RFIs). Assess and track the impact of proposed design changes and determine the effectiveness of incorporating any agreed-upon changes into the contract. (Analyze)

4. Design verification and validation
   Understand design verification and validation and understand how the design output conforms to the design input through the use of reviews, inspections, and testing. (Understand)

E. Measurement and monitoring

1. Design monitoring
   Examine design effectiveness and progress at various gates (e.g., schematic design (SD), design development (DD), construction documents (CD)) to align with the quality objectives and schedule. Address any deviations or issues discovered at each gate. (Analyze)

2. Peer reviews
Identify the necessity of peer reviews (e.g., design peer reviews and third-party consultant reviews), evaluate their input, and incorporate them into the quality documentation process. (Evaluate)

3. Quality documentation
Collaborate with the design review facilitator to document all quality-related activities, decisions, and changes, ensuring a comprehensive record for evaluating the design's adherence to quality standards. (Evaluate)

IV. Procurement (15 Questions)

A. Quality requirements

1. Quality requirements for procurement
Collaborate with engineering teams to design functional and nonfunctional quality requirements for the procurement of specific products or services. Analyze if these requirements are clearly defined, complete, correct, compliant, and coordinated. (Analyze)

2. Trade partner requirement alignment
Develop a strategy, incorporating methodologies such as first article inspection (FAI) and factory acceptance testing (FAT), to determine whether the trade partner's inspection and testing plan aligns with established quality requirements. Incorporate a rigorous review process for the inspection and test plans to ensure comprehensive evaluation. Additionally, integrate specific quality requirements for subcontractors and vendors, focusing on their scope of work, historical performance, compliance with standards, and the thoroughness of their documentation and plans. (Create)

B. Resource management

1. Pre-activity meetings
Schedule, implement, and oversee pre-activity meetings (e.g., preassembly, pre-inspection, quality kick-off) to review the project's requirements with the relevant stakeholders. (Evaluate)

2. Material control and resource identification
Incorporate material control, material traceability, and vendor requirements into the project prior to commencement of work. Evaluate the resources needed for various inspections (e.g., self-inspection and certification, third-party inspection), including those conducted at the source and as mandated by regulations. (Evaluate)

3. Resource mobilization and management
Integrate resource qualification and training, along with warehouse quality and material management into the resource management process. Assess the need for specific resources tailored to each inspection type, ensuring that these resources are available and properly allocated, and aligned with the established inspection and test plan. (Evaluate)

C. Risk management

1. Risk identification and analysis
   Analyze various risk identification and assessment tools, including failure mode and effects analysis (FMEA), fault tree analysis (FTA), and strengths-weaknesses-opportunities-threats (SWOT) analysis. Examine risks based on quantitative and qualitative data related to potential impact on the project. (Analyze)
2. Risk treatment
   Evaluate risk mitigation strategies such as risk transfer, retention, acceptance, avoidance. (Evaluate)
3. Risk monitoring
   Use risk monitoring tools (e.g., risk-based audit programs) and perform surveillance audits to regularly evaluate identified risks and monitor key factors (e.g., cost, time, safety, supplier performance), and the criticality of material and equipment. (Evaluate)

D. Quality reporting

1. Review and interpretation of quality documents
   Evaluate inspection and test reports, quality records, and manufacturers' data books to ensure traceability. Assess conformance with product specifications and quality criteria. (Evaluate)
2. Documentation and communication of quality insights
   Evaluate and document key insights from the quality management process to inform future actions. Communicate quality-related insights and determine the appropriate timing, escalation, and method to alert stakeholders about significant quality issues. (Evaluate)
3. Vendor performance evaluation
   Analyze vendor performance periodically using quality key performance indicators (KPIs) and use supplier performance incentives (e.g., awards, more wallet share, premium). (Analyze)

V. Construction (34 Questions)

A. Quality plan deployment

1. Construction plans and specifications
   Assess plans and specifications (e.g., permanent and temporary work) in relation to the project's quality plan to meet established quality standards. (Evaluate)
2. Required permits
   Understand the importance of securing all essential permits before project commencement. (Understand)

3. Construction submittals
   Review various types of construction submittals (e.g., action and deferred/delegated), taking into consideration the interdependency across trades and the project's quality plan and standards. (Apply)

4. Material evaluation
   Evaluate materials to verify their conformity to project specifications. (Evaluate)

5. Requests for information or changes
   Participate in the process to track requests for information (RFIs), field change requests (FCRs), and design change requests (DCRs) to ensure the project remains on track and that objectives adapt when necessary. (Apply)

B. Process management

1. Risk-based inspection and testing strategy
   Develop a risk-based inspection and testing strategy by identifying risks. Establish priorities for inspections, including modifications based on changes in project requirements (e.g., scope modifications), and testing in line with specifications (e.g., concrete testing) and accepted standards or regulations (e.g., welding certifications). (Create)

2. Quality management processes
   Identify inspection processes for various stages, including source, initial, intermediate, and continuous. Prioritize program inspections and tests based on their criticality, respective project stages, and potential impact on overall project quality, ensuring alignment with the project's evolving needs and scope changes. Identify the need for procedures for specialized tasks, such as welding, heat treatment, coating, and soldering. (Evaluate)

3. Quality control and monitoring
   Develop a system to address nonconformances, manage nonconformance reports (NCRs) and corrective action requests (CARs), and implement corrective and preventive action (CAPA). Design quality protocols and establish criteria for stop work notices. Establish a robust framework for maintaining quality records. Design a strategy for conducting site quality audits and create a mechanism for generating and distributing quality reports. (Create)

C. Destructive and nondestructive tests

Determine the appropriate scenarios for using destructive and nondestructive test methods and verify those performing the tests are accredited or certified. (Evaluate)

D. Compliance management

1. Laboratory accreditation
   Evaluate laboratory accreditation and conduct periodic reviews or surveillance audits based on project quality requirements. (Evaluate)

2. Compliance monitoring
   Assess compliance with applicable regulatory requirements, standards, and codes, and conformance to specifications, procedures, and drawings. Document defects and nonconformities, and collaborate with technical authorities and engineering disciplines to review, disposition, and resolve identified nonconformities. (Evaluate)

3. Regulatory body or Authority Having Jurisdiction (AHJ)
   Conduct regulatory body or AHJ reviews and audits to assess adherence to established criteria and maintain proper documentation of reviews and audits. (Analyze)

E. Quality monitoring and reporting

1. Acceptance criteria definition and implementation
   Establish criteria for accepting work at various stages, such as mock-up inspection and testing, initial inspection and testing, work-in-process inspection, punch list inspection, and final inspection, testing, and commissioning. Integrate material verification at each stage to assess if materials meet project specifications and quality standards. Apply escalation procedures to address materials that do not meet established acceptance criteria. (Evaluate)

2. Reporting and analysis
   Design periodic reports and conduct statistical analyses to identify quality trends and assess waste elimination. (Evaluate)

3. Meetings and participation
   Organize pre-inspection meetings that include quality meeting sample agenda items that address key topics, such as upcoming inspections, current quality metrics, recent findings, and open issues. (Evaluate)

4. Lessons and improvements
   Incorporate lessons learned from previous projects into the quality monitoring and reporting process. Document and communicate these lessons learned for future projects. Develop strategies to assess and manage quality risks. Engage in root cause analyses and investigations to understand underlying quality issues. Develop quality improvement programs based on identified needs and insights. (Evaluate)

5. Levels of commissioning
   Understand and determine when to use the 5-level (factory acceptance test, component verification, system construction, individual system and major equipment operation verification,

integrated system testing) and 8-level (preparation, design, pre-construction, construction, commissioning of services, pre-handover, initial occupation, post-occupancy care) commissioning frameworks. (Evaluate)

VI. Project Turnover and Closeout (11 Questions)

A. Documentation management

1. Impact assessment and compliance
   Evaluate the potential impact arising from the demobilization of quality resources. Review and assess inspection and test records related to safety critical equipment to determine if they are compliant with required standards. Confirm how final documentation will be transferred to the client (e.g., digital platforms, electronically, and hard copy). (Evaluate)

2. Required documentation
   Maintain comprehensive documentation throughout the project's duration, including records of all activities and changes. (Apply)

3. Nonconforming work
   Ensure all work not conforming to contract specifications (i.e., punch list) is closed. (Evaluate)

4. Quality deliverables verification
   Verify the completeness of all quality deliverables. (Evaluate)

5. Final certification
   Facilitate the process for obtaining final certification, ensuring inclusion and maintenance of all documentation related to each closeout milestone and stage (e.g., beneficial occupancy / use, substantial completion, and final completion). Use the current version of Issued for Construction (IFCs) documents. (Evaluate)

6. Final turnover and closeout documentation
   Audit the final turnover plan and related closeout documentation, including permit close out, operations and maintenance manuals, as-built drawings, attic stock / spare parts records, commissioning reports, and owner training materials. Confirm all quality inspection and test records are reviewed and validated for fitness for use. (Evaluate)

7. Warranties
   Implement effective communication regarding the warranty program and documentation or warranty work to all relevant stakeholders. (Evaluate)

VII. Operations and Maintenance (5 Questions)

A. Sustaining quality requirements

Prepare quality objectives required for operations and maintenance (O&M) based on customer expectations. Understand how in-contract and warranty maintenance programs are used throughout the project lifecycle. Report future O&M inspections and testing strategies needed (e.g., building component health, risk-based) to the design team. (Apply)

B. Business continuity

Provide inputs for facility shutdown plans, considering both planned (e.g., O&M-required) and unplanned events (e.g., weather or disaster-related incidents). Align the quality plan with the business continuity plan for operational resilience. Apply drills and simulations to test these plans. Monitor the quality-related impacts resulting from shutdowns. (Apply)

VIII. Quality Manager Skills (22 Questions)

A. ASQ Code of Ethics

Demonstrate appropriate behaviors and actions that align with the ASQ Code of Ethics. (Analyze)

B. Stakeholder analysis and alignment

Identify the perspectives, needs, and objectives of various internal and external stakeholders. Integrate collaboration and negotiation techniques to align the organization's strategic objectives and stakeholder expectations. (Apply)

C. Team building

1. Vision development
   Collaborate with stakeholders to adopt quality goals and a clear vision for a team or project. (Apply)
2. Stages of team development
   Identify and distinguish between the stages of team development: forming, storming, norming, performing, and adjourning. (Analyze)
3. Effective team dynamics and communication
   Employ various communication techniques (e.g., verbal, nonverbal, written, and visual) and active listening. Collaborate with multiple groups (e.g., designers, project management, and craft teams). Implement the Responsible, Accountable, Consulted, Informed (RACI) Matrix to clarify roles and responsibilities and streamline decision-making processes. Understand decision-making pitfalls (e.g., groupthink and power dynamics) to promote healthy team dynamics and achieve project objectives. Resolve conflicts through established procedures. (Apply)

D. Change management

1. Organizational culture
   Define and explain the factors that shape an organization's culture. (Apply)

2. Change management strategies
   Analyze and provide input to various change management strategies to overcome organizational roadblocks, adapt to global changes, assess the achieved change levels, and review outcomes for effectiveness. (Analyze)

E. Cost of Quality (CoQ)

1. Cost of Quality (CoQ) analysis
   Examine the methods and sources for collecting CoQ data. Interpret the CoQ data to derive meaningful insights. Report and communicate findings from the CoQ data analysis. (Analyze)

2. Improvement plans
   Incorporate CoQ as a foundation for improvement methodology, alongside Lean, Six Sigma (e.g., DMAIC), and Deming's System of Profound Knowledge (SoPK). Identify areas for improvement from CoQ data analysis, and participate in the improvement plan to address these, integrating the plan with existing business processes. Evaluate the effectiveness and applicability of CoQ tools and methodologies. (Evaluate)

3. Organizational goals
   Assess how CoQ improvement efforts align with organizational goals. Monitor and measure the progress made towards achieving organizational goals. Evaluate the steps taken to achieve organizational goals based on CoQ improvements. (Evaluate)

F. Customer satisfaction

1. Customer feedback systems
   Apply feedback collection tools, such listening posts, focus groups, complaint logs, warranty data, surveys, and interviews to capture positive and negative feedback. (Apply)

2. Customer value analysis
   Analyze a customer value analysis to determine the financial impact of customer retention and the potential consequences of customer attrition. (Analyze)

3. Customer experience enhancement
   Evaluate corrective strategies and proactive methods to enhance customer satisfaction, loyalty, and retention. (Evaluate)

G. Quality management tools

1. Quality tools
   Select the appropriate tool and evaluate the output from the seven classic quality tools (Pareto charts, cause and effect diagrams, flowcharts, control charts, check sheets, scatter diagrams, and histograms) as preventive measures, analyze root causes, and problem solve. (Evaluate)

2. Learning and best practices analysis
   Incorporate after-action reviews, and root cause analysis (RCA) meetings to critically assess project outcomes and identify lessons learned from all phases of the project, identifying both successes and areas for improvement. Analyze the critical success factors that led to best practices being implemented within the project. Use key performance indicators (KPIs) to systematically track and assess the impact of implemented improvements and evaluate strategies to reinforce and replicate these improvements across future projects. (Evaluate)
3. Knowledge management analysis
   Document the strategies and outcomes of knowledge management throughout the project. (Apply)

LEVELS OF COGNITION

BASED ON BLOOM'S TAXONOMY—REVISED (2001)

In addition to **content** specifics, the subtext for each topic in this BOK also indicates the intended **complexity level** of the test questions for that topic. These levels are based on "Levels of Cognition" (from Bloom's Taxonomy—Revised, 2001) and are presented below in rank order, from least complex to most complex.

Remember

Recall or recognize terms, definitions, facts, ideas, materials, patterns, sequences, methods, principles, etc.

Understand

Read and understand descriptions, communications, reports, tables, diagrams, directions, regulations, etc.

Apply

Know when and how to use ideas, procedures, methods, formulas, principles, theories, etc.

Analyze

Break down information into its constituent parts and recognize their relationship to one another and how they are organized; identify sublevel factors or salient data from a complex scenario.

Evaluate

Make judgments about the value of proposed ideas, solutions, etc., by comparing the proposal to specific criteria or standards.

Create

Put parts or elements together in such a way as to reveal a pattern or structure not clearly there before; identify which data or information from a complex set is appropriate to examine further or from which supported conclusions can be drawn.

Glossary

8 Wastes of Lean—The concept of 7 Wastes was given by Toyota Motor Company's Taiichi Ohno. Later, another waste was added to the original 7 and given the acronym TIMWOODS (Transportation, Inventory, Motion, Waiting, Overproduction, Overprocessing, Defects, and Skills [underutilized]).

5S—5S methodology was initially developed by Toyota Motor Company as part of its Toyota Production System (TPS) to improve workplace efficiency and organization. The *5S* stands for Sort, Set in Order, Shine, Standardize, Sustain.

A

Acceptance Criteria—The standards or requirements that construction work must meet to be considered satisfactory.

Acoustics—The science of sound and its control—often crucial in building design to minimize noise and ensure sound quality.

Aggregate—Materials such as sand, gravel, or crushed stone, used in concrete production or road construction.

Agile Retrospective—Lessons learned are conducted on tasks typically involving software-related projects, such as the implementation of software or systems supporting construction projects, where agile methodology is employed.

American Institute of Architects (AIA)—A professional organization for architects, providing resources, pursuing advocacy, and setting standards that influence building design and construction practices.

Appraisal Cost—Costs incurred to evaluate whether products or services meet quality requirements.

Asset Management—The systematic process of maintaining and operating a building or infrastructure throughout its life cycle.

Authority Having Jurisdiction (AHJ)—A governmental agency with mandates by an overarching government's laws to be the primary overseers of compliance with all local building codes and laws for a construction project.

B

B2B (business-to-business)—Transactions between businesses.

B2C (business-to-customer)—Transactions between businesses and consumers.

Backfill—Soil or other materials used to refill excavated areas.

Baseline Schedule—The approved time-phased project schedule is used as a fixed reference point to measure, monitor, and control overall project performance.

Basis of Design (BOD)—A document that outlines the design concepts, assumptions, and criteria used to develop the project's design. It serves as a reference for designers, contractors, and other stakeholders, ensuring that the project is built according to the intended design principles.

Beam—Horizontal structural members that support loads perpendicular to their length.

Beneficial Occupancy—The term "substantial completion" may also be used in reference to beneficial use; this is the stage of a construction project when the owner can use a building for its intended purpose, even if minor work remains before final completion. (Also see *Substantial Completion.*)

Bias

- **Anchoring Bias**—The tendency to rely too heavily on the first piece of information offered ("the anchor") when making decisions.
- **Confirmation Bias**—The tendency to search for, interpret, favor, and recall information in a way that confirms one's preexisting beliefs or hypotheses.
- **Overconfidence Bias**—The tendency to overestimate one's own abilities or knowledge.
- **Status Quo Bias**—The tendency to prefer things to stay relatively the same.

Bidding Process—The procedure by which organizations invite bids for projects or contracts, and evaluate and select a winning bid.

Bid Evaluation Model—A framework used to assess and compare bids based on predefined criteria.

Building Information Modeling (BIM)—A 3D modeling process that includes detailed building information.

Business Continuity Plan (BCP)—A proactive arrangement for some predetermined operations in the case of a disaster or other emergencies. In the construction industry, this plan outlines how a company will continue its essential functions during and after disruptions such as natural disasters, equipment failures, or supply chain interruptions. It includes strategies for maintaining operations, protecting assets, and ensuring the safety of personnel.

C

Certificate of Authorization (CoA)—A permit required for individuals or entities to offer professional engineering services to the public.

Certificate of Occupancy—A contract milestone describing a confirmation that the entirety of a project structure or space is suitable for habitation.

Change Control—Managing changes to a system or project in a controlled and systematic manner.

Change Control Board (CCB)—Those responsible for overseeing and managing changes to a project or system.

Change Management—A structured approach to making a change to a product, process, system, individual's role, team priorities, and organization from a current state to a desired future state. This involves processes, tools, and techniques used to manage change, including planning, validating and implementing change, and verifying effectiveness of change.

Change Order—A formal document, agreed on by both parties, that modifies the original construction contract, altering the scope of work, cost, or schedule.

Clash Detection—A process within BIM that identifies and resolves conflicts among different building systems (e.g., HVAC, plumbing, or electrical) to avoid on-site errors.

Clients—The party that commissions or contracts for the construction project; often synonymous with "owner."

Closed Competitive Tendering (or Restricted Tendering)—An invitation to bid is issued to predetermined or restricted suppliers.

Closeout—The final phase of a construction project where the physical construction is complete, documentation is finalized, and the project is officially handed over to the owner. This involves the compilation and turnover of several records generated during the performance of the work.

Commercial Client—Businesses that undertake construction projects for operational purposes or business expansion.

Commissioning—A systematic process of ensuring that all building systems perform interactively according to the design intent and the owner's operational needs.

Commissioning Plan—A document that outlines the systematic process of verifying and documenting that all building systems and components are designed, installed, tested, and operated according to the owner's project requirements. It ensures that the building functions as intended.

Communications Management Plan—A document that simplifies communication and collaboration throughout the project life cycle.

Competitive Tendering—Soliciting open bids via public notice.

Complaint Logs—Centralized system for recording and tracking customer complaints.

Compression Strength—The ability of a material (like concrete) to withstand loads that reduce its size.

Configuration Management—An essential aspect of managing changes within an organization, especially in the context of IT and software development.

Constructability Analysis—A review of the design to assess its ease of construction and identify potential challenges or inefficiencies. Typically, it is the review of design as performed by the construction contractor or by someone who represents construction.

Construction Documents—The written and graphic documents prepared or assembled for communicating the project design for construction. Includes drawings and specifications.

Construction Manager at Risk (CMAR)—A project delivery method where the construction manager commits to delivering the project within a guaranteed maximum price (GMP).

Construction Manager for Fee (CMFF)—A project delivery method where the construction manager acts as an advisor and is paid a fee without financial risk for cost overruns.

Construction Quality Control (CQC)—Processes and procedures implemented by the contractor to ensure that materials and workmanship meet the quality standards specified in the contract.

Construction Quality Managers (CQMs)—Professionals responsible for ensuring that construction projects meet specified quality standards and requirements. They develop and implement quality control plans, conduct inspections, monitor compliance, and address any quality issues that arise during the construction process. CQMs play a critical role in maintaining project quality and minimizing defects.

Consultant—A party responsible for construction oversight, quality assurance, and contract management support.

Consultant Agreement—A contract outlining the terms and scope of services provided by a consultant (e.g., engineering, architectural) to a client on a construction project.

Consultant Staffing Agreement—An agreement for providing temporary or project-specific personnel with specialized expertise for a construction project.

Contingency Fund—A dedicated amount of money included in the project budget to cover unforeseen expenses, risks, or scope changes during construction.

Contract—A legally binding agreement between parties involved in a construction project, outlining the scope of work, payment terms, timelines, and responsibilities.

Contracting—The process of creating and managing agreements for goods or services.

Contract Milestone—Specific points in time in a project marking the completion or achievement of specific tasks or making certain functions available.

Cost of Conformance—The costs associated with ensuring that products or services meet quality requirements.

Cost of Nonconformance—The costs associated with failing to meet quality requirements.

Cost of Quality (CQ)—The total cost associated with preventing failures, appraising quality, and dealing with internal and external quality failures.

Cost of Good Quality (CoGQ)—The costs associated with preventing defects and ensuring quality, including prevention and appraisal costs.

Cost of Poor Quality (CoPQ)—The costs associated with defects or failures, including rework, scrap, and warranty claims.

Cost Plus—A type of contract where the contractor is paid for the cost of the work plus a fee (usually a percentage of the cost).

Cost Plus Fixed Fee with No Maximum—A cost-reimbursement contract where the contractor receives a fixed fee regardless of the final project cost, with no upper limit on reimbursable expenses.

Cost Plus Fixed Fee with Guaranteed Maximum Price (GMP)—A cost-reimbursement contract with a fixed fee and a cap on the total reimbursable costs, beyond which the contractor is responsible.

Cost Plus Fixed Percentage—A cost-reimbursement contract where the contractor's profit is calculated as a fixed percentage of the total allowable costs of the project.

- **Costs**—The amount that must be paid or spent in order to buy or obtain something.
- **Direct Costs**—Costs that occur in a project and are attached to specific activities.
- **Fixed Costs**—Costs that do not change throughout the project life cycle.
- **Indirect Costs**—Costs that support the project, such as rent and administrative staff salaries.
- **Sunk Costs**—Investments already incurred that cannot be recovered.
- **Variable Costs**—Costs that change due to the amount of work performed.

Cradle to Cradle Certification—A product assessment and certification standard for safe, healthy, and sustainable products.

Critical Path—The sequence of project stages determining the minimum time for project completion.

Critical to Quality (CTQ) Concepts—These refer to the key characteristics of a product or service that are crucial for meeting customer requirements and quality standards.

Curing—The process of maintaining moisture in concrete to ensure proper hardening and strength development.

Customer Delight—Exceeding customer expectations to create a positive emotional response.

Customer Loyalty—The tendency of customers to repeatedly purchase from a business.

Customer Obsession—A core value where organizations prioritize customer needs above all else.

Customer Retention—The ability of a business to keep customers over a period of time.

Customer Satisfaction—The degree to which customers are happy with a product or service.

Customer Satisfaction Scores (CSAT)—A measure of customer satisfaction with a specific product or service.

Customer Value Analysis (CVA)—A method for assessing the value that customers place on products or services.

D

Deficiency—A lack of quality or a failure to meet professional standards of the product. It is identified through the quality control (QC) inspection. A deficiency (also sometimes called a defect or nonconformity) is a temporary condition that presumably can be corrected without impacting the time (schedule) or cost (budget) of the project. Deficiencies must always be bound with a time-based/schedule-based QC reinspection. If the QC inspection determines that the work has been built and cannot be corrected without affecting the schedule or budget, then it must be issued as a nonconformance and not a deficiency.

Defect (Latent)—Concealed flaws in work that exist but are hidden at the time of completion of work. Because latent defects are identified after the completion of work, they cannot be issued as a deficiency or nonconformance. They must be addressed through contract terms for the warranties of the work.

Defect (Patent)—Flaws in work that are visible or easily discoverable on inspection. The primary difference is discoverability: Patent defects are open, while latent defects are concealed. Patent defects may be issued as a deficiency only if they can be corrected without affecting the schedule or budget. However, if they cannot be corrected without negatively affecting the schedule, they should be issued as a nonconformance.

Definable Feature of Work (DFOW)—A distinct task or activity that can be identified and controlled for quality purposes.

Deflection—The degree to which a structural element bends under a load.

Delegated Design—Shifting specific design responsibilities from the project owner and architect to the contractor.

Design-Bid-Build (DBB)—A traditional project delivery method where the owner separately hires a designer and then a contractor.

Design-Build (DB)—A project delivery method where the owner contracts with a single entity to handle both design and construction.

Design, Build, Finance, and Maintain (DBFM)—A project delivery method where a private sector consortium designs, builds, finances, and maintains the infrastructure asset.

Design-Build Institute of America (DBIA)—A membership organization that advocates for and educates on the design-build project delivery method and best practices in the construction industry.

Design Criteria—Specific requirements, such as using a certain product or meeting a specific dimension.

Design Quality Managers (DQM)—Professionals responsible for ensuring that the design projects meet specified quality standards and requirements for the design. Typically, for all projects that are issued out in separate phases to different organizations (e.g., design-bid-build projects), the design quality manager is the first formally identified person assigned to ensure the project meets the quality requirements. They develop and implement quality control (QC) plans for the design phase, conduct design quality reviews, and implement QC as well as quality assurance (QA) of design on the project. The DQM position usually is a QA position that aims to establish and implement quality throughout the design phase. Even if the design and construction are performed as independent activities, the DQM must coordinate with the construction quality manager (CQM) to ensure quality compliance and address any quality issues that arise during the design process. DQMs play a critical role in maintaining project quality before the CQM is hired and the construction phase begins.

Design Review—A formal evaluation of the project's design documents to ensure compliance with requirements and identify potential issues. The formal design reviews are considered quality control reviews of design. They are of two types: within-the-discipline design review (also called intradisciplinary review) and cross-disciplinary design review (also called interdisciplinary review).

Disadvantaged Business Enterprise (DBE)—A federal certification program under the U.S. Department of Transportation for small businesses and businesses owned by historically disadvantaged groups.

DMAIC (Define, Measure, Analyze, Improve, Control)—A Six Sigma methodology for process improvement.

Durability—The ability of a construction material to withstand wear, pressure, and damage over time.

E

ENERGY STAR®—An energy efficiency program for products and materials.

Environmental, Social, and Governance (ESG)—Factors used to evaluate a company's sustainability and ethical impact.

Error Proofing (Poka-yoke)—Designing processes to prevent errors from occurring.

External Failure Costs—Costs incurred when defects are discovered after delivery to the customer.

F

Factory Witness Testing (FWT)—A process where a client or their representative observes and verifies the performance of a product or process (e.g., equipment or systems) at the manufacturer's facility before shipment. This ensures that the equipment meets specified requirements and quality standards before being installed on the construction site. FWT helps to identify and rectify any defects or issues early, reducing potential delays and costs during on-site installation. Various types of factory witness tests can be performed; the most used tests include a factory acceptance test (FAT), and first article inspection (FAI).

Fatigue—Weakening of material caused by repeated stress.

Feasibility Study—An analysis conducted to determine the viability and potential success of a construction project before significant investment is made.

Final Completion—A contract milestone describing the absolute conclusion of a project, including closure of all punch list items and the transmittal of all required documentation to an owner.

Finish—The final surface appearance and texture of materials (like concrete or wood).

First Work Inspection—An inspection conducted on the initial phase of a particular work activity to ensure that the work meets quality standards and specifications. It helps to identify and correct any potential issues early in the process.

Focus Groups—Small group discussions to explore customer experiences and expectations.

Functional Requirement—A specification that defines what a system or component should do. In the context of construction, functional requirements describe the specific functions or tasks that a built asset or its components must perform. Functional requirements typically relate directly to the operational capabilities and intended use of the facility or system. For example, a functional requirement for a building's HVAC system might be "the system must maintain a temperature of 72°F ± 2°F in all occupied spaces."

G

Gantt Charts—A project management tool used to visually represent project schedules.

General Contracting—A method where the main contractor is responsible for the entire project.

General Contractors—The primary contractor responsible for overseeing the entire construction project, coordinating subcontractors, and ensuring project completion.

General Damages (GDs)—Compensation awarded for losses that arise directly and naturally from a breach of contract, requiring proof of the actual loss suffered.

Guaranteed Maximum Price (GMP)—A cost-reimbursable contract with a ceiling price that the contractor guarantees the project cost will not exceed.

H

Heating, Ventilation, and Air Conditioning (HVAC)—Systems used to control the temperature, humidity, and air quality within a building or space. HVAC systems are essential for providing comfortable and healthy indoor environments. They include equipment such as furnaces, air conditioners, air handlers, ductwork, and control systems.

Hold Points—Specific stages in the construction process where work cannot proceed without inspection and approval.

Horizontal Construction—Construction projects that involve infrastructure built on or near the ground, such as roads, bridges, tunnels, and utility lines.

I

Indefinite-delivery, Indefinite-quantity (IDIQ)—A type of contract providing for an indefinite quantity of supplies or services for a fixed time, used when the exact needs are not known in advance.

Individual Contract—A contract between an owner and a single contractor or service provider for specific work on a construction project.

Inspection—A detailed examination to ensure that construction work meets quality standards and codes.

Inspection and Testing Procedures—Detailed procedures for examining and verifying the quality of materials and workmanship.

Inspection and Test Plan (ITP)—A document that specifies the inspections and tests to be performed at various stages of construction to ensure that the work meets quality control requirements. It details the inspection frequency, acceptance criteria, and responsibilities for each inspection and test. An ITP is typically a part of the construction quality plan.

Integrated Project Delivery (IPD)—A project delivery method that integrates people, systems, business structures, and practices into a process that collaboratively harnesses the talents and insights of all participants to optimize project results, increase value to the owner, reduce waste, and maximize efficiency through all phases of design, fabrication, and construction. IPD typically involves a multiparty agreement with shared risk and reward for all stakeholders.

Internal Failure Costs—Costs incurred when defects are discovered before delivery to the customer.

International Federation of Consulting Engineers (FIDIC)—An international body representing consulting engineers, known for publishing standard forms of contracts for construction and engineering projects.

ISO 9001—A standard related to quality management systems.

ISO 14001—A standard related to environmental management systems.

J

Job-Order Contracting (JOC)—A contracting method for recurring minor construction, repair, and renovation tasks under a single, competitively bid contract with a fixed price list.

Joint Venture Contracts—Agreements between two or more construction companies to pool resources and expertise for undertaking a specific project together, sharing risks and profits.

K

Kano Model—A theory for product development and customer satisfaction that classifies customer preferences into five categories.

Key Performance Indicators (KPIs)—Measurable values that demonstrate how effectively a company or project is achieving key business objectives. In the construction industry, KPIs might include safety incident rates, project schedule adherence, budget variance, quality metrics for construction quality, and customer satisfaction.

L

Lagging Indicator—A measure that reflects past performance or outcomes. In construction, lagging indicators might include the number of safety incidents that occurred, the final project cost variance, or the number of change orders issued. They are used to assess the overall success of a project and identify areas for improvement in future projects.

Leading Indicators—Measurable data points that signal future trends or events, acting as proactive, predictive tools to forecast the future. Leading indicators in construction can be used to predict and improve safety, quality, and project performance before incidents occur. Examples include design review metrics, supplier qualification results, quality training rates, safety training rates, near-missed incidents reported, safety audits, and project audit findings.

Lean—A methodology focused on minimizing waste and maximizing value.

Lean Construction—A methodology that focuses on minimizing waste and maximizing value throughout the construction process.

Lean Wastes (see "8 Wastes")—Non-value-adding activities or resources that consume time, money, or materials.

LEED® (Leadership in Energy and Environmental Design)—A green building rating system.

Life Cycle Management—Managing a building or infrastructure from its initial planning and design through construction, closeout, to operation and maintenance. The CCQM body of knowledge includes seven phases of a construction project life cycle.

Liquidated Damages (LDs)—Pre-agreed, estimated damages specified in a contract, payable by the contractor for failing to meet certain contractual obligations (often project deadlines).

Listening Posts—Methods of capturing informal customer feedback during interactions.

Lump Sum—A fixed price contract where the contractor agrees to complete the entire scope of work for a single, predetermined amount.

M

Material Receiving Report (MRR)—A document used to record the receipt and inspection of materials received, typically at the project site. The recipient of the goods should inspect the materials received, verify they are correct and undamaged, and then document the material's condition and receipt on the MRR.

Memorandum of Understanding (MOU)—A nonbinding agreement between two or more parties outlining their intentions and understanding for a future collaboration or project.

Method Statements—Documents detailing how a specific construction activity will be carried out safely and effectively.

Minority Business Enterprises (MBE)—Businesses owned and operated by minority group members.

Moderation meeting—A meeting to agree on a single consensus score for each bid after individual evaluations.

Multiheaded Client—A client organization with diverse stakeholders having differing needs and stakes in a project, making decision-making complex.

N

Net Promoter Score (NPS)—A measure of customer loyalty based on how likely customers are to recommend a business.

Network Diagrams—A project management tool illustrating the sequence and dependencies of project activities.

Nonconformance Report (NCR)—A document that records the condition or observation of an article of work that has been completed and does not conform to contractual specifications, specified standards, or other such requirements. A nonconformance is determined through a QC check/test that is performed as per the schedule. If the QC test is failed due to a deficiency/nonconformity that cannot be rectified without negatively impacting the schedule, then the deficiency must be issued as a nonconformance.

Nonconformances—Deviations from specified requirements, standards, or procedures in construction. Nonconformance indicates that a product, process, or service does not meet the established criteria. Examples include materials that do not meet the specified quality or dimensions; workmanship that does not comply with approved drawings or specifications; safety violations that deviate from established safety protocols; and documentation errors or omissions. Nonconformances require corrective action to rectify the deviation and prevent recurrence. They are often documented and tracked as part of a quality management system.

Nonfunctional Requirement—A specification that defines how a system or component should perform rather than what it should do. Nonfunctional requirements are performance requirements that relate to the quality attributes, constraints, or characteristics of the system or asset. Nonfunctional requirements are important because they impact the overall user experience. In construction, nonfunctional requirements might include aspects like safety, security, maintainability, reliability, performance, aesthetics, or environmental impact.

O

Open Competitive Tendering—Inviting companies to respond to a need via public notice.

Operational Expenses (OPEX)—Costs associated with day-to-day operations.

Operations and Maintenance (O&M)—The activities required to keep a building, infrastructure, or facility functioning as intended after construction is complete. This includes routine maintenance, repairs, inspections, and operational tasks necessary to ensure the asset's performance, safety, and longevity. O&M is crucial for maximizing the lifespan and efficiency of constructed assets.

Operations and Maintenance (O&M) Manuals—A comprehensive document providing guidance on how to utilize and perform upkeep on a specific piece of equipment or system. Documents that provide detailed instructions for operating and maintaining building systems and equipment. They include information on routine maintenance, troubleshooting, and repair procedures to ensure the long-term performance and reliability of the assets.

Original Equipment Manufacturer (OEM)—Companies that manufacture equipment or components used in the construction project.

Overhead—Indirect costs associated with running a business, such as office expenses, utilities, and salaries of administrative staff.

Owners—The party that holds legal title to the property and initiates the construction project.

P

Pareto Analysis—A statistical technique used to identify the most significant factors contributing to a problem.

Partnership Contract—A legal agreement forming a business partnership between two or more entities to collaborate on ventures, potentially including construction projects.

Payment applications—Documents submitted by contractors to request progress payments, detailing work completed and materials used.

Percent Completion Determination (Cost Complete vs. Work Complete)—The method used to calculate how much of the work has been finished, which can be based on costs incurred or work physically completed.

Performance Specifications—Criteria used to inform the contractor of what is to be met for proper execution of the work.

PERT Analysis—Program Evaluation and Review Technique, a statistical tool used in project management.

Plan-Do-Check-Act (PDCA)—An iterative four-step management method used for the control and continuous improvement of processes and products.

Plan-Do-Study-Act (PDSA)—A variation of PDCA, emphasizing the study phase.

Power Dynamics—The interplay of influence and authority between individuals or groups within a project.

Preconstruction Meeting—A formal meeting held before the commencement of construction activities, involving the owner, architect, engineer, general contractor, and major subcontractors. The purpose is to review the project scope, schedule, budget, safety procedures, communication protocols, and any other relevant project-related information. It aims to establish clear expectations and ensure all parties are aligned before work begins.

Preinstallation Checklists—Documents or forms used to verify that all preparations and requirements are met before installing equipment or systems.

Preinstallation Meeting/Preparatory Meeting—A meeting held before the installation of a specific system or component, involving the relevant subcontractors, suppliers, and the general contractor. It focuses on the specific installation procedures, coordination requirements, material handling, safety considerations, and quality control measures related to that particular installation. This meeting ensures that all parties involved in the task of construction installation understand the installation requirements and minimizes potential issues during the process.

Prequalification Audit—An assessment of the management systems of subcontractors and suppliers (including designers) before contracts are finalized.

Prevention Cost—Costs incurred to prevent defects from occurring in the first place.

Proactive Feedback Methods—Methods like surveys, phone calls, and customer visits.

Procurement—The acquisition of goods, services, or works from an external source.

Procurement Plan—Describes how an organization will acquire goods and services from outside the organization.

Professional Liability Insurance—Insurance that contractors should consider procuring when providing delegated design services, covering liability for the design delegated to them.

Progress Payments Based on Completion of Milestones—Partial payments are made to the contractor on the successful achievement of predefined stages or events in the project schedule.

Progress Payments Based on Percentage of Completed Work—Partial payments made to the contractor based on the assessed percentage of the total work that has been finished.

Project Brief—A document outlining the objectives, scope, and requirements of a construction project.

Project Charter—A document that formally authorizes the existence of a project and gives the project manager the authority to apply organizational resources to project activities.

Project Closure—Formalizing project completion and disseminating information to formalize completion.

Project Delivery Methods—Different project delivery methods used to organize and finance the construction projects, including design-bid-build (DBB), design-build (DB), Construction Manager at Risk (CMAR), and others.

Project Execution—The process of carrying out the work defined in the project management plan.

Project Initiation—The process of formally starting a new project or a new phase of an existing project.

Project Life Cycle—The phases of a project from initiation to closure, typically including initiation, planning, execution, monitoring and controlling, and closure.

Project Monitoring and Control—Tracking, reviewing, and regulating the progress of the project.

Project Planning—Developing a road map for the project, including defining objectives, scope, and a plan to achieve the objectives.

Project Scope Statement—A foundational document defining project objectives, deliverables, boundaries, constraints, and acceptance criteria.

Public Clients—Government agencies or departments that commission construction projects to serve public interests.

Public-Private Partnerships (P3)—Contractual agreements between a public agency and a private entity for the financing, design, construction, and/or operation of infrastructure projects.

Public Service Commission—A governmental agency overseeing public utility companies.

Punch List—A document detailing outstanding tasks or deficiencies that need to be addressed before a project is considered complete and handed over to the client. Punch list and deficiencies/nonconformities are directly related to nonconformance and to the schedule. If an article of work is identified as deficient or includes nonconformities, then the correction of those deficiencies must be verified against the schedule. Any deficiency/nonconformity that cannot be corrected without negatively impacting the schedule must not be placed on the punch list—it should be issued/elevated as a nonconformance.

Purchase Orders—Formal documents issued by a contractor to a vendor authorizing the procurement of specific materials, equipment, or services for a construction project.

Q

Qualitative Criteria—Nonnumerical factors used to evaluate bids, such as technical expertise or experience.

Qualitative Research—Research delving into motivations and opinions behind responses.

Quality Assurance (QA)—The part of quality management that provides customers and other relevant interested parties with a basis for confidence that their needs and expectations will be fulfilled. QA is process-focused; it helps establish the QA processes, as well as the quality control (QC) processes that are used by QC staff to perform their inspections and reviews. QA approaches quality by first establishing and later verifying that the quality processes are adequate to meet quality standards. QA is proactive in nature, and its primary purpose is to verify the effectiveness of the quality system (or lack of it). QA is not binary in its nature; it is to be reported on a scale (e.g., as a percentage, QA achieved 90% of its goals).

Quality Control (QC)—The part of quality management that deals with ensuring that the organization consistently meets requirements for products and services. QC is focused on product quality; it is reactive and is used to achieve quality by checking/testing the product and determining if it meets the quality specifications it is supposed to meet. QC is binary in nature, that is, the result of a QC inspection/test is either a pass or a fail. QC activities must be planned to be appropriate to the nature and complexity of processes, products, and services. For construction, the QC plan must include a more specific inspection and test plan that describes how QC testing must be performed.

Quality Management (QM)—An overarching system used to achieve quality excellence within an organization. In essence, QM encompasses both QA and QC.

Quality Management Plan (QMP)—A project-specific blueprint detailing how quality standards will be met.

Quality Management System (QMS)—A broader system for establishing policies and procedures pertaining to an entire organization's quality and business operations.

Quality Plan—A document specifying which standards, practices, resources, specifications, and sequence of activities pertain to a specific project. A quality plan issued for the design phase is the design quality plan. Likewise, the quality plan issued for the construction phase is the construction quality plan.

Quality Planning—The building block of subsequent quality management activities aimed at controlling, assuring, or improving the quality of an organization's products or services. In general terms, quality planning is connected with two different activities: ensuring there is a clear set of requirements for the product or service to be supplied, and ensuring there is a specified sequence and interaction of processes to realize the product and/or service.

Quantitative Research—Research yielding numerical data for statistical analysis.

R

Reactive Feedback Methods—Methods like complaint monitoring and warranty systems.

Regulator—A government agency or authority that enforces building codes, safety regulations, and environmental standards.

Regulatory Authorities—Entities with the power to enforce laws and regulations. They are often involved in construction projects to ensure compliance.

Regulatory Body—A group, typically an agency of a government, with mandates by an overarching government's laws or an AHJ, to control either general or specific codes, practices, standards, or laws for a construction project.

Reimbursable Costs—Project expenses incurred by the contractor that are directly reimbursed by the owner according to the contract terms.

Request for Bids (RFB)—An invitation to submit offers for goods, works, or services.

Request for Information (RFI)—Formal written request made by a contractor to the owner, architect, or engineer for clarification about plans, specifications, or other project-related information.

Request for Proposal (RFP)—A formal document used to solicit bids from contractors for a specific project.

Request for Qualifications (RFQ)—A preliminary document to evaluate and short-list potential contractors based on their qualifications.

Resource Planning—Defining and allocating the resources needed for a project.

Retainage—A percentage of payment withheld by the client until project completion to ensure all work is performed to the required standard.

Rework Costs—Expenses incurred to correct or redo work that was not performed according to specifications or quality standards.

Risk—The effect of uncertainty on an expected result. This means that risk is the potential for a deviation from what is expected. In general, the connotation of risk is that it is a negative result. However, risk can also be positive, which means a deviation from what is expected but the result is a favorable one. Positive risks are not mitigated; they must also be considered as an opportunity to enhance or exploit to achieve greater benefit.

Risk Management Plan—A component of the project management plan that describes how risk management will be structured and performed.

S

Schedule of Values (SOV)—A list of every work item on a construction job, along with each item's cost.

Scope Creep—Changes to the project scope without proper control.

Scope of Work—A document that is part of broader agreement describing the tasks and deliverables required for a project such as the work to be performed, including products, timeline, milestones, and progress reports.

Sequential Evaluation—Evaluating bids in stages, often with quality assessed before price.

Shear Strength—The ability of a material or connection to resist forces that cause sliding.

Shop Drawings—Drawings from the contractor and its design agency to the architect and client detailing how specific portions of the work will be fabricated and installed. Shop drawings corroborate the final design drawings (construction drawings), but should not be in conflict with the final design.

Single Source—Procurement where a supplier is purposefully chosen, even though others are available.

Single-Stage Bidding Procedure—Bidders submit both technical and price proposals in one envelope.

Six Sigma—A structured approach (define-measure-analyze-improve-control) that provides organizations tools to improve the capability of their business processes and solve business problems. This increase in performance and decrease in process variation results in defect reduction and improvement in profits, employee morale, and quality of products or services. Six Sigma quality is a term generally used to indicate that a process is well controlled ($\pm 6\sigma$ from the centerline in a control chart).

Slice and Package—A procedure used when a large homogeneous project is divided into smaller contracts that are bid simultaneously.

Small and Medium-Sized Enterprises (SME)—Businesses that maintain revenues, assets, or a number of employees below a certain threshold.

Specifications—A detailed description of the materials, workmanship, and quality standards required for construction.

Stakeholder—Any individual, group, or organization that has an interest in or can affect or be affected by a construction project.

Stakeholder Analysis—A strategic process used to identify, understand, and assess the interests, expectations, motivations, and influence of project stakeholders.

Stakeholder Engagement Matrix—A tool used to strategize stakeholder involvement by outlining the role of each stakeholder in the life cycle stages of a project.

Stakeholder Intent—The underlying objectives, expectations, and requirements of all parties involved in a construction project, including owners, designers, contractors, and end users. Understanding stakeholder intent is crucial for ensuring that the project meets their needs and fulfills its intended purpose.

Standard Work—Documented procedures for performing tasks consistently and efficiently.

Standstill Letter—Formal notification to unsuccessful bidders containing information about the winning bid and evaluation.

Standstill Period—A mandatory period after notifying unsuccessful bidders before awarding a contract.

Statement of Work (SOW)—A document that details the work activities, deliverables, and timeline required to complete a project.

Steering Committee—Providing strategic direction and oversight for a project, program, or organization.

Strategic Change Management—A long-term and holistic view, aligning organizational change with overall business strategy and vision.

Subcontract—A contract between a main contractor and a specialized subcontractor to perform a specific part of the larger construction project.

Subcontracting—The practice where a primary contractor hires specialized third-party companies or individuals (subcontractors) to perform specific tasks on a project.

Submittal Reviews—The process of reviewing and approving documentation, samples, or data provided by contractors and suppliers to ensure compliance with project specifications.

Substantial Completion—A contract milestone describing a condition where a significant portion of a project is fit for its intended purpose, and has been sufficiently inspected, tested, and documented to verify this conformance with the contract. The owner may occupy or utilize the work for its intended purpose despite minor remaining items. See also *Beneficial Occupancy*.

System Assurance Practices—Practices to ensure systems function as intended, often involving audits, verification, and validation.

T

Tactical Change Management—Short-term and specific initiatives, implementing particular projects or changes within the framework of the overall strategy.

Temporary Certificate of Occupancy—A contract milestone describing confirmation that either the entirety or a portion of a project structure or space is suitable for inhabitation for a specified duration of time.

Tensile Strength—The capacity of a material to resist forces that pull it apart.

Terms of Reference (TOR)—Similar to an SOW, a TOR outlines the scope of work, objectives, and responsibilities for a particular project or task, often used for consulting or advisory services.

Time and Materials (T&M) Contract—A contract where the owner pays the contractor for the actual cost of labor at agreed-on rates and the actual cost of materials used.

Tolerance—The allowable deviation from specified dimensions or quality standards.

Trade-off—Balancing competing project objectives, such as cost, schedule, and quality.

Trade Partner—A specialized contractor providing expertise in a particular trade, often involved early in the design stage. Typically, trade partners are hired in the early stages of the project due to their reliability rather than just low cost.

Tradespeople—Skilled workers specializing in a particular trade, such as carpentry, plumbing, or electrical work.

Triple Bottom Line—An accounting framework with three parts: social, environmental (or ecological), and financial.

Two-Envelope Bidding Procedure—Bidders submit technical and price proposals in separate envelopes, with technical evaluation preceding price evaluation.

Two-Stage Bidding Procedure—Bidding process involving functional specifications first and then detailed technical and financial proposals.

Two-Stage Tendering—A method with two procedures, each involving technical and financial proposals.

U

Unit Pricing Contract—A contract based on estimated quantities of specific work units and their predetermined unit rates; prices are broken into units of work, and contractors charge for each unit completed.

Unit Rates—Predetermined costs for a specific unit of work, like price per cubic yard of concrete or per linear foot of pipe.

Utility Company—A company (either private, public, or governmental) that maintains the infrastructure and supplies goods or services to the public considered an essential need. Examples are gas, electricity, and water.

V

Value Engineering—A systematic process of analyzing project functions to identify and implement cost-effective alternatives that maintain or improve project value.

Value for Money—Securing the best mix of quality and cost over the life cycle of the procured goods or services.

Value Stream—The sequence of activities required to deliver a product or service to a customer.

Value Stream Mapping—A visual tool used to analyze and improve the flow of materials and information in a value stream.

Vertical Construction—Construction projects that involve the building of structures that rise vertically, such as buildings for residential, commercial, or industrial use.

Visualization—The use of 3D models, renderings, or virtual reality to communicate and understand project design and construction.

W

Warranty—A guarantee by a contractor or manufacturer to repair or replace a defective product or to correct defective workmanship for a specific duration. These warranties can be express, clearly stated in the contract, or implied, which are legal guarantees that aren't written but are still enforceable. An explicit/express warranty is a warranty (either general or specific) for a particular component in construction listed directly in a contract. However, an implied warranty is one that is not explicitly listed in a contract but implied by common levels of quality, workmanship, and proficiency expected of a contractor.

Warranty Claims—Formal requests made by an owner or client to have repairs or replacements performed under a guarantee or warranty provided by the contractor or supplier.

Warranty Data—Information gathered from warranty claims and repair requests.

Weighting—Assigning relative importance to different evaluation criteria.

Work Breakdown Structure (WBS)—A deliverable-oriented hierarchical decomposition of the work to be executed by the project team.

Work in Place (WIP) Inspection—Inspections conducted during the construction process to verify that the work is being performed according to quality standards and specifications. These inspections are performed at work that is currently being installed.

References

Abbas, Tahir. "Stakeholder Analysis for Construction Project—Explained." November 29, 2023, https://changemanagementinsight.com/stakeholder-analysis-for-construction-project/.

Abell, Brittney, and Kristen Frisa. "5 Construction Billing Methods & When To Use Them." *Procore*, Nov 15, 2024, https://www.procore.com/library/construction-billing.

Adams, N.H. "Quality Assurance Reviews: How They Differ from Peer Reviews." *Quality Assurance*, 6 (2): 75-85 (1998), doi: https://pubmed.ncbi.nlm.nih.gov/10386330/.

Adbuildcon blog. "What Are The 5 Primary or Major Sectors of the Construction Industry." 3S Buildcon, January 21, 2020, https://3sbuildcon.in/what-are-the-5-primary-or-major-sectors-of-the-construction-industry/.

Adrian, James J. and Douglas J. Adrian. 1995. *Total Productivity and Quality Management for Construction*. Champaign, IL: Stipes Publishing LLC.

African Development Bank. 2017. *User's Guide for the Standard Prequalification Document for Procurement of Works*. Abidjan, Cote d-Ivoire: African Development Bank.

AIA Contract Documents, "Certificate of Substantial Completion vs. Final Completion: Key Construction Milestones." June 6, 2023, https://learn.aiacontracts.com/articles/substantial-completion-vs-final-completion-understanding-key-construction-milestones/.

—"Comparing Design Assist and Delegated Design in Construction Projects." August 16, 2023, https://learn.aiacontracts.com/articles/comparing-design-assist-and-delegated-design-in-construction-projects/.

—"Delegated Design, What Does It Really Mean?" January 23, 2024, https://learn.aiacontracts.com/articles/delegated-design-what-does-it-really-mean.

—"What Contractors Need to Know About Delegated Design in Construction." October 20, 2022, https://learn.aiacontracts.com/articles/what-contractors-need-to-know-about-delegated-design/.

Allen, Derek R., and Morris Wilburn. 2002. *Linking Customer and Employee Satisfaction to the Bottom Line*. Milwaukee, WI: ASQ Quality Press.

American Association of State Highway and Transportation Officials (AASHTO) Subcommittee on Construction. "Constructability Review Best Practices Guide." August 2000.

American Society for Quality (ASQ). "Insights on Excellence from ASQE." https://asq.org/quality-resources/research/asqe-insights-on-excellence/benchmarking-tool. Accessed February 25, 2026.

—"Policies." https://asq.org/about-asq/conferences-events-policies/code-of-ethics.

—"Quality Assurance vs. Quality Control: QA vs. QC." https://asq.org/quality-resources/quality-assurance-vs-control. Accessed February 25, 2026.

—"Quality Glossary." https://asq.org/quality-resources/quality-glossary. Accessed February 25, 2026.

—"Quality Resources." https://asq.org/quality-resources. Accessed February 25, 2026.

—"Seven Basic Quality Tools." https://asq.org/quality-resources/seven-basic-quality-tools. Accessed February 25, 2026.

—"What is a Gantt Chart?" https://asq.org/quality-resources/gantt-chart. Accessed February 25, 2026.

American Society of Civil Engineers (ASCE). 2012. *Quality in the Constructed Project: A Guide for Owners, Designers, and Constructors.* Reston, Virginia: American Society of Civil Engineers.

Anderson, James C., and James A. Narus. "Business Marketing: Understand What Customers Value." *Harvard Business Review* online, originally published Nov/Dec 1998, https://hbr.org/1998/11/business-marketing-understand-what-customers-value.

Anheim, Fredrik. "Importance of the Project Team to the Creation of Learning Within and Between Construction Projects." 2003. *Construction Process Improvement.* Atkin, Brian, Jan Borgbrant, and Per-Erik Josephson, eds. Blackwell Science, pp. 183–194.

Arroyo, Paz, et al. "Lean, Psychological Safety, and Behavior-Based Quality: A Focus on People and Value." 28th Annual Conference of the International Group for Lean Construction (2020).

Ashford, J.L. 1989. *The Management of Quality in Construction.* New York. E&FN Spon.

Asian Development Bank (ADB). 2018. *Guide on Bid Evaluation, June 2018.* Manila, Philippines: Asian Development Bank.

Associated Builders and Contractors, "MasterFormat CSI Codes and NAICS Codes." abc.org, https://www.abc.org/Membership/MasterFormat-CSI-Codes-NAICS-Codes/CSI-Codes (accessed September 20, 2024).

The Associated General Contractors of America/American Subcontractors Association, Inc., and Associated Specialty Contractors. 2008. *Guidelines for a Successful Construction Project.*

Association for Project Management. "10 Key Principles of Stakeholder Engagement." https://www.apm.org.uk/resources/find-a-resource/stakeholder-engagement/key-principles/ (accessed February 25, 2026).

Association of Professional Engineers of Ontario. "Advertising Information for Certificate of Authorization Holders." https://www.peo.on.ca/sites/default/files/2019-08/AdInfoCofA2015.pdf (accessed February 25, 2026).

—"Guide: Do I Need a Certificate of Authorization?" https://www.peo.on.ca/sites/default/files/2019-08/DoINeedACertificateofAuthorization.pdf (accessed February 10, 2026).

Atkins, James, B. and Grant A. Simpson. 2008. *Managing Project Risk: Best Practices for Architects and Related Professionals.* Hoboken, NJ: Wiley.

Bammer, Gabriele. "Stakeholder Engagement: Learning from Arnstein's Ladder and the IAP2 Spectrum." *Integration and Implementation Insights,* August 30, 2022, https://i2insights.org/2022/08/30/learning-from-arnsteins-ladder-and-iap2-spectrum/.

Bartholomew, Stuart H. *Construction Contracting: Business and Legal Principles, Second Edition.* Blacksburg: University Libraries at Virginia Tech, 2022. doi: https://doi.org/10.21061/constructioncontracting2e.

Bathersby Legal. "Liquidated Damages in Construction Contracts." https://www.bathersbylegal.com.au/advice/liquidated-damages-in-construction-contracts (accessed February 10, 2026).

Briefbuilder blog. "Insights: What Exactly is Requirements Management?" https://www.briefbuilder.com/what-is-requirements-management/ (accessed February 25, 2026).

Brown, Dave, and James Hamilton. "The Construction Closeout Process Explained." Procore, https://www.procore.com/library/construction-closeout (accessed May 28, 2024).

Bryce, Ian. The Gatekeeper Blog. https://www.gatekeeperhq.com/blog/author/ian-bryce (accessed February 25, 2026).

BuildingAdvisor.com. "Simple Construction Proposals." https://buildingadvisor.com/project-management/contracts/simple-construction-contracts/ (accessed February 10, 2026).

BuildingRadar.com. "Fostering a Collaborative Team Environment in Construction." https://www.buildingradar.com/construction-blog/fostering-a-collaborative-team-environment-in-construction (accessed March 3, 2026).

Burek, Paul. "Developing a complete project scope statement in 2 days." Project Management Institute (PMI), https://www.pmi.org/learning/library/project-scope-statement-skills-tools-7981 (accessed February 10, 2026).

The Business and Financial Times. "To Sole or Single or Competitively Source: Strategic Sourcing and Industrialisation Insights with Prof Douglas Boateng." October 14, 2020, https://thebftonline.com/2020/10/14/to-sole-or-single-or-competitively-source/.

Calahan Construction Services. "The Complete Guide to RFP in Construction." https://www.calahan.com/comparing-proposals-what-to-look-for-in-rfps/ (accessed February 10, 2026).

Chandler, Lori. "Integrating Quality Efforts with Contractors During the Design Phase to Reduce Duplication Efforts." *Advancing Design Quality Management*, Chicago, 2024.

Chartered Institute of Procurement & Supply (CIPS). "Sourcing & Types of Sourcing." https://www.cips.org/intelligence-hub/sourcing (accessed February 10, 2026).

Cholopray, Aaron K.D. "The Bid Evaluation Process." The Procurement ClassRoom, https://www.procurementclassroom.com/the-bid-evaluation-process/ (accessed February 10, 2026).

Chertoff, Gregory H., and Navid Ansari. "Design Delegation—Legal Definitions, Practical Considerations and the Need for Clarity." Peckar & Abramson, PC, https://www.agc.org/sites/default/files/pdf/Handout%20on%20Delegated%20Design.pdf (accessed February 11, 2026).

Chung, H.W. 1999. *Understanding Quality Assurance in Construction: A Practical Guide to ISO 9000*. London: E&FN Spon.

Clear People, "Lessons Learned Guidelines & Examples. How to Automate Them for Project Management." Atlas blog, https://www.clearpeople.com/blog/lessons-learned-guide (accessed February 25, 2026).

Cockfield, Richard W. "Scope Management." *PM Network*, 1(3): 12-15.

Coleman, Lance B. 2020. *The ASQ Certified Quality Auditor Handbook, Fifth Edition*. Milwaukee, WI: ASQExcellence.

Construction Innovation Hub. *CPQP Measurement System Analysis Guideline*. Cambridge, UK: Construction Innovation Hub, https://constructioninnovationhub.org.uk/media/v2tbt24h/6-cpqp-msa-final.pdf (accessed April 16, 2025).

Construction Management Association of America (CMAA). 2015. *An Owner's Guide to Project Delivery Methods*. CMAA.

The Construction Specifications Institute. 2011. *Construction Specifications Practice Guide*. Wiley.

Contracts Counsel. "Time and Materials Contract: A Basic Guide." https://www.contractscounsel.com/t/us/time-and-materials-contract (accessed February 10, 2026).

Creasey, Tim. "Change vs. Change Management." Prosci blog, March 3, 2021 (updated October 28, 2025), https://www.prosci.com/blog/change-vs-change-management.

—"A Communications Plan is Not Change Management." Prosci blog, January 2, 2020 (updated August 11, 2025), https://www.prosci.com/blog/a-communications-plan-is-not-change-management.

Cunnningham, Jean. "The Eight Wastes of Lean." Lean Enterprise Institute, January 18, 2020, https://www.lean.org/the-lean-post/articles/the-eight-wastes-of-lean/.

Defense Acquisition University. "Quality Assurance Surveillance Plan." https://www.dau.edu/acquipedia-article/quality-assurance-surveillance-plan-qasp (accessed February 27, 2026).

DeFeo, Joseph A. 2016. *Juran's Quality Handbook*, 7th ed. McGraw-Hill.
Deming, W. Edwards. 1982 *Out of the Crisis*. Cambridge, MA: Massachusetts Institute of Technology.
Design Council. 2013. "Design Review—Principles and Practice." Updated May 2019, https://www.designcouncil.org.uk/fileadmin/uploads/dc/Documents/Design%2520Review_Principles%2520and%2520Practice_May2019.pdf.
Dixon, Ian. "Partnering Tip: The Zipper Approach," Partnering Essentials." September 24, 2019, https://www.partneringessentials.com/news-articles/2019/9/23/the-zipper-approach.
Dobberstein, Parker. "Construction Closeout: What to Expect and How to Prepare." Neumann Monson Architects blog, May 9, 2022, https://neumannmonson.com/blog/construction-closeout.
Dupslaff, Ben. "A Better Checklist." March 11, 2024. CSI Milwaukee Chapter Meeting.
—"Creating Effective Checklists in Design and Construction." *Advancing Construction Quality*. Phoenix, 2023.
—"Implement Your Business Case and Succeed in Business Improvement." Deconstrategy. November 20, 2024. https://deconstrategy.substack.com/p/implement-your-business-case-and.
—"On Definable Features of Work." Deconstrategy. May 15, 2024, https://deconstrategy.com/definable-features-of-work.
—"There Is No Perfect Tool." Deconstrategy. May 15, 2024, https://deconstrategy.com/there-is-no-perfect-tool/.
—"Weekly Tactics: Focus, MEP, and Lessons Learned." Deconstrategy, November 17, 2024, https://deconstrategy.substack.com/p/weekly-tactics-focus-mep-and-lessons.
DXP Marketing, "What is Factory Inspection Testing (FAT) & How Does It Work?" DXP, December 21, 2023, https://www.dxpe.com/what-is-factory-acceptance-test-protocol-purpose/.
EcoSys Team. "Project Controls: Key Elements, Benefits and Challenges." Hexagon, August 1, 2019, https://aliresources.hexagon.com/enterprise-project-performance/project-controls.
Englund, Randall L., and Alfonso Bucero. "Influencing Others: Negotiating and Achieving Desire Results with Stakeholders." Paper presented at PMI® Global Congress 2013—North America, New Orleans, LA. Newtown Square, PA: Project Management Institute.
Ewenstein, Boris, Wesley Smith, and Ashvin Sologar. "Changing Change Management." McKinsey & Company, July 1, 2025, https://www.mckinsey.com/featured-insights/leadership/changing-change-management.
Finity, Johnny. "6 Types of Construction Projects: Key Differences for Owners and Contractors." Procore, August 7, 2025, https://www.procore.com/library/construction-project-types#4-institutional-projects.
—"Payment Applications 101: A Guide to Pay Apps in Construction." August 6, 2023, https://www.fieldwire.com/blog/a-guide-to-pay-apps-in-construction.
Flanagan, Roger, and George Norman. 1993. *Risk Management and Construction*. Wiley.
Flyvbjerg, Bent and Dan Gardner. *How Big Things Get Done: The Surprising Factors That Determine the Fate of Every Project, From Home Renovations to Space Exploration and Everything in Between*. New York: Currency, 2023.
Furterer, Sandra L., and Douglas C. Wood. 2021 *The ASQ Certified Manager of Quality/Organizational Excellence Handbook, Fifth Edition*. Milwaukee, WI: ASQExcellence.
Gandhi, Gourav. "8 Steps to a Successful Construction Closeout Process." *Rdash.io* https://rdash.io/blog/8-steps-to-a-successful-construction-closeout-process/ (accessed September 20, 2024).

Ganiyu, Abideen Adekunle. "Subcontracting Strategy in Industrial Projects: Evidence from Saudi Arabia." *LAUTECH Journal of Civil and Environmental Studies*. 7. 10.36108/laujoces/1202.70.0170.

Gannett Fleming, Inc. "Construction project management handbook: September 2009." 2009.

Garcia, Ernest and James Hamilton. "Construction Closeout Documents: What's Included and Why." Procore, March 13, 2025, https://www.procore.com/library/construction-closeout-documents.

Gawande, Atul. *The Checklist Manifesto: How to Get Things Right*. New York: Picador, 2009.

Generation.org. "Collaboration Across Cultures: Fostering Teamwork in a Diverse Workforce." https://www.generation.org/news/collaboration-across-cultures-fostering-teamwork-in-a-diverse-workforce. Accessed March 3, 2026.

Giusti, Michael, and Mary Carroll-Coelho. "General Conditions in Construction: A Contractor's Guide." Procore, June 20, 2024, https://www.procore.com/library/general-conditions-construction.

Global Sources. "Sole vs Single vs Multiple Sourcing: Choosing the Right Supply Chain Strategy." Updated December 30, 2025, https://www.globalsources.com/knowledge/sole-sourcing-vs-single-sourcing-vs-multiple-sourcing/.

GOAL/QPC. 2018. *The Memory Jogger 2*, 2nd ed. Goal/QPC.

Gonzalez, Edward. "Successful Project Closeout: The Comprehensive Guide." Buildr.com, March 10, 2022, https://buildr.com/blog/project-closeout/.

Gould, F. E., and N.E. Joyce. 2020. *Construction Project Management*, 5th Ed. New York: Pearson.

Government Commercial Function. "Bid Evaluation: Guidance Note—May 2021." Civil Service and Cabinet Office, May 2021, https://assets.publishing.service.gov.uk/government/uploads/system/uploads/attachment_data/file/987130/Bid_evaluation_guidance_note_May_2021.pdf.

Green, David. "Equipment Reliability: Paving the Way to Sustainable Equipment Performance." Click Maint blog, July 21, 2023, https://www.clickmaint.com/blog/equipment-reliability.

Griffith, Alan. 2011. *Integrated Management Systems for Construction: Quality, Environment and Safety*. New York: Pearson.

Hammer, Michael and Steve Stanton. "How Process Enterprises Really Work." *Harvard Business Review*, November/December 1999, https://hbr.org/1999/11/how-process-enterprises-really-work.

Hare, Rachelle. "Ultimate Guide to Construction Contracts in Australia (2025)." Blaze Business & Legal, January 4, 2025, https://blazebusinessandlegal.com.au/construction-contracts/.

Harrington, H. James. 2006. *Change Management Excellence: The Art of Excelling in Change Management*. Paton Press.

Harris, Georgia L., ed. 2019. *NISTIR 8250: Calibration Procedures for Weights and Measures Laboratories*. Gaithersburg, MD: National Institute of Standards and Technology, doi: 10.6028/NIST.IR.8250.

Hart, Roger. 1994. *Quality Handbook for the Architectural, Engineering, and Construction Community*. Milwaukee, WI: ASQC Quality Press.

Hayes, Adam. "Subcontracting: How It Works, Benefits, Definition, and Taxation." Investopedia, February 25, 2025, https://www.investopedia.com/terms/s/subcontracting.asp.

Hayes, Bob E. 2008. *Measuring Customer Satisfaction and Loyalty: Survey Design, Use, and Statistical Analysis Methods*. Milwaukee, WI: ASQ Quality Press.

Heslin, Kevin, Ryan Orr, Chris Brown, and Ed Rafter. "Improve Project Success Through Mission Critical Commissioning." Uptime Institute. May 4, 2015, https://journal.uptimeinstitute.com/improve-project-success-through-mission-critical-commissioning/.

Hitech BIM Services. "What Does SD, DD, & CD Mean in the Construction World?" January 20, 2021, https://www.hitechbimservices.com/blog/what-is-sd-dd-cd-mean-in-construction-world.php.

Hoffnagle, Elena, Kate Mertz, and Sarah Mostafa. "Assessing and Tracking Vendor Performance." Harvard Kennedy School Government Performance Lab, 2022.

Howarth, Tim, and Tim Greenwood. 2018. *Construction Quality Management: Principles and Practice*, 2nd Ed. New York: Routledge.

Howarth, Tim, and Paul Watson. 2011. *Construction Quality Management: Principles and Practice*, 1st Ed. New York: Routledge.

Inglesis, Pavlo. "7 Types of Construction Audits." Quality In Construction, June 16, 2025, https://qualityinconstruction.com/quality-audits-in-a-construction-project/.

Jain, Ayush. "How to Define Non-Functional Requirements of a Project?" Mindbowser, https://www.mindbowser.com/non-functional-requirements-of-a-project/ (accessed February 11, 2026).

Jain, Vinay. "Types of Construction Contracts and Their Comparisons." TheConstructor.org, https://theconstructor.org/construction/types-of-construction-contracts-comparison/14268/?amp=1 (accessed February 10, 2026).

Jamble, Riya, and Dinna Dragonetti. "Construction Bidding Process: A Comprehensive Guide." SelectHub, January 9, 2025, https://www.selecthub.com/construction-management/construction-bidding/construction-bidding-process/.

Jha, Kumar Neeraj, and K.C. Iyer. "Critical Factors Affecting Quality Performance in Construction Projects." *Total Quality Management & Business Excellence* 17, no. 9 (2006): 1155–1170, doi: 10.1080/14783360600750444.

Juran, Joseph M., and A. Blanton Godfrey, eds. 1999. *Juran's Quality Handbook*, 5th ed. New York: McGraw-Hill.

Juran, Joseph M., and Frank M. Gryna. 1988. *Juran's Quality Control Handbook*, 4th ed. New York: McGraw-Hill.

Kakabadse, Andrew, and Nada Kakabadse. "Trends in Outsourcing: Contrasting USA and Europe." *European Management Journal* 20, no. 2 (2002): 189–198, https://www.sciencedirect.com/science/article/abs/pii/S0263237302000294.

Kargina, Mariya. "Cultural Differences in Global Virtual Teams: Mapping Knowledge and Identifying Research Directions." *Journal of Management & Organization* 31, no. 6 (2025): 3054–74. https://doi.org/10.1017/jmo.2023.41.

Kelly, John and Steven Male. 1993. *Value Management in Design and Construction: The Economic Management of Projects.* London: E&FN Spon.

KGDR Consultants, Inc. "Efficient Management of Utilities in Construction Projects: Principles and Best Practices." LinkedIn, January 23, 2024. https://www.linkedin.com/pulse/efficient-management-utilities-construction-projects-8wghc/.

Killough, Dawn. "Guide to Progress Payments on Construction Projects." Procore, January 15, 2025, https://www.procore.com/library/progress-payments.

—"Construction Warranty Types." Procore, June 11, 2024, https://www.procore.com/library/construction-warranty.

Koutsogiannis, Anastasios. "5 Major Changes Coming to the Construction Industry in the Future." ConstructConnect, September 21, 2017, https://www.constructconnect.com/blog/5-major-changes-coming-construction-industry-future.

Kubal, Michael T. 1994. *Engineered Quality in Construction: Partnering and TQM*. New York: McGraw-Hill.

Laman, Scott A., ed. 2022. *The ASQ Certified Quality Engineer Handbook, Fifth Edition.* Milwaukee, WI: ASQExcellence.

LetsBuild. "Project Deliverables in Construction." LetsBuild.com blog, https://www.letsbuild.com/blog/project-deliverables-in-construction (accessed July 18, 2024).

—"Construction Site Inspection: Why It's Important to Your Construction Project and Business." LetsBuild.com blog, April 4, 2024, https://www.letsbuild.com/blog/construction-site-inspection-why-its-important-to-your-construction-project-and-business.

Lindfors, Christian. "Improving Project Efficiency through Process Transparency in Management Information Systems." *Construction Process Improvement*, 2003, 211–224.

Los Alamos National Laboratory. "Sustainable Design Guide." 2002.

Luck, Ian. "What is the Relationship Between Customer Retention and CLV?" CustomerGaige, February 14, 2024. https://customergauge.com/blog/relationship-between-customer-retention-and-clv.

Mamlouk, M. S., and J.P. Zaniewski. 2021. *Materials for Civil and Construction Engineers.* Upper Saddle River, NJ: Pearson Prentice Hall.

Marsden, Paul. 2019. *Digital Quality Management in Construction.* New York: Routledge.

McShane-Vaughn, Mary, ed. 2022. *The ASQ Certified Six Sigma Black Belt Handbook, Fourth Edition.* Milwaukee, WI: ASQExcellence.

Medina, Hildy. "RFIs: A Contractor's Guide to Requests for Information," Procore. June 18, 2024, https://www.procore.com/library/rfi-construction.

Metropolitan Transit Authority (MTA) Construction & Development. "MTA General Provisions for the Design-Bid-Build Contract." October 2025, https://www.mta.info/document/114766.

Mosaic Project Services Pty Ltd. "Project Controls: A Definition." Mosaic's PMKI Free Library, January 18, 2014, https://mosaicprojects.com.au/WhitePapers/WP1093_Project_Controls.pdf.

—"V&V = the Verification and Validation of Deliverables." Mosaic's PMKI Free Library, June 12, 2017, https://mosaicprojects.com.au/WhitePapers/WP1098_V_and_V.pdf.

Munro, Roderick A., Govind Ramu, and Daniel J. Zrymiak, eds. 2022. *The ASQ Certified Six Sigma Green Belt Handbook, Third Edition.* Milwaukee, WI: ASQExcellence.

National Academies of Sciences, Engineering, and Medicine. 2021. *Practices for Closing Out Highway Projects from Substantial Completion to Final Payment.* Washington, DC: The National Academies Press.

National Institute of Building Sciences. "Reliability-Centered Maintenance." Whole Building Design Guide, https://www.wbdg.org/resources/reliability-centered-maintenance-rcm. Accessed February 25, 2026.

—"Whole Building Design and the WBDG®." Whole Building Design Guide, https://www.wbdg.org/whole-building-design. Accessed February 25, 2026.

National Procurement and Tender Administration (NPTA). 2009. *Standard Evaluation Criteria Handbook for Prequalification and Bidding.* Georgetown, Guyana: NPTA.

Nelson, Charles. 2017. *Managing Quality in Architecture: Integrating BIM, Risk & Design Process,* 2nd Edition. New York: Routledge.

New York City Department of Buildings. "Temporary Certificate of Occupancy." https://www.nyc.gov/site/buildings/property-or-business-owner/temporary-certificate-of-occupancy.page (accessed May 6, 2024).

NFU Mutual. "Construction-Client Duties: Health and Safety Guidance Note." https://www.nfumutual.co.uk/globalassets/business/rms/rms-business-section/cdm-client-duties.pdf (accessed March 3, 2026).

Oakland, John and Marton Marosszeky. 2017. *Total Construction Management: Lean Quality in Construction Project Delivery.* New York: Routledge.

Oberg, Mats. "Application of Integrated Life Cycle Design to Housing: Coordination of the Design and Building Process for Optimal Building Performance." in *Construction*

Process Improvement, Brian Atkin, Jan Borgbrant, and Per-Erik Josephson, eds. 2003. Blackwell Science.

Oden, Rod. "The Relationship Between Building & Fire Codes." Locksmith Ledger International, October 1, 2007, https://www.locksmithledger.com/business-technical-aids/article/10229599/the-relationship-between-building-fire-codes.

Office of Justice Programs. "Fact Sheet: Sole Source Justification." U.S. Department of Justice, April 2022, https://www.ojp.gov/sites/g/files/xyckuh241/files/media/document/Sole_Source_FactSheet_C.pdf.

Optimove. "Customer Acquisition vs. Retention Costs." https://www.optimove.com/resources/learning-center/customer-acquisition-vs-retention-costs (accessed March 1, 2025).

Overvest, Marijn. "5 Popular Procurement Methods in 2026." Procurement Tactics website, https://procurementtactics.com/procurement-methods/ (accessed February 25, 2026).

—"Sole Sourcing vs. Single Sourcing—6 Crucial Differences." Procurement Tactics website, Last updated November 7, 2025, https://procurementtactics.com/sole-sourcing-vs-single-sourcing/.

Partida, Devin. "First Article Inspection 101: What You Need to Know." *Quality Magazine* online, December 4, 2020, https://www.qualitymag.com/articles/96312-first-article-inspection-101-what-you-need-to-know.

Patterson, Jack, Danny Morton, and Katy Griffin. 2008. *Quality and Construction: How to Develop a Quality Management System for Your Construction Firm.* Vista, California: BNi Publications.

Pennington, Alisha. "SWOT Analysis: The Most Overlooked Business Tool, and How to Use It." Forbes Business Council, February 24, 2022, https://www.forbes.com/councils/forbesbusinesscouncil/2022/02/24/swot-analysis-the-most-overlooked-business-tool-and-how-to-use-it/.

Pettee, Stephen R. "Constructability Reviews." Construction Management Association of America (CMAA). https://www.cmaanet.org/sites/default/files/2018-04/Constructiblity%20Reviews%20-%20a%20primer%206-6-12%20versionFINAL.pdf (accessed 2/10/26).

Pilot, Simha. "Back to Basics: What Is a Fault Tree Analysis?" *Quality Progress*, March 2002, 35(3): 120.

PlanRadar. "5 Key Factors Contributing to Budget Overruns in Construction Projects and Strategies to Mitigate Them." PlanRadar Blog, July 24, 2023, https://www.planradar.com/ae-en/construction-project-budget-overruns-strategies/.

Project Management Institute (PMI). 2017. *A Guide to the Project Management Body of Knowledge (PMBOK® Guide)*, 7th Ed.

—*A Guide to the Project Management Body of Knowledge (PMBOK® Guide)*, 5th Ed.

—"Projects and the Project Lifecycle." https://www.pmi.org/about/what-is-a-project. Accessed March 3, 2026.

Program on Negotiation (PON) staff. "The Anchoring Effect and How it Can Impact Your Negotiation." Program on Negotiation at Harvard Law School, December 3, 2025, https://www.pon.harvard.edu/daily/negotiation-skills-daily/the-drawbacks-of-goals/.

Prosci. "Avoid These 5 Change Management Obstacles." March 16, 2022 (updated August 8, 2025), https://www.prosci.com/blog/avoid-these-change-management-obstacles.

Quick Response Fire Supply (QRFS). "What Is a Fire Marshal? And Exactly What Is an AHJ?" https://blog.qrfs.com/290-what-is-a-fire-marshal-and-exactly-what-is-an-ahj/ (accessed February 11, 2026).

Ramu, Govind. "A Tailored Approach: A Framework for Thorough, Value-Add Process Auditing." *Quality Progress*, Sep. 2024 57 (9): 30-37.

—"Demystifying Cost of Quality." Standard Issues, *Quality Progress*, Sep 2021, 54(9):56-59.

—"Expert Answers: December 2021." *Quality Progress*, Dec. 2021, 54 (12): 8-9.

—"Lineup of Players." *Six Sigma Forum*, Nov. 2016 16 (1): 23-25.

—2022. *The ASQ Certified Six Sigma Yellow Belt Handbook, Second Edition*. Milwaukee, WI: ASQExcellence.

Reyes, Jaydee. "What is Commissioning? A Comprehensive Guide," SafeCulture website, March 27, 2024, https://safetyculture.com/topics/commissioning.

Rogers, Cephas B. "The Spider Chart: A Unique Tool for Performance Appraisal." World Conference on Quality and Improvement, May 1995 (49): 16–22.

Rumane, Abdul Razzak. 2017. *Quality Management in Construction Projects*, 2nd ed. Boca Raton, FL: CRC Press.

—2023. *Risk Management Applications Used to Sustain Quality in Projects: A Practical Guide*. Boca Raton, FL: CRC Press. https://doi.org/10.1201/9781003245612.

—2024. *Total Quality Management: Applications and Concepts for Construction Projects*. Boca Raton, FL: CRC Press. https://doi.org/10.1201/9781003451464.

Russell, James P., and Terry Regel. 2000. *After the Quality Audit: Closing the Loop on the Audit Process*. Milwaukee, WI: ASQ Quality Press.

Saha, Anupam. "Difference Between Certification Audit and Surveillance Audit." Cert-Pro, September 25, 2025, https://certpro.com/difference-between-certification-audit-and-surveillance-audit/.

Schroeder, Jason. "Organizing Your Meetings." Elevate Construction Podcast, Ep. 1108, June 21, 2024.

—"Winning with Stakeholders in Construction: A Guide to Success." Elevate Construction Blog, January 3, 2024, https://elevateconstructionist.com/winning-with-stakeholders-in-construction-a-guide-to-success/.

Schwallier, David. "The Significance of Organizational Culture in the Construction Industry." Associated Builders of Wisconsin, https://www.abcwi.org/wisconsin-contractor-blog/the-significance-of-organizational-culture-in-the-construction-industry/ (accessed February 25, 2026).

Sears, S. Keoki, Glenn A. Sears, Richard H. Clough, Jerald L. Rounds, and Robert O. Segner. 2015. *Construction Project Management*, 6th Ed. Hoboken, NJ: Wiley.

Sheldon, Wade, and James Dougherty. "How to Set Up a Team with Trade Partners and Subcontractors for Success." CBR Group, https://www.crbgroup.com/insights/construction/trade-partners-and-subcontractors (accessed February 10, 2026).

Silvestrini, Rachel T., and Sarah E. Burke. 2017. *The Certified Quality Engineer Handbook, Fourth Edition*. Milwaukee, WI: ASQ Quality Press.

Singh, Vishal, Ning Gu, and Xiangyu Wang. "A Theoretical Framework of a BIM-Based Multi-Disciplinary Collaboration Platform." *Automation in Construction* 20, no. 2 (2011): 134–144, https://www.sciencedirect.com/science/article/abs/pii/S0926580510001408?via%3Dihub.

Sison, Marianne and Trevor Greenberg. "5 Phases of Project Management Life Cycle and Why They Matter." December 4, 2025, https://project-management.com/project-management-phases/#what-are-the-5-phases-of-project-management.

Smith, Larry W. "Stakeholder Analysis a Pivotal Practice of Successful Projects." Paper presented at Project Management Institute Annual Seminars & Symposium, Houston, TX (2000). Newtown Square, PA: Project Management Institute.

Snitchler, Hubbard, and Eric Parzianello. "Joint Venture Agreement vs Partnership Agreement: What's the Difference?" Hubbard Snitchler & Parzianello, May 02, 2023, https://www.hspplc.com/blog/2023/may/joint-venture-agreement-vs-partnership-agreement/.

Sofio, Filipa. "Requirements in Management in Civil Engineering and Construction." Valispace, January 19, 2023, https://www.valispace.com/requirements-management-in-civil-engineering-and-construction/.

Steinkamp, Ronald. "Payment Application Best Practices for Construction Projects." Armanino, December 22, 2022, https://www.armanino.com/articles/payment-application-best-practices-for-construction-projects/.

Thomas, Willis H., Raymond W. Lam, David J. Nutt, and Michael E. Thas. 2011. *The Basics of Project Evaluation and Lessons Learned*, 1st ed. Boca Raton, FL: CRC Press.

Trinidad, Janine, Anthresia McWashington, and Kristen Frisa. "Ultimate Guide to Submittals in Construction." Procore, July 25, 2025, https://www.procore.com/library/construction-submittals.

Trinidad, Janine, and Trey Strange. "AHJs in Construction: Gaining Approval from Government Authorities." Procore, October 3, 2025, https://www.procore.com/library/ahjs-in-construction.

Tronier, Ryan. "The Eisenhower Matrix—Prioritizing Tasks and Do-To Lists." Asana, December 29, 2025, https://asana.com/resources/eisenhower-matrix.

Turtledove, Nati. "A Friendly Reminder: Quality Assurance and Peer Reviews." Nati Turtledove blog, February 20, 2023, https://nati-turtledove.com/2023/02/20/a-friendly-reminder-quality-assurance-and-peer-reviews/.

United Nations (UN). "Sustainability." https://www.un.org/en/academic-impact/sustainability (accessed February 11, 2026).

U.S. General Services Administration. *GSA Commissioning Guide.* U.S. General Services Administration, September 2020, https://www.wbdg.org/FFC/GSA/gsa_commissioning_guide_2020.pdf.

United States Army Corps of Engineers (USACE). 2020. *Construction Quality Management for Contractors.*

—*United Facilities Guide Specifications.* Division 1—General Requirements—Section 01 45 00.05 20—Design & Construction, January 2026.

Usmani, Fahad. "Stakeholder Analysis: Identify and Engage Key Players in Project Management." PM Study Circle, September 2, 2024, https://pmstudycircle.com/stakeholder-analysis/.

Vogwell, Deborah. "Stakeholder Management." Paper presented at PMI® Global Congress 2003—EMEA, The Hague, South Holland, The Netherlands. Newtown Square, PA: Project Management Institute.

West Chester University, "Tuckman's Stages of Group Development." https://www.wcupa.edu/coral/tuckmanStagesGroupDelvelopment.aspx (accessed February 25, 2026).

Wharton Online Insights. "What Is Status Quo Bias and How Does It Affect the Workplace?" Wharton Executive Education, https://executiveeducation.wharton.upenn.edu/thought-leadership/wharton-online-insights/status-quo-bias/ (accessed February 25, 2026).

Wilson, Shauna, and Paul Russell. 2024. *Remote Auditing Fundamentals, Second Edition.* Milwaukee, WI: Quality Press.

Wood, Douglas C., and ASQ Quality Management Division. 2013. *Principles of Quality Costs, Fourth Edition.* Milwaukee, WI: ASQ Quality Press.

The World Bank. 2023. *Evaluating Bids and Proposals Including Use of Rated Criteria for Procurement of Goods, Works, and Non-Consulting Services*, 2nd Ed. Washington, D.C.: The World Bank.

The World Bank with Global Water Security & Sanitation Partnership (GWSP). "Appendix 1: Construction Supervision and Quality Assurance Plan (Sample Framework)." March 2021. https://openknowledge.worldbank.org/server/api/core/bitstreams/f08df5ed-da64-52e8-a578-0ee92d241654/content.

Younis, Jourdan. "The Role of Consultants in Construction Projects: How to Improve Your ROI." Alpin, https://www.alpinme.com/the-role-of-consultants-in-construction-projects-how-to-improve-your-roi/ (accessed February 10, 2026).

Zou, Patrick X. W., Guomin Zhang, and Jian Zuo. "Understanding the Key Risks in Construction Projects in China." *International Journal of Project Management* 25, no. 6 (2007): 601–614. https://doi.org/10.1016/j.ijproman.2007.03.001.

Standards and Guidelines

American Society of Heating, Refrigerating and Air-Conditioning Engineers (ASHRAE). *ASHRAE Guideline 0-2019: The Commissioning Process.* Atlanta: ASHRAE, 2019.

American Society of Heating, Refrigerating and Air-Conditioning Engineers (ASHRAE). *ASHRAE Guideline 1-1996: The HVAC Commissioning Process.* Atlanta: ASHRAE, 1996.

ASQ. *ASQ TR2:2024: Cost of Quality—Guidelines for Development, Implementation and Monitoring to Improve Quality Performance.* Milwaukee, WI: ASQ, 2024.

International Organization for Standardization. *ISO 10004:2018: Quality Management—Customer Satisfaction—Guidelines for Monitoring and Measuring.* Geneva: ISO, 2018.

International Organization for Standardization. *ISO 10014:2021: Quality Management Systems—Managing an Organization for Quality Results—Guidance for Realizing Financial and Economic Benefits.* Geneva: ISO, 2021.

International Organization for Standardization. *ISO/TS 10020:2022: Quality Management Systems—Organizational Change Management—Processes.* Geneva: ISO, 2022.

International Organization for Standardization. *ISO/IEC 17025:2017: General Requirements for the Competence of Testing and Calibration Laboratories.* Geneva: ISO, 2017.

International Organization for Standardization. *ISO 19011:2018: Guidelines for Auditing Management Systems.* Geneva: ISO, 2018.

International Organization for Standardization. *ISO 31000:2018: Risk Management—Guidelines.* Geneva: ISO, 2018.

International Organization for Standardization. *ISO 9000:2015: Quality Management Systems—Fundamentals and Vocabulary.* Geneva: ISO, 2015.

International Organization for Standardization. *ISO 9001:2015: Quality Management Systems—Requirements.* Geneva: ISO, 2015.

International Organization for Standardization. *ISO 9004:2018: Quality Management—Quality of an Organization—Guidance to Achieve Sustained Success.* Geneva: ISO, 2018.

Project Management Institute. *A Guide to the Project Management Body of Knowledge (PMBOK® Guide).* PMI, 2025.

Organizations and Programs

American Association of State Highway and Transportation Officials (AASHTO), https://transportation.org/.

American National Standards Institute (ANSI), https://www.ansi.org/.

American Society for Quality (ASQ) Certified Construction Quality Manager (CCQM) certification, www.asq.org.

Construction Industry Institute (CII), https://www.construction-institute.org/.

Department of Small Business and Supplier Diversity (SWaM), Commonwealth of Virginia, https://www.virginia.gov/agencies/department-of-small-business-and-supplier-diversity/.

Energy Star Certification, https://www.energystar.gov/buildings/resources-audience/licensed-professionals/cost-free-verification.

Institute for Sustainable Infrastructure (ISI), Envision Sustainability Professional (ENV SP), https://sustainableinfrastructure.org/.

International WELL Building Institute (IWBI) WELL certification, https://www.wellcertified.com/.

Lean Construction Institute, https://leanconstruction.org/.

National Institute of Standards and Technology (NIST), a National Voluntary Laboratory Accreditation Program (NVLAP), https://www.nist.gov/.

U.S. Green Building Council (USBGC) Leader in Energy and Environmental Design (LEED) certification, https://www.usgbc.org/leed.

About the Editors

Govind Ramu, CMQ/OE, CCQM, CQE, CSSBB, CQA, CSQE, CSQP, and CRE, is a LEAD Distinguished Scholar from the Stanford Graduate School of Business and a Mechanical Engineer from Bangalore University, India. He has over 30 years of international quality management experience from various industry sectors, including construction. He is an associate member of the International Academy for Quality (IAQ). He was the convener for the ISO technical committee for ISO 10014.

Ramu is an ASQ Fellow, recipient of the ASQ Crosby Medal, ASQ LA Simon Collier Quality Award, ISO Excellence Award, and DCD Distinguished Service Award, and holds eight ASQ certifications. He is author of *The ASQ Certified Six Sigma Yellow Belt Handbook, Second Edition* (ASQExcellence, 2022), coeditor of *The ASQ Certified Six Sigma Green Belt Handbook, Third Edition* (ASQExcellence, 2022), and contributed to *The Lean Handbook* (ASQ Quality Press, 2012) and the American standard *ASQ TR 2:2024, Cost of Quality: Guidelines for Development, Implementation and Monitoring to Improve Quality and Performance* (ASQ, 2024). He organized the Body of Knowledge for ASQ's certification Construction Quality Manager (CCQM).

Tariq Masud, CMQ/OE, CCQM, CQE, CQA, PMP, CMC, ENV-SP, is a Registered Professional Engineer in the U.S. and Canada with diverse experience in project and construction management, risk, and quality management. Masud is also certified by the Project Management Institute (PMI) and holds multiple ASQ certifications as well as credentials with other professional associations for consulting, sustainability, and procurement and supplier management.

Masud has over 30 years of experience in international quality and risk management and is an accomplished conference and webinar speaker. He has presented at the ASQ World Conference on Quality and Improvement, as well as at the Quality Show, and at several design and construction world conferences. Masud is also the 2026 Chair for the Chicago's NEI Chapter of ASQ, and is actively engaged with PMI as well as with the Institute for Management Consulting (IMC).

Index

C

D

G

H

I

N

O

P

Q

V

W

Z

www.ingramcontent.com/pod-product-compliance
Lightning Source LLC
LaVergne TN
LVHW080645210926
848485LV00003B/330
9781636941912